Geodynamics Series

Geodynamics Series

1. Dynamics of Plate Interiors
   A. W. Bally, P. L. Bender, T. R. McGetchin, and R. I. Walcott (Editors)

2. Paleoreconstruction of the Continents
   M. W. McElhinny and D. A. Valencio (Editors)

3. Zagros, Hindu Kush, Himalaya, Geodynamic Evolution
   H. K. Gupta and F. M. Delany (Editors)

4. Anelasticity in the Earth
   F. D. Stacey, M. S. Patterson, and A. Nicholas (Editors)

5. Evolution of the Earth
   R. J. O'Connell and W. S. Fyfe (Editors)

6. Dynamics of Passive Margins
   R. A. Scrutton (Editor)

7. Alpine-Mediterranean Geodynamics
   H. Berckhemer and K. Hsü (Editors)

8. Continental and Oceanic Rifts
   G. Pálmason, P. Mohr, K. Burke, R. W. Girdler, R. J. Bridwell, and G. E. Sigvaldason (Editors)

9. Geodynamics of the Eastern Pacific Region, Caribbean, and Scotia Arcs
   Rámon Cabré, S. J. (Editor)

10. Profiles of Orogenic Belts
    N. Rast and F. M. Delaney (Editors)

11. Geodynamics of the Western Pacific-Indonesian Region
    Thomas W. C. Hilde and Seiya Uyeda (Editors)

12. Plate Reconstruction From Paleozoic Paleomagnetism
    R. Van der Voo, C. R. Scotese, and N. Bonhommet (Editors)

13. Reflection Seismology: A Global Perspective
    Muawia Barazangi and Larry Brown (Editors)

14. Reflection Seismology: The Continental Crust
    Muawia Barazangi and Larry Brown (Editors)

15. Mesozoic and Cenozoic Oceans
    Kenneth J. Hsü (Editor)

16. Composition, Structure, and Dynamics of the Lithosphere-Asthenosphere System
    K. Fuchs and C. Froidevaux (Editors)

17. Proterozoic Lithospheric Evolution
    A. Kröner (Editor)

18. Circum-Pacific Orogenic Belts and Evolution of the Pacific Ocean Basin
    J. W. H. Monger and J. Francheteau (Editors)

19. Terrane Accretion and Orogenic Belts
    Evan C. Leitch and Erwin Scheibner (Editors)

20. Recent Plate Movements and Deformation
    K. Kasahara (Editor)

21. Geology of the USSR: A Plate-Tectonic Synthesis
    L. P. Zonenshain, M. I. Kuzmin, and L. M. Natapov   B. M. Page (Editor)

22. Continental Lithosphere: Deep Seismic Reflections
    R. Meissner, L. Brown, H. Dürbaum, W. Franke, K. Fuchs, F. Seifert (Editors)

23. Contributions of Space Geodesy to Geodynamics: Crustal Dynamics
    D. E. Smith, D. L. Turcotte (Editors)

24. Contributions of Space Geodesy to Geodynamics: Earth Dynamics
    D. E. Smith, D. L. Turcotte (Editors)

25. Contributions of Space Geodesy to Geodynamics: Technology
    D. E. Smith, D. L. Turcotte (Editors)

26. Structure and Evolution of the Australian Continent
    Jean Braun, Jim Dooley, Bruce Goleby, Rob van der Hilst and Chris Klootwijk (Editors)

27. Mantle Dynamics and Plate Interactions in East Asia
    M.F.J. Flower, S.-L. Chung, C.-H. Lo, and T.-Y. Lee

The Core-Mantle Boundary Region

Michael Gurnis
Michael E. Wysession
Elise Knittle
Bruce A. Buffett
Editors

Geodynamics Series   Volume 28

American Geophysical Union
Washington, D.C.

Published under the aegis of the AGU Books Board.

QE
509.3
.C67
1998

Library of Congress Cataloging-in-Publication Data

The core-mantle boundary region / Michael Gurnis ... [et al.],
  editors.
          p.   cm. -- (Geodynamics Series ; v. 28)
            Includes bibliographical references.
            ISBN 0-87590-530-7
            1. Core-mantle boundary. I. Gurnis, Michael, 1959-
      II. Series.
      QE509.3.C67   1998
      551.1' 1--dc21   98-7870
                                                                 CIP

ISSN 0277-6669

ISBN 0-87590-530-7

Cover image courtesy of Jeremy Bloxham, Harvard University.

Figures, tables, and short excerpts may be reprinted in scientific books and journals if the source is
properly cited.

Authorization to photocopy items for internal or personal use, or the internal or personal use of
specific clients, is granted by the American Geophysical Union for libraries and other users registered
with the Copyright Clearance Center (CCC) Transactional Reporting Service, provided that the base
fee of $01.50 per copy plus $0.50 per page is paid directly to CCC, 222 Rosewood Dr., Danvers, MA
01923.0277-6669/98/$01.50+0.5. This consent does not extend to other kinds of copying, such as
copying for creating new collective works or for resale. The reproduction of multiple copies and the
use of full articles or the use of extracts, including figures and tables, for commercial purposes requires
permission from the American Geophysical Union.

Copyright 1998
Published by the American Geophysical Union
2000 Florida Avenue, N.W.
Washington, D.C. 20009

Printed in the United States of America.

39130589

CONTENTS

Preface
Michael Gurnis, Michael E. Wysession, Elise Knittle, and Bruce A. Buffett　　vii

Introduction
Michael Gurnis, Michael E. Wysession, Elise Knittle, and Bruce Buffett　　1

STRUCTURE OF THE LOWERMOST MANTLE AND D"
Deep subduction and aspherical variations in P-wavespeed at the base of Earth's mantle
Rob D. van der Hilst, Sri Widiyantoro, Kenneth C. Creager, and Thomas J. McSweeney　　5

Global analysis of shear wave velocity anomalies in the lower-most mantle
Xian-Feng Liu and Adam M. Dziewonski　　21

PKP and PKKP precursor observations: Implications for the small-scale structure
of the deep mantle and core
Peter M. Shearer, Michael A. H. Hedlin, and Paul S. Earle　　37

Illuminating the base of the mantle with diffracted waves
Raul W. Valenzuela and Michael E. Wysession　　57

PHYSICAL AND CHEMICAL PROPERTIES OF THE CORE MANTLE REGION
The temperature contrast across D"
Quentin Williams　　73

Elastic constants and anisotropy of $MgSiO_3$ perovskite, periclase, and SiO_2 at high pressure
Lars Stixrude　　83

Investigating causes of D" anisotropy
J.-M. Kendall and P. G. Silver　　97

The solid/liquid partitioning of major and radiogenic elements at lower mantle pressures:
Implications for the core-mantle boundary region
Elise Knittle　　119

Is there a thin electrically conducting layer at the base of the mantle ?
J. P. Poirier, V. Malavergne, and J. L. Le Mouël　　131

CORE MANTLE COUPLING AND THE GEOMAGNETIC FIELD
Electromagnetic core-mantle coupling II: Probing deep mantle conductance
Richard Holme　　139

Free oscillations in the length of day: Inferences on physical properties near the core-mantle boundary
Bruce A. Buffett   153

Interpreting the paleomagnetic field
David Gubbins   167

A one-dimensional map of B_S from torsional oscillations of the Earth's core
Stephen Zatman and Jeremy Bloxham   183

DYNAMICS OF THE CORE AND LOWER MANTLE
Numerical dynamo modeling: Comparison with the Earth's magnetic field
Weijia Kuang and Jeremy Bloxham   197

Geodynamically consistent seismic velocity predictions at the base of the mantle
I. Sidorin and Michael Gurnis   209

Three-dimensional simulations of mantle convection with a thermo-chemical basal boundary layer: D" ?
Paul J. Tackley   231

INTERPRETATIONS OF LOWER MANTLE STRUCTURE, CHEMISTRY, AND DYNAMICS
The EDGES of the mantle
Don L. Anderson   255

The D" discontinuity and its implications
Michael E. Wysession, Thorne Lay, Justin Revenaugh, Quentin Williams, Edward J. Garnero, Raymond Jeanloz, and Louise H. Kellogg   273

Seismic wave anisotropy in the D" region and its implications
Thorne Lay, Quentin Williams, Edward J. Garnero, Louise Kellogg, and Michael E. Wysession   299

Ultralow velocity zone at the core-mantle boundary
Edward J. Garnero, Justin Revenaugh, Quentin Williams, Thorne Lay, and Louise H. Kellogg   319

PREFACE

The core-mantle boundary (CMB) is the largest density interface within the Earth's interior, and the change in material properties is as significant as that between the solid Earth and the hydrosphere. The two giant heat engines responsible for plate tectonics and the geodynamo dynamically interact at this boundary. The CMB is as dynamic as the Earth's outer skin, and seismological observations show that dense mantle dregs, anti-continents, raft around the mantle's base and are underlain by giant molten oceans. The mantle dregs are sheared and have developed strong seismic anisotropy. The CMB dynamically interacts with the planet's surface; old oceanic lithosphere has been imaged down to the mantle's base where it accumulates. Reversal pathways of the geomagnetic field may be controlled by molten oceans at the CMB. The molten oceans – the seismological ultra low velocity zone – may be the source of hot-spot volcanism at the Earth's surface. The complex magnetohydrodynamics of the geodynamo within the fluid outer core of the Earth has been simulated on a computer, and the models show a surprising similarity to the observed geomagnetic field, including the observed westward drift and episodes of flux expulsion.

Like the interdisciplinary plate tectonics revolution which brought order to our view of the lithosphere, deciphering the state, history, and dynamics of the deep interior will likewise be interdisciplinary. The principal disciplines contributing toward an understanding of the core-mantle region include seismology, high-pressure and high-temperature mineral physics, geodynamics, geomagnetism, magnetohydrodynamics, geodesy, and geochemistry. In an effort to promote dialog between these fields, AGU's Committee on Studies of the Earth's Deep Interior organized a special session at the Fall 1996 meeting entitled, "Theoretical and observational constraints on the core-mantle boundary region". A diverse range of exciting, provocative, and interdisciplinary presentations were followed by spirited discussion. We selected a number of the most exciting papers and invited their authors to contribute to this volume. In addition, we solicited papers from individuals who were unable to present papers at the special session.

This field is rapidly moving forward. Hardly a month passes without the publication in the major scientific journals of several new and stimulating results on the nature of the core-mantle region. We hope that the present volume will contribute to an interdisciplinary understanding of a region of the Earth that plays a fundamental role in its dynamics and evolution. Finally, we would like to thank the many reviewers of the manuscripts published here. Their dedicated service helps maintain the high editorial standards expected for an AGU publication.

Michael Gurnis
Michael E. Wysession
Elise Knittle
Bruce A. Buffett
Editors

INTRODUCTION

Michael Gurnis
Michael E. Wysession
Elise Knittle
Bruce A. Buffett

The solid lower mantle plays a fundamental role in the thermal and chemical evolution of the planet. Within the fluid outer core the geomagnetic field is generated. Two giant heat engines, one powering tectonic processes the other the geodynamo, dynamically interact at the largest of all density interfaces within the planet. The mantle side of this interface, D", is extraordinarily complex with a myriad of fine structure. Thermal and chemical heterogeneity, anisotropy, and melting atop the core-mantle boundary (CMB) may all be required in order to explain observed seismological fine structure. The picture emerging is one in which the lower boundary of the mantle is as complex as the mantle's top boundary. One of a number of plausible scenarios of this region has anti-continents and molten anti-oceans rafting around the base of the mantle on geologic time scales, perhaps in harmony with the motions of the lithosphere, periodically giving rise to super-plumes, and modulating reversals of the geomagnetic field.

Recent advances in seismology have provided amazing insights into the geophysical processes at work in the lower mantle and at the core-mantle boundary. Lateral variations in compressional wave speeds have never been detected in the fluid outer core and none are expected (Liu and Dziewonski). However, the lateral variations of seismic wave speeds in the mantle are refined in two new inversions presented by Liu and Dziewonski and van der Hilst and colleagues. They corroborate earlier inversions showing that the velocity anomalies in the lowermost 500 km of the mantle are very long wavelength. These lateral variations in seismic velocity and radial gradients just above the CMB lead to large changes in the waveforms and travel times of body waves diffracting long distances along the CMB. Valenzuela and Wysession show that seismic constraints from diffracted waves are largely consistent with other inversions of large-scale structure

— slow beneath the central Pacific and fast beneath the circum-Pacific. A marked advance has been the identification in some regions of the continual transport of subducted lithosphere from the surface to the CMB, as highlighted by van der Hilst and colleagues, ending the debate over whether or not slabs can penetrate into the lower mantle. But continual does not mean continuous: the slabs seem to break up into smaller segments below a depth of about 2000 km. Moreover, at the very base of the mantle, there is a poor correlation between P- and S-wave velocities, suggesting the presence of significant amounts of either chemical heterogeneity or anisotropy.

In addition to long wavelength, small-scale structure is also inferred in the deep mantle. Seismic waves passing through the core have been known to have precursors to PKP and PKKP phases which Shearer and colleagues argue is better modeled with small-scale velocity perturbations (with ~ 8-km scale lengths) distributed throughout the mantle, perhaps evidence of incomplete mantle mixing and small-scale chemical heterogeneity throughout the mantle.

Another fine scale feature of the lower mantle is the ~3% increase is seismic velocity 250 km above the CMB which has been the subject of considerable previous study. Wysession and colleagues argue that the presence of the discontinuity (often called the D" discontinuity) does not correlate with the large-scale seismic structure of the lower mantle. The D" discontinuity is traditionally interpreted as a chemical or phase transition, and to this list Wysession and colleagues add the possibility that the cold thermal anomaly associated with ancient slabs leads to increased seismic velocity. The dynamics of a chemical layer at the base of the mantle are addressed in separate papers by Tackley and Sidorin and Gurnis. Sidorin and Gurnis find that the intermittent character and impedance contrast of the seismic discontinuity at 300 (±

1

about 150) km above the CMB are difficult to explain with a simple chemical layer and argue that rather dramatic shifts in composition would be required to match the observed seismic and likely dynamic properties of this layer. Tackley also has significant problems interpreting the D" discontinuity as a simple chemical layer. He finds in models with a high-viscosity lower mantle that stable thermochemical mega-plumes can form and may provide an explanation for the structure imaged seismically beneath Africa by van der Hilst and colleagues.

The D" may be anisotropic to seismic shear waves as horizontally polarized waves may travel faster in places than vertically polarized waves. Lay and colleagues review constraints on seismic wave anisotropy in D" and argue that it is most likely present in the circum-Pacific areas which have higher-than-average lower mantle shear velocity. Kendall and Silver also argue that anisotropy is present largely beneath the Caribbean and Alaskan regions and may be due to a shape-preferred orientation, analogous to the horizontal layering of slower-velocity material. They suggest that such a structure could result from the horizontal inclusions of former oceanic crust within laterally accumulated ancient slabs. Stixrude draws theoretical constraints on the likely magnitudes of anisotropy of major mineralogic components of the lowermost mantle at the pressures of the CMB. All phases undergo significant pressure-induced shifts in their elastic anisotropy, with the changes of magnesiowuestite and SiO_2 being particularly notable.

One of the most exciting discoveries of the past decade has been the ultralow velocity zone (ULVZ), discussed by Garnero and colleagues. Involving velocity reductions of at least 10% in a layer that extends up to 40 km above the CMB, the ULVZ may be the result of partial melting, and therefore provides a vital constraint on the melting curve and thus the geochemistry of the lowermost mantle. ULVZs may correlate with the surface hot spot distribution but anti-correlate with previously subducted oceanic lithosphere, suggesting that the thickness of the layer is coupled with the pattern of whole-mantle convection. On the other hand, Anderson assesses whether plumes need be derived from the lowermost mantle, and concludes that the present state of geochemical and geodynamic data does not require that D" be the source of plumes. Instead, he proposes that upper mantle/tectonic processes explain midplate volcanism.

The mineral physics of the ULVZ is addressed in various papers. Knittle experimentally studied melting of perovskite at high pressures and argues that the melted phase is more dense than the residual solid and if this melt accumulated it would lead to a molten layer at the base of the mantle, perhaps consistent with the ULVZ. Moreover, she shows that U may

be partitioned into the melt while K would be left behind in the solid. Melting in the lower mantle could have profound implications for the chemical evolution of the mantle and the distribution of heat-producing elements. The possibility that the lowermost mantle may contain a thin layer of high electric conductivity is treated by Poirier and colleagues. Such a layer could have a controlling influence on the propagation of the geomagnetic field through this region of the planet. While not precluding the existence of such a layer, they conclude that simple partial melting and core-mantle boundary reactions may each not suffice to generate such a layer. If the ULVZ actually represents a zone of partial melt is not certain. If the region is molten and the composition of D" can be determined, powerful constraints on the geotherm will emerge. In an effort to better understand this region, Williams reviews our understanding of the mantle and core and estimates a 1000-2000 K superadiabatic temperature increase across D".

Historical observations of changes in the Earth's magnetic field provide important clues about dynamical processes in the fluid core. Fluid velocities inferred at the top of the core from changes in the magnetic field have been used in recent studies to calculate changes in the angular momentum of the core. Comparisons with the measured changes in the angular momentum of the mantle afford new opportunities to learn about the dynamics of the core and its interaction with the mantle. Holmes applies inverse theory to explain observed variations in the length of day under the assumption that the electromagnetic coupling fully explains the transfer of angular momentum between the core and mantle. This study explores the range of electrical conductivity profiles in the lower mantle that are needed to successfully explain the observations with electromagnetic coupling. Fluid motions in the core that contribute to changes in the angular momentum are constrained by the influences of rotation to occur as fluid cylinders aligned with the axis of rotation. Motions of adjacent cylinders are coupled by the radial component Bs of the magnetic field, which permits waves in the core known as torsional oscillations. Zatman and Bloxham attribute fluctuations in the fluid velocities at the top of the core to torsional oscillations and infer the strength of the radial field that best explains the changes in the fluid velocities. Such studies open new lines of inquiry which demand improvements in the dynamical description of the fluid core. Buffett takes a step in this direction by developing a theoretical model to describe changes in the length of day. The model is based on torsional oscillations in the fluid core, and includes the effects of electromagnetic, topographic and gravitational coupling to the mantle. Distinctive features of the different coupling mechanisms provide testable predictions for future investigations.

Paleomagnetic measurements provide a valuable record of the behavior of the Earth's magnetic field over geological time. Although these measurements do not resolve features in the field with the same detail possible using historical data, the longer record is essential to establish the average structure of the field. Features of the field that persist over millions of years suggest a controlling influence of the mantle on the dynamics of the core. Gubbins reviews current attempts to relate paleomagnetic observations to the dynamics of the dynamo process. Interactions between the core and the mantle may arise from topography, as well as lateral variations in electrical conductivity or heat flow at the core-mantle boundary. A convincing case is made that interpretations of the paleomagnetic field hold important insights about the core-mantle boundary region.

One of the most exciting areas of deep Earth research is the numerical solution of geodynamo in a spherical geometry. Although the parameter range thought to exist within the outer core is still not accessible to numerical models, Kuang and Bloxham achieve a dynamical regime that is similar to that of the geodynamo. They find similarities between the model and the observed field, including field strength and the secular variation of the field, particularly the westward drift and episodes of flux expulsion.

Deep Subduction and Aspherical Variations in *P*-Wavespeed at the Base of Earth's Mantle

Rob D. van der Hilst

Massachusetts Institute of Technology, Department of Earth, Atmospheric, and Planetary Sciences, Cambridge, Massachusetts

Sri Widiyantoro[1]

Research School of Earth Sciences, Australian National University, Canberra, Australia

Kenneth C. Creager, and Thomas J. McSweeney[2]

Geophysics Program, University of Washington, Seattle, Washington

We review some recent results of tomographic imaging that are relevant for deep subduction and the structure of the lowermost mantle. Tomography reveals long, narrow structures of higher than average seismic wavespeed in the lower mantle that are continuous to seismogenic slabs in the upper mantle and whose geographical distribution correlates very well with locations at the surface of plate convergent zones in the past 120 Ma. Near 2000 km depth these long linear features disintegrate into smaller segments. However, in several regions narrow downwellings seem to continue to the base of the mantle where they spread out to form long wavelength structure, thus supporting speculations of a relationship between deep subduction and structural complexity in the lowermost mantle. Upwellings seem to be less ubiquitous than downwellings but are not yet well imaged by the *P*-wave data used. These observations confirm many previous results based on residual sphere analysis and are consistent with whole mantle convection with significant internal heating owing to radioactive decay. Despite whole mantle overturn, it appears that convective flow is significantly distorted in the upper mantle transition zone and also in a transitional interval between approximately 1800 and 2300 km depth. The deep change in planform of

---

[1] Now at: Department of Geophysics and Meteorology, Bandung Institute of Technology, Bandung, Indonesia.

[2] Now at: IRIS, Data Management Center, Seattle, WA 98105.

The Core-Mantle Boundary Region
Geodynamics 28
Copyright 1998 by the American Geophysical Union.

INTRODUCTION

Numerous analyses of seismic body waves, both compressional and shear waves, have indicated a significantly enhanced level of structural heterogeneity in the lowermost several hundreds of kilometers of the Earth's lower mantle. Indeed, it has been argued that the amplitude and the scale of the lateral variations in seismic wavespeeds in the core mantle boundary region are comparable to those observed in the lithospheric part of the

convection is not necessarily due to changes in state, but joint inversion of P- and S-wave data begin to indicate that it coincides with a change in the ratio between bulk sound and shear wavespeed. This behavior of the elastic moduli, which is reported elsewhere, suggests widespread chemical heterogeneity near base of Earth's mantle. The lowermost 800 km or so of the mantle thus contains important clues for understanding the composition and evolution of Earth. Unfortunately, it is poorly sampled by direct P-waves, but data coverage can be improved by using core phases. Our preliminary study shows that the current P model explains much of the PKP differential travel time residuals that have been used to study heterogeneity just above the core mantle boundary, but that it underestimates their amplitude. Joint inversion of the differential times along with our direct P data is expected to reduce this apparent bias in heterogeneity level in our current wavespeed model and it fully exploits the advantages of differential residuals without having to make assumptions about the location of the anomalies.

Earth's upper mantle. This need not be surprising since these regions represent the thermal (and perhaps chemical) boundary layers of convection in Earth's mantle. For excellent reviews of this subject we refer to *Jordan et al.* [1989], *Lay* [1989], *Lay et al.* [1990], *Wysession* [1996b], and *Loper and Lay* [1995].

Our understanding of the nature, origin, and evolution of the heterogeneity in the CMB region hinges critically on the issue as to the scale of mantle flow, and, in particular on the existence of significant thermal boundary layers elsewhere in Earth's mantle, for instance at 660 km depth. Although still controversial [*Lay*, 1994; *Loper and Lay*, 1995], several investigators [e.g., *Davies and Gurnis*, 1986; *Wysession*, 1996a,b; *Kendall and Silver*, 1996] have speculated that at least part of the aspherical structure above the CMB can be attributed to the deep subduction of slabs of former oceanic lithosphere. This interpretation implies significant mass exchange between upper and lower mantle associated with downwellings and is inconsistent with convection models that invoke layered convection with some flux between the layers by means of plumes to explain the isotope data [e.g., *Kellogg and Wasserburg*, 1990; *O'Nions and Tolstikhin*, 1994; *Allègre et al.*, 1995].

The postulated relationship between deep subduction and aspherical structure in the lowermost mantle is supported by the strong correlation between the locations of subduction in the geologic past (in particular in the Mesozoic) and the distribution of high wavespeed anomalies in the lower mantle according to tomographic images of long wavelength structure [*Richards and Engebretson*, 1992]. The tomographic model used for that study [*Su and Dziewonski*, 1992] represents the effects of dynamic processes averaged over long periods of time and does not reveal the trajectories of mantle

flow in sufficient detail to demonstrate that slabs are indeed connected to the CMB region. However, the main conclusions drawn from long wavelength imaging are substantiated by evidence from recent tomographic imaging, which strongly supports con-vection models in which lowermost mantle structure is causally related to the deep subduction of former oceanic lithosphere.

The tomographic model from which examples are drawn here [*Van der Hilst et al.*, 1997; *Widiyantoro*, 1997] is based on data associated with first arriving body waves. The major disadvantage of such tomography with respect to studies of the lowermost mantle is the uneven data coverage; there are large regions where sampling is inadequate, especially in the southern hemisphere. In future studies we aim to improve this situation by incorporating phases other than P; for example the core phase PKP and the core diffracted (P_{diff}) and reflected phases (PcP). Despite the extensive processing by *Engdahl et al.* [1998] we expect that bulletin data are too noisy for this purpose. Instead, we aim to use high-quality data determined by waveform cross correlation techniques, which can be supplemented by our routinely processed phase data after additional quality control and under application of a proper weighting scheme.

The construction of differential travel time residuals, for instance PcP-P or PKP_{DF}-PKP_{AB}, helps to extract struc-tural signal pertinent to aspherical wavespeed variations in the lowermost mantle. Indeed, it is often assumed that differential PKP residuals are only sensitive to structure just above the CMB. We show that heterogeneity maps based on differential data determined from waveform analysis of the different PKP arrivals are consistent with our current P-wave model. A joint inversion of the different classes of phase data does not rely on an explicit isolation of structural signal and is expected to produce better constraints on lower mantle heterogeneity than either data set alone. In turn, the improved model will enable the calculation of more accurate mantle corrections which may help extract the subtle signal pertinent to CMB topography, heterogeneity, if any, of the outer core, and isotropic and anisotropic structure of the inner core.

EVIDENCE FOR DEEP SUBDUCTION

Van der Hilst et al. [1997] and *Grand et al.* [1997] reported recent advances in the imaging of the aspherical structure of the Earth's (lower) mantle. They showed, for the first time, that tomographic inversions of travel time data now begin to show agreement on structural features on length scales of 500-1000 km, even less in the best sampled regions. The *P*-wave model [*Van der Hilst et al.*, 1997; *Widiyantoro*, 1997], hereinafter referred to as P97, is based on a large data set of reported *pP*- and *P*-wave arrival times and earthquake locations originally published by the International Seismological Centre (ISC) and the U. S. Geological Survey's National Earthquake Information Center (NEIC) but extensively reprocessed by Engdahl and co-workers [*Engdahl et al.*, 1998]. The technique used is very similar to that described by *Widiyantoro and Van der Hilst* [1996] in their regional study of Indonesia. The *S*-wave study [*Grand*, 1994] - the current model, presented in *Grand et al.* [1997], is hereinafter referred to as S97 - is based on carefully processed travel times extracted from a range of shear wave arrivals. For the interpretation of the data, both studies use local basis functions (constant wavespeed blocks, or cells, with dimensions of the order of $2° \times 2° \times 150$ km) for the parametrization of the model space, which enables the detection of spatial variations in wavespeeds on a scale length that is significantly smaller than when global basis functions, such as spherical harmonics, are used for the representation of structure. The potential resolution of variations in wavespeed over relatively short distances allows the mapping of mantle structure in well sampled regions of the mantle in a detail that was not previously possible.

P97 and S97 reveal long, narrow linear features of faster-than-average seismic wavespeeds in the mid mantle beneath the Americas and beneath southern Asia (Figure 1). These linear structures are detected to at least 1800 km depth. Once identified, these structural features can also be recognized clearly in long wavelength models of shear wavespeed such as those published by *Su et al.* [1994], *Li and Romanowicz* [1996],

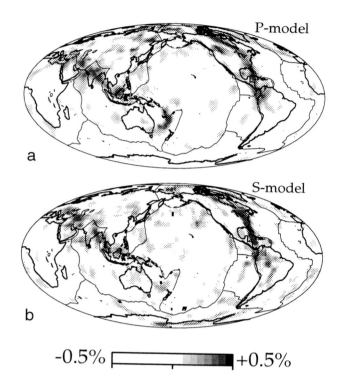

Figure 1. Lateral variation in compressional (top) and shear (bottom) wavespeeds in the lower mantle at about 1350 km depth (After *Van der Hilst et al.* [1997] and *Grand et al.* [1997]). For display purposes the diagrams only depict the mantle region where the wavespeed is higher-than-average.

Masters et al. [1996], in the *P*-wavespeed model by *Bolton and Masters* [1996], and in earlier block-model inversions using the original set of ISC travel time residuals [*Inoue et al.*, 1990; *Vasco et al.*, 1994]. This robustness of the observations along with results of resolution tests with synthetic data calculated from known aspherical test models (e.g., Figure 2A,B in *Van der Hilst et al.* [1997]) supports the existence of the long, linear structures in the lower mantle.

Displays of wavespeed variations at several depths demonstrate that the long, narrow structure in the lower mantle beneath the Americas can be traced over a large depth interval (Plate 1). In the uppermost part of the lower mantle (700 km, bottom panel in Plate 1) the fast anomalies form a narrow structure that extends with only few interruptions from the Hudson Bay region in northern Canada across central America to as far south as northern Chile in South America. Comparison with *S* wave maps at the same depth [*Grand et al.*, 1997] indicates that variations in amplitude and sign

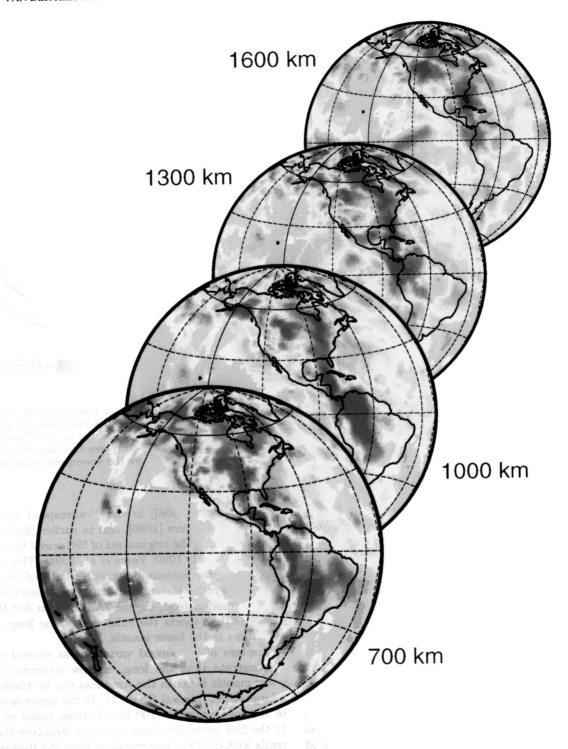

Plate 1. Lateral variation in seismic wavespeed at different depth levels beneath north, central, and south America after the model published by *Van der Hilst et al.* [1997] and *Widiyantoro* [1997]. Mantle regions with insufficient data coverage are depicted in gray.

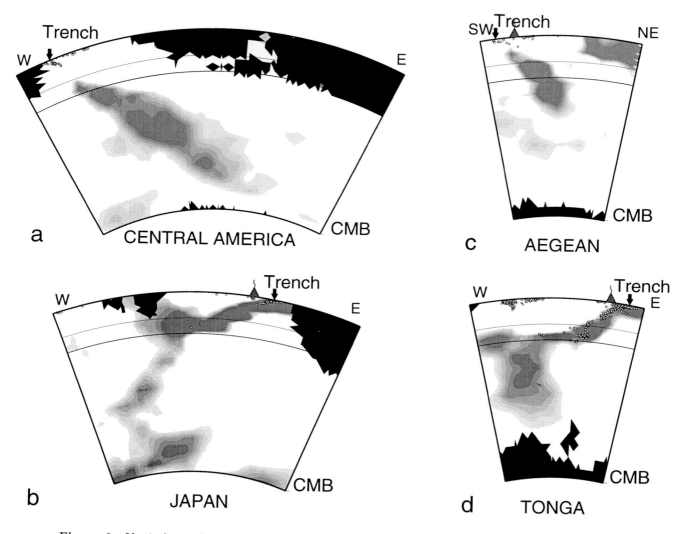

Figure 2. Vertical mantle sections from the Earth's surface to the core mantle boundary across (a) Middle America, depicting the lower mantle structure known as the 'Caribbean' anomaly, (b) central Japan and east Asia, (c) southeastern Europe, across the convergent margin between the African and Eurasian plates, and (d) northern Tonga. For display purposes the diagrams only depict the mantle region where the wavespeed is higher-than-average. Mantle regions with insufficient data coverage are depicted in black.

of the wavespeed perturbations along the strike of the linear feature are probably real. This may justify the interpretation of even smaller scale structure than has so far been discussed in literature. At larger depth, the anomalies become slightly broader (note that in Plate 1 the dimensions of the maps are adjusted to account for the decreasing radius within Earth) and the geometry changes, in particular in the far north where alternations of high and low wavespeed are visible [also in S97]. Beneath southernmost South America the high wavespeed anomaly is not visible beneath about 1300

km depth due to a deflection of the structure at a shallower depth in the mantle [*Grand*, 1994; *Engdahl et al.*, 1995; *Grand et al.*, 1997]. Vertical mantle sections further illustrate the continuity of high wavespeed trajectories from the upper mantle to very large depths in the lower mantle (Figure 2).

The continuity of the deep high wavespeed anomalies to subduction zones delineated by seismicity in the upper mantle and the strong spatial correlation with locations of known convergent margins in the Mesozoic [*Richards and Engebretson*, 1992; *Grand*, 1994; *Van der*

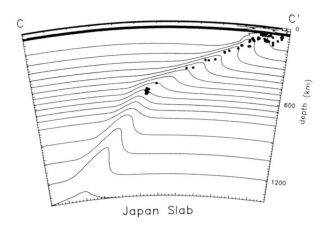

Figure 3. Cross section CC' from *Creager and Jordan* [1986]. The contours connect points of constant *P* wavespeed. The location of the cross section is almost the same as that used for Figure 2b. The increase in dip angle in the top of the lower mantle is in excellent agreement with inferences from tomographic imaging (Figure 2b).

Hilst et al., 1997; *Grand et al.*, 1997] provide strong support for the ability of slabs to sink across the upper mantle transition zone and the 660 km discontinuity into the lower mantle. Indeed, the recent tomographic models confirm inferences pertinent to the fate of slabs from studies based on residual sphere analysis for selected arc segments [*Jordan*, 1977; *Creager and Jordan*, 1984, 1986; *Fischer et al.*, 1988]. Compare, for instance, the result of tomographic imaging (Figure 2b) with inferences from residual sphere analysis for central Japan (Figure 3). For other regions, for instance Java and Izu Bonin, the residual spheres were less diagnostic of unhindered slab penetration into the lower mantle and the tomography indicates severe deformation and even deflection of the slab [e.g., *Van der Hilst et al.*, 1991; *Fukao et al.*, 1992; *Widiyantoro and Van der Hilst*, 1996]. Flow in the lower mantle induced by conductive cooling across an interface without mass exchange does not provide a plausible explanation for the observations described above. For a review of arguments against the significance of thermal coupling across a boundary layer at around 660 km depth we refer to *Jordan et al.* [1989]. Here we only make some simple inferences from the cross sections displayed in Figure 2.

The time scale for dynamic processes in the upper mantle is significantly shorter than in the higher viscosity lower mantle. Yet, there is an excellent geographical match between the fast anomalies in the upper- and

lower mantle. Many lower mantle anomalies seem to be continuous to rapidly developing, subduction related structures in the upper mantle. Consider, for instance, the structure beneath Tonga (Figure 2d) and the Marianas [*Creager and Jordan*, 1986; *Van der Hilst et al.*, 1991]. Barring coincidence, the excellent match of lower mantle structures with transient features in the upper mantle is hard to explain by separate flow regimes coupled by heat exchange alone [*Jordan et al.*, 1989]. Moreover, thermal coupling of flow regimes on either side of an interface effectively decouples lower mantle flow from the motion of plates and is not likely to produce the observed geometry (e.g., constant dip angle) of the high wavespeed anomalies, for instance for Central America and southeastern Europe (Figures 2a,c). Also, for several subduction systems the time constraints provided by the known tectonic history and the inferred depth to the leading edge of the high wavespeed slab rule out a controlling role of a very slow physical process such as thermal conduction, unless the tectonic reconstructions are in error or flow in the lower mantle is much faster than expected from its high viscosity.

The seismological observations described above can best be explained by some form of whole mantle flow with substantial flux across the upper mantle transition zone into the lower mantle. We have not yet used our images to quantify the amount of mass exchange but they suggest that it should be at least the volume involved in the slabs of subducted lithosphere. While this 'slab flux' is likely to be a lower bound of the actual mass exchange [*Puster and Jordan*, 1997], it is up to two orders of magnitude higher than expected from mantle flow models used to explain noble gas data [e.g., *Hofmann*, 1997].

COMPLEX FLOW ACROSS THE UPPER MANTLE TRANSITION ZONE

The conclusion that the seismic data can best be explained by large scale convective flow does not mean that the upper mantle transition zone has no effect on the flow pattern. On the contrary: flow across the upper mantle transition zone can be complex, with local deflection and kinking of the slab between 500 and 1000 km in depth, in particular in some segments of the western Pacific arc system [e.g., *Zhou and Clayton*, 1990; *Van der Hilst* et al., 1991; *Fukao et al.*, 1992; *Van der Hilst*, 1995; *Thoraval et al.*, 1995; *Widiyantoro and Van der Hilst*, 1996; *Castle and Creager*, 1997].

It has been argued from seismic observations and numerical and analog modeling of mantle flow that rapid lateral motions of the convergent margin can be causally related to changes in the morphology of the slab when the down-wellings interact with changes in physical properties at or near the 660 km discontinuity (e.g., a 30-fold jump in viscosity; an endothermic, iso-chemical phase change in mantle silicates; possible changes in rheology; perhaps a slight increase in intrinsic density). Several feed back relationships have been investigated [*Zhong and Gurnis*, 1995] but the complex geodynamical system is not yet fully understood. A more complete discussion of this issue is beyond the scope of this paper and for an overview of the arguments and modeling results we refer to *Gurnis and Hager* [1988], *Van der Hilst and Seno* [1993], *Zhong and Gurnis* [1995], *Griffiths et al.* [1995], *Christensen* [1996], and *Olbertz et al.* [1997].

While the effect of the 660 km discontinuity on down-wellings is still debated, its role in stratifying mantle flow far away from convergent margins is perhaps even more uncertain. Owing to limited data coverage our technique does not constrain upper mantle structure beneath such regions, but *Katzman et al.* [1997] provide compelling evidence for upper mantle heterogeneity beneath the western Pacific, which they interpret in terms of convective rolls above the 660 km discontinuity.

TRANSITION TO THE LOWERMOST 800 KM OF THE MANTLE

What is the ultimate fate of the slabs that sink, somehow, across the upper mantle transition zone? The new class of high resolution models suggests that linear features that are related to subduction of former oceanic lithosphere persist to at least 1800 km depth and that some of them continue all the way to the base of the mantle. For instance, the high wavespeed anomaly beneath central America (Figure 2a), which was first associated with subduction more than two decades ago [*Jordan and Lynn*, 1974], provides support for the postulation by *Kendall and Silver* [1996] that piling up of slab material atop the CMB in that region results in an effective anisotropic medium that causes the shear wave splitting that they observed. The deep anomaly as inferred from P97 is very similar to the structure deduced from S97, although the correlation between variations in P and S wavespeed seems to break down just above the CMB. Another region beneath which narrow high wavespeed anomalies form a continuous trajectory from the Earth's surface to the CMB is central Japan [Figures 2b, 3]. The anomaly is relatively narrow in the mid mantle but spreads out at the base of the mantle

to form a large scale high wavespeed anomaly above the CMB.

These examples suggest that some downwellings can reach the CMB, but this may not be the fate of all slabs. Indeed, in P97 the heterogeneity pattern of the mid mantle does not seem to connect in a simple way to the large scale anomalies above the CMB. This is illustrated in Plate 2, which depicts the lateral variation in P-wavespeed according to P97 at 1300, 2100, and 2700 km depth, respectively. The long, linear structures that characterize the mid-mantle seem to disintegrate across a depth interval of roughly 1800-2300 km, in which the pattern of heterogeneity seems to be dominated by a distribution of - in map view - more equidimensional structures. This change in the planform of mantle structure roughly coincides with a change in heterogeneity spectrum [*Su and Dziewonski*, 1992; *Liu and Dziewonski*, 1994, this issue] but the inferred depth interval over which the changes occur is somewhat model dependent which renders the inference of this transition zone in the lower mantle tentative.

Even though a change in physical parameters or composition may not be required to explain the observed change in heterogeneity pattern [*Bercovici et al.*, 1989; *Glatzmaier et al.*, 1990] it is interesting to note that the transition zone as speculated upon by *Van der Hilst et al.* [1997] coincides with the depth range where the proportionality between variations in P- and S-wavespeed begins to break down [*Bolton and Masters*, 1996; *Robertson and Woodhouse*, 1996; *Su and Dziewonski*, 1997]. A joint inversion of P- and S- data for global variations in bulk sound and shear wavespeed by *Kennett et al.* [1998] confirms most previous conclusions and shows that the contribution of changes in bulk modulus to the variations in seismic wavespeed decreases gradually with increasing depth in the lowermantle and reaches a minimum at about 2000 km depth. The amplitudes of variations in bulk modulus increase again towards the base of the mantle but at those depths there appear to be large regions where the ratio of bulk sound and shear wavespeed anomalies can no longer be explained by thermal perturbations alone. The inferred behavior of the elastic moduli is suggestive of large scale chemical heterogeneity in the lowermost mantle [*Robertson and Woodhouse*, 1996; *Su and Dziewonski*, 1997; *Kennett et al.* 1998].

The presence of widespread chemical heterogeneity near the base of the mantle may also explain the high frequency (1 Hz) precursors to the core phase PKP_{DF}, which are thought to be caused by small (10s of km) randomly distributed scatters [*Haddon and Cleary*, 1974].

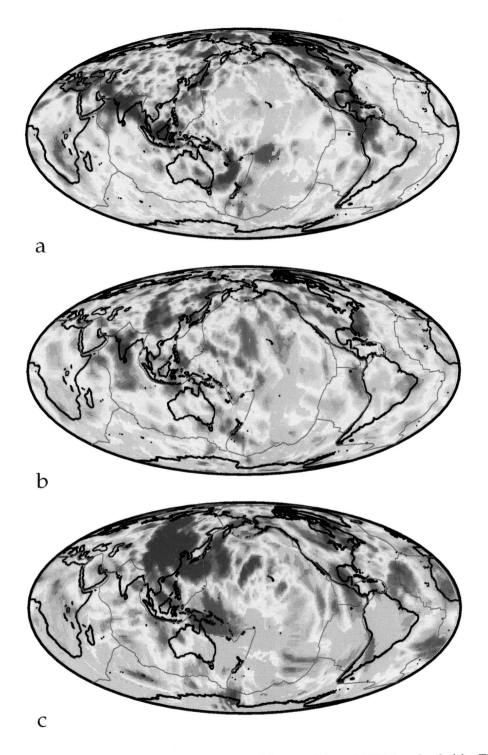

Plate 2. Lateral variation in *P*-wavespeed at 1350 (a), 2100 (b), and 2750 km depth (c). The gray patches indicate regions where sampling by the *P*-paths used in our study is unsufficient to constrain variations in wavespeed.

Hedlin et al. [1997] and *Shearer et al.* [this issue] provide compelling evidence that such scatterers can exist up to at least several hundreds of km above the CMB, perhaps throughout the lower 1000 km of the mantle.

HOW ABOUT UPWELLINGS?

Low wavespeed anomalies are not as well imaged by our tomographic technique as the fast anomalies associated with downwellings. Large scale upwellings in the deep mantle are difficult to detect in the mid-mantle if they become more focused and cylindrically shaped as they ascend [e.g., *Bercovici et al.*, 1989]. There may also be a natural bias towards the imaging of high wavespeed anomalies since slow areas can be overlooked by first arriving waves owing to diffraction [*Wielandt*, 1987; *Nolet and Moser*, 1993]. For practical purposes, however, another important complication for plume imaging is that upwellings are usually not located in regions where many earthquakes occur. As a result, the effective data coverage by first arriving P waves is typically significantly worse than that of the mantle regions associated with downwellings. Mantle structure beneath intraplate regions is better mapped by tomographic studies that exploit structural information from direct and multiply reflected body waves and from surface waves [e.g., *Su et al.*, 1994; *Grand*, 1994; *Masters et al.*, 1996]. Even though we are not as confident in the low wavespeed anomalies as in the fast structures, we can make some first order observations from the current whole mantle P-wave model.

Some large scale features stand out owing to slower-than-average P-wave propagation. In agreement with inferences from shear wave data [*Su et al.*, 1994; *Grand et al.*, 1997], the P-wave images suggest that beneath southern Africa a slow anomaly extends upwards from the core mantle boundary to a depth of about 800 km. A possible continuation to even shallower depths can not be resolved with the data used. Beneath the Society Islands in the southwest Pacific our results suggest the presence of a continuous conduit of slower-than-average wavespeeds from Earth's surface to a depth of about 2500 km. This structure is not well resolved, but the observations could not all be attributed to smearing in radial direction since in our imaging we do not use P-waves with such small ray parameters that they would travel sub-vertically from a 2500 km depth to Earth's surface. Interestingly, the low wavespeed feature does not seem to connect to the pronounced slow anomaly above the CMB, which is located vertically below the Ontong Java plateau, i.e. much further to the west. We remark, however, that within the current model un-

certainties these structures may well represent different parts of a single, large slow anomaly in the lowermost mantle beneath the southwest Pacific.

P-WAVESPEED ABOVE THE CMP FROM TOMOGRAPHY

At this stage of model development we chose not to discuss the images of the structure above the CMB in detail or compare them with independent wavespeed variations from independent studies. We realize that the current model is severely influenced by uneven sampling, as is evident from the large gray patches in Plate 2c, and that construction of better P-wave models is possible. In fact, we are in the process of adding complementary data in order to reduce the sampling problem and thus improve the lower mantle part of our model [*Kárason et al.*, 1997]. A detailed comparison with other results is not so meaningful either; previously published P-wave models constrain structure at much longer wave lengths (e.g., the model by *Wysession* [1996a] based on P_{diff} differential residuals) and the comparison with S-wave models is complicated by the effect of chemical heterogeneity and, perhaps, seismic anisotropy. We do note, however, that the amplitude of anomalies in the lowermost mantle is generally lower in P97 than suggested by other studies, which probably indicates that too much damping was used to constrain the solution. Despite these uncertainties, several robust features in the images can be addressed.

A pronounced, well resolved structure of higher-than-average compressional wavespeed is located in the lowermost mantle beneath east Asia. This high wavespeed anomaly has also been reported by previous investigators of heterogeneity in the lowermost mantle. Mounting evidence from seismic imaging for deep subduction [e.g., *Creager and Jordan*, 1986; *Kamiya et al.*, 1989; *Van der Hilst et al.*, 1997] suggests that this structural heterogeneity above the CMB is the result of deep subduction of former oceanic lithosphere beneath island arcs in the west Pacific and southeast Asia. Figure 2b suggests that the downwelling is locally continuous from Earth's surface to the lowermost mantle and spreads out to form long wavelength structure atop the CMB. The patches of high velocity beneath the Americas (Plate 2c) are consistent with inferences from long wave length models [e.g., *Su et al.*, 1994] and possibly represent remnants of previously subducted lithosphere (e.g., Figure 2a). Other robust observations are the pronounced slow anomalies in the lowermost mantle vertically below southern Africa and the Ontong Java plateau in the west Pacific. The former is less well constrained by the data used to construct P97 (note the gray areas in Plate 2c) than in *S* wave studies [*Su et al.*, 1994; *Grand et al.*, 1997]. The latter is in excellent agreement with previous *P*-wave tomography [*Inoue et al.*, 1991; *Vasco et al.*, 1994], *S*-wave tomography [*Su et al.*, 1994; *Masters et al.*, 1996; *Li and Romanowicz*, 1996], and with inferences from imaging with *ScS-S* differential travel times [*Wysession et al.*, 1994], anomalous core diffractions [*Wysession*, 1996a], core refracted shear waves *SKS* [*Garnero and Helmberger*, 1993; *Liu and Dziewonski*, 1994, this issue], and compressional *PKP* waves [*Garnero et al.*, 1993; *Song and Helmberger*, 1997]. As mentioned above, this pronounced low wavespeed anomaly seems confined to the lowermost 300 or so km of the mantle, whereas further to the east, beneath the southwestern Pacific Society Islands, a (poorly constrained) low velocity conduit is continuous from a depth of approximately 2500 km to Earth's surface. A possible connection between two slow anomalies just above the CMB beneath the south Pacific and the conduit to shallower depths is not resolved by the data used.

Variations in *P* wavespeed seem to differ significantly from shear heterogeneity in some regions above the CMB. Perhaps the most significant disparity between *P*- and *S*- anomalies occurs beneath the Alaska and the Bering sea; P97 suggests that *P* wavespeed is lower than average whereas most shear wave studies reveal faster than average wave propagation [e.g. *Lay et al.*, 1997]. A joint inversion of *P*- and *S*-bulletin data reveals anomalous ratios between bulk sound and shear wavespeed in this region, indicating variations in bulk composition [*Kennett et al.*, 1998].

ADDITIONAL CONSTRAINTS FROM LATER PHASES SUCH AS PKP

The sampling of structure in the lowermost mantle by compressional waves can be improved significantly by the addition of travel time data from phases such as the core reflections (*PcP*) or arrivals at distances in and beyond the *P*-wave shadow zone, such as core diffractions (*P*$_{\text{diff}}$) and refractions (*PKP*). For these phases the bulletin data are rather noisy, despite the careful processing by *Engdahl et al.* [1998]. Therefore, for the extraction of structural signal pertinent to the lowermost mantle from these phases we will initially rely on waveform data. Here we present results of a reconnaissance study that should lead to the incorporation of *PKP* data in our global tomography.

Ray paths of the phase *PKP*$_{\text{AB}}$ bottom in the shallow regions of the outer core and are almost core grazing (Figure 4). As a result, *PKP*$_{\text{AB}}$ paths are signif-

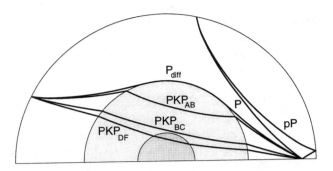

Figure 4. Path geometry and nomenclature of the compressional phases that have been or will be used in our whole mantle tomography. The *P* and *pP* phases have been used in the model discussed here (P97). Differential times for the various *PKP* branches and *P*$_{\text{diff}}$ have been used by *Kárason et al.* [1997].

icantly longer near the base of the mantle than the steeper paths associated with the phase *PKP*$_{\text{BC}}$, which bottom in the lower regions of the outer core, or with *PKP*$_{\text{DF}}$, which pass through the inner core (Figure 4). If heterogeneity in the outer core can be neglected and if large *PKP*$_{\text{DF}}$ residuals owing to inner core anisotropy (as used by, e.g., *Morelli et al.*, 1986; *Shearer and Toy*, 1991; *Creager*, 1992; *Song and Helmberger*, 1993; *Tromp*, 1993; *Song*, 1996; *Creager*, 1997] are omitted from the data set, differential residual times formed by pairing the *AB* branch with the *BC* or *DF* branch (referred to as *AB-BC* times and *AB-DF* times, respectively) are expected to be most sensitive to structure in the lowermost mantle [e.g. *Sylvander and Souriau*, 1996; and many others].

Figure 5a shows about 1150 observed *PKP*$_{\text{AB}}$-*PKP*$_{\text{DF}}$ and *PKP*$_{\text{AB}}$-*PKP*$_{\text{BC}}$ travel time residuals [*McSweeney*, 1995; *McSweeney et al.*, 1997] with respect to the global average according to model *ak135* [*Kennett et al.*, 1995] plotted at the core entrance and exit points of the *PKP*$_{\text{AB}}$ paths. The *PKP*$_{\text{BC}}$ and *PKP*$_{\text{DF}}$ entrance and exit points are located nearly directly beneath sources and receivers and fill in gaps in *P* wave sampling beneath South America, Tonga-New Hebrides, Indonesia, the Philippines, Japan, China, Europe, North America and parts of Africa. The data used in Figure 5a do not include times from rays that traverse the inner core parallel to the spin axis so they do not contain the large (> 5s) signals produced by inner core anisotropy. However, heterogeneity in the inner core can have a significant effect on *AB-DF* residuals [*Creager*, 1997], which will have to be accounted for in tomographic inversions based on these data.

In order to assess the compatibility of the structural signal in the *P* and the *PKP* data we computed travel

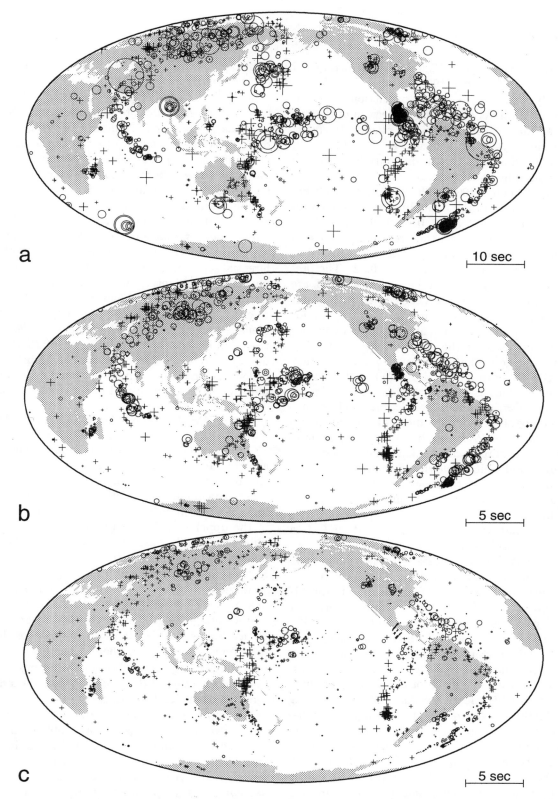

Figure 5. (a) PKP_{AB}-PKP_{DF} and PKP_{AB}-PKP_{BC} differential travel time residuals determined from waveform cross correlation and projected at the entry and exit intersections of the PKP_{AB} path with the core mantle boundary. Pluses are positive residuals, circles are negative. (b) Predictions from the P wave model by *Van der Hilst et al.* [1997] using the same paths as in (a). (c) Same as (b) except predictions are calculated from the lowermost 900 km of the mantle. Note the difference in symbol scale. For a detailed description, see *McSweeney* [1995] and *McSweeney et al.* [1997].

times by tracing the *PKP* paths used by *McSweeney et al.* [1997] and *McSweeney* [1995] through model P97. Subsequently, we compared model predictions with observations: qualitatively by visual inspection of the differential residuals in map view and quantitatively by computing the correlation coefficient between the maps. The contribution of structure in different depth ranges was investigated by considering 5 intervals bounded by interfaces at 0, 660, 1400, 2000, 2600, and 2889 km depth. For structure in the individual layers, the correlation coefficients are 0.20, 0.13, 0.16, 0.19, and 0.14, respectively. Differential travel time residuals calculated for the whole earth, i.e., all five layers, are displayed in Figure 5b; the correlation coefficient is 0.32. The predictions for the bottom 900 km, i.e., layers 4 and 5, are plotted in Figure 5c and show a correlation with observations of 0.24. Notice that the symbol scale used for the model predictions (Figures 5b and 5c) is two times as large as that used for the observations (Figure 5a). Our analysis suggests that the current *P* model underestimates the size of the residuals by up to half an order of magnitude, which is in accord with the findings by *Song and Helmberger* [1997]. This is due to (1) damping of the tomographic solution and (2) the complementary nature of *P* and *PKP* sampling in the deep mantle. *PKP* ray paths sample many regions that are not well sampled by *P* and which have therefore acquired only a small anomaly, if any, in the inversion.

There are two main points we want to make on the basis of this comparison. First, there is a remarkable agreement, in map view, between observed *PKP* data and the predictions from P97, which, we stress, was constructed without the core phases. This is evident from the spectacular match of predicted and observed residuals along the equator: eastward from Brazil into the Atlantic, westward from South America into easternmost Pacific, the east-west gradients beneath the central Pacific, and the anomalies beneath the Indian ocean. Also for central Asia and north America the model predictions agree remarkably well with the observations. *PKP*AB does not sample much of the pronounced high wavespeed anomaly beneath east Asia, but the sampling improves further to the north and the agreement between model and data is excellent. Just west of the date line in the northern and central Pacific, east-west gradients over relatively small distances are revealed by both the observations and the predictions from P97, which suggests that such relatively small scale structure and the associated steep velocity gradients are real. There are also disagreements. For instance, for the region east of Australia the gradients are reversed, and for the Indian Ocean southeast of Africa the sign of the observed anomalies is opposite of the P97 predictions. However,

both regions are poorly sampled by the *P* paths used for the tomography (Plate 2c).

Second, the maps displayed in Figures 5b and 5c and the correlation coefficients computed for different depth intervals suggest that structure in a significant fraction of the lower mantle can contribute to the signal contained in the differential travel time residuals. Apparently, some structure, for instance deep slabs, can be sampled differentially by the *PKP* branches owing to the slight difference in take off angle between the *PKP*AB path and either the *PKP*BC or the *PKP*DF path. Even though we have overestimated its effect by using ray theory, which ignores the effective width of rays, differential sampling may violate the assumptions underlying conventional interpretations of differential residuals. However, it is not a problem in tomographic inversions since the residuals are back projected along all path segments involved [*Van der Hilst and Engdahl*, 1991]. We therefore expect that the inclusion of *PKP* data (and other core data) will improve the constraints on aspherical structure in the entire lower mantle, although most improvements are expected in the lowermost 800 km.

SUMMARY AND CONCLUSIONS

There is increasing consensus from seismological and geodynamical studies that slabs of subducted lithosphere sink deep into Earth's lower mantle and that present-day mantle convection is predominated by some form of whole mantle flow. The pattern of heterogeneity in the upper mantle is determined by ocean-continent differences and by narrow slabs of subducted lithosphere delineated by seismicity. In the mid-mantle, long linear features are detected to at least 1800 km depth and are continuous to seismogenic slabs in the upper mantle beneath major convergence zones. The linear structures may disintegrate at even larger depth but there is evidence that some slabs sink to the base of the mantle and connect to structural heterogeneity atop the CMB, providing strong support for a relationship between aspherical structure at the base of the mantle and material recycled from Earth's surface.

We may have overlooked slow anomalies as a result of uneven data coverage and a combination of diffraction and wavefront healing, but the current images seem to indicate that the aspherical structure in the Earth's lower mantle is predominated by downwellings. This asymmetry in the planform of mantle flow is consistent with thermal con-vection that is driven primarily by internal heating (decay of radio-isotopes) and conductive cooling through the top thermal boundary layer, with basal heating due to cooling of the core amounting to less than 20% of the total energy required for convection [*Schubert et al.*, 1980, *Loper*, 1985; *Davies*, 1988; *Sleep*, 1990; *Davies and Richards*, 1992].

In this scenario of essentially mantle wide flow we identify two depth intervals (between 500-800 km and 1800-2300 km in depth) where the planform of mantle heterogeneity related to convection changes.

Under certain conditions (which depend on thermal structure of the slab, rate of subduction, age of lithosphere, subduction history, trench migration, membrane strength, etc.) the upper mantle transition zone can distort flow trajectories and cause a transient layering that prevents slabs from sinking into the deeper mantle. Along with the increase in viscosity [*Bunge et al.*, 1996] such transient processes may filter out significant amounts of small scale and rapidly varying structures, which may explain the difference between the heterogeneity spectrum of the upper and lower mantle.

At larger depth the character of heterogeneity changes again: between approximately 1800 and 2300 km depth the planar, sheet like slabs seem to break down into laterally more confined downwellings that, eventually, spread out to form long wavelength structure at the base of the mantle. In the scenario of 'penetrative convection' as proposed by *Silver et al.* [1988] this transition in the planform and spectrum of heterogeneity could indicate the depth where the intrinsic density of the ambient mantle increases owing to changes in bulk chemistry so that downwellings lose their excess negative (thermal) buoyancy. The observations could also be indicative of resistance to flow due to deep phase changes in the mantle silicates, for instance stishovite [*Kingma et al.*, 1995], but such transformations are still enigmatic [*Kesson, personal communication*, 1997]. If the change in the planform of heterogeneity does not mark the maximum penetration depth of slabs there must be significant lateral flow within the sheet-like structures towards and into the more localized and equidimensional downwellings (which may be detectable by other means, such as the analysis of birefringence of *S* waves that bottom in this depth interval).

Support for changes in bulk chemistry in the bottom 800 km or so of Earth's mantle comes from joint inversions of *P*- and *S*-data, which indicate that the ratio of variations in bulk- and shear wavespeed is anomalous and inconsistent with expected thermal effects [*Bolton and Masters*, 1996; *Robertson and Woodhouse*, 1996; *Su and Dziewonski*, 1997; *Kennett et al.*, 1998]. The conclusive interpretation of these increasingly robust observations awaits better and more mineral physics data pertinent to the physical con-ditions of the lowermost mantle. However, widespread heterogeneity is also invoked to explain the high frequency precursors to the *PKP* arrivals [*Hedlin et al.*, 1997; *Shearer et al.*, this

issue], and it may represent primitive mantle that survived convective mixing in a mantle with depth dependent viscosity [*Gurnis and Davies*, 1986; *Loper*, 1985; *Davies and Richards*, 1992]. This may go towards reconciling whole mantle flow with inferences from noble gas data (see *Hofmann* [1997] for a review), but the volume of the lowermost 800 km of the mantle is too small to encompass all geochemical reservoirs.

The depth interval between approximately 2000 km and the CMB thus contains many clues to a better understanding of the composition, dynamics, and the thermal and chemical evolution of Earth's mantle. Differential travel times have traditionally been used to isolate structural signal from the CMB region but the relation to heterogeneity in the mantle above is perhaps of as much fundamental interest as the boundary layer structure itself. Unfortunately, in the current global models based on *P*-wave data poor sampling prohibits a more detailed description of the variations in *P* wavespeed in the bottom 800 km of the mantle. Advances can be made by inclusion of core phases. Preliminary results indicate that *PKP* data are compatible with our *P*-wave model. However, it also shows that structure in a large part of the lower mantle can contribute to differential *PKP* signal. Models for heterogeneity above the core mantle boundary based on differential residuals alone can thus be contaminated by structural signal from sources elsewhere in the mantle. A joint inversion of the different classes of phase data will produce better constraints on lower mantle structure than either data set alone. In turn, the improved model enables the calculation of more accurate mantle corrections which may help the extraction from waveforms of structural signal pertinent to more detailed structure of the base of the mantle, CMB topography, and subtle heterogeneity, if any, in the outer core, and may create a clear observation window for further investigations of the inner core.

Note added to proof. Recently, we have successfully incorporated a range of *PKP* and P_{diff} differential travel time residuals in inversions for variations in compressional wavespeed in Earth's mantle [*Kárason et al.*, 1997]. The definition and resolution of structure in the bottom 500 or so km of the mantle has improved dramatically, but the conclusions of this paper do not need to be revised. The new model reduces the variance of the differential times significantly without degrading the fit to the *P* data (as was to be expected from the complementary nature of sampling). The wavespeed variations above the CMB are consistent with independent models. However, in accord with the observations made above we detect sharp lateral gradients in some parts of the core mantle boundary region. The change in planform of heterogeneity from the mid mantle to the CMB region is even more pronounced than inferred from the model used here, and we now begin to resolve the deep slow anomaly beneath Africa and also the connection between the slow anomalies above the CMB and the low wavespeed conduit that continues to shallow depths beneath the southern Pacific.

Acknowledgments: We thank Hrafnkell Kárason for discussions, access to his preliminary whole mantle model based on P, pP, PKP, and P_{diff} data, and for producing Figure 4. Constructive reviews by Peter Shearer, Xiaodong Song, and Michael Wysession improved the manuscript. We thank Michael Gurnis and the co-editors for organizing this monograph, and Bethany Matsko and Frederik Simons for their assistance with the LaTeX preparation of the final manuscript. This research was supported by NSF grant EAR9628087.

REFERENCES

Allègre, C. J., M. Moreira, and T. Staudacher, ^{4}He/^{3}He dispersion and mantle convection, *Earth Planet. Sci. Lett.*, 136, 629-646, 1995.

Bercovici, D., G. Schubert, and G. Glatzmaier, 3-dimensional spherical models of convection in the Earth's mantle, *Science*, 244, 950-955, 1989.

Bolton, H., and G. Masters, A region of anomalous $d\ln V_s/d\ln V_p$ in the deep mantle, *(abstr.) EOS Trans. Am. Geophys. Un.*, 1996.

Bunge, H.-P., M.A. Richards, and J.R. Baumgardner, The effect of viscosity stratification on mantle convection, *Nature*, 379, 436-438, 1996.

Castle, J.C. and K.C. Creager, Topography of the 660 km seismic discontinuity beneath Izu-Bonin: implications for tectonic history and slab deformation (submitted).

Christensen, U.R., The influence of trench migration on slab penetration into the lower mantle, *Earth Planet. Sci. Lett.*, 140, 27-39, 1996.

Creager, K.C., Anisotropy of the inner core from differential travel times of the phases PKP and $PKIKP$, *Nature*, 356, 309-314, 1992.

Creager, K.C., Anisotropic structure and rotation of the inner core, *(abstr.) EOS Trans. Am. Geophys. Un.*, 78, F458, 1997.

Creager, K.C., and T.H. Jordan, Slab penetration into the lower mantle, *J. Geophys. Res.*, 89, 3031-3049, 1984.

Creager, K.C.,and T.H. Jordan, Slab penetration into the lower mantle below the Mariana arc and other island arcs of the northwest Pacific, *J. Geophys. Res.*, 91, 3573-3589, 1986.

Davies, G.F., Ocean bathymetry and mantle convection, 1, Large-scale flow and hot spots, *J. Geophys. Res.*, 93, 10447-10480, 1988.

Davies, G.F., and M. Gurnis, Interaction of mantle dregs with convection: lateral heterogeneity at the core-mantle boundary, *Geophys. Res. Lett.*, 13, 1517-1520, 1986.

Davies, G.F., and M.A. Richards, Mantle convection, *J. Geol.*, 100, 151-206, 1992.

Engdahl, E.R., R.D. van der Hilst, and J. Berrocal, Imaging of subducted lithosphere beneath South America, *Geophys. Res. Lett.*, 22, 2317-2320, 1995.

Engdahl, E.R., R.D. van der Hilst, and R.P. Buland, Global teleseismic earthquake relocation from improved travel times and procedures for depth determination, *Bull. Seism. Soc. Am.* (in press, 1998).

Fischer, K., T.H. Jordan, and K.C. Creager, Seismic constraints on the morphology of deep slabs, *J. Geophys. Res.*, 93, 4773-4783, 1988.

Fukao, Y., M. Obayashi, H. Inoue, and M. Nenbai, M., Subducting slabs stagnant in the mantle transition zone, *J. Geophys. Res.*, 97, 4809-4822, 1992.

Garnero, E.J., and D.V. Helmberger, Travel times of S and SKS: implications for three-dimensional lower mantle structure beneath the central Pacific, *J. Geophys. Res.*, 98, 8225-8241, 1993.

Garnero, E.J., S.P. Grand, and D.V. Helmberger, Low P-wave velocity at the base of the mantle, *Geophys. Res. Lett.*, 20, 1843-1846, 1993.

Glatzmaier, G.A., G. Schubert, and D. Bercovici, Chaotic, subduction-like downflows in a spherical model of convection in the Earth's mantle, *Nature*, 347, 274-277, 1990.

Grand, S.P., Mantle shear structure beneath the Americas and surrounding oceans, *J. Geophys. Res.*, 99, 11591-11621, 1994.

Grand, S.P., R.D. van der Hilst, and S. Widiyantoro, High resolution global tomography: a snapshot of convection in the Earth, *Geol. Soc. Am. Today*, 7, 1-7, 1997.

Griffiths, R.W., Hackney, R., and R.D. van der Hilst, A laboratory investigation of effects of trench migration on the descent of subducted slabs, *Earth Planet. Sci. Lett.*, 133, 1-17, 1995.

Gurnis, M., and G.F. Davies, The effect of depth-dependent viscosity on convective mixing in the mantle and the possible survival of primitive mantle, *Geophys. Res. Lett.*, 13, 541-544, 1986.

Gurnis, M. and B.H. Hager, Controls on the structure of subducted slabs, *Nature*, 335, 317-321, 1988.

Haddon, R.A.W., and J.R. Cleary, Evidence for scattering of seismic PKP waves near the mantle-core boundary, *Phys. Earth Planet. Inter.*, 8, 211-234, 1974.

Hedlin, M., P.M. Shearer, and P.S. Earle, Seismic evidence for small-scale heterogeneity throughout the Earth's mantle, *Nature*, 1997.

Hofmann, A., Mantle geochemistry: the message from oceanic volcanism, *Nature*, 385, 219-229, 1997.

Inoue, H., Y. Fukao, K. Tanabe, and Y. Ogata, Whole mantle P-wave travel time tomography, *Phys. Earth Planet. Inter.*, 59, 294-328, 1990.

Jordan, T.H. and W.S. Lynn, A velocity anomaly in the lower mantle, *J. Geophys. Res.*, 79, 2679-2685, 1974.

Jordan, T.H., Lithospheric slab penetration into the lower mantle beneath the sea of Okhotsk, *J. Geophys. Res.*, 43, 473-496, 1977.

Jordan, T.H., A.L. Lerner-Lam, and K.C. Creager, Seismic imaging of boundary layers and deep mantle convection, In *Mantle Convection*, edited by W.R. Peltier, pp. 98-201, Gordon and Breach Scientific Publishers, 1989.

Kamiya, S., T. Miyatake, and K. Hirahara, How deep can we see the high velocity anomalies beneath the Japanese island arcs? *Geophys. Res. Lett.*, 15, 828-831, 1988.

Kárason, H., R.D. van der Hilst, and K.C. Creager, Improving seismic models of global P-wavespeed by the inclusion of core phases, (abstr.) EOS Trans. Am. Geophys. Un., 78, F17, 1997.

Katzman, R., L. Zhao, and T.H. Jordan, High resolution, 2D vertical tomography of the central Pacific mantle using ScS reverberations and frequency-dependent travel times, J. Geophys. Res. (in press, 1998).

Kellogg, L., and G.J. Wasserburg, The role of plumes in Helium fluxes, Earth Planet. Sci. Lett., 99, 276-289, 1990.

Kendall, J.-M., and P.G. Silver, Constraints from seismic anisotropy on the nature of the lowermost mantle, Nature, 381, 409-412, 1996.

Kennett, B.L.N., E.R. Engdahl, and R. Buland, Constraints on seismic velocities in the Earth from travel times, Geophys. J. Int., 122, 108-124, 1995.

Kennett, B.L.N., Widiyantoro, S., and R.D. van der Hilst, Joint seismic tomography for bulk-sound and shear wave speed, J. Geophys. Res. (in press, 1998).

Kingma, K.J., R.E. Cohen, R.J. Hemley, and H.K. Mao, Transformation of stishovite to a denser phase at lower mantle pressures, Nature, 374, 243-245, 1995.

Lay, T, Structure of the core-mantle transition zone: A chemical and thermal boundary layer, EOS Trans. Am. Geophys. Un., 70, 49, 54-55, 58-59, 1989.

Lay, T., T.J. Ahrens, P. Olson, J. Smyth, and D. Loper, Studies of Earth's Deep Interior: goals and trends, Phys. Today, 43, 44-52, 1990.

Lay, T. The fate of descending slabs, Ann. Rev. Earth Planet. Sci., 22, 33-61, 1994.

Lay, T., E.J. Garnero, C.J. Young, and J.B. Gaherty, Scale length of shear velocity heterogeneity at the base of the mantle from S wave differential travel times, J. Geophys. Res., 102, 9887-9909, 1997.

Li, X.-D. and B. Romanowicz, Global mantle shear-velocity model developed using nonlinear asymptotic coupling theory, J. Geophys. Res., 101, 22245-22272, 1996.

Liu, X.-F., W.-j. Su, and A.M. Dziewonski, Improved resolution of lowermost mantle shear wave velocity structure obtained using SKS-S data, (abstr.) EOS Trans. Am. Geophys. Un., 75, 232, 1994.

Liu, X.-F., W.-j. Su, and A.M. Dziewonski, Global analysis of shear wave velocity anomalies in the lowermost mantle, (this volume).

Loper, D.E., A simple model of whole mantle convection, J. Geophys. Res., 90, 1809-1836, 1985.

Loper, D., and T. Lay, The core-mantle boundary region, J. Geophys. Res., 100, 6397-6420, 1995

Masters, G., S. Johnson, G. Laske, and H. Bolton, A shear-velocity model of the mantle, Phil. Trans. R. Soc. Lond. A, 354, 1385-1411, 1996.

McSweeney, T.J., Seismic constraints on core structure and dynamics, Ph. D. thesis, University of Washington, Seattle, 1995.

McSweeney, T.J., K.C. Creager, and R.T. Merrill, Depth extent of inner-core seismic anisotropy and implications for geomagnetism, Phys. Earth Planet. Inter., 101, 131-156, 1997.

Morelli, A., A.M. Dziewonski, and J.H. Woodhouse, Anisotropy of the inner core inferred from PKIKP travel times, Geophys. Res. Lett., 13, 1545-1548, 1986.

Nolet, G. and T.-J. Moser, Teleseismic delay times in a 3-dimensional Earth and a new look as the S-discrepancy, Geophys. J. Int., 104, 185-195, 1993.

Olbertz, D., M.J.R. Wortel, and U. Hansen, Trench migration and subduction zone geometry, Geophys. Res. Lett., 24, 221-224, 1997.

O'Nions, K., and I. Tolsthikin, Behavior and residence times of lithophile and rare gas tracers in the upper mantle, Earth Planet. Sci. Lett., 124, 131-138, 1994.

Puster, P., and T.H. Jordan, How stratified is mantle convection, J. Geophys. Res., 102, 7625-7646, 1997.

Richards, M.A., and D.C. Engebretson, Large-scale mantle convection and the history of subduction, Nature, 355, 437-440, 1992.

Robertson, G.S., and J.H. Woodhouse, Ratio of relative S to P velocity heterogeneities in the lower mantle, J. Geophys. Res., 101, 20041-20052, 1996.

Rodgers, A., and J. Wahr, Inference of core-mantle boundary topography from ISC PcP and PKP traveltimes, Geophys. J. Int., 115, 991-1011, 1993.

Schubert, G., D. Stevenson, and P. Cassen, Whole planet cooling and the radiogenic heat source content of Earth and Moon, J. Geophys. Res., 85, 2531-2538, 1980.

Shearer, P.M, and K.M. Toy, PKP(BC) versus PKP(AB) differential travel times and aspherical structure in the Earth's inner core, J. Geophys. Res., 96, 2233-2247, 1991.

Shearer, P.M., M.A.H. Hedlin, and P.S. Earle, PKP and PKKP precursor observations: implications for the small scale structure of the deep mantle, (this volume).

Silver, P.G., R.W. Carlson, and P. Olson, Deep slabs, geochemical heterogeneity, and the large-scale structure of mantle convection, Rev. Earth Planet. Sci., 16, 477-541, 1988.

Sleep, N.H., Hotspots and mantle plumes, J. Geophys. Res., 95, 6715-6736, 1990.

Song, X., Anisotropy in the central part of the inner core, J. Geophys. Res., 101, 16089-16097, 1996.

Song, X., and D.V. Helmberger, Anisotropy of Earth's inner core, Geophys. Res. Lett., 20, 2591-2594, 1993.

Song, X., and D.V. Helmberger, PKP differential travel times: implications for 3-D lower mantle structure, Geophys. Res. Lett., 24, 1863-1866, 1997.

Su, W.-j., and A.M. Dziewonski, Predominance of long-wavelength heterogeneity in the mantle, Nature, 352, 121-126, 1992.

Su, W. -j., R.L. Woodward, and A.M. Dziewonski, Degree 12 model of shear velocity heterogeneity in the mantle, J. Geophys. Res., 99, 6945-6981, 1994.

Su, W.-j., and A.M. Dziewonski, Simultaneous inversion for 3D variations in shear and bulk velocity in the mantle, Phys. Earth Planet. Inter., 100, 135-156, 1997.

Sylvander, M., and A. Souriau, P-velocity structure of the core-mantle boundary region inferred from PKP(AB)-PKP(BC) differential travel times, Geophys. Res. Lett., 23, 853-856, 1996.

Thoraval, C., P. Machetel, and A. Cazenave, A locally layered convection inferred from dynamic models of the Earth's mantle, Nature, 375, 777-780, 1995.

Tromp, J., Support for anisotropy of the Earth's inner core from free oscillations, Nature, 366, 678-681, 1993.

Van der Hilst, R.D., and E.R. Engdahl, On ISC pP and PP data and their use in delay time tomography of the Caribbean region, Geophys. J. Int., 106, 169-188, 1991.

Van der Hilst, R.D., E.R. Engdahl, W. Spakman, and G. Nolet, Tomographic imaging of subducted lithosphere below northwest Pacific island arcs, *Nature*, 353, 37-42, 1991.

Van der Hilst, R.D., and T. Seno, Effects of relative plate motion on the deep structure and penetration depth of slabs below the Izu-Bonin and Mariana island arcs, *Earth Planet. Sci. Lett.*, 120, 375-407, 1993.

Van der Hilst, R.D., Complex morphology of subducted lithosphere in the mantle beneath the Tonga trench, *Nature*, 374, 154-157, 1995.

Van der Hilst, R.D., S. Widiyantoro, and E.R. Engdahl, Evidence for deep mantle circulation from global tomography, *Nature*, 386, 578-584, 1997.

Vasco, D.W., L.R. Johnson, R.J. Pulliam, and P.S. Earle, Robust inversion of IASP91 travel time residuals for mantle *P* and *S* velocity structure, *J. Geophys. Res.*, 99, 13727-13755, 1994.

Widiyantoro, S., Ph.D. Thesis, Australian National University, Canberra, 1997.

Widiyantoro, S., and R.D. van der Hilst, Structure and evolution of lithospheric slab beneath the Sunda arc, Indonesia, *Science*, 271, 1566-1570, 1996.

Widiyantoro, S., and R.D. van der Hilst, Mantle structure beneath Indonesia inferred from high-resolution tomographic imaging, *Geophys. J. Int.*, 130, 167-182, 1997.

Wielandt, E., On the validity of the ray approximation for interpreting delay times, In *Seismic Tomography*, edited by G. Nolet, pp. 85-98, D. Reidel Publishing Company, Dordrecht, The Netherlands, 1987.

Wysession, M.E., Continents of the core, *Nature*, 381, 373-374, 1996.

Wysession, M.E., Imaging cold rock at the base of the mantle: the sometimes fate of slabs? In *Subduction: Top to Bottom*, edited by G.E. Bebout, D.W. Scholl, S.H. Kirby, and J.P. Platt, American Geophysical Union, *Geophysical Monograph Series, Am. Geophys. Un.*, 96, 369-384, 1996.

Wysession, M.E., L. Bartkó, and J. Wilson, Mapping the lowermost mantle using core-reflected shear waves, *J. Geophys. Res.*, 99, 13667-13684, 1994.

Zhong, S. and M. Gurnis, Mantle convection with plates and mobile, faulted plate margins, *Science*, 267, 838-843, 1995.

Zhou, H.-W., and R.W. Clayton, *P* and *S* wave travel time inversion for subducting slabs under island arcs of the northwest Pacific, *J. Geophys. Res.*, 95, 6829-6851, 1990.

R. D. van der Hilst, Massachusetts Institute of Technology, Department of Earth, Atmospheric, and Planetary Sciences, Rm 54-514, Cambridge MA 02319, U.S.A. (email: hilst@mit.edu)

S. Widiyantoro, Research School of Earth Sciences, Australian National University, Canberra ACT 0200, Australia.

K.C. Creager, and T.J. McSweeney, Geophysics Program, University of Washington, Seattle WA 98195, U.S.A.

Global Analysis of Shear Wave Velocity Anomalies in the Lower-Most Mantle

Xian-Feng Liu and Adam M. Dziewonski

Department of Earth and Planetary Sciences, Harvard University, Cambridge, MA 02138-2902

The lateral variation of seismic wave velocity in the outer core, if any, is too small to be detected seismically. The differential travel time anomalies of $S - SKS$ (including diffracted S) should be, therefore, mainly sensitive to shear wave velocity anomalies in the lower-most mantle. Because the differential travel times of $S - SKS$ can be observed within a distance range $85^0 < \Delta < 130^0$, in which ray paths are quite different from $ScS - S$, ray path coverage is greatly improved by using both $S - SKS$ and $ScS - S$. In this study, the deviations of arrival times of these phases from the standard spherically symmetric Earth model PREM (Dziewonski and Anderson, 1981) are observed by matching the recorded seismogram with a synthetic that is calculated by normal mode summation with a cut-off period $T = 8$ s. The source mechanisms are taken from the Harvard CMT catalogue. The data include about 1,000 earthquakes of the period 1989-1994 with $M_w \geq 5.5$. We compare our observations of differential travel time residuals $ScS - S$ and $SS - S$ with an earlier study by Woodward and Masters (1991), where the difference was read from waveforms alone. The agreement between results obtained by these two different methods is good. Our observations of $S - SKS$ agree with the data published in an earlier study by Garnero and Helmberger (1993) for ray paths bottoming under the eastern Pacific region. Examination of the global $ScS - S$, $S - SKS$ and $SKKS - SKS$ data sets in both frequency domain and space domain, before they are used in inversion for a 3-D structure, tells us that the velocity anomalies in the lowermost mantle are dominated by spherical harmonics of degree 2 and 3 and that this anomalous structure extends from the core-mantle boundary to at least 500 km above it.

1. INTRODUCTION

Bullen (1940, 1942) divided the Earth along its radius in a series of shells designated by letters from A (crust) to G (inner core). In this nomenclature the lower mantle was the region D. However, it was known from the work of Jeffreys (1939) that the slowness of the P-waves beyond 90° had unusually low gradient, corresponding to a flattening of the velocity distribution in the last 150 km, or so, of the lower mantle. Bullen (1950), having recognized the statistical significance of this feature, divided the D region into D', covering most of the lower mantle, and D'' representing the deepest 180 km.

This term has been modified by the usage during the last decade or so because of the work of Lay and Helmberger (1983), and a number of following studies, who proposed a discontinuity in the lowermost mantle. In the recent literature, D'' came to mean the lowermost 250-300 km of the mantle, in which the precursors to

The Core-Mantle Boundary Region
Geodynamics 28
Copyright 1998 by the American Geophysical Union.

ScS and PcP are said to originate. The question of the semantic differences, which may lead to a real confusion, is discussed at some length by Dziewonski et $al.$ (1997) in a section $Where$ is D''?, who suggest: "In view of our reluctance to call this region D''... we propose the term 'lower-most mantle' (LMM)".

Knowledge of the seismic velocity structure at the base of the mantle is of great importance in Earth sciences. Identification of the origin of the velocity anomalies in the deepest 150–300 km of the mantle, a region called here LMM, is important to the understanding of the style of mantle convection and plume formation. The seismic velocity discontinuity between LMM and the rest of the mantle, if it exists, implies the presence of radial heterogeneity in composition or a phase change some 200 km above the core–mantle boundary (CMB).

There are two different approaches to studying lower-most mantle structure. In the first approach, seismologists focus on the observation of signals reflected or scattered from internal structure and interpret it as a radially symmetric structure, not included in the reference Earth model. The amplitude of the observed signal is dependent on the gradient of the structure, according to classic ray theory. In the other approach, which relies on interpretation of travel time anomalies, the signal is equal to the integral of the anomalies along the entire ray path. This approach tends to average out small (less than the wavelength) scale structure and has the advantage of recovering large-scale structure. On the other hand, the first approach has an advantage in locating rapid changes in seismic wave speed. It is easy to understand why these two approaches would arrive at different answers. What is disturbing, however, is that, unlike the situation in the upper mantle — where modeling results seem to converge — models of the LMM appear disjoined at present: compare this report with that of Lay et $al.$ (1997).

The first approach indicates that the LMM is characterized by the presence of scatterers (Cleary and Haddon, 1972) of 10–100 km dimensions (Doornbos, 1976; Bataille et $al.$, 1990), although Hedlin et $al.$ (1997) propose that small scatterers are uniformly distributed through the lower mantle. Lay and Helmberger (1983), Young and Lay (1987, 1990), and Gaherty and Lay (1993) proposed the existence of a radial discontinuity, on the order of 2–3%, in shear velocity some 200–300 km above the CMB in various parts of the world. Weber (1993) demonstrates rapid variation of the presence or absence of reflections from D''. He arrives at the conclusion that representation of the D'' discontinuity by a simple, nearly horizontal boundary is not valid, and

that focusing and defocusing of wave energy by 3-D heterogeneity may play an important role in generating the observed waveform complexity.

On the other hand, global tomographic models (Dziewonski, 1984; Tanimoto, 1990; Dziewonski and Woodward, 1992; Su et $al.$, 1994; Masters et $al.$, 1996; Li and Romanowicz, 1996) show anomalies that are dominated by the gravest spherical harmonics. In particular, Su et $al.$ (1994) have shown that in the power spectrum of the velocity anomalies in the lower-most mantle, degrees 2 and 3 are by far the largest and that a shift in the spectrum begins some 1,000 km above the CMB. This is also clearly visible in the pattern of data such as $ScS - S$ residuals (Woodward and Masters, 1991b) and the data that we have collected here. The level of heterogeneity is also high: $2 - 3\%$ at the base of the mantle. This anomaly is too large to be the result of low-pass filtering a signal dominated by high-wavenumber energy.

However, the resolution tests performed by Su et $al.$ (1994) indicate that the radial resolution of their model is rather poor in the lower-most mantle and that distinct properties of the LMM region could be averaged with those of the mantle above it. It is not surprising that models inverting for LMM structure using $ScS - S$ travel time residuals would smear structure in the lower-most mantle, since the deepest turning depth of S phases in observable $ScS - S$ residuals is some 1000 km above the CMB. To improve the resolution in the LMM, we need data that turn near or at the CMB and are highly sensitive to its properties.

The P- and S-wave phases diffracted at the CMB satisfy this requirement, in principle, and have been used in the context of laterally heterogeneous structure by Wysession et $al.$ (1992). In particular, SH waves can be observed at distances as large as 130°, without being dominated by very long period energy. Other phases that can be considered are SKS and $SKKS$, even though parts of their ray paths are in the outer core. By forming the travel time differences $S - SKS$ we can obtain a datum which has a sensitivity kernel with a strong maximum near the CMB and significantly reduced sensitivity to the mantle structure above, say, 1500 km depth. Also, the effects of errors in hypocentral coordinates and origin time are nearly completely eliminated by differencing the arrival times.

Garnero and Helmberger (1993) measured the travel times of S and SKS from deep focus Fiji-Tonga events recorded in North America, and compared their observations of differential travel times with the predictions of three tomographic Earth models: MDLSH of Tanimoto (1990), model $S12$ of Su et $al.$ (1992, 1994) and model $SH.10c.17$ of Masters *et al.* (1992, 1996). The fact that these three models give significantly different predictions is a strong indication that $S - SKS$ differential travel times with a good global path distribution would be an important addition to the data set.

Souriau and Poupinet (1991) measured differential travel times of $SKKS$ and SKS ($t_{SKKS-SKS}$) selecting paths to obtain a good sampling of the core. They read the arrival times from seismograms recorded by globally distributed digital stations (broadband GEOSCOPE, NARS and intermediate period GDSN) in the epicentral distance range $83^0 < \Delta < 130^0$. They concluded that lateral heterogeneities in the outer core, if any, are very small because their $SKKKS - SKKS$ residual variations are small. The low level of scatter in $t_{SKKKS-SKKS}$ residuals compared with $t_{SKKS-SKS}$ indicates that the lateral variation of shear wave velocity in the LMM may explain most of the residual variations. Their simultaneous analysis of $SKKKS - SKS$ and $SKKS - SKS$ travel time residuals indicates that large-scale structure dominates the LMM.

Following the encouraging results of Souriau and Poupinet (1991) and Garnero and Helmberger (1993) we undertake a large scale project of measuring the travel times of seismic phases from digital recordings of FDSN stations from which data are currently available.

2. TRAVEL TIME MEASUREMENTS

Picking the onset of a phase is the traditional way of reading the arrival times of body waves. If the data are low-pass filtered long-period seismograms, this may be highly inaccurate. Woodward and Masters (1991a) avoided this difficulty by measuring the differential travel times $SS - S$ through cross-correlation of the Hilbert transformed SS waveform with the S waveform. The time at which the maximum correlation occurs gives the desired measurement. This works well when a differential travel time measurement is sought, assuming that there is no superposition of other phases on either of the two waveforms. If, however, the latter is true or the 'absolute' arrival times need to be measured, another approach is necessary.

Su and Dziewonski (1991) measured the absolute travel times of S and SS by cross-correlating the observed SS waveform, for example, with that predicted by a synthetic seismogram for known seismic source parameters and a spherically symmetric reference Earth model. This involves an intermediate step of a comparison with the synthetic seismogram, which is useful even if differential travel times are desired in a particu-

lar experiment. The synthetic allows us to estimate the quality of the signal, or signals. In the case of a minor interference from other weaker phases, the fact that all of them are automatically present in the synthetic tends to reduce the potential measurement error.

Su and Dziewonski (1991) used long-period signals; their synthetic seismograms, obtained by summation of normal modes, had a cut-off period of 32 s. This removes much of the available energy, and here we extend the analysis to four times higher frequency by introducing a cut-off period of 8 seconds. This is close enough to the microseism band that not much is lost in terms of the signal-to-noise ratio. However, this increases the number of the normal modes in summation by a factor of 16 or so. A similar technique to cross-correlate the synthetic seismogram with the recorded one has also been used in other studies (Wysession *et al.*, 1994).

We implement an efficient method for the computation of synthetic seismograms by using the algorithm described in Dziewonski and Woodhouse (1983), which calls for the evaluation of 'excitation' functions that depend only on the source depth. A sum of the contributions of all overtones with the same angular order, ℓ, is obtained first. Summation over ℓ, which involves multiplication by the spherical harmonic functions and depends on the receiver position, is done next. In this way, all synthetic seismograms for a particular earthquake are calculated simultaneously for all the available station–components. The savings are proportional to ℓ_{max}: in the most time consuming part of the computations, the number of operations is proportional to ℓ_{max} rather than the naive ℓ_{max}^2. With ℓ_{max} exceeding 1500, the savings are substantial. It takes approximately 5 minutes on a Sparc-10 workstation to compute 200 seismograms, each of which contains contributions from over 160,000 normal modes and has, on the average, duration of 30 minutes. It should be noted that the method applies only in the case of a spherically symmetric Earth; to compute seismograms for a laterally heterogeneous Earth we must deal with one mode at a time.

Figure 1 is a copy of a computer screen display showing the interactive program for measuring the differences of the observed arrival times with respect to PREM (Dziewonski and Anderson, 1981). The event is a deep earthquake in the Bonin Islands region and the data are from station Eskdalemuir (ESK) in Scotland at a distance of 100.7°. The system response is flat to ground velocity from 12 to 360 s, and a $\cos^2$ taper is applied from 12 to 8 s. All three (rotated) components of motion are shown, with the recorded data at

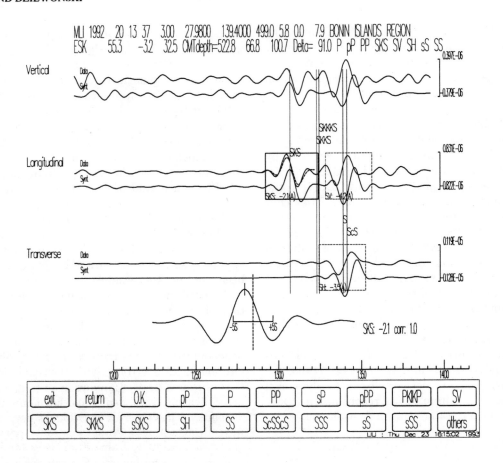

Figure 1. A copy of the computer screen display showing an example of the interactive program for determination of differences in arrival times between the recorded waveform (top) and a synthetic (bottom) computed for an earthquake with a known source mechanism. For details see text.

the top and the synthetic below. In this particular example, the operator has measured differences between the observed and computed phases for SKS and SV on the longitudinal component and SH on the transverse component.

The cross-correlogram between the observed SKS waveform, delineated by the box, and the synthetic is shown just above the time scale. It has a peak at -2.1 s, meaning that the observed pulse arrived earlier than predicted by PREM. The cross-correlation coefficient at the peak is rounded off to 1.0, indicating excellent correlation; this is confirmed by superimposing the shifted synthetic on the observed seismogram — the traces are easier to distinguish in color. The quality of the observation is rated A; B and C are used in less satisfactory cases.

SV and SH are both earlier with respect to PREM than SKS: -4.2 s and -3.7 s, respectively; we leave it to the reader to infer whether this 0.5 s difference

should be interpreted as evidence of anisotropy.

The panel at the bottom of Figure 1 shows 20 options that an operator might choose; among them are 16 different seismic phases. This list can be extended by pointing at 'others'. Once a phase is chosen, the vertical line indicates its arrival time as predicted by PREM and the label identifies the phase. The operator selects the time window for the given phase and the cross-correlogram is automatically computed and the peak correlation and the time shift printed. The operator either accepts the measurement, with an appropriate quality grade, or rejects it. The measurement is then entered into the permanent data base.

3. DATA

Figure 2 shows the distribution of stations used in this study; at least one observation from the period

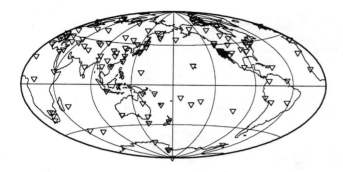

Figure 2. Distribution of 160 digital seismic stations used in this study.

1989–1994 has been contributed by each station included. The stations represent the efforts of many networks: GDSN (USGS), IRIS/IDA, IRIS/USGS, Terrascope, GEOSCOPE, MedNet, and CDSN. The coverage today is vastly superior to that available in the early 1980's, when the first tomographic studies were undertaken.

As part of our project of systematic gathering of travel time data, we collect measurements for, roughly, 15 phases. For some of them, such as SH, we have collected close to 30,000 measurements. Others, like $SKKS$, for example, have about 2,300 readings. In what follows, we shall use our measurements of $SS - S$ and $ScS - S$ to compare with the results of Woodward and Masters (1991a, b), and will discuss $S - SKS$ and $SKKS - SKS$ in greater detail, in the context of inversion for a 3-D Earth model with enhanced resolution in the lower-most mantle.

3.1. Differential Travel Times of SS and S

Woodward and Masters (1991a) measured $SS - S$ differential travel times by comparing S with Hilbert-transformed SS within the distance range from 50° to 100°. We extend the distance range up to $\Delta = 130°$ by including the diffracted S phase; mostly, SH. For distances above 75°, $SS - S$ differential travel time measurements are sensitive to the structure of the upper mantle as well as that of the lower-most mantle.

Figure 3 shows a comparison of our $SS - S$ with those of Woodward and Masters (1991a). Woodward and Masters (1991a) have made 5,462 observations using the GDSN data of 1976-1986; our map is based on 5,709 measurements made on seismograms from the years 1989 through 1994. The two studies agree well for areas with good coverage. For example, both maps show positive residuals for mid-points below the East, South and North Pacific Ocean, North-East Asia and

South Asia, and negative residuals below the Central Pacific Ocean and stable continental areas. There are some differences, however. For example, the $SS - S$ times with reflection points under the Arabian Peninsula and the Caucasus are fast in (a) but slow to normal in (b). In general, we find the results of this comparison satisfactory; we infer that our measurement philosophy does not introduce a systematic bias into the data base. Figure 3(c) demonstrates that the anomalies associated with the S leg of $SS - S$ may be important. The areas in Figure 3(c) where there are large differences with respect to Figure 3(b) correspond to the regions with highly anomalous travel times of $ScS - S$ and $S - SKS$, to be described next.

3.2. Differential Travel Times of ScS and S

$ScS - S$ differential travel times are ideal for imaging lower-mantle velocity structure. These differences can be best observed at distances of $45° - 75°$. For distances less than $45°$, ScS either overlaps with SS or surface waves. Woodward and Masters (1991b) used waveform cross correlation to measure $ScS - S$ differential travel times in this distance range by taking advantage of the shape similarity between S and ScS phases.

Figure 4 shows the global distribution of our $ScS - S$ measurements as compared with that of Woodward and Masters (1991b); the measurements agree with each other, in general. For mid-points below the Central Pacific Ocean, both studies observe positive residuals of $ScS - S$. For mid-points below the Caribbean Sea, the North Pacific Ocean and North-East Asia, dominantly negative residuals of $ScS - S$ appear in both sets of measurements.

3.3. Differential Travel Times of S and SKS

Figure 5 shows the ray geometries of S, diffracted S, SKS and $SKKS$. S and SKS have very similar ray paths in the upper mantle, but very different ray paths in the lower mantle. Beyond 100° or so, S is diffracted. The diffracted S wave is especially sensitive to structure near the CMB; the differential travel times of $S - SKS$ are therefore very sensitive to the structure of the lower-most mantle. The ray paths of SKS and $SKKS$, however, are not far away from each other even near the CMB. Thus, the $SKKS - SKS$ residuals carry information about the horizontal gradient of velocity anomalies in the lower-most mantle.

Although we observe the SV as well as the SH phase, we use the SH in our estimates of differential travel times of $S - SKS$. There are two reasons; first, our experience indicates that observations of diffracted SV

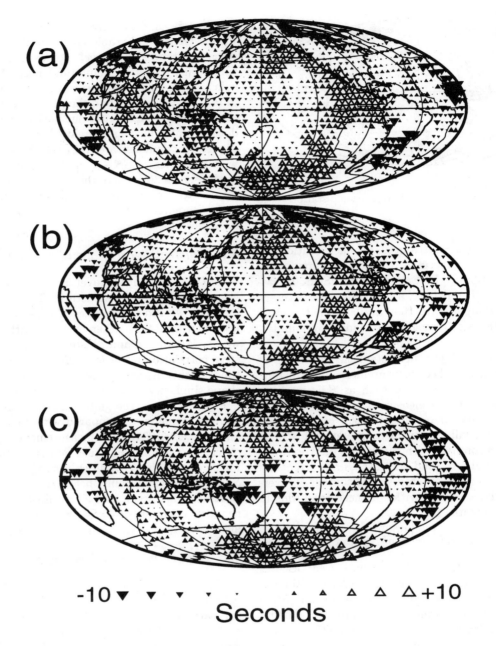

Figure 3. Comparison of 5^0 cap averaged $SS - S$ measurements. A minimum of one point per cap was required to present a cap average. (a) Measurements by Woodward and Masters(1991a); (b) $SS - S$ (c) $SS - S_{diffracted}$ Despite the difference in measurement methods, pass-band of the signal, differences in station coverage, and completely different sets of sources, the maps in (a) and (b) show great similarity.

phases at large distances ($\Delta > 105°$) are rare. The reason is that the diffracted SV decays quickly with distance due to the conversion to P-wave energy, transmitted into the core as phases such as $SdKS$ or $SPdKS$ (Choy, 1977). Therefore, the SV arrivals are too small to measure reliably at large epicentral distances. SH, on the other hand, can be observed at distances up to $\Delta = 130°$, and sometimes beyond. The second reason

ScS-S Measurements

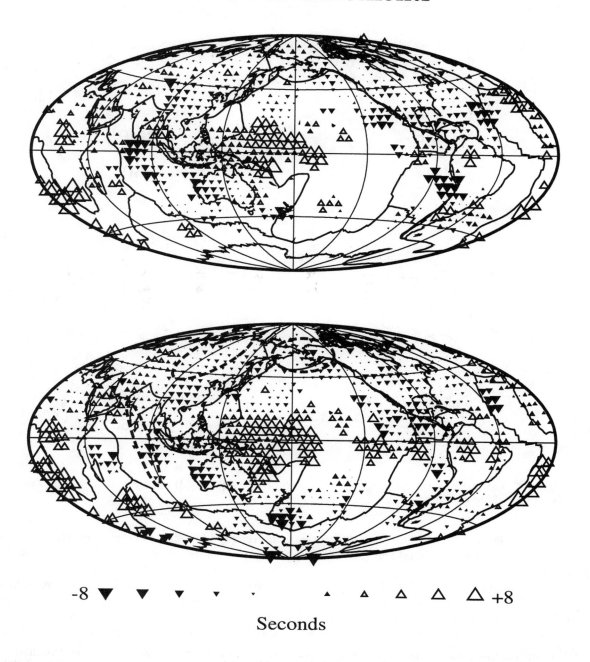

Figure 4. Comparison of cap averaged $ScS - S$ measurements of Woodward and Masters (1991a) (top) with those obtained in this study (bottom). A minimum of one point per cap was required to present a cap average. Despite the difference in measurement methods, pass-band of the signal, differences in station coverage, and completely different sets of sources, the two maps show great similarity. The dashed line ovals (bottom) indicate regions where Lay *et al.* (1997) performed their "pair average residual" experiment.

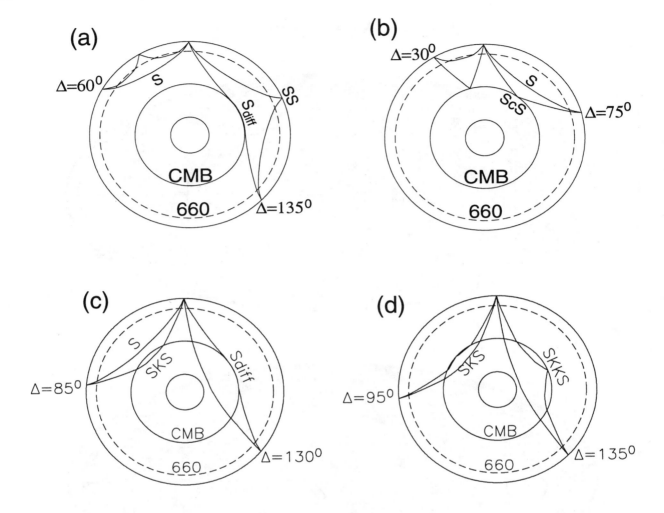

Figure 5. Ray paths of (a) $SS - S$, (b) $ScS - S$, (c) $S - SKS$ and (d) $SKKS - SKS$ at the ends of their range of observations.

for measuring SH is that, from $84° < \Delta < 90°$ interference occurs between the SV signal and that of SKS or $sSKS$ phases.

Young and Lay (1990) suggested from their observations of S and ScS bottoming under Alaska that there might be SV/SH anisotropy near the CMB. If such anisotropy at the base of the mantle were widespread, using $SH - SKS$ instead of $SV - SKS$ could introduce a systematic bias. We compare in Figure 6 the travel time residuals of SH and SV in the distance range $85° < \Delta < 105°$. We find that the SH residuals correlate very well with the SV residuals. It also appears that the global distribution of travel time differences $t_{SV} - t_{SH}$ is random. When $t_{SV} - t_{SH}$ is plotted at mid-points on the surface of the globe, there is

no readily identifiable systematic pattern of positive or negative values, although the global distribution of the mid-points for which both measurements are available is far from satisfactory. This should not be interpreted as a statement contradicting the available evidence for the presence of anisotropy in LMM, or anywhere else along the ray path. We simply conclude that lateral heterogeneity near the CMB is the statistically dominant effect, and that we are justified in using SH measurements in forming the differential travel times $S_H - SKS$.

We obtained almost 5,000 measurements of $S - SKS$ by applying our measurement method to over 1,000 events from 1989 to 1994 with $M_w \geq 5.5$. Since the data coverage is uneven, we average the $S - SKS$ residuals in 5° radius caps to obtain a more informative display of

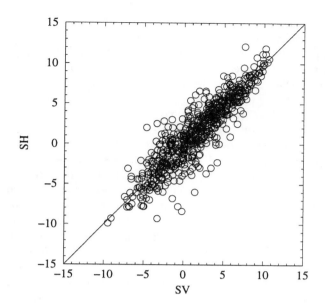

Figure 6. Correlation of 663 SV and SH measurements for the distance range from 85° to 105°; the inference is that isotropic lateral heterogeneity is the principal cause of travel time anomalies in this distance range.

the observations. Figure 7(b) shows our cap averaged measurements at the mid-points between sources and receivers. In the limited number of regions for which such a comparison can be made, the pattern of our observations agrees with previous reports. For example, Lay and Young (1990) reported negative residuals of $S - SKS$ for mid-points below Alaska and this agrees with the anomalies in Figure 7(b). In addition, positive residuals beneath the Central Pacific were reported by Garnero and Helmberger (1993) and Vinnik *et al.* (1995), and our observations show the same sign anomaly.

Jordan and Lynn (1974) showed that it is possible to determine which of the phases in a set of differential travel time measurements plays the dominant role. They plot the absolute residuals of a particular phase against the appropriate differential travel times; the phase that shows better correlation is most likely the dominant phase. In Figure 8 we show on the left δt_S vs. δt_{S-SKS} and on the right $-\delta t_{SKS}$ vs. δt_{S-SKS}. The correlation on the left is much stronger, although the scatter plot on the right is not entirely random either. We conclude that S is the dominant phase in the differential travel times of $S - SKS$; this points to lateral heterogeneity in the deep mantle as the principal source of the anomalies in the S-wave travel times. The fact that $-\delta t_{SKS}$ shows some correlation with $S - SKS$

is the consequence of the very large wavelength of the anomaly in LMM; the paths of S and SKS (see Figure 5) are close to each other on the scale of the wavelengths of the second and third spherical harmonics.

The damped least-square spherical harmonic fit to the residuals shown in Figure 7(b) yields a power spectrum in which degrees 2 and 3 contain by far the most energy. An independent study using a similar, but significantly smaller, data set by Kuo and Wu (1997) confirms our result. Our results on $S - SKS$ are compatible with those of Woodward and Masters (1991b) on $ScS - S$. The overall geographical pattern of the residuals and their spectral characteristics are similar, confirming the inference of Woodward and Masters (1991b) of a particularly high amplitude of anomalies in the lower-most mantle. The travel times of the diffracted S-waves, which are used extensively here, are particularly sensitive to the properties near or at the CMB,

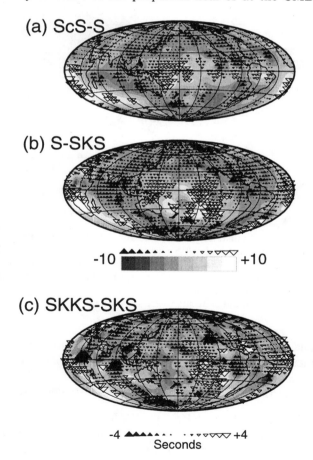

Figure 7. Comparison of cap averages with spherical harmonic expansion of (a) $ScS - S$, (b) $S - SKS$ and (c) $SKKS - SKS$ travel time measurements.

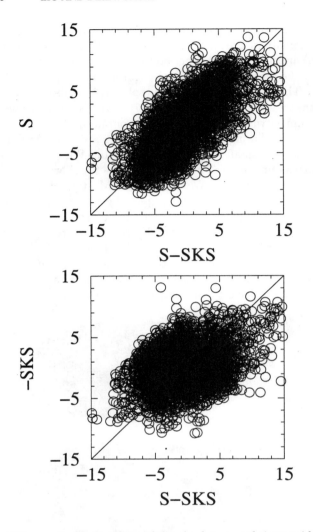

Figure 8. Scatter plots of the absolute travel time residual of S (top) and SKS (bottom) against the differential travel time $S-SKS$. The inference is that S is the dominant phase; the weaker, but noticeable, correlation of $-\delta t_{SKS}$ is the consequence of the very large wavelength of the anomalies in the lower-most mantle.

and we expect that including $S-SKS$ will substantially increase the resolution in this region.

3.4. Differential Travel Times of SKKS and SKS

The core phases SKS and $SKKS$ have been extensively used in the past to study the compressional velocity structure of the outer core (Souriau and Poupinet, 1990, 1991; Choy, 1977; Lay and Young, 1990; Hales and Roberts, 1970, 1971; Schweitzer and Müller, 1986; Garnero *et al.*, 1988). In several of these papers an anomalous gradient in the outermost liquid core was

introduced to explain the observations. Here, we shall primarily address the question of the contribution of these phases to modeling of the lateral heterogeneity. Because the measurement of $SKKS$ is more difficult than that of SKS, we have determined only slightly more than 2,000 $SKKS-SKS$ differential travel times. Nevertheless, we believe ours to be the largest set of such data ever assembled.

Figure 7(c) shows the geographical distribution of the cap averaged $SKKS-SKS$ residuals. Schweitzer and Müller (1986) and Lay and Young (1990) reported positive residuals for mid-points below the Central Pacific and the southern Pacific, in agreement with our observations. The constant offset of the $SKKS-SKS$ residuals with respect to PREM has been noted before (Souriau and Poupinet, 1991; Schweitzer and Müller, 1986).

There are two immediate inferences that can be made by comparing Figure 7(c) with Figure 7(b). First, the level of the anomalies is substantially lower in Figure 7(c) (the scale range is ±4 s *vs.* ±10 s). Second, the pattern of the residuals is more complex. The very high degree of correlation between δt_{SKKS} and δt_{SKS} explains the change in the pattern. This is shown in Figure 9, where the two quantities are plotted against each other. Except for a few outliers, the two core phases appear to be correlated about as well as the SV and SH in Figure 6. The constant offset of $SKKS$ with respect to SKS is clearly seen. The proximity of the SKS and $SKKS$ ray paths and their relatively steep incidence angles in the lower mantle make the difference between their residuals sensitive to the horizontal derivative of the lateral heterogeneity in the deepest 1000 km or so of the mantle. The fact that the r.m.s. amplitude of $SKKS-SKS$ is some 2.5 times smaller than that of $S-SKS$ indicates that while the absolute level of the residuals is high, the horizontal gradient of the heterogeneity pattern is small. This is consistent with the inference of Souriau and Poupinet (1991), who concluded that the spectrum of heterogeneity in the LMM is dominated by large wavelengths.

3.5. Absolute Travel Times

Absolute travel times of S and SS have been used in previous studies (Su *et al.*, 1994; Masters *et al.*, 1996). Using absolute travel times in the inversion can improve the global coverage of the whole data set even though absolute travel times are quite sensitive to earthquake mis-location errors. The uncertainty due to mislocations can be mitigated by using a very large data set. Using the method that we introduced earlier, we

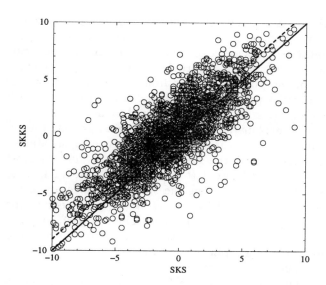

Figure 9. Scatter plot of δt_{SKKS} against δt_{SKS}. Note the constant offset ~1 second of $SKKS$ observations with respect to SKS. This probably means that the velocity in the outermost liquid core in PREM is slightly slower than in the real Earth. Dashed line: the best linear fit to the $SKKS$ vs. SKS measurements.

have collected about $30,000$ S travel times and $12,000$ SS travel times. We include this large data set in our inversion.

3.6. Summary of Data Used

Here we summarize the data set used in the inversion for our new earth model $S16U6L8$:

- 4,601 $S - SKS$ differential travel time residuals;

- 3,455 $ScS - S$ differential travel time residuals;

- 6,356 $SS - S$ differential travel time residuals;

- 2,216 $SKKS - SKS$ differential travel time residuals;

- 27,657 S absolute travel time residuals;

- 11,848 SS absolute travel time residuals;

- 7,400 paths of Love and Rayleigh mantle wave waveforms;

- 9,700 paths of SH body wave waveforms.

We correct both our travel time and waveform data using the crustal model of Mooney *et al.* (1997). Ellipticity corrections (Dziewonski and Gilbert, 1976) and

bathymetry/topography corrections (ETOPO5, National Geophysical Data Center, 1986) are also applied to the theoretical predictions of our travel time measurements.

4. ANALYSIS OF THE DATA

It is generally believed that heterogeneities in seismic velocity structure are dominated by large scale structure, although there are some differences of opinions (e.g., Davies, 1996; Lay *et al.*, 1997), based on regional studies. This question is very important because it justifies our limited (degree 16) parameterization of the Earth's structure. Fortunately, it can be answered by simply analyzing our original measurements.

Since the data coverage is uneven, we use spherical splines (Parker, 1994) to interpolate the travel time residuals of $ScS - S$, $S - SKS$ and $SKKS - SKS$ from degree 0 to degree 36. Figure 7 shows a comparison between the 5° cap averaged measurements and the spherical harmonic expansions of $ScS - S$, $S - SKS$ and $SKKS - SKS$ travel time residuals with respect to PREM. Figure 10 shows their power spectra. Both $ScS - S$ and $S - SKS$ measurements are dominated by degree 2 and 3 spherical harmonics. This result leads to a conclusion that this structure cannot be limited to the to the lowermost 200 km, or so. ScS, from $ScS - S$, and S, from $S - SKS$, have different ray geometries, therefore ScS spends much less time in the bottom 200 km than, say, S_{diff} at 110°. Therefore, for the ScS spectrum to have a similar power to that of S_{diff}, the radial extent of the LMM degree 2–3 anomaly must be considerable: 500 km or more. However, smaller scale structure (such as degrees 5 and 6) show different behavior and this part of the spectrum of $ScS - S$ and $S - SKS$ is not equally well correlated. The dominating degree 2 pattern in our $S - SKS$ residuals agrees with the recent study of Kuo and Wu (1997).

The finding of a recent study (Lay, *et al.*, 1997), who used data from regions beneath Alaska, Eurasia and India, that the heterogeneity in the LMM is decoupled from the overlying mantle is in sharp contrast with our data. The similar pattern in our global data set $ScS - S$ and $S - SKS$, which are sensitive to different regions in the lower mantle leads to the conclusion that the large scale heterogeneity continues from the middle mantle to the LMM. Lay *et al.* also introduce a method, which we call "Pair Residual Average", to assess the scale of heterogeneities in their data. In "Pair Residual Average" method, a moving average is computed for the absolute value of the residual difference as a function of distance of separation between the mid-points for each pair of observations. This method essentially provides

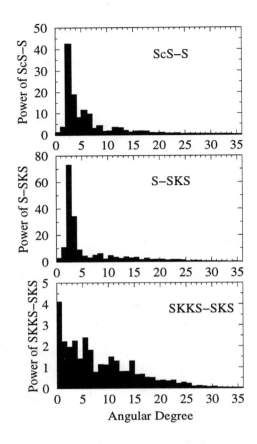

Figure 10. Power spectra of (a) $ScS - S$, (b) $S - SKS$ and (c) $SKKS - SKS$ travel time measurements.

an alternative way to analyze the scale of heterogeneity in space domain, instead of frequency domain. The strength of heterogeneity in their sparse data set in different regions is not dropping off precipitously at the shorter scales. This behavior suggests that the heterogeneity is not dominated by large scale structure which, as they point out, does not agree with other studies using spherical harmonics expansion (Su *et al.*, 1994; Masters *et al.*, 1996; Li and Romanowicz, 1996; Dziewonski *et al.*, 1997; Kuo and Wu, 1997). We apply "Pair Residual Average" method to our global data set of $ScS - S$ and $S - SKS$ and obtain a dramatically different result. Figure 11 shows our result. The analysis has been performed in two ways. In Figure 11a, we use all the data, similarly as Lay *et al.*. In Figure 11b we use cap averaged data. There are, of course fewer of those, and a slight smoothing has been introduced, but the low-pass filter is all-pass to degree 18, so it should not affect the results at distances above 10°. There is a small difference in the absolute value of the average residual,

related to the fact that the cap averaged data are less noisy, but the shape of the curves in both cases unmistakenly points to the dominating effect of the large scale heterogeneity: in Figure 11b, based on cap-averaged data, the plateau extends from 60° to 120°, which is consistent with our result shown in Figure 10, in which the spectrum is clearly dominated by degrees 2 and 3.

One explanation of this discrepancy is that the regions Lay *et al.* have selected in their analysis are not characteristic of the global properties of the lowermost mantle; see Figure 4 (bottom) and Figure 11 in Kuo and Wu (1997) showing the preferential location of the regions considered by Lay *et al.* with respect to the high velocity regions in LMM. This, in part, explains the low r.m.s. values of their differences between the data pairs shown in their Figure 14.

Figure 7(c) shows the spherical spline expansion of our measurements of $SKKS - SKS$ in spherical harmonics from degree 0 to degree 36. The largest term comes from degree 0 (Figure 10, bottom) which corresponds to a systematic deviation of $SKKS - SKS$ from PREM. The power spectrum, which is plotted using a more than tenfold smaller scale, decreases much less rapidly with the increasing angular order than either $S - ScS$ or $S - SKS$; this may also be caused by a significantly higher relative noise level.

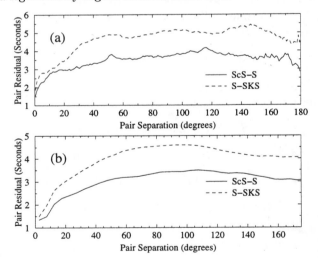

Figure 11. Spatial characteristic of the residual patterns for $ScS - S$ and $S - SKS$ — a repetition of an experiment by Lay *et al.* (1997, Figure 14) but using our global data set. (a) raw measurements; (b) 5° cap average data, which diminishes the random errors and guarantees more even geographical distribution. The absolute value of the difference between each pair of residuals is plotted as a function of the separation distance between their midpoints, and a moving average is computed for a distance of (a) 1° interval, and (b) 5° interval.

5. DERIVATION OF THE SHEAR VELOCITY MODELS OF THE MANTLE

The detailed presentation of our new earth model $S16U6L8$ is given in separate reports (Liu, 1997; Liu and Dziewonski, 1998). This model is parameterized separately in the upper and the lower mantle using B-spline basis (de Boor, 1978) functions. The resolution test shows that $S16U6L8$ has indeed improved the resolution in the lower-most mantle. The shear wave velocity anomalies of $S16U6L8$ are dominated by long wavelength structure, and have a very similar pattern to the previous degree 12 model $S12\_WM13$. However, the root-mean-square amplitude of velocity anomalies in $S16U6L8$ increases dramatically in D'' and peaks at the core-mantle boundary (CMB). These features exist in other three-dimensional Earth models, and may indicate existence of strong compositional heterogeneities in the lower-most mantle and core-mantle reactions are important. The maximum amplitude of slow velocity anomalies (-5%) in our model $S16U6L8$ exceeds that of fast anomalies ($+3\%$), a finding which may result from vigorous reactions between core material and lower mantle material.

Model Coefficients of $S16U6L8$. The new model $S16U6L8$ is available by anonymous ftp at saf.harvard.edu/pub/S16U6L8 or visiting our web page http://www.seismology.harvard.edu/projects/3D/. The postscript file of a few cross-sections of major subduction zones are also available at the same site.

6. DISCUSSION

D'' must be a thermal boundary layer because of the large difference in temperature between the core and the mantle. For any kind of convection, there will be large temperature variations within this region. Theoretical studies (Stevenson, 1994) point out that thermal and compositional fluctuations have different characteristics in the power spectrum of anomalies. Lateral variations in temperature at the CMB must be zero: because of the rapid flow of the low viscosity core liquid, the CMB is forced to be an isotherm. Temperature heterogeneity will also be low well above the boundary layer, due to the development of large radial motions characteristic of high Rayleigh number convection.

On the other hand, the power spectrum resulting from compositional fluctuations need not be zero due to the slowness of compositional diffusion, and may even peak at the CMB if core-mantle reactions are important. Therefore, the sharp increase in the amplitudes of velocity anomalies in D'' revealed in our study argues strongly for both thermal and compositional contributions to the heterogeneity of that region. The study by Su and Dziewonski (1997) indicates that the compositional heterogeneity in the lower-most mantle is very strong.

Laboratory experiments (Knittle and Jeanloz, 1989) show that liquid iron (core material) reacts chemically with $(Mg, Fe)SiO_3$ perovskite (lower mantle material) at the conditions of the lower-most mantle (pressure $\geq$ 70 GPa and temperature $\geq$ 3,700 K) to produce $MgSiO_3$, SiO_2 and the iron alloys FeO and FeSi (Jeanloz, 1990). The overall effect of these reactions would be to reduce seismic wave velocities due to the presence of the metallic alloys (Wysession et al., 1992). Core-mantle reactions may contribute to the extremely slow velocity anomalies below central Pacific and Africa.

The pattern of $ScS - S$ differential travel time residuals in Figure 7(a) is very similar to $S - SKS$ in Figure 7(b), in which S waves have turning depths between 2700 km and the CMB, i.e. they bottom within the 'classical' D'' region. The pattern is still very coherent. This confirms that there is no obvious change in the velocity pattern in D'' region, and that the velocity anomaly dominated by degrees 2 and 3 extends over a substantial range of depths, as shown by Su et al. (1994, Figure 6).

We conclude that the lower-most mantle of the Earth is a complex region in which different phenomena, on different horizontal and radial scales take place. Large-scale heterogeneity is continuous across the 'classical' boundary between the D' and D'' regions, but is strongly enhanced in close proximity to the CMB. This effect is most likely associated with large scale flow in the mantle — as supported by its correlation with the geoid (e.g., Hager et al., 1985). The nature of the 'classical' D'' anomaly is not clear, but it could be related to the thermal boundary layer at the CMB. The fact that its expression, both in terms of thickness and gradient, does not seem to change much with location (Dziewonski, et al., 1997) suggests that differences in the thermal gradient across D'' are not highly variable and that a significant part of the observed heterogeneity must be compositional in nature.

The following conclusions may be drawn from our work on mantle S-velocity structure:

- The new model $S16U6L8$, by using $S - SKS$ and $SKKS - SKS$ differential travel times improves resolution in the deepest 600 km of the mantle.

- In the lower-most mantle, degrees 2 and 3 dominate. This large scale structure begins at about 2000 km depth and increases to the core-mantle boundary.

- The large scale structure in D'' continues into the mantle above.

- The sharp increase in the amplitude of lateral heterogeneity in D'' argues strongly for both thermal and chemical origins of lateral variations in D''. The fact that the root-mean-square amplitude of the model peaks at the CMB indicates that core-mantle reactions may be important.

Acknowledgments. We thank Wei-jia Su for help in various stages of this project and, in particular, the graphics. We also thank Meredith Nettles for her help during the preparation of the manuscript. We thank Babara Romanowicz and the other anonymous reviewer for their thoughtful and helpful reviews. This research was supported by NSF grant EAR92-19361.

REFERENCES

Bataille, K., R. S. Wu, and S. M. Flatté, Inhomogeneities near the core-mantle boundary evidenced by seismic wave scattering – A review, *Pure and Appl. Geophys.*, **132**, 151-174, 1990.

Bolton, H. F. Long period travel times and the structure of the mantle, *Ph.D Thesis*, University of California, San Diego, 1996.

Bullen, K. E., The problem of the Earth's density variation, *Bull. Seism. Soc. Am.*, **30**, 235-250, 1940.

Bullen, K. E., The density variation of the Earth's central core, *Bull. Seism. Soc. Am*, **32**, 19-29, 1942.

Bullen, K. E., An Earth model based on compressibility–pressure hypothesis, *Month. Not. R. Astr. Soc., Geophys. Suppl.*, **6**, 50-59, 1950.

Choy, G. L., Theoretical seismograms of core phases calculated by frequency-dependent full wave theory, and their interpretations. *Geophys. J. R. Astron. Soc.*; **65**: 55-70, 1977.

Cleary, J. R., The velocity at the core-mantle boundary, from observations of diffracted S. *Bull. Seism. Soc. Am.*, **59**: 1399-1405, 1969.

Cleary, J. R. and R. A. W. Haddon, Seismic wave scattering near the core-mantle boundary: a new interpretation of precursors to PKP, *Nature*, **240**, 549-551, 1972.

Davies, J. H., A discussion to *A shear-velocity model of the mantle* by Masters *et al.* (1996), *Phil. Trans. R. Soc. Lond. Ser. A*, **354** 1411, 1996.

de Boor, C., *A practical guide to splines, Applied Mathematical Sciences*, **27**, Springer, New York, 1978.

Doornbos, D. J., Characteristics of lower mantle heterogeneities from scattered waves, *Geophys. J. Roy. Astron. Soc.*, **44**, 447-470, 1976.

Dziewonski, A.M., Mapping the lower mantle: Determination of lateral heterogeneity in P velocity up to degree and order 6, *J. Geophys. Res.*, **89**, 5929-5952, 1984.

Dziewonski, A. M. and D. L. Anderson, Preliminary reference Earth model. *Phys. Earth Planet. Inter.*, **25**: 297-356, 1981.

Dziewonski, A. M., T.-A. Chou and Woodhouse, J. H., Determination of earthquake source parameters from waveform data for studies of global and regional seismicity, *J. Geophys. Res.*, **86**: 2825-2852, 1981.

Dziewonski, A. M., W.-J. Su, and R. L. Woodward, Grand structures of the Earth's Interior, *EOS Trans. AGU Fall Meeting Suppl.*, **72**, 451, 1991.

Dziewonski, A. M., A. M. Forte, W.-J. Su and R. L. Woodward, Seismic tomography and geodynamics, *Relating Geophysical Structures and Processes: The Jeffreys Volume*, Geophysical Monograph 76, **IUGG Volume 16**, pp 67-105, edited by K. Aki and R. Dmowska, 1993.

Dziewonski, A. M. and F. Gilbert, The effect of small, aspherical perturbations on travel times and re-examination of the corrections for ellipticity, *Geophys. J. R. Astron. Soc.*, **44**, 7-18, 1976.

Dziewonski, A. M., X.-F. Liu and W.-J. Su, Lateral heterogeneity in the lowermost mantle, *Earth's Deep Interior: the Doornbos Memorial Volume*, edited by D. J. Crossley, Gordon and Breach Science Publishers, The Netherlands, 11-49, 1997.

Dziewonski, A. M. and J. H. Woodhouse, Studies of the seismic source using normal mode theory, Proceedings of the International School of Physics, 'Enrico Fermi" Course LXXXV, in: *Earthquakes: Observations, Theory and Interpretation* (H. Kanamori and E. Boschi, eds.), North Holland, 45-137, 1983.

Dziewonski, A. M., and R. L. Woodward, Acoustic imaging at the planetary scale, *in Acoustical Imaging*, **19**, H. Ermert and H.-P. Harjes, eds, Plenum Press, New York, 785-797, 1992.

Fukao, Y., Obayashi, M., Inoue, H., and Nenbai, M., Subducting slabs stagnant in the mantle transition zone, *J. Geophys. Res.* **97**, 4809-4822, 1992.

Garnero, E. and D. Helmberger, Travel times of S and SKS: implications for 3-D lower mantle structure beneath the central Pacific. *J. Geophys. Res.*, **98**: 8225-8241, 1993.

Garnero, E. J., and D. V. Helmberger, Seismic detection of a thin laterally varying boundary layer at the base of the mantle beneath the central-Pacific, *Geophys. Res. Letters*, **23**, 977-980, 1996.

Garnero, E., D. Helmberger and G. Engen, Lateral variations near the core-mantle boundary. *Geophys. Res. Letters*, **6**, 609-612, 1988.

Gaherty, J. B. and T. Lay, Investigation of laterally heterogeneous shear velocity structure in D'' beneath Eurasia. *J. Geophys. Res.*, **97**: 417-435, 1993.

Hales, A. L. and J. L. Roberts, Shear velocities in the lower mantle and the radius of the core, *Bull. Seism. Soc. Am.*, **60**: 1427-1436, 1970.

Hales, A. L. and J. L. Roberts, The velocities in the outer core, *Bull. Seism. Soc. Am.*, **61**: 1051-1059, 1971.

Hager, B. H., R. W. Clayton, M. A. Richards, R. P. Comer and A. M. Dziewonski, Lower mantle heterogeneity, dynamic topography and the geoid, *Nature*, **313**, 541-545, 1985.

Hedlin, M. A. H., P. M. Shearer and P. S. Earle, Seismic evidence for small-scale heterogeneity throughout the Earth's mantle, *Nature*, **387**, 145-150, 1997.

Jordan, T. H. and W. S. Lynn, A velocity anomaly in the lower mantle, *J. Geophys. Res.*, **79**, 2679-2685, 1974.

Kind, R. and G. Müller, The structure of the outer core from *SKS* amplitudes and travel times, *Bull. Seism. Soc. Am.*, **67**: 1541-1554, 1977.

Knittle, E. and R. Jeanloz, Simulating the core-mantle boundary: an experimental study of high-pressure reactions between silicates and liquid iron, *Geophys. Res. Letters*, **16**, 609-612, 1989.

Kuo, B.-Y. and K.-Y. Wu, Global shear velocity heterogeneities in the D'': Inversion from $Sd - SKS$ differential travel times, *J. Geophys. Res.*, **102**, 11775-11788, 1997.

Lay, T. and D. V. Helmberger, A lower mantle S-wave triplication and the shear velocity structure of D'', *Geophys. J. R. astr. Soc.*, **75**, 799-837, 1983.

Lay, T. and C. J. Young, The stably-stratified outermost core revisited, *Geophys. Res. Letters*, **17**: 2001-2004, 1990.

Lay, T., Garnero, E. J., Young, C. J. and J. B. Gaherty, Scale lengths of shear velocity heterogeneity at the base of the mantle from S wave differential travel times, *J. Geophys. Res.*, **102**: 9887-9909, 1997.

Li, X.-D. and B. Romanowicz, Comparison of global waveform inversions with and without considering cross-branch modal coupling, *Geophys. J. Int.*, **121**, 695-709, 1995.

Li, X.-D. and B. Romanowicz, Global mantle shear-velocity model developed using nonlinear asymptotic coupling theory, *J. Geophys. Res.*, **101**, 22245-22272, 1996.

Li, X.-D. and T. Tanimoto, Waveforms of long-period body waves in a slightly aspherical Earth model, *Geophys. J. Int.*, **112**, 92-102, 1993.

Liu, X.-F., The three-dimensional shear wave velocity structure of the Earth's lower-most mantle, *Ph.D Thesis*, Harvard University, 1997.

Liu, X.-F., W.-J. Su and A. M. Dziewonski, Improved resolution of the lower-most mantle shear wave velocity structure obtained using $SKS - S$ data (abstract), *EOS Trans. AGU Spring Supplement*, **75**, 232-233, 1994.

Liu, X.-F. and A. M. Dziewonski, Lower-most mantle shear wave velocity structure (abstract), *EOS Trans. AGU Fall Supplement*, **75**, 663, 1994.

Liu, X.-F. and A. M. Dziewonski, Lower-most mantle shear wave velocity structure, *in preparation*, 1998.

Liu, X.-F. and J. Tromp, Uniformly valid body-wave ray theory, *Geophys. J. Int.*, **127**, 461-491, 1996.

Masters, G., H. Bolton and P. Shearer, Large-scale 3-dimensional structure of the mantle (abstract), *EOS Trans. AGU.*, **73**, 201, 1992.

Masters, G., S. Johnson, G. Laske and H. Bolton, A shear-velocity model of the mantle, *Phil. Trans. R. Soc. Lond. A*, **354** 1385-1411, 1996.

Mooney, W., G. Laske, and G. Masters, A new global crustal model at 5 x 5 degrees: CRUST-5.1, *J. Geophys. Res.*, in press, 1997.

Jeanloz, R., The nature of the Earth's core, *Annu. Rev. Earth Planet. Sci.*, **18**: 357-86, 1990.

Johnson, S., G. Masters, P. Shearer, G. Laske and H. Bolton, A shear velocity model of the mantle (abstract), *EOS Trans. AGU Fall Supplement*, **75**, 475, 1994.

Parker, R. P., Geophysical Inverse Theory, *Princeton University Press*, 1994.

Schweitzer, J. and G. Müller, Anomalous difference travel

times and amplitude ratios of SKS and $SKKS$ from Tonga-Fiji events, *Geophys. Res. Letters*, **13**: 1529-1532, 1986.

Souriau, A. and G. Poupinet, A latitudinal pattern in the structure of the outermost liquid core, revealed by the travel times of SKKS-SKS seismic phases, *Geophys. Res. Letters*, **17**: 2005-2007, 1990.

Souriau, A. and G. Poupinet, A study of the outermost liquid core using differential travel times of the SKS, $SKKS$ and $S3KS$ phases. *Phys. Earth Planet. Inter.*, **68**: 183-199, 1991.

Stevenson, D. J., Limits on lateral density and velocity variations in the Earth's outer core. *Geophys. J. R. Astron. Soc.*, **88**, 311-319, 1987.

Stevenson, D. J., Relative Importance of Thermal and Compositional Fluctuations in D'' *EOS Trans. AGU Fall Supplement.*, **75**: 662, 1994.

Su, W.-J. and A. M. Dziewonski, Predominance of long-wavelength heterogeneity in the mantle, *Nature*, **352**: 121-126, 1991.

Su, W.-J. and A. M. Dziewonski, Simultaneous inversion for 3-D variations in shear and bulk velocity in the mantle, *Phys. Earth Planet. Inter.*, **100**: 134-156, 1997.

Su, W.-J, The Three dimensional shear wave velocity structure of the Earth mantle, *Ph.D Thesis*, Harvard University, 1993.

Su, W.-J., R. L. Woodward and A. M. Dziewonski, Degree-12 model of shear velocity heterogeneity in the mantle, *J. Geophys. Res.*, **99**: 6945-6980, 1994.

Tanimoto, T., Long-wavelength S-wave velocity structure throughout the mantle, *Geophys. J. Int.*, **100**: 327-336, 1990.

Vinnik, L., B. Romanowicz, Y. Le Stunff and L. Makayeva, Seismic anisotropy in the D'' layer, *Geophys. Res. Lett.*, **22**, 1657-1660, 1995.

Weber, M., P and S wave reflections from anomalies in the lowermost mantle, *Geophys. J. Int.*, **115**, 183-210, 1993.

Woodhouse, J. H. and A. M. Dziewonski, Mapping the upper mantle: Three dimensional modelling of Earth structure by inversion of seismic waveforms, *J. Geophys. Res.*, **89**, 5953-5986, 1984.

Woodhouse, J. H. and A. M. Dziewonski, Seismic modeling of the Earth's large-scale three dimensional structure, *Philos. Trans. R. Soc. Lond. A*, **328**, 291-308, 1989.

Woodward, R. L. and G. Masters, Global upper mantle structure from long-period differential travel times. *J. Geophys. Res*, **96**: (6351-6377), 1991a.

Woodward, R. L. and G. Masters, Lower-mantle structure from $ScS - S$ differential travel times. *Nature*, **352**: 231-233, 1991b.

Wysession, M. E., E. A. Okal and C. R. Bina, The structure of the core-mantle boundary from diffracted waves, *J. Geophys. Res.*, **97**, 8749-8764, 1992.

Wysession, M. E., L. Barkó and J. B. Wilson, Mapping the lowermost mantle using core-reflected shear waves, *J. Geophys. Res.*, **99**, 13667-13684, 1994.

Wysession, M. E., R. W. Valenzuela, A.-N. Zhu and L. Barkó, Investigating the base of the mantle using differential travel times, *Phys. Earth Planet. Inter.*, **92**, 67-84, 1995.

Young, C. J. and T. Lay. Evidence for a shear velocity discontinuity in the lower-most mantle beneath India and the Indian Ocean, *Phys. Earth Planet. Inter.*, **49**: 37-53, 1987a.

Young, C. J. and T. Lay, The core-mantle boundary, *Annu. Rev. Earth Planet. Sci.*, **15**, 25-46, 1987b.

Young, C. J. and T. Lay, Multiple phase analysis of the shear velocity structure in the D'' region beneath Alaska, *J. Geophys. Res.*, **95**: 17385-17402, 1990.

X.-F. Liu and A. M. Dziewonski, Department of Earth and Planetary Sciences, 20 Oxford Street, Harvard University, Cambridge, MA 02138-2902. (E-mails: liu@seismology.harvard.edu; dziewons@seismology.harvard.edu).

PKP and PKKP Precursor Observations: Implications for the Small-scale Structure of the Deep Mantle and Core

Peter M. Shearer and Michael A.H. Hedlin

Cecil H. and Ida M. Green Institute of Geophysics and Planetary Physics, Scripps Institution of Oceanography, University of California, San Diego, La Jolla, California

Paul S. Earle

Department of Earth and Planetary Sciences, University of California, Los Angeles, Los Angeles, California

Precursors to the seismic core phases PKP and $PKKP$ result from scattering off small-scale heterogeneity in the deep mantle or core. We apply stacking techniques to high-frequency data from the global seismic networks to image details of the time and range dependence of the precursor wavefields. Results for PKP suggest that small-scale velocity perturbations with ~8-km scale length and ~1% rms amplitude are distributed throughout much of the lower mantle, with no evidence for a concentration of heterogeneity near the core-mantle boundary. This supports geochemical evidence and numerical convection experiments that suggest incomplete mixing and small-scale chemical heterogeneity are present in the mantle. The $PKKP$ precursor observations can be modeled by short wavelength topography on the core-mantle boundary (~300 m rms height), but exhibit a range dependence that is difficult to explain with simple models. The outer core appears free of scattering, but the inner core is a possible contributor of scattered energy to the $PKKP$ precursors.

1. INTRODUCTION

At Earth's surface, heterogeneous structures exist at a wide range of sizes, from the ocean/continent differences at large scales to the complexities of local geology at small scales. Seismic surveys have shown that small-scale heterogeneity can be observed to extend through the crust and, in some cases, into the uppermost mantle. However, it becomes increasingly difficult to resolve small-scale features as one probes deeper into Earth's interior. Whole-mantle seismic tomography studies are currently producing maps of velocity heterogeneity at scales of thousands of kilometers, and some have begun to image subducting slabs in the middle to upper

mantle. However, the resolution of these models, even in areas with favorable ray coverage, is no better than hundreds of kilometers in the upper mantle and much less in the lower mantle or in the vicinity of the core-mantle boundary.

The extent to which smaller-scale structure may be present within the deep Earth remains largely unknown. Mantle convection simulations often predict relatively narrow plumes of upwelling material, but these have so far not been imaged in global tomography studies. The eventual fate of subducting lithosphere is also not well understood. It is possible that processes of thermal and chemical diffusion result in a relatively homogenous mantle at small scales. Alternatively, incomplete mixing may prevent local homogenization of the mantle and small-scale features may be common throughout the mantle, particularly in boundary layers near major discontinuities such as the core-mantle boundary

The Core-Mantle Boundary Region
Geodynamics 28
Copyright 1998 by the American Geophysical Union.

(CMB). The limited resolution of travel-time tomography studies prevents direct discrimination between these models. The observed wavenumber spectrum of velocity heterogeneity in global tomography studies at long wavelengths suggests that mantle heterogeneity is dominated by very large-scale anomalies [e.g, *Su and Dziewonski*, 1991], but the extent to which this result can be extrapolated to shorter wavelengths is uncertain. Subducting slabs are quite narrow in the upper mantle and may persist as relatively sharp features in the mid-mantle [e.g., *Grand et al.*, 1997; *van der Hilst et al.*, 1997]. In the lowermost mantle, large lateral velocity gradients have recently been identified just above the core-mantle boundary [e.g., *Garnero and Helmberger*, 1996].

At even smaller scales (tens of kilometers or less), travel-time studies are not suited to resolving deep heterogeneities, but such small-scale heterogeneity is not invisible to seismology, as it causes scattering of high-frequency waves. While a deterministic mapping of the exact form of such heterogeneity is unlikely to be achieved, a statistical characterization of its properties can be made from observations of the scattered wavefield. The high-frequency coda that follows direct P and S arrivals is an example of such a scattered wavefield, and studies of seismic codas have led to a number of advances in scattering theory and in the characterization of crustal heterogeneity and attenuation. However, it is difficult to use P and S coda observations to constrain small-scale heterogeneity in the deep mantle and core because of the uncertainties involved in removing the strong scattering signal contributed by the crust and uppermost mantle.

Fortunately, there are some ray geometries that are uniquely sensitive to scattering in the deep Earth [for a review, see *Bataille et al.*, 1990]. This was first recognized in the classic study of *Cleary and Haddon* [1972], who found that the observed onset time of precursors to PKP could be explained by scattering from small-scale heterogeneities in the vicinity of the core-mantle boundary. Modeling work has shown that these precursors could be caused either by short wavelength CMB topography or by volumetric velocity heterogeneity in the lowermost mantle [e.g., *Bataille and Flatté*, 1988]. Precursors to $PKKP$ have also been identified [*Doornbos*, 1974, 1980; *Chang and Cleary*, 1978, 1981] and attributed to scattering from the CMB. Together PKP and $PKKP$ precursors provide important constraints on near-CMB structure at very short scales (∼10 km).

However, the data do not justify focusing exclusively on the core-mantle boundary as the source of all deep scattered energy. Although some scattering of short-period energy must be occurring near the core-mantle boundary, it does not follow that the CMB is the only source of deep scattering or the only region in the deep mantle and core where small-scale structure may be present. It is also largely unknown if the small-scale velocity perturbations are widely distributed or if they might be concentrated in a few areas of anomalously strong scattering. Possible lateral variations in PKP scattering strength have been noted by *Bataille and Flatté* [1988] and *Hedlin et al.* [1995], and localized sources of S-to-P scattering have recently been identified at 1600 to 1800 km depth in the mantle [*Kaneshima and Helffrich*, 1997].

In this paper, we examine both PKP and $PKKP$ precursors, applying a new stacking technique to records from the global seismic networks in order to resolve the time and range dependence of the precursor wavefields. The resulting images represent a more complete description of the globally averaged properties of the precursors than is available from previous studies, and enable analyses that address the strength, scale length, and depth extent of the scattering regions. We find that average PKP precursor amplitudes are best fit with models in which small-scale heterogeneity is distributed throughout the mantle, rather than concentrated in the core-mantle boundary region. Our observations of $PKKP$ precursors suggest small-scale topography on the CMB, but the precursors exhibit a range dependence that is difficult to explain with scattering from the CMB alone. Thus other scattering source regions, such as the inner-core boundary, should also be considered.

Many of our results were described in *Hedlin et al.* [1997] and *Earle et al.* [1997]. Our goal in this paper is to provide more details regarding our analyses, discuss how they relate to other studies, and suggest possible future directions for research.

2. PKP PRECURSORS

For over 50 years it has been noted that $PKPdf$ at source-receiver ranges between 120° and 145° is often preceded by high-frequency precursors that arrive up to about 20 s before the main arrival. The precursors are typically emergent in character and stronger in records from longer ranges. The first satisfactory explanation for the origin of these arrivals was described by *Haddon* [1972] and *Cleary and Haddon* [1972], who showed that the observed precursor onset times could be explained if scattering of seismic energy occurs at the core-mantle boundary. Their idea is illustrated in Figure 1, which shows the ray paths and travel-time curves for the different branches of PKP in the mantle and core. These

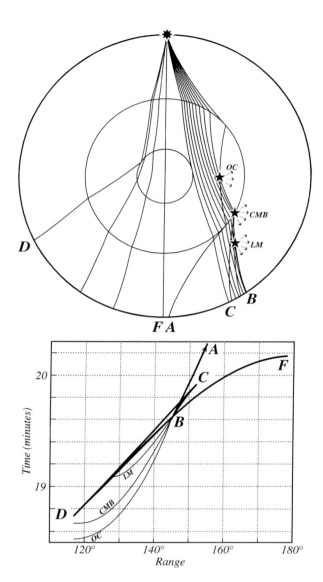

Figure 1. Ray paths and travel time curves for the different branches of *PKP*. *PKPab* and *PKPbc* turn in the outer core, *PKPcd* is reflected off the inner core boundary, and *PKPdf* penetrates the inner core. Scattering of *PKPab* or *PKPbc* waves can produce precursory arrivals to *PKPdf* at ranges less than the *B* caustic at 145°. Hypothetical scattering points in the outer core (OC), core-mantle boundary (CMB) and lower mantle (LM) are shown as the stars in the top figure; corresponding minimum times for scattered arrivals are plotted below. These curves define the expect onset times for precursor wavefields resulting from scattering at various depths. In general, deeper sources of scattering will produce earlier onset times for the precursors. Although we have depicted *PKP* to *P* scattering on the receiver side of the path, precursors can also result from *P* to *PKP* scattering on the source side.

branches result from the *P* velocity drop at the CMB and the spherical geometry.

PKPab is the retrograde branch that turns in the middle of the outer core. *PKPbc* is the prograde branch between about 145° and 155° that bottoms in the lower part of the outer core. The *b* caustic at about 145° marks the minimum range for rays that turn in the outer core. *PKPcd* is the retrograde branch that reflects off the inner core boundary; *PKPdf* is the prograde branch that turns within the inner core. At ranges between 120° and 145°, the first arriving phase is expected to be *PKPdf*, followed closely by *PKPcd*. However, if scattering deflects *PKPbc* or *PKPab* rays to ranges shorter than 145°, this can produce precursors to *PKPdf* (indicated by the labeled curves in Figure 1). These scattered ray paths require the presence of laterally heterogeneous structures; they are not predicted by ray tracing through spherically symmetric models.

Assuming a reference 1-D velocity model, it is possible to compute those points inside the Earth where single scattering could produce *PKPdf* precursors. In Figure 2, we plot cross-sections of these scattering volumes at source-receiver ranges of 125°, 130°, 135° and 140°. The possible source region includes a large fraction of the outer core, extending almost down to the inner core boundary. However, outer-core scattering would produce precursors up to a minute before *PKPdf*, much earlier than those observed. In contrast, scattering restricted to the CMB and overlying mantle produces minimum times that show good agreement with the onset time of the observed precursors [e.g., *Cleary and Haddon*, 1972]. At 125°, the possible scattering source region extends only about 200 km up into the lower mantle. However, at longer ranges an increasingly large fraction of the mantle could be involved in the scattering.

2.1. Stacking PKP precursors

A number of studies have documented *PKP* precursor observations and compared their amplitudes to the main *PKPdf* arrivals [e.g., *Haddon and Cleary*, 1974; *King et al.*, 1974; *Husebye et al.*, 1976; *Haddon*, 1982; *Bataille and Flatté*, 1988]. However, these reports do not necessarily provide an unbiased estimate of average *PKP* precursor amplitudes because they typically are based on a limited number of records and may tend to preferentially use those source-receiver paths for which the precursors are seen most clearly. Further, they have not generally examined the time dependence of precursor amplitudes, information that is critical for resolving the depth extent of mantle scattering.

We begin our analysis by extracting 21,945 broadband waveforms from the online waveform archive of

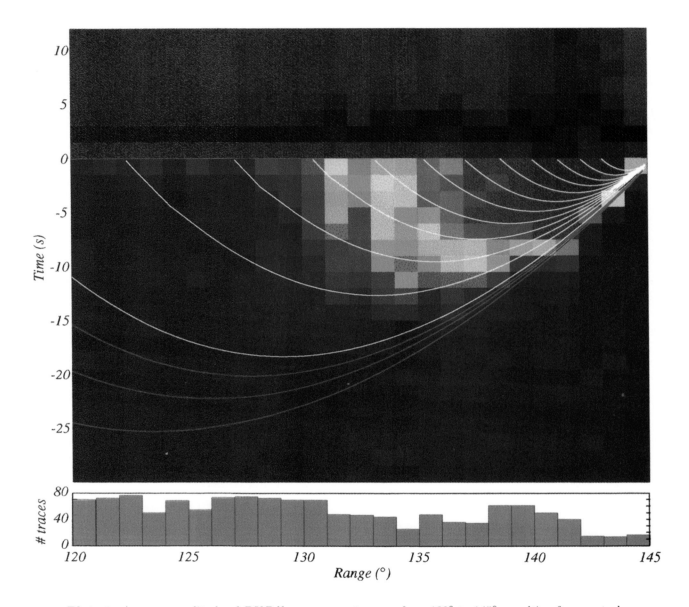

Plate 1. Average amplitude of $PKPdf$ precursors at ranges from 120° to 145°, resulting from a stack of 1600 short-period seismograms. The colors indicate the strength of the precursors, ranging from dark blue (zero amplitude) through green, yellow, red and brown (highest amplitudes). The onset of $PKPdf$ is shown by the edge of the brown region at zero time. The histogram indicates the number of seismograms included in each 1° range bin. The precursors exhibit increasing amplitude with both range and time. The curved lines indicate the theoretical precursor onset times for scattering at various depths. The earliest white curve shows the minimum arrival times for scattering at the CMB, with the later white curves corresponding to 200 km depth intervals above the CMB (the last curve represents scattering 2200 km above the CMB). The red curves show onset times for outer core scattering at 200 km intervals below the CMB; the earliest red curve is for scattering 600 km below the CMB. The observed onset times generally agree with the CMB scattering curve, suggesting that no observable scattering originates from the outer core and that some scattering must originate near the CMB.

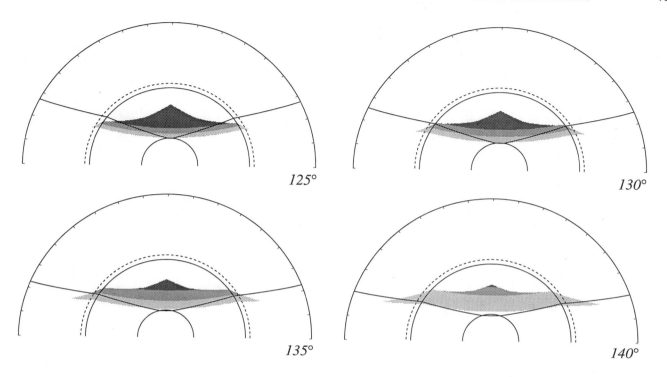

Figure 2. Cross-sections of possible scatterer source volumes at 125°, 130°, 135° and 140° source-receiver ranges. The *PKPdf* ray path is also shown. The dashed line`is at a radius 200 km above the CMB. Precursor times at 0 to −10 s relative to *PKPdf* are plotted in light gray, times from -10 to -20 s are shown as medium gray, and times earlier than -20 s are dark gray. Scattering from the outer core can produce earlier arrivals than mantle or CMB scattering. At longer ranges, an increasing fraction of the mantle can contribute to the precursors.

the Incorporated Research Institutions for Seismology (IRIS) between 1989 and 1994 at source-receiver ranges from 118° to 145°. After bandpass filtering the records to frequencies between 0.7 and 2.5 Hz, we select only those traces with stable and low levels of noise (relative to *PKPdf*), reducing our dataset to 1600 high-quality seismograms. To avoid biasing our estimates of average precursor amplitudes, we include records regardless of the presence or absence of visible precursors.

The precursors are not coherent between different seismograms, so we compute the envelope function of each trace prior to stacking. We then adjust the level of each trace to account for the average power level of the noise as measured from a time window prior to the precursors. Finally, we align the records on the first break of the *PKPdf* arrival, apply a range adjustment to correct for the effect of variable source depths, and stack the envelope functions into 1° bins in source-receiver range and 1.5 s time bins. We use a weighting scheme that emphasizes those records with the highest signal-to-noise ratios. Standard error estimates are obtained for the stacked traces by using a bootstrap method [e.g.,

Efron and Tibshirani, 1991] in which we repeat the stack many times for random subsets of the data.

The resulting image of the precursor wavefield is displayed in Plate 1, together with a histogram showing the total number of traces that were averaged in each 1° range bin. The main *PKPdf* arrival is aligned at zero time; the precursors appear at negative times and exhibit increasing amplitude both with range and time. The curves show minimum travel time curves for single scattering at various depths computed using the PREM velocity model [*Dziewonski and Anderson*, 1981]. The earliest of the white curves indicates the expected precursor onset times for scattering at CMB; the series of later curves represent minimum times for scattering at 200 km intervals above the CMB (up to 2200 km above the CMB for the last curve). The red curves show the expected precursor onsets for scattering in the outer core at 200 km intervals below the CMB.

The agreement between the onset of the observed precursor wavefield and the CMB scattering curve, first noted by *Cleary and Haddon* [1972], is strong evidence that the maximum depth of scattering is close to the

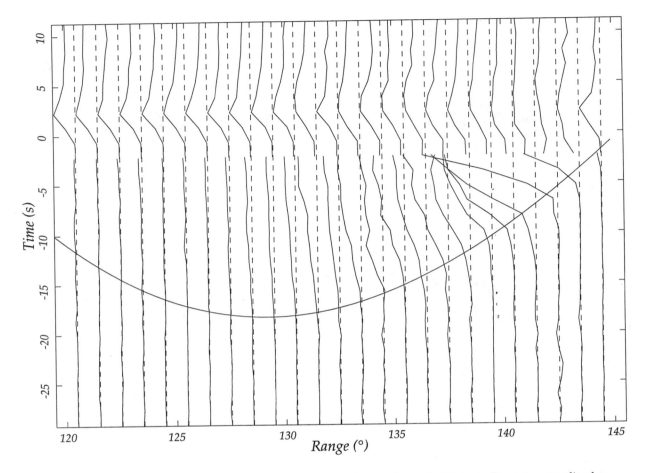

Figure 3. An alternative plot of the *PKPdf* precursor image shown in Plate 1. Precursor amplitudes have been multiplied by 10 at times before -2 s to show them on the same plot as the *PKPdf* arrivals. The curve shows the theoretical onset time (for the PREM model) for scattering at the CMB. Note the emergent onsets and gradual increase in precursor amplitudes within each range bin.

CMB. Scattering from significant depths within the outer core would result in arrivals at earlier times than are seen in the data. This constraint is best seen in our stack at ranges less than 135°; we attribute the apparent early arrivals at longer ranges (seen in Figure 3) to uncertainties in picking the *PKPdf* onset, the coarseness of our binning scheme, and the effects of three-dimensional velocity variations. Although some scattering must be coming from the vicinity of the CMB, the upper limit of the scattering source region is harder to constrain because the later arrivals could have originated from a variety of scattering depths in the mantle. The curves shown in Plate 1 and Figure 3 display only the minimum times for scattering at certain depths; scattering at these depths will also result in later arrivals. For example, scattering at the CMB alone will produce a precursor wavefield that arrives at all the times between the onset time and the *PKPdf* arrival.

Resolving the depth dependence of the scattering requires analysis of the time dependence of the precursor amplitudes, not simply their onset times.

This ambiguity in determining the minimum depth of scattering was recognized in many of the older studies, with some suggesting a scattering region that extends at least 600 to 900 km above the CMB [*Doornbos and Vlaar*, 1973; *Doornbos*, 1976], while others finding that the data could be explained with much thinner layers within the D″ region at the base of the mantle [*Haddon and Cleary*, 1974] or by short-wavelength CMB topography [*Doornbos*, 1978]. While not explicitly rejecting the possibility of scattering higher in the mantle, recent analyses have tended to focus on the CMB region as the source of the scattering. *Bataille and Flatté* [1988] examined 130 seismograms with clear *PKP* precursors and found that their amplitudes could be explained equally well by 0.5 to 1.0% rms velocity perturbations in a 200 km thick layer at the base of the mantle, or by CMB topography with rms height of about 300 m.

2.2. Modeling PKP precursors

Assuming a reference seismic velocity model and single scattering, it is possible to identify those points within the Earth's interior that could produce an arrival at a given time-distance point in the precursor wavefield. Some cross-sections of these scattering source regions are plotted in Figure 2. A plan view at the CMB of the scattering points for a source-receiver distance of 135° is shown in Figure 4. In this case, we assume that both the source (indicated with the star) and the receiver (indicated with the triangle) are located along the equator. Precursory arrivals could result either from scattering near the source or near the receiver in the shaded areas. Arrival times relative to $PKPdf$ are contoured; these contours are truncated by the minimum range of the b caustic. Because of the geometrical focusing of energy at the caustic, the strongest precursors are scattered close to the caustic, indicated by the thick shaded lines in the figure.

The time contours plotted in Figure 4 are part of three-dimensional isotime surfaces that extend up into the mantle. However, each source-receiver pair samples only a small fraction of the CMB and overlying mantle. The total coverage of our data set is shown in Figure 5 which plots the theoretical $PKPdf$ entry and exit points at the CMB. Due to the relatively steep mantle ray paths, the points on the CMB are clustered near the sources and receivers. Coverage is best in the western Americas, Eurasia, and the southwest Pacific; coverage is sparse beneath the Atlantic, Indian and mid-Pacific oceans.

To model our precursor observations, we apply single-scattering theory to a medium with random velocity variations. This is the approach used by most previous analyses of PKP precursors and is described in *Haddon and Cleary* [1974], *Aki and Richards* [1980], *Wu and Aki* [1985a,b], and *Bataille and Flatté* [1988]. For our application we assume that the random velocity perturbations are characterized by an exponential autocorrelation function:

$$N(\mathbf{r}) = e^{-|r|/a}$$

where a is termed the correlation length. More loosely, we will refer to a as the scale length, as it provides a rough measure of the typical separation between the high and low velocity regions. If the material behaves as a Poisson solid (P and S scale equally), and considering only single scattering, then the average scattered power

as a function of scattering angle may be expressed as:

$$< |\Phi_s(\theta)|^2 > = \frac{2k^4 a^3 \tilde{\alpha}^2 V A^2}{\pi r^2} \; \frac{\frac{1}{4}\left(\cos\theta + \frac{1}{3} + \frac{2}{3}\cos^2\theta\right)^2}{\left(1 + 4k^2 a^2 \sin^2\frac{\theta}{2}\right)^2}$$

where A is the incident wave amplitude, V is the volume of scattering, $\tilde{\alpha}$ is the rms velocity perturbation ($\delta\alpha/\alpha_0$), k is the wavenumber (ω/α_0), r is the scatterer-receiver distance, and θ is the scattering angle. To use this equation for PKP scattering from mantle heterogeneity, we use geometrical ray theory to compute A and replace the $1/r^2$ factor with the appropriate scatterer-to-receiver geometrical spreading factor. The transmission coefficients at the core-mantle boundary are also included in these terms. We sum contributions from both near-source and near-receiver scattering and from the ab and bc branches of PKP. The scattering angle is defined by the difference in direction between the incident and scattered ray paths.

Ray theory is well suited to this problem because of the relatively high frequencies that are involved. High amplitudes are predicted in the vicinity of the b caustic, but these are smoothed by numerical integration over distances that span the caustic. We experimented with applying smoothing filters to the ray theoretical amplitudes to prevent an abrupt cutoff in the amplitudes at the b caustic and did not see a significant change in our results. A more sophisticated synthetic seismogram study by *Cormier* [1995] has also shown that simple ray theory is adequate for examining PKP precursors.

Using PREM as a reference velocity model, the only free parameters in the scattering model are the scale length a, the amplitude of the velocity anomalies $\tilde{\alpha}$ and the volume of the scattering. To compare with our observations, we normalize our synthetic wavetrain to the predicted peak amplitude of $PKPdf + PKPcd$. The largest source of uncertainty in this calculation is the assumed level of inner core attenuation, which has a large effect on $PKPdf$ amplitudes at high frequencies. We use the inner core Q value of 360 obtained by *Bhattacharyya et al.* [1993]. Choosing a different value for Q would change the amplitude and range dependence of $PKPdf$ and the predicted precursor-to-$PKPdf$ amplitude ratio. This would affect our estimates of the best a and $\tilde{\alpha}$ values to fit our data, but not our conclusions, discussed below, concerning the depth extent of the scattering (as these are primarily derived from the time dependence in the individual precursor amplitudes at fixed source-receiver range).

To model our observations, we assume that the scattering properties are homogeneous within spherical shells of 20 km thickness and compute the precursor

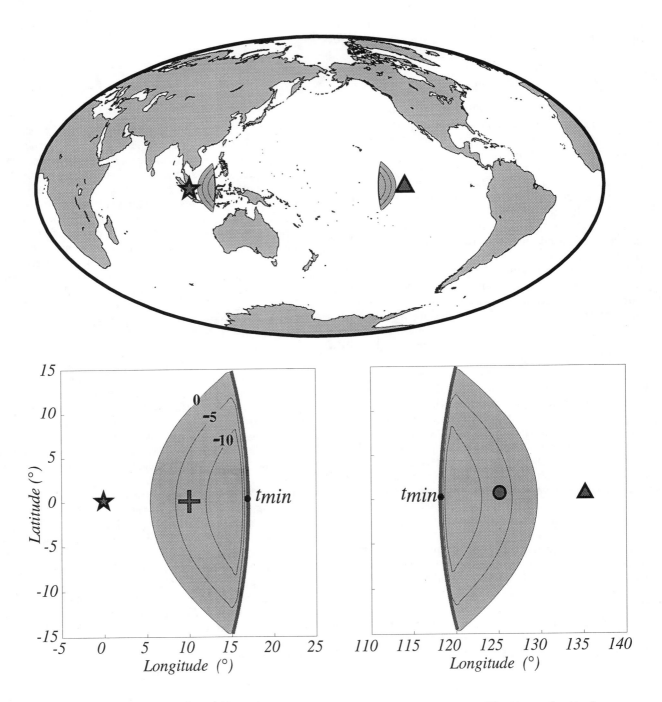

Figure 4. A map showing CMB scattering points at 135° source-receiver range. The star and triangle indicate the source and receiver locations, respectively. Scattering of *PKPbc* at the CMB within the cresent shaped regions will produce precursors to *PKPdf*. Closeups of these scattering source regions are shown below; the contours show the arrival times relative to *PKPdf*. The cross and circle respectively show the *PKPdf* entry and exit points on the CMB. Due to geometrical focusing, the strongest precursors are generated by scattering along the *B* caustic, the position of which is indicated by the heavy line in the lower plots.

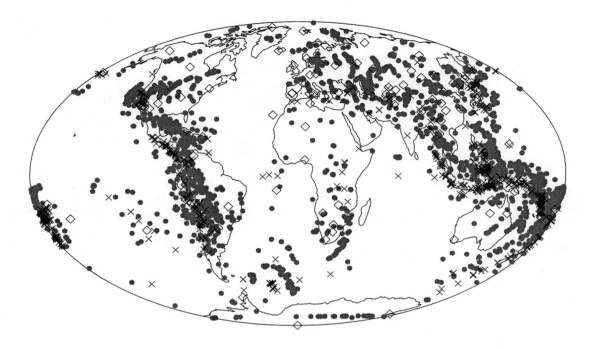

Figure 5. Map showing the $PKPdf$ entry and exit points on the CMB for the 1600 seismograms in the stack of PKP precursors. Source locations are shown as crosses, receiver locations as diamonds.

wavefield for each shell separately. The effect of scattering over different depth intervals can then be obtained by summing (in power) the appropriate number of these response kernels. We find that the predicted scattering envelope at each depth has roughly the form of a step function. No energy arrives prior to the minimum possible time, as defined by the curves shown in Plate 1. The predicted power level then jumps sharply and remains at a fairly constant level until the $PKPdf$ arrival. The earliest energy comes from scattering where the b caustic intersects the vertical source-receiver plane (the points labeled t_{min} in Figure 4). Subsequent arrivals also come primarily from the b caustic but from ray paths that are deflected out of this plane. To first order, a change in the scattering depth shifts the onset time of the response kernel and scales the amplitude but does not have a drastic effect on the shape.

A cartoon that illustrates the time dependence in the precursor amplitudes that might be expected from this type of behavior is shown in Figure 6. Scattering that is restricted to the core-mantle boundary should produce impulsive precursors of relatively constant amplitude. In contrast, scattering that extends uniformly up into the mantle should produce a wedge-shaped precursor amplitude profile. In principle, a rough measure of the scattering strength at a given depth in the mantle could be obtained from the slope at the appropriate time along the precursor power curve. Comparing Figures 3

and 6, it would appear that the continuous increase in observed precursor amplitudes with time favors models in which the scattering is distributed throughout the lower mantle. However, the comparison is complicated by the fact that the response kernels must first be convolved with realistic source-time functions and this greatly reduces the differences between the model predictions. To properly compare our synthetic precursors with the observations, we use an empirical source-time function derived from the $PKPdf$ pulses in the stack.

At a source-receiver range of 135.5°, Figure 7 shows the kernels for energy scattered from random velocity perturbations (8 km scale length) at 7 depths ranging from the CMB up to 1200 km above. The square root of a sum of these energy kernels can be used to model the data amplitudes; this is shown below for models of scattering from the CMB only, the lowermost 200 km of the mantle, and the entire mantle. In each case the rms velocity is adjusted to give the best least-squares fit to the observations, but is assumed uniform over the entire depth extent of the scattering. In this case, we find that the whole-mantle scattering model provides a better fit to the observations than those models in which scattering is restricted to the lowermost mantle. The latter models predict a steeper onset to the precursor amplitudes and a flatter slope immediately prior to $PKPdf$ than are seen in the data. This misfit would appear even worse, except that we permit a small time shift

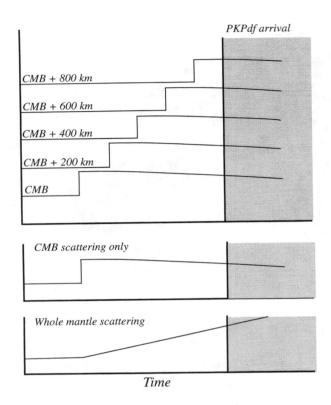

Figure 6. A cartoon illustrating the behavior of the theoretical kernels for *PKP* precursors for scattering at various depths in the mantle. In general, scattering restricted to the vicinity of the CMB should produce impulsive precursors, whereas whole mantle scattering should produce a gradual increase in precursor amplitude with time.

between data and predictions to adjust for uncertainties in the reference velocity model. The three different models all agree at about -11 s, but, for the CMB models, the slope of the curve at this point is steeper than the observations.

These differences are rather subtle and our results would not be convincing if they were obtained from this range bin alone. However, similar outcomes are found through independent fitting of a number of different range bins. This is illustrated in Figure 8, which plots the best fits of the three different models across 11 range bins between 130° and 141°. Only the whole mantle model provides an adequate overall fit to the data; the other models do not reproduce the emergent onset of the precursors and their steady increase in amplitude until the main *PKPdf* arrival. This is confirmed by χ^2 tests of the significance of the data fits; only the whole mantle model fits the data within the standard errors [*Hedlin et al.*, 1997].

Our preferred model of whole-mantle scattering has a scale length of 8 km and rms velocity perturbations that vary between different range bins with $\tilde{\alpha}$ values from

0.8 to 2.0%. Scale lengths between 4 and 12 km also can produce satisfactory fits to the data, but require a greater range dependence in the velocity perturbations. To test if our conclusions might be specific to the exponential autocorrelation model, we repeated our calculations for a Gaussian model. The fits were very similar to those obtained from the exponential model and continued to favor the whole mantle scattering model. However, the required rms velocity perturbations were somewhat less, ranging from 0.8 to 1.4%.

Our results do not require any enhanced velocity heterogeneity near the CMB or short wavelength topography on the CMB to explain the observed precursors. However, we cannot exclude the possibility that

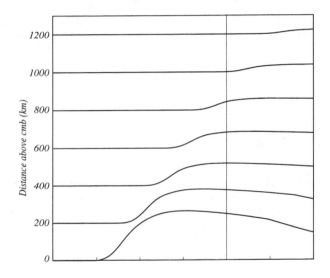

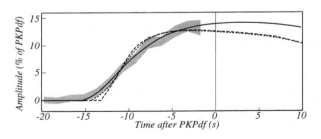

Figure 7. Scattering energy kernels at 135.5° range for random mantle velocity perturbations at 7 depths ranging from the CMB to 1200 km above. The kernels in this figure have been calculated assuming an exponential autocorrelation function with an 8 km scale length, and have been convolved with an empirical source-time function derived from the data stack. The whole mantle model, computed by summing kernels at 20 km intervals, produces the solid curve in the lower plot. This fits the data (shown with 1 σ error bounds) much more closely than the curves that result from models that allow scattering only in the lower 200 km of the mantle (short dashes) and in the lowermost 20 km of the mantle (long dashes).

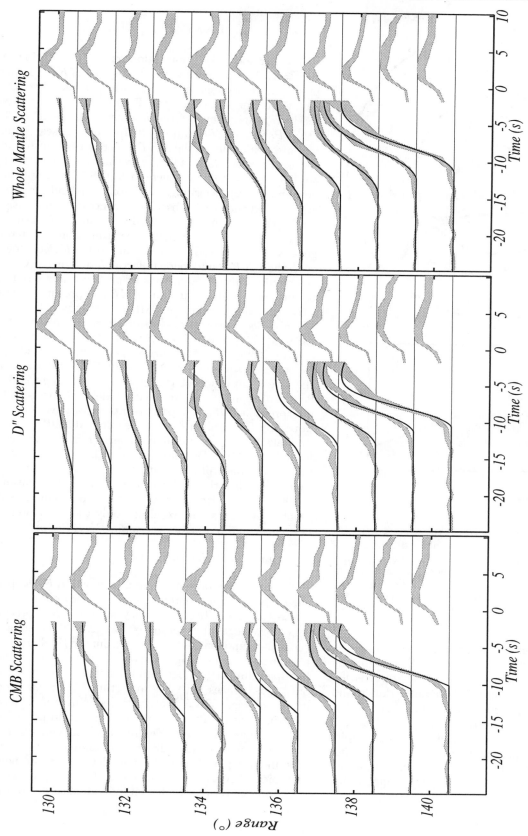

Figure 8. Observed average *PKPdf* precursor amplitudes at ranges between 130° and 141° compared to theoretical predictions for three different models of mantle heterogeneity. The data and their standard errors are indicated with the gray shaded regions. The left panel shows the best fit to the data achieved with a model in which scattering is restricted to the core-mantle boundary, the center panel indicates the fit when scattering is permitted within the lowermost 200 km of the mantle, and the right panel shows the the fit obtained from a model in which the scattering is uniform throughout the mantle. Only the whole mantle model provides an adequate overall fit to the data.

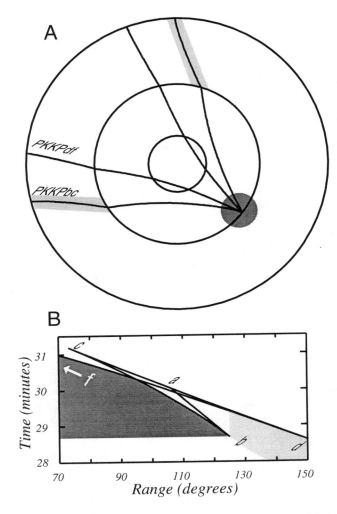

Figure 9. Ray paths (a) and travel times curves (b) for *PKKP*, showing the effect of scattering along different parts of the ray paths. Backscattering near the CMB can produce precursors in the dark shaded region. Forward scattering in the mantle can deflect *PKKPab* and *PKKPbc* to longer ranges, producing *PKKPdf* precursors in the light shaded region.

some level of increased scattering might occur near the CMB. To place an upper limit on the CMB topography we computed the maximum rms velocity perturbation within the lowermost 20 km of the mantle that would produce precursor amplitudes that did not exceed the maximum amplitude of our observations, and then converted this contrast to the equivalent rms topography on the CMB [using equation 8 of *Bataille and Flatté*, 1988]. For a scale length of 8 km we found that the smallest upper bound was 300 m in rms topography, provided by the 124° to 125° bin.

3. PKKP PRECURSORS

Another source of information on deep scattering in the Earth is provided by precursors to *PKKP*. These precursors have been observed within at least two different distant intervals (see Figure 9). At source-receiver ranges beyond the *b* caustic at about 125°, *PKKPdf* precursors can result from mantle scattering and are analogous to the *PKP* precursors discussed in the previous section. These were observed by *Doornbos* [1974], who detected them in NORSAR (Norwegian Seismic Array) recordings of Solomon Islands events, and showed that their observed slownesses were consistent with scattering from the deep mantle. At ranges less than 125°, forward scattering from mantle heterogeneity does not produce precursors to *PKKPbc*. However, the *PKKP* underside reflection point on the CMB is a maximum time point with respect to changes in the bouncepoint position, making *PKKPbc* precursors sensitive to short wavelength CMB topography.

PKKPbc precursors were observed by *Chang and Cleary* [1978, 1981] from Novaya Zemlya explosions recorded by the LASA array in Montana at about 60° range. Although suggestive of CMB topography, the large amplitude of the observed precursors is difficult to explain with realistic models. More readily interpretable *PKKPbc* precursor observations were obtained by *Doornbos* [1980], who examined NORSAR array data from several different source regions at ranges from 80° to 110°. Doornbos showed that these observations could be explained with a rough CMB with horizontal scale lengths of 10 to 20 km and depth variations of 100 to 200 m. However, this result is based on a small number of events and examined only the time interval between 1 and 5 s before the main *PKKPbc* arrival.

3.1. Stacking PKKP precursors

To obtain a more complete image of *PKKPbc* precursors, we implemented a stacking procedure similar to that described previously for the *PKP* precursors. We began by obtaining over 25,000 broadband vertical-component seismograms from the IRIS archives at source-receiver ranges between 80° to 120°. After bandpass filtering to between 0.7 and 2.5 Hz, we select only those traces with low and stationary noise properties and clear *PKKP* arrivals, reducing our data set to 1856 records. Theoretical CMB reflection points for these data are plotted in Figure 10. Identifying *PKKP* precursors is more difficult than finding *PKP* precursors, in large part due to the much lower signal-to-noise in typical *PKKP* arrivals. Clear *PKKP* precursors are almost never observable on individual seismograms.

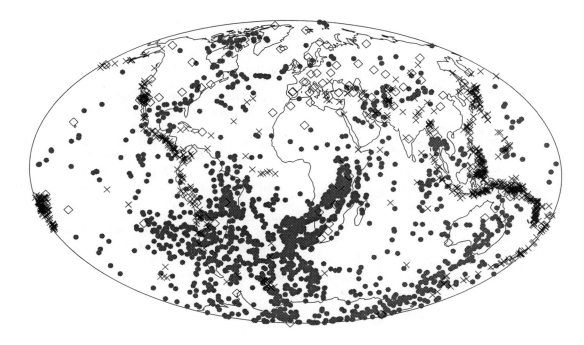

Figure 10. Map showing the *PKKP* CMB bouncepoints for the 1856 seismograms in the stack of *PKKP* precursors. Source locations are shown as crosses, receiver locations as diamonds.

To image the precursors, we applied stacking methods similar to that used in the *PKP* analysis, but with a coarser binning scheme (4° by 4 s) to reduce the noise levels in the stack. Figure 11 shows the results obtained using a variety of different stacking techniques. At ranges between 80° and 112°, precursory energy is seen up to 60 s before the main *PKKPbc* arrival, with a generally later onset at longer ranges. The precursor wavefield is emergent, with amplitudes gradually increasing up to the main arrival. It is likely that the precursor wavefield extends to ranges less than 80°; however, attempts to image it at these ranges were unsuccessful due to a lack of suitable *PKKPbc* arrivals to use as reference phases.

3.2. Modeling PKKP precursors

To interpret these observations, we apply Kirchhoff theory [e.g. *Kampfmann and Müller*, 1989] to models of small-scale CMB topography. The response for a specified topography at a given range is calculated by propagating the wavefield to the CMB with ray theory, evaluating the surface Kirchhoff integral, and then propagating the solution back to the surface. We include the effects of the reflection and transmission coefficients at the CMB and inner-core boundary (ICB), inner core

attenuation (assumed Q=360), and scattering between all branches of *PKKP*. We assume random CMB topography, parameterized by a Gaussian autocorrelation function with horizontal correlation length λ and rms amplitude σ. In order to compare to the power stack of the data (Figure 11, bottom panel), we stack synthetic seismograms (each derived from a different realization of the CMB topography but having identical statistical properties) using the same stacking method. We then convolve with an empirical source-time function (derived from *PKKPbc* in the data stack) and align on the *PKKPbc* onset.

Values of λ from 7 to 10 km and σ values from 250 to 350 m match many of the characteristics of the observed waveforms. Our preferred model has a horizontal scale length of 8 km and rms amplitude of 300 m and produces the fit shown in Figure 12. This is not too far from the 100 to 200 m topography obtained by *Doornbos* [1980] in his study of *PKKPbc* precursors. However, there is a problem with our fit to the observations, in that the synthetics predict that precursor energy levels should grow with range, whereas no such trend is apparent in the data. This is why our model underpredicts the precursors at small ranges and overpredicts them at long range. The increase in precursor amplitude with range is a robust feature of the CMB topography models that we considered; we could not find

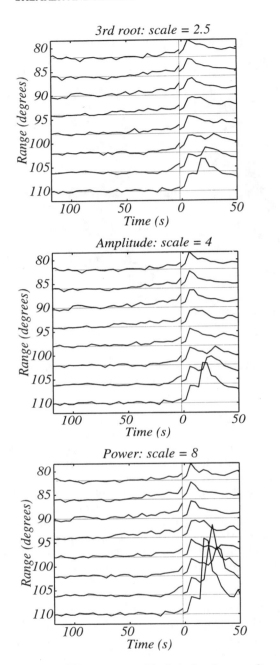

3rd root: scale = 2.5

Amplitude: scale = 4

Power: scale = 8

Figure 11. PKKP precursor stacks, showing results of 3 different processing methods. The top plot is a stack of the cube root of the envelope functions of the traces, the middle plot is a simple amplitude stack, and the bottom plot is a stack of the power in each trace (the square of the amplitude). The scale changes to the left of the vertical line in each plot to enhance the visibility of the precursors.

a way to remove this trend by using different values of λ and σ.

Our preferred CMB topography model comes close to explaining the observations, but, given the mismatch in

the amplitudes with range, it is not clear how much confidence should be accorded to the CMB topography model. What could be causing this systematic misfit to the data? We considered several possibilities:

(1) There is a problem with our Kirchhoff calculation. To check this possibility, we performed a simple calculation in which point scatterers are assumed to be present immediately below the CMB and experimented with a variety of different scattering angle dependencies. As in our Kirchhoff calculation, the resulting precursor/$PKKPbc$ amplitude ratios increased with range. Thus, we do not think it likely that the predicted range dependence is a result of a problem in our implementation of the Kirchhoff method.

(2) The assumption of Gaussian topography is too limiting; a different form of random topography might explain the observations. While we cannot eliminate this possibility with certainty, the test that we applied for (1) suggests that this is unlikely.

(3) Multiple scattering at the CMB is involved; Kirchhoff or Born approaches do not produce reliable results. In the case of transmitted PKP, this idea was tested by

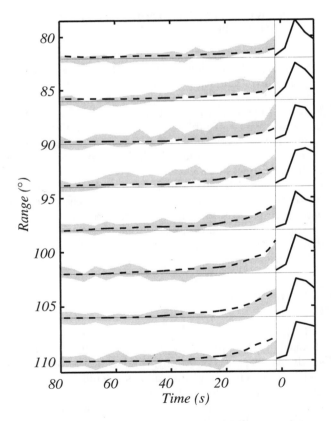

Figure 12. The fit to the observed $PKKP$ precursor power levels obtained by our preferred model of random small-scale CMB topography. The shaded regions show the ±2 standard error limits on the stack, computed using a bootstrap resampling method.

Doornbos [1988] who found that first-order scattering theory produces reasonably accurate results for CMB topography of 500 m or less. However, this remains a possibility that should be further studied.

(4) Lateral variations in CMB topography produce the observed range dependence. This idea is suggested by the fact that we could fit the data almost perfectly if we allowed a different topography amplitude for each range bin. Since the different range bins sample different parts of the CMB, this is a possibility. However, it seems improbable that such variations would result in a smooth change in precursor amplitude with range.

(5) The precursors result not from CMB topography, but from back-scattering off volumetric heterogeneities in the lowermost mantle (above and near the CMB bouncepoints). This could produce precursors if the heterogeneities are located close to the CMB. Backscattering requires relatively large impedance variations to produce significant precursors, but perhaps these are possible given the large velocity anomalies that have been observed in the vicinity of the CMB. However, a preliminary calculation indicates that such a mechanism would result in a precursor/$PKKPbc$ power ratio that grows with range in the same way that the CMB topography models predict.

(6) The precursors result from forward scattering from velocity heterogeneity in the outer core. Although the outer core is generally considered to be laterally homogeneous, such scattering would produce $PKKPbc$ precursors and it is conceivable that a model of this type could be constructed that would mimic the range dependence in our data. However, if such scattering were present, it should also be observable in PKP precursors. As shown in Plate 1, there is no evidence for outer-core scattering in PKP data stacks.

(7) The precursors are caused by scattering from the inner core. This would not be as surprising as outer-core scattering because observations have indicated that both lateral heterogeneity and anisotropy are present within the inner core. Ray tracing calculations show that energy single-scattered from the $PKKPdf$ ray-path at the ICB or within the inner core can arrive at times that are consistent with the observed precursors (Figure 13). However, very large scattering angles are required to explain the onset of the precursors, in some cases exceeding 90°. Some of these peculiar ray paths are illustrated in Figure 14. Another potential difficulty is that the predicted amplitude of the direct $PKKPdf$ arrival is extremely small, due to inner-core attenuation and a steep CMB reflection angle. Thus there would not appear to be enough energy available to scatter to generate precursors of the required amplitude. This model would work only if some of the observed P wave attenuation within the inner core were actually due to scattering [*Earle and Shearer*, 1997]. In any case, the large scattering angles and the lack of direct $PKKPdf$ in the observations suggest that multiple scattering theory will be necessary to test this idea.

Additional work will be required to obtain a definitive answer regarding the problem with the range dependence in the precursor amplitudes. However, even if the precursors are derived in part from sources other than the CMB, it is still possible to use our Kirchhoff calculation to place upper limits on CMB topography. Using the precursor/$PKKP$ power ratio at 106°, a conservative upper limit on σ is 315 m for $\lambda = 10$ km and 630 m for $\lambda = 20$ km [*Earle and Shearer*, 1997].

4. DISCUSSION

The lowermost mantle within the D″ region is known to be anomalous in many respects [reviews include *Young and Lay*, 1987a; *Loper and Lay*, 1995]. Travel time patterns and tomographic inversions show much stronger lateral S velocity perturbations in the lowermost mantle than in the mid-mantle [e.g., *Woodward and Masters*, 1991; *Su et al.*, 1994; *Masters et al.*, 1996; *Lay et al.*, 1997]. Recent analyses of the SKS family of waveforms have suggested P velocity anomalies of 5 to 10% within a thin layer above the CMB [*Garnero et al.*, 1993; *Garnero and Helmberger*, 1995, 1996]. Additional arrivals indicative of velocity discontinuities at a variety of depths above the CMB have been identified by a number of authors [*Lay and Helmberger*, 1983; *Lay*, 1986, 1989; *Young and Lay*, 1987b, 1989, 1990; *Weber and Davis*, 1990; *Houard and Nataf*, 1992; *Weber*, 1993; *Vidale and Benz*, 1993; *Kendall and Shearer*, 1994; *Yamada and Nakanishi*, 1996]. Shear-wave splitting indicative of anisotropy has recently been identified in the lowermost mantle [*Vinnik et al.*, 1989, 1995; *Lay and Young*, 1991; *Kendall and Silver*, 1996; *Matzel et al.*, 1996; *Garnero and Lay*, 1997].

The large change in composition, temperature and seismic velocities across the CMB may produce chemical and thermal boundary layers within the D″ region. Thus it would not be surprisingly if the CMB region was also anomalous with respect to generating high-frequency scattered energy. However, this has not been demonstrated from analysis of PKP precursors; our results, discussed above, indicate that the globally averaged data can be explained better with models of uniform scattering throughout the mantle than with scattering concentrated near the CMB. This evidence for small-scale seismic velocity perturbations within the mantle supports models of compositional heterogeneity due to incomplete mixing during convection [e.g., *Davies*, 1984, 1990; *Gurnis and Davies*, 1986; *Christensen and Hofmann*, 1994; *Schmalzl et al.*, 1996]. Mantle heterogeneity at a range of scale lengths is also indicated by geochemical studies, which require a variety

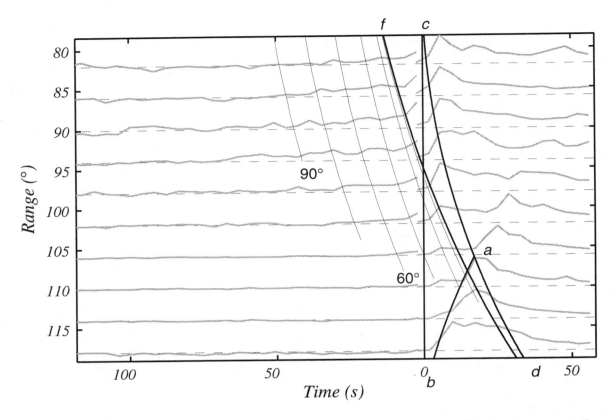

Figure 13. Predicted precursor onset times for scattering at the inner-core boundary, shown at 15° increments in the maximum scattering angle. Precursor onsets that agree with those seen in the data are possible, but require scattering angles of 90° or more. The data stack and travel times curves for the different branches of $PKKP$ are plotted for comparison. Note that the direct $PKKPdf$ arrival cannot be seen in the data stack.

of different source chemistries [e.g., *Zindler and Hart*, 1986; *Carlson*, 1994]. A possible source for 8-km scale length material is detached pieces of subducted oceanic crust that have become entrained in the mantle flow field.

Regardless of structures higher in the mantle, PKP precursor observations can be used to focus on the CMB region by excluding the later arrivals in the precursor wavetrain. Our PKP waveform stack shows that some scattering must be originating from the vicinity of the CMB, but that no scattering is detectable from the outer core (although scattering from the uppermost ~50 km cannot be ruled out). CMB topography is not required to explain the data, as volumetric heterogeneity in the lowermost mantle can account for the observed signal. $PKPdf$ precursor amplitudes limit topography on the CMB to no more than ~300 m at 8-km scale length.

Our analysis of $PKKPbc$ precursors has so far produced mixed results. We obtain a clear image of the average precursor wavefield, but have not succeeded in

explaining the range dependence in the precursor to $PKKPbc$ energy ratios with simple models of CMB topography. Our overall best-fitting CMB topography model has an 8-km horizontal scale length and rms amplitude of 300 m (identical to the upper limit on CMB topography suggested by our PKP analysis). However, this model overpredicts precursor amplitudes at long range and underpredicts the precursors at short range; we have been unable to identify with certainty the source of this mismatch. It is possible that scattering from the vicinity of the inner-core boundary also contributes to the $PKKP$ precursor wavefield. Thus, the extent to which $PKKP$ precursor data require CMB topography remains somewhat uncertain, although we can use our results to place upper bounds on the amplitude of the topography.

Observations of topside reflections off the CMB have generally been found to be consistent with the absence of small-scale topography along the interface, and have been used to place upper limits on its size. *Menke* [1986] modeled amplitude variations of PcP and found that

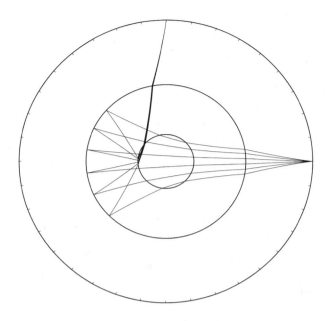

Figure 14. Possible ray paths for $PKKPdf$ single-scattering at the inner-core boundary (ICB) at a source-receiver distance of 90°. Here the source is shown at the top; the scattering occurs at the second ICB crossing along the ray path. To match the observations of precursors up to about 50 s before $PKKPbc$, scattering angles of 90° or more are required at the ICB.

CMB topography is less than a few hundred meters at horizontal scale lengths of 2 to 50 km. *Neuberg and Wahr* [1991] found that PcP travel-time variations limited CMB topography to less than 2 to 3 km at wavelengths of 50 to 400 km. *Vidale and Benz* [1992] stacked PcP and ScP arrivals across short-period stations in northern California and found that the simple pulse shapes limited CMB topography to less than 500 m at scale lengths of 50 to 200 km. These results are consistent with our inferred upper limits of ∼300 to 600 m at horizontal scale lengths of 10 to 20 km.

The possibility that the inner core could cause scattering of seismic energy deserves further study. A number of recent studies have found evidence for both lateral heterogeneity and anisotropy within the inner core [e.g., *Poupinet et al.*, 1983; *Morelli et al.*, 1986; *Woodhouse et al.*, 1986; *Shearer and Toy*, 1991; *Creager*, 1992; *Vinnik et al.*, 1994; *Shearer*, 1994; *Song and Helmberger*, 1995; *Su and Dziewonski*, 1995; *Tromp*, 1993; *Tanaka and Hamaguchi*, 1997]. Significant changes in $PKPdf$ travel times are sometimes observed to occur over shorter spatial scales than can be explained by simple anisotropy or low-order heterogeneity models [e.g., *Shearer*, 1994]. $PKPdf$ rays that traverse the inner core at angles close to the rotation axis

produce early arrivals indicative of anisotropy, but also appear anomalously weak in amplitude [*Creager*, 1992; *Souriau and Romanowicz*, 1996]. The weak amplitude of these arrivals could be indicative of differences in intrinsic attenuation, but, alternatively, might result from scattering of high-frequency energy due to small scale structures at or near the core-mantle boundary.

More research is needed to help to resolve the depth extent of scattering sources in the deep Earth. In addition to PKP and $PKKP$, there are other phases that may provide additional constraints on scattering. For example, high-frequency $Pdiff$ can be modeled with multiple scattering within the D″ region [*Bataille and Lund*, 1996]. If small-scale heterogeneity is present throughout the mantle, as our PKP precursor analysis suggests, it should produce scattered arrivals for a number of different seismic phases. However, identification of such deep-mantle scattering may be difficult due to interference from strong scattering in the crust and upper mantle. A comprehensive analysis of scattering from a variety of phases and ray geometries would help to isolate the contributions from different depths in the Earth. It is likely that lateral variations in scattering strength are also present; we are currently experimenting with back-projection techniques in an effort to image scattering sources more directly.

Acknowledgments. Our research would not have been possible without the ready accessibility of high-quality data from the global seismic networks; in particular we would like to thank Rick Benson for maintaining the FARM archive at the IRIS DMC. Ken Creager and Michael Wysession provided constructive reviews. This work was supported by National Science Foundation grants EAR93-15060, EAR95-07994, EAR96-14350, and EAR96-28020.

REFERENCES

Aki, K., and P.G. Richards, *Quantitative Seismology, Theory and Methods, Volume 2*, W.H. Freeman, San Franciso, 1980.

Bataille, K. and S.M. Flatté, Inhomogeneities near the core-mantle boundary inferred from short-period scattered PKP waves recorded at the global digital seismograph network, *J. Geophys. Res.*, *93*, 15,057–15,064, 1988.

Bataille, K., and F. Lund, Strong scattering of short-period seismic waves by the core-mantle boundary and the P-diffracted wave, *Geophys. Res. Lett.*, *23*, 2413–2416, 1996.

Bataille, K., R.-S. Wu, and S.M. Flatté, Inhomogeneities near the core-mantle boundary evidenced from scattered waves: a review, *Pageoph.*, *132*, 151–173, 1990.

Bhattacharyya, J., P.M. Shearer, and G. Masters, Inner core attenuation from short-period PKP(BC) versus PKP(DF) waveforms, *Geophys. J. Int.*, *114*, 1–11, 1993.

Carlson, R.W., Mechanisms of Earth differentiation: Consequences for the chemical structure of the mantle, *Rev. Geophys.*, *32*, 337-361, 1994.

Chang, A.C., and J.R. Cleary, Precursors to PKKP, *Bull. Seismol. Soc. Am.*, *68*, 1059–1079, 1978.

Chang, A.C., and J.R. Cleary, Scattered PKKP: Further evidence for scattering at a rough core-mantle boundary, *Phys. Earth Planet. Inter.*, *24*, 15–29, 1981.

Christensen, U.R., and A.W. Hofmann, Segregation of subducted oceanic crust in the convecting mantle, *J. Geophys. Res.*, *99*, 19,867–19,884, 1994.

Cleary, J.R., and R.A.W. Haddon, Seismic wave scattering near the core-mantle boundary: a new interpretation of precursors to PKP, *Nature*, *240*, 549–551, 1972.

Cormier, V.F., Time-domain modeling of PKIKP precursors for constraints on the heterogeneity in the lowermost mantle, *Geophys. J. Int.*, *121*, 725–736, 1995.

Creager, K.C., Anisotropy of the inner core from differential travel times of the phases *PKP* and *PKIKP*, *Nature*, *356*, 309–314, 1992.

Davies, G.F., Geophysical and isotopic constraints on mantle convection: an interim synthesis, *J. Geophys. Res.*, *89*, 6017–6040, 1984.

Davies, G.F., Mantle plumes, mantle stirring and hotspot chemistry, *Earth Planet. Sci. Lett.*, *99*, 94–109, 1990.

Doornbos, D.J., Seismic wave scattering near caustics: observations of PKKP precursors, *Nature*, *247*, 352–353, 1974.

Doornbos, D.J., Characteristics of lower mantle heterogeneities from scattered waves, *Geophys. J. R. Astron. Soc.*, *44*, 447–470, 1976.

Doornbos, D.J., On seismic-wave scattering by a rough core-mantle boundary, *Geophys. J. R. Astron. Soc.*, *53*, 643–662, 1978.

Doornbos, D.J., The effect of a rough core-mantle boundary on PKKP, *Phys. Earth Planet. Inter.*, *21*, 351–358, 1980.

Doornbos, D.J., Multiple scattering by topographic relief with application to the core-mantle boundary, *Geophys. J.*, *92*, 465–478, 1988.

Doornbos, D.J., and N.J. Vlaar, Regions of seismic wave scattering in the Earth's mantle and precursors to PKP, *Nature Phys. Sci.*, *243*, 58–61, 1973.

Dziewonski, A.M., and D.L. Anderson, Preliminary reference Earth model, *Phys. Earth Planet. Inter.*, *25*, 297–356, 1981.

Earle, P.S., and P.M. Shearer, New observations of PKKP precursors: constraints on small-scale topography on the core-mantle boundary, *Science*, *277*, 667–670, 1997.

Efron, B., and R. Tibshirani, Statistical data analysis in the computer age, *Science*, *253*, 390–395, 1991.

Garnero, E.J., S.P Grand, and D.V. Helmberger, Low P-velocity at the base of the mantle, *Geophys. Res. Lett.*, *20*, 1843–1846, 1993.

Garnero, E.J., and D.V. Helmberger, A very slow basal layer underlying large-scale low-velocity anomalies in the lower mantle beneath the Pacific: evidence from core phases, *Phys. Earth Planet. Inter.*, *91*, 161–176, 1995.

Garnero, E.J., and D.V. Helmberger, Seismic detection of a thin laterally varying boundary layer at the base of the mantle beneath the central-Pacific, *Geophys. Res. Lett.*, *23*, 977–980, 1996.

Garnero, E.J., and T. Lay, Lateral variations in lowermost mantle shear wave anisotropy beneath the north Pacific and Alaska, *J. Geophys. Res.*, *102*, 8121–8135, 1997.

Grand, S.P., R.D. van der Hilst, and S. Widiyantoro, Global seismic tomography: a snapshot of convection in the Earth, *GSA Today*, *7*, 1–7, 1997.

Gurnis, M., and G.F. Davies, Mixing in numerical models of mantle convection incorporating plate tectonics, *J. Geophys. Res.*, *91*, 6375–6395, 1986.

Haddon, R.A.W., Corrugations on the CMB or transition layers between inner and outer cores, *Trans. Am. Geophys. Union*, *53*, 600, 1972.

Haddon, R.A.W., Evidence for inhomogeneities near the core-mantle boundary, *Phil. Trans. R. Soc. Lond. A*, *306*, 61-70, 1982.

Haddon, R.A.W., and J.R. Cleary, Evidence for scattering of seismic PKP waves near the mantle-core boundary, *Phys. Earth Planet. Inter.*, *8*, 211–234, 1974.

Hedlin, M.A.H., P.M. Shearer, and P.S. Earle, Imaging CMB scatterers through migration of GSN PKPdf precursor recordings, *EOS Trans. AGU (Fall meeting supplement)*, *76*, F402, 1995.

Hedlin, M.A.H., P.M. Shearer, and P.S. Earle, Waveform stacks of PKP precursors: evidence for small-scale heterogeneity throughout the mantle, *Nature*, *387*, 145–150, 1997.

Houard, S., and H-C. Nataf, Further evidence for the 'Lay discontinuity' beneath northern Siberia and the North Atlantic from short period P-waves recorded in France, *Phys. Earth Planet. Inter.*, *72*, 264–275, 1992.

Husebye, E.S., D.W. King, and R.A.W. Haddon, Precursors to PKIKP and seismic wave scattering near the mantle-core boundary, *J. Geophys. Res.*, *81*, 1870–1882, 1976.

Kampfmann, W., and G. Müller, PcP amplitude calculations for a core-mantle boundary with topography, *Geophys. Res. Lett.*, *16*, 653–656, 1989.

Kanshima, S., and G. Helffrich, Detection of lower mantle scaterers northeast of the Mariana subduction zone using short-period array data, *J. Geophys. Res.*, submitted, 1997.

Kendall, J.-M., and P.M. Shearer, Lateral variation in D″ thickness from long-period shear wave data, *J. Geophys. Res.*, *99*, 11,575–11,590, 1994.

Kendall, J.-M., and P.G. Silver, Constraints from seismic anisotropy on the nature of the lowermost mantle, *Nature*, *381*, 409–412, 1996.

King, D.W., R.A.W. Haddon, and J.R. Cleary, Array analysis of precursors to PKIKP in the distance range 128° to 142°, *Geophys. J. R. Astron. Soc.*, *37*, 157–173, 1974.

Lay, T., Evidence for a lower mantle shear velocity discontinuity in *S* and *sS* phases, *Geophys. Res. Lett.*, *13*, 1493–1496, 1986.

Lay, T., Structure of the core-mantle transition zone, *EOS Trans. AGU*, *70*, 49–55, 1989.

Lay, T., and D. V. Helmberger, A lower mantle *S*-wave triplication and the shear velocity structure of D″, *Geophys. J. R. Astron. Soc.*, *75*, 799–837, 1983.

Lay, T., and C.J. Young, Analysis of SV waves in the core's penumbra, *Geophys. Res. Lett.*, *18*, 1373–1376, 1991.

Lay, T., E.J. Garnero, C.J. Young, and J.B. Gaherty, Scale-lengths of shear velocity heterogeneity at the base of the mantle from S wave differential travel times, *J. Geophys. Res.*, *102*, 9887–9909, 1997.

Loper, D.E., and T. Lay, The core-mantle boundary region, *J. Geophys. Res.*, *100*, 6397–6420, 1995.

Masters, G., S. Johnson, G. Laske, and H. Bolton, A shear-velocity model of the mantle, *Phil. Trans. R. Soc. Lond.*, *354*, 1285–1411, 1996.

Matzel, E., M.K. Sen, and S.P. Grand, Evidence for anisotropy in the deep mantle beneath Alaska, *Geophys. Res. Lett.*, *23*, 2417–2420, 1996.

Menke, W., Few 2-50 km corrugations on the core-mantle boundary, *Geophys. Res. Lett.*, *13*, 1501–1504, 1986.

Morelli, A., A.M. Dziewonski, and J.H. Woodhouse, Anisotropy of the inner core inferred from *PKIKP* travel times, *Geophys. Res. Lett.*, *13*, 1545–1548, 1986.

Neuberg, J., and J. Wahr, Detailed investigation of a spot on the core-mantle boundary using digital PcP data, *Phys. Earth Planet. Inter.*, *68*, 132–143, 1991.

Poupinet, G., R. Pillet, and A. Souriau, Possible heterogeneity of the earth's core deduced from *PKIKP* travel times, *Nature*, *305*, 204–206, 1983.

Schmalzl, J., G.A. Houseman, and U. Hansen, Mixing in vigorous, time-dependent three-dimensional convection and application to Earth's mantle, *J. Geophys. Res.*, *101*, 21,847–21,858, 1996.

Shearer, P.M., Constraints on inner core anisotropy from PKP(DF) travel times, *J. Geophys. Res.*, *99*, 19,647–19,659, 1994.

Shearer, P.M., and K.M. Toy, *PKP(BC)* versus *PKP(DF)* differential travel times and aspherical structure in the Earth's inner core, *J. Geophys. Res.*, *96*, 2233–2247, 1991.

Song, X., and D.V. Helmberger, Depth dependence of anisotropy of Earth's inner core, *J. Geophys. Res.*, *100*, 9805–9816, 1995.

Souriau, A., and B. Romanowicz, Anisotropy in inner core attenuation: a new type of data to constrain the nature of the solid core, *Geophys. Res. Lett.*, *23*, 1–4, 1996.

Su, W.-J., and A.M. Dziewonski, Predominance of long-wavelength heterogeneity in the mantle, *Nature*, *352*, 121–126, 1991.

Su, W.-J., and A.M. Dziewonski, Inner core anisotropy in three dimensions, *J. Geophys. Res.*, *100*, 9831–9852, 1995.

Su, W.-J., R.L. Woodward, and A.M. Dziewonski, Degree 12 model of shear velocity heterogeneity in the mantle, *J. Geophys. Res.*, *99*, 6945–6980, 1994.

Tanaka, S., and H. Hamaguchi, Degree one heterogeneity and hemispherical variation of anisotropy in the inner core from PKP(BC)-PKP(DF) times, *J. Geophys. Res.*, *102*, 2925–2938, 1997.

Tromp, J., Support for anisotropy of the Earth's inner core from free oscillations, *Nature*, *366*, 678–681, 1993.

van der Hilst, R.D., and E.R. Engdahl, Evidence for deep mantle circulation from global tomography, *Nature*, *386*, 578–584, 1997.

Vidale, J.E., and H.M. Benz, A sharp and flat section of the core-mantle boundary, *Nature*, *359*, 627–629, 1992.

Vidale, J. E., and H. M. Benz, Seismological mapping of fine structure near the base of the Earth's mantle, *Nature*, *361*, 529–532, 1993.

Vinnik, L.V., V. Farra, and B. Romanowicz, Observational evidence for diffracted SV in the shadow of the Earth's core, *Geophys. Res. Lett.*, *16*, 519–522, 1989.

Vinnik, L., B. Romanowicz, and L. Breger, Anisotropy in the center of the inner core, *Geophys. Res. Lett.*, *21*, 1671–1674, 1994.

Vinnik, L.V., B. Romanowicz, Y. Le Stunff, and L. Makeyeva, Seismic anisotropy of the D″ layer, *Geophys. Res. Lett.*, *22*, 1657–1660, 1995.

Weber, M., P and S wave reflections from anomalies in the lowermost mantle, *Geophys. J. Int.*, *115*, 183–210, 1993.

Weber, M, and J.P. Davis, Evidence of a laterally variable lower mantle structure from P- and S-waves, *Geophys. J. Int.*, *102*, 231–255, 1990.

Woodhouse, J. H., D. Giardini, and X.-D. Li, Evidence for inner core anisotropy from free oscillations, *Geophys. Res. Lett.*, *13*, 1549–1552, 1986.

Woodward, R.L., and G. Masters, Lower-mantle structure from ScS-S differential travel times, *Nature*, *352*, 231–233, 1991.

Wu, R., and K. Aki, Scattering characteristics of elastic waves by an elastic heterogeneity, *Geophysics*, *50*, 582–595, 1985a.

Wu, R., and K. Aki, Elastic wave scattering by a random medium and the small-scale inhomogeneities in the lithosphere, *J. Geophys. Res.*, *90*, 10,261–10,273, 1985b.

Yamada, A., and I. Nakanishi, Detection of P-wave reflector in D″ beneath the south-western Pacific using double-array stacking, *Geophys. Res. Lett.*, *13*, 1553–1556, 1996.

Young, C.J., and T. Lay, The core-mantle boundary, *Ann. Rev. Earth Planet. Sci.*, *15*, 25–46, 1987a.

Young, C.J., and T. Lay, Evidence for a shear velocity discontinuity in the lower mantle beneath India and the Indian Ocean, *Phys. Earth Planet. Inter.*, *49*, 37–53, 1987b.

Young, C. J., and T. Lay, The core shadow zone boundary and lateral variations of the P velocity structure of the lowermost mantle, *Phys. Earth Planet. Inter.*, *54*, 64–81, 1989.

Young, C. J. and T. Lay, Multiple phase analysis of the shear velocity structure in the D″ region beneath Alaska, *J. Geophys. Res.*, *95*, 17,385–17402, 1990.

Zindler, A., and S. Hart, Chemical geodynamics, *Annu. Rev. Earth Planet. Sci.*, *14*, 493–571, 1986.

P. Earle, Dept. of Earth and Space Science, U.C. Los Angeles, Los Angeles, CA 90024-1567. (e-mail: pearle@moho.ess.ucla.edu)

M. Hedlin and P. Shearer, IGPP 0225, U.C. San Diego, La Jolla, CA 92093-0225. (e-mail: mhedlin@ucsd.edu; pshearer@ucsd.edu)

Illuminating the Base of the Mantle with Diffracted Waves

Raul W. Valenzuela[1] and Michael E. Wysession

Department of Earth and Planetary Sciences, Washington University, St. Louis, Missouri

A review is presented of the state of knowledge about the core-mantle boundary (CMB) obtained from diffracted waves. Because diffracted waves spend a considerable portion of their paths traveling through the lowermost mantle, these studies have better CMB coverage than most other techniques, span a unique range of frequencies and CMB length scales, and sample geographic regions inaccessible to other seismic phases. Diffracted waves place constraints on CMB properties such as the seismic velocity heterogeneity of the D″ layer, the velocity gradients in D″, anelastic attenuation, and seismic anisotropy. We discuss the most common techniques that rely on diffracted waves, and include results from recent studies based on measurements of ray parameters, decay constants, and differential travel times between *PKP* and *Pdiff* phases. Measurements of globally-available shear wave slownesses in D″ indicate variations of ±3.75% relative to PREM. D″ shear wave velocities are up to 3% slow beneath the north and northeastern Pacific Ocean, and change from fast to slow towards the northeast under central Asia. The possibility for azimuthal anisotropy exists in D″ beneath eastern Siberia. Other regions with good coverage are under western Africa and the east central Atlantic Ocean, beneath the central Atlantic, and under the southwestern Pacific. Locations of fast and slow velocity anomalies at the CMB are consistent with the model of cold paleoslabs ponding at the CMB and forcing D″ rock laterally to form hot aggregates. Decay constants of *SHdiff* waves were measured in two regions at the CMB and used to constrain radial velocity structure. Observations under eastern Siberia favor a model with a discontinuous velocity increase followed by a gradual velocity decrease with increasing depth. The modeled D″ discontinuity is about 210 km above the CMB with a velocity increase of ~3.8 - 5.4%. The modeled D″ discontinuity under the east central Pacific Ocean is ~185 km above the CMB with a velocity jump of ~3.4%, underlain by a negative velocity gradient. These models are consistent with the presence of a thermal/chemical boundary layer in D″, and either a phase change or a chemical discontinuity at the top of D″.

1. INTRODUCTION

Seismology provides some of the most effective tools to study the lowermost mantle. *Gutenberg and Richter* [1939] and *Jeffreys* [1939] obtained seismic velocity models from the inversion of travel times that indicated a reduced velocity gradient for both *P* and *S* waves in the lowermost mantle relative to the mantle immediately above. *Bullen* [1950] named this region of reduced velocity gradients D″ and hypothesized that it is inhomogeneous, either because of compositional changes or because of a nonadiabatic thermal structure. Current knowledge of the base of the mantle indicates that it could act as a thermal boundary layer [*Stacey and Loper*, 1983; *Lay and Helmberger*, 1983; *Doornbos et al.*, 1986; *Lay*, 1989; *Poirier*, 1991; *Loper and Lay*, 1995] as well as a chemical boundary layer [*Davies and Gurnis*,

The Core-Mantle Boundary Region
Geodynamics 28
Copyright 1998 by the American Geophysical Union.

1986; *Lay*, 1989; *Christensen and Hofmann*, 1994; *Weber*, 1994], and that the top of D″ could mark the depth of a phase transition [*Nataf and Houard*, 1993; *Stixrude and Bukowinski*, 1990; *Wang et al.*, 1992; *Wysession*, 1996c].

In this paper we summarize the contributions from diffracted-wave studies to our understanding of the core-mantle boundary. The travel times, amplitudes, and waveforms of diffracted waves are affected by the velocity structure at the base of the mantle. Specifically, diffracted wave studies help to place constraints on the heterogeneity of the D″ layer, the velocity gradients in D″, velocity contrasts across the CMB, anelastic damping, and anisotropy [*Alexander and Phinney*, 1966; *Phinney and Alexander*, 1966, 1969; *Sacks*, 1966, 1967; *Teng and Richards*, 1969; *Bolt et al.*, 1970; *Mondt*, 1977; *Doornbos*, 1978, 1983; *Doornbos and Mondt*, 1979a, 1979b, 1980; *Okal and Geller*, 1979; *Mula and Müller*, 1980; *Mula*, 1981; *Bolt and Niazi*, 1984; *Doornbos et al.*, 1986; *Van Loenen*, 1988; *Wysession and Okal*, 1988, 1989; *Vinnik et al.*, 1989, 1995; *Bataille et al.*, 1990; *Lay and Young*, 1991; *Wysession et al.*, 1992, 1993; *Souriau and Poupinet*, 1994; *Maupin*, 1994; *Valenzuela Wong*, 1996; *Kendall and Silver*, 1996; *Bataille and Lund*, 1996; *Hock et al.*, 1996; *Garnero and Lay*, 1997; *Kuo and Wu*, 1997; *Ritsema et al.*, 1997a,b].

Techniques for the study of the CMB based on diffracted waves include the following: (1) time domain measurements of the ray parameter, $dt/d\Delta$ (t is time; Δ is epicentral distance); (2) time domain measurements of the amplitude decrease as the ray propagates along the CMB; (3) frequency domain determination of the decay constant (amplitude decrease); (4) extended *Pdiff* codas as an indication of CMB scattering; and (5) comparisons between phase pairs such as *SHdiff* - *SVdiff*, *Sdiff* - *SKS*, *Sdiff* - *SKKS*, and *PKP* - *Pdiff*.

Core-diffracted waves, observed in the core shadow zone, behave as non-geometrical rays traveling across the base of the mantle (Figure 1). Spending a considerable portion of their paths at the CMB, they provide a powerful means of sampling D″. *Pdiff* and *SHdiff* are easy to identify because they are the first compressional and transverse arrivals, respectively. Using the Preliminary Reference Earth Model (PREM) [*Dziewonski and Anderson*, 1981], the first diffracted arrivals from a surface earthquake are expected at epicentral distances of $\Delta = 98.4°$ (*Pdiff*) and $\Delta = 102.7°$ (*Sdiff*), assuming infinite frequency. Diffraction occurs at shorter epicentral distances for typical intermediate and long-period waves, as well as for deep earthquakes. Diffracted waves are observed after traveling distances along the CMB in excess of 70° for *Pdiff* and 50° for *Sdiff*, enabling good spatial coverage. Core-diffracted waves display the following characteristics: (1) for a D″ layer with a non-zero velocity gradient, dispersion of the diffracted wave occurs because its different frequencies sample regions of different velocities; (2) as the ray propagates along the CMB its amplitude decreases in a frequency-dependent manner, with high frequencies decaying much faster; (3) amplitudes decrease due to propagation of energy back up into the mantle as well as the leakage of *Pdiff* and *SVdiff* energy into the

core as *PdiffKP* and *SVdiffKS* (*SHdiff* energy can also leak into the core by converting into *SVdiff* in the presence of anisotropy, or if the rigidity at the top of the core is not zero); (4) effects of anelastic attenuation further diminish the high frequency components of the signal; and (5) the velocity gradient in D″ plays a significant role in affecting amplitudes. Many of these complexities can be accounted for by comparing data with synthetic seismograms computed using methods that incorporate the effects of diffraction such as normal mode summation and reflectivity.

Due to the high degree of lateral heterogeneity in D″, studies of diffracted waves show large regional variations. Much work has been done using profiles of seismograms to several stations from a single earthquake, with the analysis drawn from the change in the diffracted phase as distances increase. In order to reliably provide information about the ray parameter at the base of the mantle, these profiles need to have (1) a distance interval in excess of 15° to minimize the possibility of bias from local variations of the crustal structure under the stations and from different upswing paths through the mantle, and (2) an azimuthal range of less than 20° to help ensure that the downswing mantle paths are the same, and that a single region of the CMB is sampled [*Alexander and Phinney*, 1966; *Wysession and Okal*, 1989; *Wysession et al.*, 1992, 1995]. While these criteria, as well as the global distribution of sources and receivers, limit the number of useful profiles, they allow us to obtain reliable long-wavelength information (at a scale of ~1000 km) for many regions of the D″ by sampling well-defined strips across the CMB. Given existing global seismicity distributions and that *Pdiff* and *Sdiff* are best recorded at distances of 100° - 140°, best coverage is in the South Atlantic Ocean, southern Africa, and eastern North America [*Wysession*, 1996b]. A large amount of global digital data is available, and additional sampling is achieved with portable arrays such as the IRIS PASSCAL Missouri-to-Massachusetts (MOMA) experiment [*Wysession et al.*, 1996].

In short, diffracted waves are valuable for the study of the CMB because they (1) can spend a significant portion of their total path length within D″ (accruing large travel-time anomalies), (2) provide observations within a frequency range and length scale not covered by other seismic phases, and (3) increase our geographical coverage of the CMB by sampling regions inaccessible to other phases.

2. DIFFRACTED RAY PARAMETERS (SLOWNESSES)

Studies of the diffracted wave ray parameter (or slowness) aim in determining the average velocity of a given region of the lowermost mantle. Diffracted waves theoretically have a constant ray parameter and therefore a linear travel time slope because all arrivals bottom at the same depth (Figure 2). In practice the slowness varies due to both waveform frequency effects and CMB lateral heterogeneity.

Measurements of *Pdiff* slownesses date back to the start of the century with observations by *Gutenberg* [1914], *Macelwane* [1930] and *Krumbach* [1931, 1934]. While

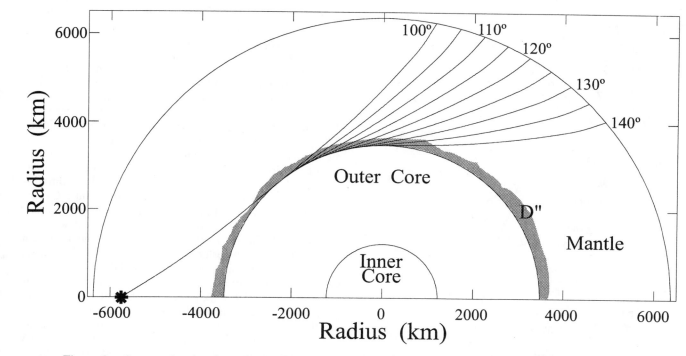

Figure 1. Cartoon showing the path of diffracted waves in the mantle and at the core-mantle boundary (CMB) for 600-km deep earthquake. Diffracted arrivals are observed in the core shadow zone. In this figure diffracted waves are observed at epicentral distances between 100° and 140°. A significant portion of the travel times of diffracted waves is spent in D″, providing a good probe of the region.

some doubted the existence of core-diffraction, preferring instead models of refraction within a very slow CMB layer [*Lehmann*, 1953, 1958], the phenomenon of diffraction was firmly established by the time of *Gutenberg* [1960]. The late 60's saw the establishment of a theoretical basis for diffracted waves [*Alexander and Phinney*, 1966; *Phinney and Alexander*, 1966, 1969; *Sacks*, 1966, 1967; *Teng and Richards*, 1969], and the determination of CMB velocities from diffracted wave slownesses was presented by Doornbos and *Mondt* [1979a] and *Mula and Müller* [1980]. Early studies of diffracted wave ray parameters assumed a globally homogeneous layer [*Doornbos and Mondt*, 1979b; *Mula and Müller*, 1980; *Doornbos*, 1983], while later work using *SHdiff* [*Van Loenen*, 1988; *Wysession and Okal*, 1988], *Pdiff* [*Wysession and Okal*, 1989; *Souriau and Poupinet*, 1994; *Hock et al.*, 1996], and both *Pdiff* and *SHdiff* [*Wysession et al.*, 1992] confirmed a laterally heterogeneous D″.

We present the results here of 8 years of global digital *SHdiff* profile data [*Valenzuela Wong*, 1996; *Valenzuela and Wysession*, 1998]. A similar study was carried out by *Hock et al.* [1996] using global digital *Pdiff* profiles. A search of the NEIC (National Earthquake Information Center) digital waveform catalog produced 122 useful events that provided a total of 161 profiles. Most of these profiles contain (1) at least 4 stations (the mean was 4.6; the maximum was 12), (2) azimuthal variations of less than 20°, and (3) distance ranges of more than 15° along the CMB. Ray

parameters are determined from a linear regression through the peak times of the pulses with the instrument responses removed. Times are corrected for ellipticity, and for mantle path heterogeneity along the upswing paths using the mantle tomographic model of *Woodward et al.* [1993]. The latter correction is an attempt to correct for the different paths the diffracted waves take through the mid- and upper mantle on their paths up from the CMB. It is likely that these tomographic corrections underestimate the true travel-time perturbations accrued within the upper mantle, and an exact quantification of the existent uncertainties is not possible. The average velocity of the CMB regions sampled are determined based upon comparisons with the slownesses of synthetic counterparts [*Wysession et al.*, 1992] computed using reflectivity [*Kennett*, 1983]. Plate 1 shows the resulting map of the lateral changes in slowness relative to PREM [*Dziewonski and Anderson*, 1981]. The absolute values of most of the observed slownesses range between 8.2 and 8.8 s/°, and correspond to a heterogeneity of ±3.75% relative to PREM. Less reliable profiles (with fewer stations, a greater azimuthal range, or shorter epicentral distance range) show a slightly greater spread of about ±5%. This increase may represent true structure, as longer profiles will have a smoothing effect, possibly averaging out differing positive and negative anomalies. The precision of the linear regression is reliable to within ±0.7%, and though the absolute accuracy of the slownesses is impossible to assess given the

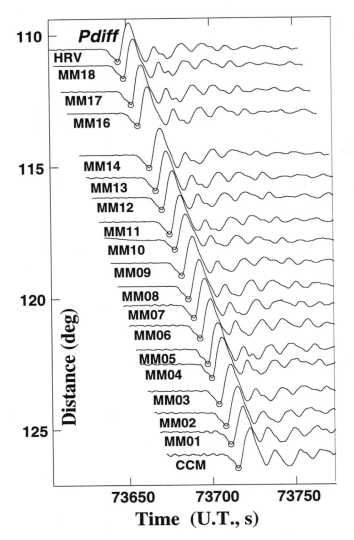

Figure 2. An example of a profile of core-diffracted *Pdiff* phases, recorded by the Missouri-to-Massachusetts (MOMA) IRIS PASS-CAL stations from an $M_w = 7.7$ earthquake in the Loyalty Islands on May 16, 1995. The *Pdiff* arrivals have been band-pass filtered for periods between 2 and 100 s, with the instrument responses removed. Small circles represent the times of the arrival peak maxima, used in determining the ray parameter (the slope through these times). Narrower filters (25 - 50 s) were used to limit the effects of dispersion. An average velocity for the base of the mantle can be obtained by comparing the ray parameter (or slowness) to that obtained from synthetic seismograms, once corrections are made for mantle path heterogeneity.

uncertainties of the tomographic mantle path corrections, we estimate it to be no more than twice this value. The lateral variations of ±4% we find in *Sdiff* slownesses are of the same range as current tomographic models of shear velocities in D″ [*Masters et al.*, 1996; *Dziewonski et al.*, 1996].

The largest number of observations sample the CMB beneath the northern and northeastern Pacific Ocean (Plate 1). There is a transition from fast shear velocities beneath Alaska to slow velocities beneath the Pacific. Another densely sampled region is found beneath central and northeastern Asia. In both the NE Pacific and Asian regions there is a pattern of alternating slow and fast regions at a scale of thousands of km. The same alternation was also observed in the region beneath the NE Pacific in preliminary studies of *Sdiff* profiles from over 40 earthquakes to the 20 stations in the MOMA array [*Wysession et al.*, 1997]. Predominantly slow D″ velocities are found beneath the southwestern Pacific, with evidence for strong slow velocity anomalies beneath western Africa and the northern Atlantic Ocean. Other D″ regions with fast velocities include Central America, the central Atlantic, western Asia, and western Australia. The locations of the slow features under western Africa and the southwestern Pacific Ocean coincide with the locations of the Great African Plume and the Equatorial Pacific Plume Group (EPPG), seen in tomographic shear velocity models [*Su et al.*, 1994]. Fast D″ velocities beneath the Pacific rim and slow velocities beneath western Africa and the mid-Pacific agree with geodynamic models suggesting that cold, seismically fast, paleoslabs reach the CMB beneath the locations of paleotrenches, and in the process laterally sweep ambient D″ rock into large hot, seismically slow, aggregates beneath western Africa and the mid-Pacific that may be the underlying source for the large numbers of Pacific and African plate hotspots [*Dziewonski et al.*, 1993, *Wysession*, 1996a; *Grand et al.*, 1997].

Several previous studies mapped the lateral velocity structure of D″ from the slowness of *Pdiff* waves [*Wysession and Okal*, 1989; *Wysession et al.*, 1992; *Souriau and Poupinet*, 1994; *Hock et al.*, 1996]. *Wysession et al.* [1992] found fast velocities beneath north-central Asia, southeast Asia, and the Arctic Ocean and northern Canada, and slow velocities under the Mediterranean, the northeast Atlantic, Alaska and the North Pacific, the east central Pacific, and northwestern South America. *Souriau and Poupinet* [1994] report the regions under Kamchatka, northern Asia, northeast Pacific, northwestern South America and the Caribbean to be fast and the regions beneath the southwest Pacific, the south Atlantic, New Guinea, the Philippines, and the Mediterranean to be slow. They find a ± 2.5% range for D″ *P* velocities. Where *Wysession et al.* [1992] obtained observations of both *Pdiff* and *Sdiff* ray parameters they found fast *S* velocities and slow *P* velocities underneath Alaska and Canada, with the opposite for north-central Asia.

These differences between *P*-wave and *S*-wave velocities, often quantified as the material Poisson ratio, may be a result of chemical rather than thermal heterogeneities at the base of the mantle. *Zandt and Ammon* [1995] used variations in the Poisson ratio to parameterize the composition of the continental crust, and measurements of the Poisson ratio in the lower mantle have been made from global data sets of *P* and *S* waves [*Bolton and Masters*, 1996]. A variability of

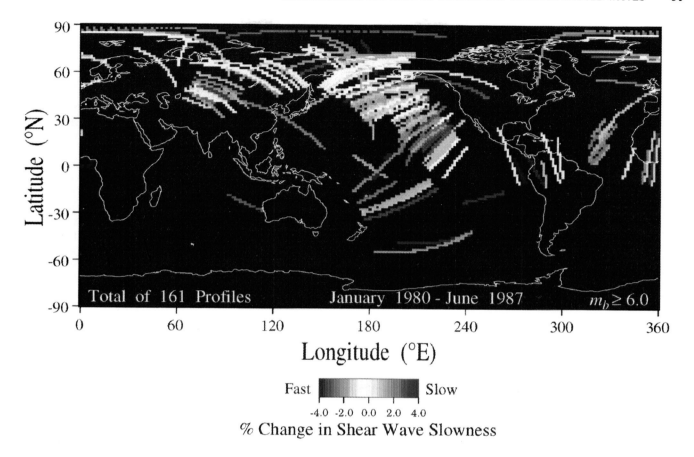

Fast ▬▬▬▬▬ Slow

-4.0 -2.0 0.0 2.0 4.0

% Change in Shear Wave Slowness

Plate 1. Shear wave slowness in D″ relative to PREM from 161 *Sdiff* profiles. Earthquakes are from 1/80 - 6/87. Variations in slowness are ±3.75% relative to PREM for most profiles. Best coverage is under the northern and northeastern Pacific Ocean, with predominantly slow velocities. SW-NE and NW-SE trending paths that intersect under eastern Siberia may indicate the presence of azimuthal anisotropy. Good coverage is also obtained under central and northeastern Asia, where velocities are fast in the southwest, become slow towards the northeast and then are close to PREM or slightly fast farther to the northeast. Other areas with coverage include slow velocities under western Africa, fast velocities under the central Atlantic Ocean, and slow velocities under the southwestern Pacific Ocean.

5% in the inferred Poisson ratio was found by *Wysession et al.* [1992] from *Pdiff* and *Sdiff* slownesses (0.295 - 0.310). However, a much larger range (~20%) was found using diffracted wave slownesses from the Missouri-to-Massachusetts array [*Wysession et al.*, 1997]. Using identical earthquake-station geometries for *Pdiff* and *Sdiff*, the Poisson ratio varied from 0.31 beneath the mid-eastern Pacific to 0.25 beneath Alaska. The observation of fast shear velocities coincident with slow *P* velocities beneath Alaska is consistent with many seismic studies, and is hard to explain with only thermal variations [*Wysession et al.*, 1993].

As variations in seismic velocity can be produced by thermal or chemical anomalies, slowness measurements can be used to place constraints on the thermal structure and chemical composition of the lowermost mantle. *Wysession et al.* [1992, 1993] used a third-order Birch-Murnaghan equation of state to estimate that a 1% variation in D″ seismic velocities can be produced by lateral variations of approximately 200 °C for *P* waves and 300° C for *S* waves. The seismic velocities are also sensitive to the relative amounts of Fe and Mg: a 5% variation from Mg/(Mg+Fe) = 0.90 would change *P* and *S* velocities by almost 1%. Changes in the partitioning between oxides and silicates were less important. It is likely that large velocity anomalies result from a combination of thermal and chemical effects. For instance, the D″ region beneath a mantle plume could have hotter temperatures and a thicker layer of dense iron-rich chemical dregs, with both contributing to lower velocities.

It has also been observed that the diffracted wave ray parameter varies as a function of frequency [*Okal and Geller*, 1979; *Souriau and Poupinet*, 1994]. This is because the diffracted wave undergoes a form of dispersion similar to that

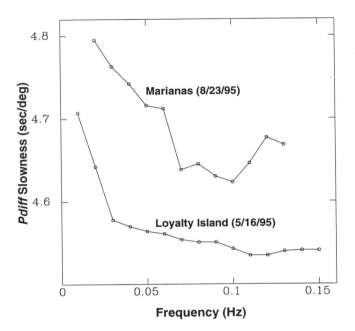

Figure 3. Two examples of the geometrical dispersion of *Pdiff* waves at the CMB, taken from the MOMA array for earthquakes on August 23, 1995 (Marianas) and May 16, 1995 (Loyalty I.). Since longer wavelengths travel through more of the mantle, with a lower average velocity, they travel more slowly. The increase in slowness above 0.1 Hz for the Marianas event is the result of a strong negative velocity gradient. This feature is not seen in the Loyalty I. event, showing a difference in D″ structure for regions sampled by the two earthquakes.

of surface waves at Earth's surface, with high frequencies traveling close to the boundary and low frequencies propagating through and therefore sampling a much larger portion of the mantle. Figure 3 shows the effect of frequency on the *Pdiff* ray parameters for two earthquakes recorded across the MOMA array. Because the velocity decreases moving up into the mantle above D″, the slowness increases moving toward longer periods. *Souriau and Poupinet* [1994] found this to be the case sampling digital GEOSCOPE *Pdiff* station-pair slownesses at dominant periods of 15, 25 and 35 s, with the lowest slownesses at 15 s periods, and the highest slownesses at 35 s periods. They estimate that *Pdiff* waves at these periods (15, 25, 35 s) sample D″ layers at thicknesses of roughly 150, 250 and 350 km, and find evidence for a negative gradient over this range because the observed decrease in slowness with decreased period is less than for synthetics computed using PREM, which has a zero gradient in D″. The dispersion curves in Figure 3 suggest differences in Pacific D″ velocity structure for the regions sampled by the two different earthquakes. The current example does not go to low enough periods to be able to resolve a narrow ultra-low velocity zone of the form described by *Garnero et al.* [1998].

3. AMPLITUDE DECAY OF DIFFRACTED WAVES

3.1. Time Domain Measurements

Measurements of the amplitudes of diffracted waves were first done in the time domain to constrain the radial velocity structure as well as the amount of anelastic attenuation in the lowermost mantle. Early work determined amplitude-distance curves for short-period *P* and *Pdiff* waves [*Carpenter et al.*, 1967; *Cleary*, 1967; *Bolt*, 1972; *Booth et al.*, 1974; *Ansell*, 1974], but demonstrated the difficulty of separating crustal and upper mantle effects from the CMB signal. Models showed velocities that were either close to constant [*Ansell*, 1974] or gradually decreasing down through D″ [*Bolt*, 1972]. *Mula* [1981] obtained amplitude-distance curves for several earthquakes and used comparisons with synthetic seismograms to find models with either a zero velocity gradient or a slightly positive gradient in D″. *Mula* [1981] also concluded that Q values less than 250 were unlikely. *Ruff and Helmberger* [1982] used Soviet nuclear tests recorded by North American WWSSN stations to sample D″ beneath northern Greenland and the Arctic Canadian Islands and found a decreasing velocity gradient (with increasing depth) at the top of D″ followed by an increasing velocity gradient at the bottom of D″. They corrected arrivals for receiver effects, greatly reducing the scatter in amplitude values, and discounted the existence of a thin low Q zone at the base of D″ in this region. *Ruff and Lettvin* [1984] used deep earthquakes in Tonga and Japan to sample two patches of D″ under the Pacific Ocean, concluding that a thermal boundary layer could explain their observations, contrary to *Ruff and Helmberger* [1982].

3.2. Frequency Domain Measurements

More information about D″ can be found by examining the amplitude decay of core-diffracted waves across a wide range of frequencies. Early studies of the velocity structure at the base of the mantle determined the diffracted wave decay constant, $\gamma(\omega)$, a measure of the amplitude decay as a function of frequency [*Alexander and Phinney*, 1966; *Phinney and Alexander*, 1966, 1969]. The amplitude decay of diffracted waves is largely controlled by the velocity gradient across D″, providing a tool to constrain radial D″ velocity structure. The theoretical basis behind the amplitude decay of core-diffracted waves is established in *Alexander and Phinney* [1966], *Chapman and Phinney* [1972], *Doornbos and* Mondt [1979a], *Mula* [1981], and *Van Loenen* [1988]. In the mid-70's and early 80's, attempts to construct a global 1-D model of the velocity structure in D″ reached conflicting conclusions. *Mondt* [1977], *Doornbos and Mondt* [1979b], and *Doornbos* [1983] suggested negative *P* and *S* velocity gradients in the bottom 50 - 100 km of D″, but *Mula and Müller* [1980], and *Mula* [1981] found positive velocity gradients at the base of D″. The discrepancy was likely due to different regional sampling of D″.

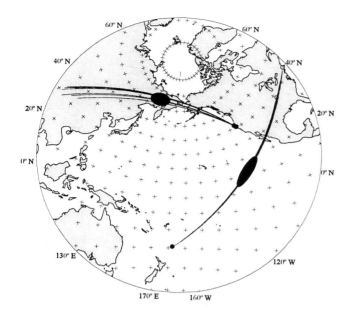

Figure 4. Regions at the base of the mantle sampled by the decay constant studies of *Valenzuela Wong* [1996]. Ray paths are shown for *SH* and *SHdiff* waves from five earthquakes in northern California recorded in Tibet. The epicentral distances vary roughly between 93° and 101°. The base of the mantle sampled under eastern Siberia is indicated by a dark oval. The distance sampled at the CMB is ~6.7°. The second region studied is under the east-central Pacific Ocean, as indicated by the other dark oval. The ray paths of *SHdiff* waves go from the Kermadec Islands to the MOMA array in the northeastern United States, at distances 104.8° - 120.4°. The azimuthal window is ~1.3°, providing a coherent CMB sampling.

An accurate assessment of the amplitude decay of diffracted waves as a function of frequency and distance requires several broadband seismometers closely spaced but spanning a large distance, best served with portable broadband arrays such as the NARS array in Western Europe [*van Loenen*, 1988] or various IRIS PASSCAL arrays. We discuss here results from two examples [*Valenzuela Wong*, 1996; *Valenzuela et al.*, 1998]: the Tibetan Plateau Passive Seismic Experiment [*Owens et al.*, 1993], and the Missouri-to-Massachusetts (MOMA) Broadband Experiment [*Wyssession et al.*, 1996]. We determine the shear wave radial velocity structure of D″ under eastern Siberia using Tibetan Plateau data [*Owens et al.*, 1993] and under the east-central Pacific Ocean using data from MOMA. The source to station geometries of these studies, as well as the regions sampled at the CMB are shown in Figure 4. The shear wave decay constant beneath eastern Siberia (Figure 5a) was determined by averaging the results of five northern California earthquakes recorded in Tibet. A Kermadec Islands earthquake recorded by MOMA was used to evaluate the shear wave decay constant under the east central Pacific (Figure

5b). Reliable measurements of the decay constant were obtained in the 0.020 - 0.071 Hz frequency range (50 to 14 s) under eastern Siberia, and at 0.016 - 0.045 Hz (63 to 22.4 s) beneath east-central Pacific. The different frequency bands are due to differences in distance: the Kermadec-MOMA data are at a greater distance and energy is shifted to longer periods. In the California-Tibet case we used Tibetan recordings of seismograms from along-strike earthquakes in Alaska in order to help remove the effects of mantle and crustal structure beneath the Tibetan array.

We used forward-modeling with reflectivity synthetic seismograms to test a large variety of possible D″ velocity structures including (1) near-zero gradients like that of PREM, (2) a gradual decrease of the velocity throughout D″ with increasing depth, (3) a gradual velocity increase, and (4) a discontinuous velocity increase at the top of D″. Models with a discontinuous increase in velocity were initially tested because of their prevalence in other studies [*Wyssession et al.*, 1998], but were extensively explored because they did the best job of fitting the amplitude data. In order to find the best match between data and synthetics we varied the following parameters: (1) the amount and style of the velocity decrease throughout D″, (2) the thickness of D″, (3) the velocity increase at the discontinuity, (4) the velocity gradient above D″ and (5) anelastic attenuation.

The basic idea behind the decay constant is that different velocity gradients will focus or defocus the seismic energy in different ways. A positive gradient will bend energy back toward the surface, generating a larger decay constant. A negative gradient will keep energy trapped near the CMB, cause large *Sdiff* amplitudes, and therefore smaller decay constants. So a simple D″ model with a negative velocity gradient (Figures 5+6) will have low decay constant. The data decay constant curve in Figure 5b shows the effect of both the negative gradient and the D″ discontinuity. The curve has low decay constants at high frequencies (the right side of Figure 5b), where the waves travel closer to the CMB and strongly "feel" the negative velocity gradient. The middle part of the curve has large decay constants due to intermediate wavelengths sampling the high velocities at the top of D″. At even lower frequencies (left side) the decay constants are low again because waves sample the slower velocities found above the discontinuity. The resolution of this method is much better at the CMB and decreases upward, so the negative velocity gradients right at the CMB are well-constrained, but the form of the "discontinuity" is not and could be gradual rather than discontinuous.

A different situation holds in Figure 5a, where many of the data arrive at distances less than 100°, and are direct *S* (not *Sdiff*). The high decay constants at low frequencies are a result of energy that is refracted into D″ by the negative velocity gradient beneath the velocity increase. It is this energy, channelled away from returning to the surface, that causes the large high frequency *Sdiff* amplitudes and therefore smaller decay constants at greater distances, as in Figure 5b. This phenomenon is less pronounced at longer

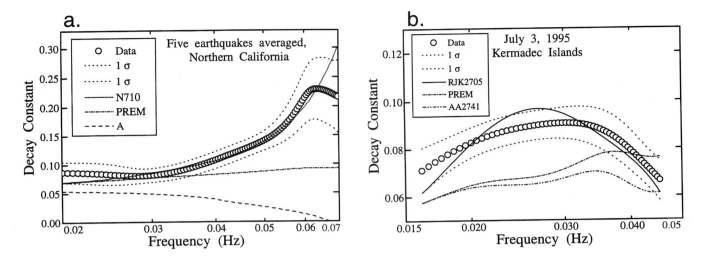

Figure 5. Decay constants as a function of frequency from observations and synthetic seismograms in the study by *Valenzuela Wong* [1996]. The decay constant, γ(ω), measures decay per degree of epicentral distance. The γ(ω) in (a) was determined from earthquakes at shorter epicentral distances and consequently the data are reliable up through higher frequencies. Open circles represent observations and thin dashed lines are the corresponding 1-σ standard deviations. (a) γ(ω) in D″ under eastern Siberia as sampled by earthquakes in Northern California. (b) γ(ω) under the east central Pacific Ocean as determined from an earthquake in the Kermadec Islands. IN both cases, γ(ω) is best fit by models with a discontinuous increase atop D″ underlain by a negative velocity gradient. For (a) model N710 the discontinuity is 185 - 235 km above the CMB, and for (b) model RJK2705 the discontinuity is about 185 km above the CMB. PREM does not provide a satisfactory fit to the *Sdiff* amplitudes, and nor do models without a discontinuity, like models A and AA2741.

wavelengths due to the larger Fresnel zones that average a broader range of the velocity structure.

The preferred models (Figure 6) for both regions show a velocity increase at the top of D″ followed by a gradual decrease with increasing depth. The fits of the model frequency-domain decay constants to that of the data are shown in Figure 5. Observation of the velocity increase atop D″ provides important independent corroboration of the results of many studies that are done with different data and techniques: triplicated *PdP* and *SdS* waves bottoming at the base of the mantle [e. g., *Weber*, 1993; *Lay et al.*, 1997; *Sherbaum et al.*, 1997; *Kendall and Shearer*, 1994; for reviews see *Loper and Lay*, 1995; *Weber et al.*, 1996; *Wysession et al.*, 1998]. Our best-fitting model has a discontinuity beneath eastern Siberia at about 185 - 235 km above the CMB and an *S*-wave velocity increase of 3.8 - 5.4%: there is a trade-off between discontinuity height and velocity increase. While the globally averaged height for the discontinuity estimates is about 250 - 260 km above the CMB for both *P* and *S* velocities [*Wysession et al.*, 1998], the D″ discontinuity in this region beneath eastern Siberia (65N, 155E) also shows a greater depth in other studies. For example, it is at the western edge of the region beneath Alaska modeled by *Young and Lay* [1990] with a discontinuity 243 km above the CMB, and near the binned-data location (63N, 164E) modeled by *Kendall and Shearer* [1994] with a discontinuity 235 km above the CMB.

We model the *S*-wave velocity increase under the east central Pacific (13N, 232E) to be ~3.4% and 185 km above the core-mantle boundary. *Garnero et al.* [1993] find a 2.4% shear wave velocity discontinuous increase located 180 km above the CMB for a nearby region (~10N, ~210E) slightly to the west, though *Kendall and Shearer* [1994] find a shallower discontinuity 319 km above the CMB for the same region (13N, 207-212E). This discrepancy may point to the difficulties in using 1-D modeling to represent highly heterogeneous structure, and highlights previous observations of large changes in apparent discontinuity height over short lateral distances [*Wysession et al.*, 1998].

The amplitude decay data strongly suggest a negative gradient at the base of D″, also supporting an increasingly negative gradient. While the D″ discontinuity likely has non-thermal contributions of a chemical [*Christensen and Hofmann*, 1994; *Weber*, 1994] or mineral phase nature [*Stixrude and Bukowinski*, 1990; *Wang et al.*, 1992], the velocity decrease at the base of D″ is an expected feature of a thermal boundary layer [*Stacey and Loper*, 1983; *Lay and Helmberger*, 1983; *Doornbos et al.*, 1986; *Poirier*, 1991]. Support for a negative velocity gradient in D″ also comes from the studies of free oscillations [*Kumagai et al.*, 1992], *SHdiff* slownesses as a function of frequency [*Souriau and Poupinet*, 1994], and diffracted *SV* waves [*Ritsema et al.*, 1997a]. The latter study modeled *SH-SKS* differential travel times, *SV/SKS* amplitude ratios, and *Sdiff* waveforms. It is possible the low velocities detected by these *Sdiff* studies are related to the ultra-low velocity layer identified in many parts of D″ [*Garnero et al.*, 1998], and suggested to be a result of partial melting [*Williams and Garnero*, 1996].

4. ANELASTIC ATTENUATION

The effects of anelastic attenuation in D″ can mirror in some ways the effects of varying velocity gradients on diffracted wave amplitudes, and it is difficult to separate them [*Young and Lay*, 1987]. *Doornbos and Mondt* [1979a] showed that a very low Q at the base of the mantle can significantly lower diffracted wave amplitudes. There are reasons to expect high attenuation in D″, as a rapid temperature increase across a D″ thermal boundary layer, perhaps in excess of 1000 K [*Williams and Jeanloz*, 1990], could approach the melting point of D″ constituents. The suggestion of partial melting in D″ [*Williams and Garnero*, 1996; *Holland and Ahrens*, 1997] lends support to this. Seismic evidence has been mixed, as some early models support a distinct low Q_μ (Q for *S* waves) zone within D″ [*Anderson and Hart*, 1978a, 1978b], and *Widmer et al.* [1991] note a systematic increase in shear attenuation with increasing depth in the lower mantle. But other models have shown no distinction between Q in D″ and the rest of the lower mantle [*Sailor and Dziewonski*, 1978]. *Durek and Ekström* [1996] compared models with and without a separate D″ and found that a model with a separate low Q_μ in D″ does not significantly improve the fit to their observations. Mentioned earlier, the studies of *Mula* [1981] and *Ruff and Helmberger* [1982] argued against the existence of a thin low Q zone at the base of the mantle. There are also thermochemical reasons for D″ materials to be far from their melting temperatures and Q values to be high [*Poirier*, 1991]. We stress the difficulties of determining deep-Earth Q values due to trade-offs with focussing effects from lateral and radial velocity variations, and the fact that some data and methods are more sensitive to Q than others.

We attempted to model the amplitude decay of the California-Tibet data with changes in attenuation, but were unable to do so. Starting with a PREM velocity structure, we found that increasing the attenuation within D″ caused an increase in the decay constant that was roughly consistent with the data values, but we were unable to match the particular decay distribution as a function of frequency in the manner possible by altering the velocity gradients. While the decay constant is fairly insensitive to low values of attenuation ($Q_\mu > 150$), models with very high attenuation ($Q_\mu < 80$) do a poor job of fitting gamma. This supports previous suggestions that the diffracted wave amplitude decay is affected more by velocity gradients than by anelastic attenuation [*Doornbos and Mondt*, 1979a], but it is possible that there is a contribution from anelastic attenuation. *Souriau and Poupinet* [1994] interpreted the decay of *Pdiff* amplitudes as the result of attenuation, and used the lack of correlation between the attenuation coefficients and ray parameter of GEOSCOPE station pairs to suggest that D″ heterogeneities have a chemical rather than thermal origin.

5. DIFFERENTIAL TRAVEL TIMES

The studies of core-diffracted profile ray parameters, previously discussed, are differential travel-time studies of a sort, using the difference in time between successive *Pdiff* or *Sdiff* arrivals. Other studies of diffracted waves uses their differential travel times with core phases such as *PKP* and *SKS*. The loss of high frequencies during core-diffraction and resulting change in appearance of the diffracted phases not only makes it hard to pick onset times, but also hard to compare with other phases. As a result, diffracted phase onset arrival times are not incorporated in tomographic inversions for mantle structure, and diffracted phases are also rarely used in differential travel time studies.

Differential travel-time studies are powerful tools for examining the CMB region because of their ability to suppress source and receiver effects, as well as upper mantle structure common to both phases [*Wysession et al.*, 1995]. However, most such studies have used *ScS-S*, *ScS-S-SdS*, *PcP-P*, *PcP-P-PdP*, *SKS-S*, *SKKS-SKS*, *SKKS-SKS-S*, *SPdiffKS-SKS*, *PKP(AB)-PKP(DF)*, or *PKP(AB)-PKP(BC)*. There are, however, some important exceptions. In *Garnero and Helmberger* [1993], the *SKS-S* times included *SVdiff* phases to distances as large as 123°, providing greater coverage of the CMB. The mantle tomographic model SKS12/WM13 of *Dziewonski et al.* [1996] uses large numbers of *SKS-Sdiff* travel times to provide excellent shear-wave coverage of the base of the mantle, though data in that study are filtered at periods greater than 35 s, so the loss of *Sdiff* high frequencies is not a problem.

Wysession [1996a] used the differential travel times of *PKP(DF)* and *Pdiff* to obtain a map with very good lateral coverage of the large-scale *P*-velocity variations at the base of the mantle. Because the very different waveforms of *PKP* and *Pdiff* precluded direct comparisons, reflectivity synthetic counterparts were created with a PREM structure. The differential travel-time residuals, relative to PREM, were then found through the circuit PKP_{data} - PKP_{synth} + $Pdiff_{synth}$ - $Pdiff_{data}$. An example is shown in Figure 7. A total of 543 good differential times, shown distributed along their CMB paths in Plate 2, were obtained from 8 years of global digital data. These paths were then inverted to provide the large-scale vertically-averaged D″ *P* velocity map shown in *Wysession* [1996a]. The large-scale coherency of the velocity anomalies displayed in *Wysession* [1996a] are evident in the path anomalies shown here in Plate 2.

In a similar technique, *Kuo and Wu* [1997] use a careful analysis of 340 *SHdiff-SKS* differential travel times to invert for the shear-wave velocity structure of D″. The differential travel-time residuals, even after correcting for upper mantle anisotropy and mantle path heterogeneity, have the enormous range of -16 to +18 s, demonstrating the large degree of heterogeneity in D″. *Kuo and Wu* [1997] achieve a 56%

variance reduction of the data misfit, very good for studies of the CMB. Their fast and slow regions correlate with both the inferred paleoslab and major hotspot distributions at the 99% confidence levels. Because of the long distances covered in D″ by *Pdiff* and *Sdiff*, inversions of large numbers of *PKP-Pdiff* and *SKS-Sdiff* differential travel times go a long ways toward providing complete maps of the large-scale structure of the base of the mantle.

Ritsema et al. [1997a] modeled *SH(SHdiff) - SKS* differential travel times simultaneously along with the amplitude ratios of *SV(SVdiff)/SKS* and the waveforms of *Sdiff* in order to constrain the vertical velocity structure at the base of the mantle beneath the mid-Pacific. The 338 data (83° - 125°) included both direct *S* as well as *Sdiff* arrivals. The combination of all three types of observations provides a strong constraint on the shear velocity gradient, found to be strongly negative. *Ritsema et al.* [1997a] find a 3% *S*-velocity reduction at the base of D″ relative to PREM, with significant small-scale (100-500 km) heterogeneity (0.5%-1%) required to explain scatter in the observations.

6. D″ ANISOTROPY FROM DIFFRACTED WAVES

An additional form of differential travel time is that of *SVdiff-SHdiff*, used in the determination of D″ anisotropy (splitting also occurs for pre-diffracted arrivals at distances like 95° that bottom within D″) . There is a growing number of observations of seismic anisotropy within D″, with important implications for either the structural fabric [*Kendall and Silver*, 1996] or mineralogical behavior [*Stixrude*, 1998] of the base of the mantle (for a more complete discussion, see *Lay et al.* [1998]). Perturbations to diffracted *S* waves have played an important part of this line of study. An early indication of D″ anisotropy [*Vinnik et al.*, 1989], found *SVdiff* arrivals from southwest Pacific earthquakes recorded in the eastern United States that were many seconds later than *SHdiff* arrivals. Splitting between *SHdiff* and *SVdiff* was also observed by *Lay and Young* [1991]. *Kendall and Silver* [1996] showed that as distances increased and the *ScS* phases changed to *Sdiff* phases, the amount of *SV-SH* splitting beneath Central America increased.

Difficulties with *SHdiff/SVdiff* studies exist because isotropic radial velocity structures in D″ can causes significant differences between the *SHdiff* and *SVdiff* waveforms, the latter usually being of higher frequency and complexity. *Maupin* [1994] demonstrated this by incorporating anisotropic effects into synthetic seismograms. One solution is to track the propagation of *SHdiff* and *SVdiff* separately across a linear array from the same earthquake, as is done with MOMA data [*Wysession et al.*, 1997]. *Vinnik et al.* [1997] use a complementary approach, using a range of earthquake distances from seismic stations to examine anisotropy beneath the western Pacific with *Sdiff* phases. They find that while *SV* velocity is about 10% slower than *SH* velocity, there is a very large variation in *SH* such that it is sometimes slower than *SV* (most observations of D″ anisotropy find *SV* slower than *SH* [*Lay et al.*, 1998]).

One possibility for this is the presence of azimuthal anisotropy, suggested by *Bolt and Niazi* [1984] from the observation of an unusually large *Sdiff* profile slowness (9.48 ± 0.26 s/°) under the Arctic Ocean. The presence of azimuthal anisotropy in D″ was introduced on theoretical grounds by *Doornbos et al.* [1986]. In examining *Sdiff* ray parameters [*Valenzuela Wong*, 1996], we find the interesting observation that paths extending from the southwest to the northeast under eastern Siberia and the Arctic Ocean have slownesses that are very close to or slightly slower than PREM values, whereas profiles trending northwest to southeast are fast. Heterogeneities existing along the non-shared segments of the two different sets of profiles may be responsible for the difference, but could also result from azimuthal anisotropy. This azimuthal dependence for Pacific D″ has been observed in individual *Sdiff* phases [*Ritsema et al.*, 1997b; *Wysession et al.*, 1997] as well as *ScS* travel times [*Winchester and Creager*, 1997; *Russell et al.*, 1997]. Better azimuthal sampling of the CMB will be required before azimuthal anisotropy can be further understood.

7. PDIFF CODAS AND CMB SCATTERING

Diffracted waves provide information as to small-scale lateral heterogeneities (~10 to 100 km) at the CMB. *Bataille*

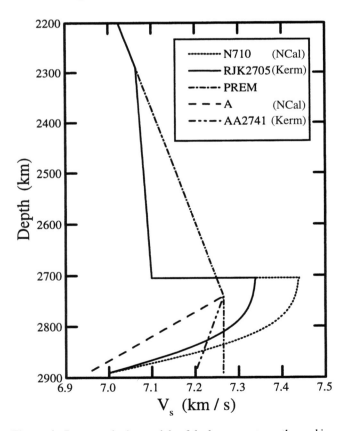

Figure 6. *S*-wave velocity models of the lowermost mantle used in the *Sdiff* amplitude analysis of Figure 5.

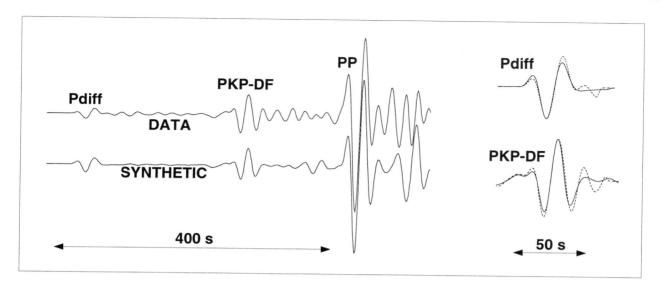

Figure 7. An example of the analysis of *PKP-Pdiff* differential travel times, taken from *Wysession* [1996a]. Because of the significant differences between the *PKP(DF)* and *Pdiff* waveforms, the differential times are determined through a circuit including PREM [*Dziewonski and Anderson*, 1981] reflectivity synthetic seismograms. The data (dashed) and synthetic *Pdiff* phases (solid) are aligned, and the same is done for *PKP(DF)*, with the data *PKP-Pdiff* times obtained by incorporating the theoretical PREM *PKP-Pdiff* times. Mantle heterogeneity and inner core anisotropy perturbations to the *PKP* arrival times are computed using models of *Vasco et al.* [1994] and *Song and Helmberger* [1993] and removed, and the remaining residuals are interpreted as the result of D″ velocity heterogeneities. In this example for a KONO record of an August 5, 1982, Vanuatu earthquake, the *PKP-Pdiff* residual is +0.8 s, suggesting a slightly slow path through D″.

et al. [1990] and *Bataille and Lund* [1996] observed a long tail of high-frequency energy following the *Pdiff* arrival for up to three minutes. They argued that this long *Pdiff* coda originates in D″ because (1) the frequency content and the particle motion for any short (10 s) window within the long tail are similar to those in the window containing the first arrival, and (2) the complexity of the coda increases as arrivals are recorded at increasing distances, which means that the ray paths sample deeper into the lower mantle and go from grazing the top of D″ to traveling completely within D″. *Bataille et al.* [1990] and *Bataille and Lund* [1996] suggest that scattering of *Pdiff*, caused by small-scale (~10 to 100 km) volume heterogeneities within D″ and/or topographic relief of the CMB, gives rise to the long tail. *Bataille and Lund* [1996] proposed a model, based on the theory of energy transfer, where the *P*-wave is trapped by a CMB low velocity zone and interacts with heterogeneities. As it progresses along the CMB, some of the energy leaks to the core, and some is scattered back to the surface. The scattering of *Pdiff* energy remains relatively unexamined.

8. CONCLUSIONS

Diffracted waves shed light on several properties of the CMB. (1) The lateral velocity structure of the lowermost mantle is quantified using *Pdiff* and *Sdiff* ray parameters and from differential travel times involving diffracted and core phases. Velocity anomalies of up to ±5% are found for D″ *S* velocities and up to 2.5% for *P* velocities. The large-scale patterns of these velocity heterogeneities are consistent with findings from recent tomographic models, though the diffracted waves sample many regions unsampled or poorly sampled by data used in tomographic studies. These patterns also correlate well with the history of paleosubduction, with seismically fast D″ regions underlying paleotrenches, and seismically slow regions beneath the Pacific and west African hotspot superplumes. (2) The radial velocity structure of the base of the mantle is examined using the amplitude decay of diffracted waves as well as the dispersion (change in ray parameter as a function of frequency). Results show strong evidence for a negative velocity gradient at the base of the mantle, consistent with a strong thermal boundary layer at the CMB, but also for a velocity increase at the top of D″. The cause for such a velocity increase is still unknown. (3) Relative differences between *SHdiff* and *SVdiff* have led to the identification of transverse isotropy in some regions of D″ and azimuthal anisotropy in others. (4) Extended *Pdiff* codas suggest strong seismic scattering at the base of the mantle, and are indicative of the scale of heterogeneity above the CMB.

Understanding the seismic lateral structure, radial structure, anisotropy and scattering has provided constraints on

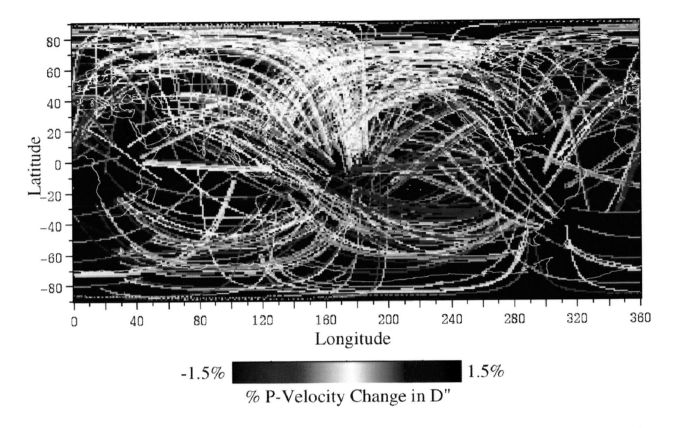

Plate 2. A map of the 543 *PKP-Pdiff* travel time residuals from *Wysession* [1996a], distributed along their respective *Pdiff* path segments in D″. There is a strong geographical coherence at these very long length scales, and the significant cross-coverage allows for an inversion of average D″ *P* velocity (shown in *Wysession* [1996a]). There is significant agreement with the *Sdiff* profile slownesses shown in Plate 1, though there are noticeable differences that may be the product of D″ chemical heterogeneity.

the thermochemistry and geodynamics of the CMB region. Study of the phenomenon of core diffraction is still in its infancy, and many discoveries are yet to be made.

Acknowledgments. We thank Karen Fischer, Tim Clarke, Ghassan Al-Eqabi and Patrick Shore for help with the Missouri-to-Massachussetts deployment, and Tom Owens for providing data from the Tibetan Plateau deployment. We thank the staff of IRIS PASSCAL and DMC programs for much assistance. We thank Tim Clarke for providing reflectivity programs, Patrick Shore for computer support, Doug Wiens, Thorne Lay, and Ed Garnero for discussions and suggestions, and Matt Neustadt, Jessica Butler, Brian Murray and Andy Hovland for help with data reduction. The research presented in this paper was funded by the National Science Foundation through grants NSF-EAR-9205368 and NSF-EAR-9319324, and by The David and Lucile Packard Foundation.

REFERENCES

Alexander, S. S., and R. A. Phinney, A study of the core-mantle boundary using *P* waves diffracted by the earth's core, *J. Geophys. Res., 71*, 5943-5958, 1966.
Anderson, D. L. and R. S. Hart, Attenuation models of the Earth, *Phys. Earth Planet. Int., 16*, 289-306, 1978a.
Anderson, D. L. and R. S. Hart, Q of the Earth, *J. Geophys. Res., 83*, 5869-5882, 1978b.
Ansell, J. H., Observation of the frequency-dependent amplitude variation with distance of P waves from 87° to 119°, *Pure Appl. Geophys., 112*, 683-700, 1974.
Bataille, K., and F. Lund, Strong scattering of short-period seismic waves by the core-mantle boundary and the P-diffracted wave, *Geophys. Res. Lett., 23*, 2413-2416, 1996.
Bataille, K., R. S. Wu, and S. M. Flatté, Inhomogeneities near the core-mantle boundary evidenced from scattered waves: A review, *Pure Appl. Geophys., 132*, 151-174, 1990.
Bolt, B. A., M. Niazi and M. R. Somerville, Diffracted ScS and the shear velocity at the core boundary, *Geophys. J. R. astr. Soc., 19*, 299-305, 1970.

Bolt, B. A., The density distribution near the base of the mantle and near the Earth's center, *Phys. Earth Planet. Int., 5*, 301-311, 1972.

Bolt, B. A., and M. Niazi, S velocities in D″ from diffracted SH-waves at the core boundary, *Geophys. J. R. astron. Soc., 79*, 825-834, 1984.

Bolton, H., and G. Masters, A region of anomalous *dlnVs/dlnVp* in the deep mantle (abstract), *Eos Trans. AGU, 77*(46), Fall Meeting Supp., F697, 1996.

Booth, D. C., P. D. Marshall, and J. B. Young, Long and short period P-wave amplitudes from earthquakes in the range 0°-114°, *Geophys. J. R. astr. Soc., 39*, 523-537, 1974.

Bullen, K. E., An Earth model based on a compressibility-pressure hypothesis, *Mon. Not. R. Astr. Soc., Geophys. Suppl., 6*, 50-59, 1950.

Carpenter, E. W., P. D. Marshall, and A. Douglas, The amplitude-distance curve for short period teleseismic P-waves, *Geophys. J. R. astr. Soc., 13*, 61-70, 1967.

Chapman, C. H., and R. A. Phinney, Diffracted seismic signals and their numerical solution, *Methods in Computational Physics, 12*, 165-230, 1972.

Christensen, U. R., and A. W. Hofmann, Segregation of subducted oceanic crust in the convecting mantle, *J. Geophys. Res., 99*, 19,867-19,884, 1994.

Cleary, J., Analysis of the amplitudes of short-period P waves recorded by long range seismic measurements stations in the distance range 30° to 102°, *J. Geophys. Res., 72*, 4705-4712, 1967.

Davies, G. F., and M. Gurnis, Interaction of mantle dregs with convection: lateral heterogeneity at the core-mantle boundary, *Geophys. Res. Lett., 13*, 1517-1520, 1986.

Doornbos, D. J., On seismic wave scattering by a rough core-mantle boundary, *Geophys. J. R. astron. Soc., 53*, 643-662, 1978.

Doornbos, D. J., Present seismic evidence for a boundary layer at the base of the mantle, *J. Geophys. Res., 88*, 3498-3505, 1983.

Doornbos, D. J., and J. C. Mondt, Attenuation of *P* and *S* waves diffracted around the core, *Geophys. J. R. astron. Soc., 57*, 353-379, 1979a.

Doornbos, D. J., and J. C. Mondt, *P* and *S* waves diffracted around the core and the velocity structure at the base of the mantle, *Geophys. J. R. astron. Soc., 57*, 381-395, 1979b.

Doornbos, D. J., and J. C. Mondt, The interaction of elastic waves with a solid-liquid interface, with applications to the core-mantle boundary, *Pure Appl. Geophys., 118*, 1293-1309, 1980.

Doornbos, D. J., S. Spiliopoulos, and F. D. Stacey, Seismological properties of D″ and the structure of a thermal boundary layer, *Phys. Earth Planet. Int., 41*, 225-239, 1986.

Durek, J. J. and G. Ekström, A radial model of anelasticity consistent with long-period surface-wave attenuation, *Bull. Seism. Soc. Am., 86*, 144-158, 1996.

Dziewonski, A. M., and D. L. Anderson, Preliminary reference earth model, *Phys. Earth Planet. Inter., 25*, 297-356, 1981.

Dziewonski, A. M., A. M. Forte, W.-J. Su, and R. L. Woodward, Seismic tomography and geodynamics, in *Relating Geophysical Structures and Processes: The Jeffreys Volume*, ed. by K. Aki and R. Dmowska, AGU, Washington, D. C., 67-105, 1993.

Dziewonski, A. M., G. Ekström, and X.-F. Liu, Structure at the top and bottom of the mantle, in *Monitoring a Comprehensive Test Ban Treaty*, Kluwer Academic Publishers, pp. 521-550, 1996.

Garnero, E. J., and D. V. Helmberger, Travel times of *S* and *SKS*: Implications for three-dimensional lower mantle structure beneath the central Pacific, *J. Geophys. Res., 98*, 8225-8241, 1993.

Garnero, E. J., and T. Lay, Lateral variations in lowermost mantle shear wave anisotropy beneath the north Pacific and Alaska, *J. Geophys. Res., 102*, 8121-8135, 1997.

Garnero, E. J., D. V. Helmberger, and S. Grand, Preliminary evidence for a lower mantle shear wave velocity discontinuity beneath the central Pacific, *Phys. Earth Planet. Int., 79*, 335-347, 1993.

Garnero, E. J., J. Revenaugh, Q. Williams, T. Lay, and L. Kellogg, Ultralow velocity zone at the core-mantle boundary, in *Observational and Theoretical Constraints on he Core-Mantle Boundary Region*, ed. by M. Gurnis, M. Wysession, B. Buffett and E. Knittle, AGU, Washington, D. C., this volume, 1998.

Grand, S. P., R. D. van der Hilst, and S. Widiyantoro, Global seismic tomography: A snapshot of convection in the Earth, *GSA Today, 7*, 1-7, 1997.

Gutenberg, B., Ueber Erdbebenwellen VIIA, *Nachr. Ges. Wiss. Göttingen, Math.-Phys. Kl.*, 166-218, 1914.

Gutenberg, B., The shadow of the Earth's core, *J. Geophys. Res., 65*, 1013-1020, 1960.

Gutenberg, B., and C. F. Richter, On seismic waves, *Beitr. Geophys., 54*, 94-136, 1939.

Hock, S., M. Roth, and G. Müller, Long-period ray parameters of the core diffraction *Pdiff* and mantle heterogeneity, *EOS Trans. AGU, 77 (46)*, Fall Meeting suppl., F680-681, 1996.

Holland, K. G., and T. J. Ahrens, Melting of $(Mg,Fe)_2SiO_4$ at the core-mantle boundary of the Earth, *Science, 275*, 1623-1625, 1997.

Jeffreys, H., The times of P, S, and SKS and the velocities of P and S, *Mon. Not. R. Astron. Soc., 4*, 498-533, 1939.

Kendall, J. M., and P. M. Shearer, Lateral variations in D″ thickness from long-period shear-wave data, *J. Geophys. Res., 99*, 11,575-11,590, 1994.

Kendall, J. M., and P. G. Silver, Constraints from seismic anisotropy on the nature of the lowermost mantle, *Nature, 381*, 409-412, 1996.

Kennett, B. L. N., *Seismic Wave Propagation in Stratified Media*, Cambridge University Press, Cambridge, MA, 1983.

Krumbach, G., *Die instrumentallen Aufzeichnungen des Erdbeben vom 26. Juni 1924*, Veröff. Reichsanstalt f. Erd-bebenforsch., Jena, Heft 16b, 9 pp., 24 pl., 1931.

Krumbach, G., *Die Ausbreitung von Erdbebenwellen in grossen Herdentfernungen bei dem Südseebeben vom 26. Juni 1924*, Veröff. Reichsanstalt f. Erdbebenforsch., Jena, Heft 16a, 68 pp., 1934.

Kumagai, H., Y. Fukao, N. Suda, and N. Kobayashi, Structure of the D″ layer inferred from the Earth's free oscillations, *Phys. Earth Planet. Inter., 73*, 38-52, 1992.

Kuo, B. Y., and K. Y. Wu, Global shear velocity heterogeneities in the D″ layer: Inversion from *Sd-SKS* differential travel times, *J. Geophys. Res., 102*, 11775-11788, 1997.

Lay, T., Structure of the core-mantle transition zone: a chemical and thermal boundary layer, *Eos Trans. AGU, 70*, 49, 54-55, 58-59, 1989.

Lay, T., and D. V. Helmberger, A lower mantle *S*-wave triplication and the velocity structure of D″, *Geophys. J. R. astron. Soc., 75*, 799-837, 1983.

Lay, T., and C. J. Young, Analysis of seismic *SV* waves in the core's penumbra, *Geophys. Res. Lett., 18*, 1373-1376, 1991.

Lay, T., E. J. Garnero, C. J. Young, and J. B. Gaherty, Scale-lengths of shear velocity heterogeneity at the base of the mantle from *S* wave differential travel times, *J. Geophys. Res., 102*, 9887-9910, 1997.

Lay, T., E. J. Garnero, Q. Williams, R. Jeanloz, B. Romanowicz, L.

Kellogg, and M. E. Wysession, Seismic wave anisotropy in the D″ region and its implications, in *Observational and Theoretical Constraints on he Core-Mantle Boundary Region*, ed. by M. Gurnis, M. Wysession, B. Buffett and E. Knittle, AGU, Washington, D. C., this volume, 1998.

Lehmann, I., On the shadow zone of the earth's core, *Bull. Seism. Soc. Am., 43*, 291-306, 1953.

Lehmann, I., On amplitudes of P near the shadow zone, *Ann. geofis. Rome, 11*, 153-156, 1958.

Loper , D. E., and T. Lay, The core-mantle boundary region, *J. Geophys. Res., 100*, 6397-6420, 1995.

Macelwane, J. B., The South Pacific earthquake on June 26, 1924, *Gerlands Beitr. Geophys., 28*, 165-227, 1930.

Masters, G., S. Johnson, G. Laske, and H. Bolton, A shear-velocity model of the mantle, *Phil. Trans. R. Soc. Lond. A, 354*, 1385-1411, 1996.

Maupin, V., On the possibility of anisotropy in the D″ layer as inferred from the polarisation of diffracted S waves, *Phys. Earth Planet. Int., 87*, 1-32, 1994.

Mondt, J. C., *SH* waves: Theory and observations for epicentral distances greater than 90 degrees, *Phys. Earth Planet. Inter., 15*, 46-59, 1977.

Mula, A. H., Amplitudes of diffracted long-period P and S waves and the velocities and Q structure at the base of the mantle, *J. Geophys. Res., 86*, 4999-5011, 1981.

Mula, A. H., and G. Müller, Ray parameters of diffracted long period P and S waves and the velocities at the base of the mantle, *Pure Appl. Geophys., 118*, 1270-1290, 1980.

Nataf, H.-C., and S. Houard, Seismic discontinuity at the top of D″: A world-wide feature?, *Geophys. Res. Lett., 20*, 2371-2374, 1993.

Okal, E. A., and R. J. Geller, Shear-wave velocity at the base of the mantle from profiles of diffracted SH waves, *Bull. Seismol. Soc. Am., 69*, 1039-1053, 1979.

Owens, T. J., G. E. Randall, F. T. Wu, and R. Zeng, PASSCAL instrument performance during the Tibetan Plateau Passive Seismic Experiment, *Bull. Seism. Soc. Am., 83*, 1959-1970, 1993.

Phinney, R. A., and S. S. Alexander, P wave diffraction theory and the structure of the core-mantle boundary, *J. Geophys. Res., 71*, 5959-5975, 1966.

Phinney, R. A., and S. S. Alexander, The effect of a velocity gradient at the base of the mantle on diffracted P waves in the shadow, *J. Geophys. Res., 74*, 4967-4971, 1969.

Poirier, J.-P., *Introduction to the Physics of the Earth's Interior*, Cambridge University Press, Cambridge, England, 264 pp., 1991.

Ritsema, J., E. Garnero, and T. Lay, A strongly negative shear velocity gradient and lateral variability in the lowermost mantle beneath the Pacific, *J. Geophys. Res., 102*, 20,395-20,411, 1997a.

Ritsema, J., T. Lay, E. Garnero, and H. Benz, Variable anisotropy in the D″ region beneath the Pacific, Atlantic and Indian Oceans, *Eos Trans. AGU, 78*(46), Fall Meeting suppl., F10, 1997b.

Ruff, L. J., and D. V. Helmberger, The structure of the lowermost mantle determined by short-period P-wave amplitudes, *Geophys. J. R. astr. Soc., 68*, 95-119, 1982.

Ruff, L., and E. Lettvin, Short period P-wave amplitudes and variability of the core shadow zone boundary (abstract), *Eos Trans. AGU, 65*, 999, 1984.

Russell, S., and T. Lay, D″ velocity structure beneath the central Pacific, *Eos Trans. AGU, 78*(46), Fall Meeting suppl., F16, 1997.

Sacks, S., Diffracted waves studies of the earth's core: 1. Amplitudes, core size, and rigidity, *J. Geophys. Res., 71*, 1173-1181, 1966.

Sacks, S., Diffracted P wave studies of the Earth's core, 2, Lower mantle velocity, core size, lower mantle structure, *J. Geophys. Res., 72*, 2589-2594, 1967.

Sailor, R. V. and A. M. Dziewonski, Measurements and interpretation of normal mode attenuation, *Geophys. J. R. astr. Soc., 53*, 559-581, 1978.

Scherbaum, F., F. Kruger, and M. Weber, Double beam imaging: Mapping lower mantle heterogeneities using combinations of source and receiver arrays, *J. Geophys. Res., 102*, 507-522, 1997.

Song, X., and D. V. Helmberger, Anisotropy of Earth's inner core, *Geophys. Res. Lett., 20*, 2591-2594, 1993.

Souriau, A. and G. Poupinet, Lateral variations in P velocity and attenuation in the D″ layer, from diffracted P waves, *Phys. Earth Planet. Int., 84*, 227-234, 1994.

Stixrude, L., and M. S. T. Bukowinski, Fundamental thermodynamic relations and silicate melting with implications for the constitution of D″, *J. Geophys. Res., 95*, 19,311-19,325, 1990.

Stixrude, Elastic constants and anisotropy of $MgSiO_3$ perovskite, periclase, and SiO_3 at high pressure, in *Observational and Theoretical Constraints on he Core-Mantle Boundary Region*, ed. by M. Gurnis, M. Wysession, B. Buffett and E. Knittle, AGU, Washington, D. C., this volume, 1998.

Stacey, F. D., and D. E. Loper, The thermal boundary-layer interpretation of D″ and its role as a plume source, *Phys. Earth Planet. Inter., 33*, 45-55, 1983.

Su, W., R. L. Woodward, and A. M. Dziewonski, Degree 12 model of shear velocity heterogeneity in the mantle, *J. Geophys. Res., 99*, 6945-6981, 1994.

Teng, T. L., and P. G. Richards, Diffracted P, SV and SH waves and their shadow boundary shifts, *J. Geophys. Res., 74*, 1537-1555, 1969.

Valenzuela Wong, R., Lateral and radial velocity structure at the base of the mantle from diffracted shear waves, Ph. D. thesis, 255 pp., Washington University, St. Louis, Missouri, USA, 1996.

Valenzuela, R. W., and M. E. Wysession, Lateral variations at the base of the mantle from profiles of digital *Sdiff* data, *J. Geophys. Res.*, in preparation, 1998.

Valenzuela, R. W., M. E. Wysession, G. I. Al-eqabi, P. J. Shore, K. M. Fischer, and T. J. Clarke, T. J. Owens, The radial structure of the base of the mantle from broadband array *Sdiff* waves, *J. Geophys. Res.*, in preparation, 1998.

Van Loenen, P. M., S velocity at the base of the mantle from diffracted *SH* waves recorded by the NARS array, M.Sc. Thesis, Department of Theoretical Geophysics, Utrecht, 1988.

Vasco, D. W., L. R. Johnson, R. J. Pulliam, and P. S. Earle, Robust inversion of IASP91 travel time residuals for mantle P and S velocity structure, earthquake mislocations, and station corrections, *J. Geophys. Res., 99*, 13,727-13,755, 1994.

Vinnik, L. P., V. Farra, and B. Romanowicz, Observational evidence for diffracted SV in the shadow of the earth's core, *Geophys. Res. Lett., 16*, 519-522, 1989.

Vinnik, L., B. Romanowicz, Y. Le Stunff, and L. Makeyeva, Seismic anisotropy in the D″ layer, *Geophys. Res. Lett., 22*, 1657-1660, 1995.

Vinnik, L., L. Breger, and B. Romanowicz, Anisotropic structures at the base of the Earth's mantle, *Nature*, submitted, 1997.

Wang, Y., F. Guyot and R. C. Liebermann, Electron microscopy of (Mg, Fe)SiO3 perovskite: Evidence for structural phase transitions and implications for the lower mantle, *J. Geophys. Res.*, 97, 12,327-12,347, 1992.

Weber, M., *P* and *S* wave reflections from anomalies in the lowermost mantle, *Geophys. J. Int.*, 115, 183-210, 1993.

Weber, M., Lamellae in D″? An alternative model for lower mantle anomalies, *Geophys. Res. Lett.*, 21, 2531-2534, 1994.

Weber, M., J. P. Davis, C. Thomas, F. Krüger, F. Scherbaum, J. Schlittenhardt, and M. Körnig, The structure of the lowermost mantle as determined from using seismic arrays, in *Seismic modelling of the Earth's Structure*, Edited by E. Boschi, G. Ekström, and A. Morelli, pp. 399-442, Istit. Naz. di Geophys., Rome, 1996.

Widmer, R., G. Masters and F. Gilbert, Spherically symmetric attenuation within the Earth from normal mode data, *Geophys. J. Int.*, 104, 541-553, 1991.

Williams, Q., and E. J. Garnero, Seismic evidence for partial melt at the base of Earth's mantle, *Science*, 273, 1528-1530, 1996.

Williams, Q. and R. Jeanloz, Melting relations in the iron-sulfur system at ultra-high pressures: Implications for the thermal state of the Earth, *J. Geophys. Res.*, 95, 19,299-19,310, 1990.

Winchester, J. P., and K. C. Creager, Azimuthal anisotropy and abrupt transitions from slow to fast anomalies in the velocity structure of D″, *Abstracts 9th Ann. IRIS Workshop*, 1997.

Woodward, R. L., A. M. Forte, W.-J. Su, and A. M. Dziewonski, Constraints on the large-scale structure of the Earth's mantle, in *Evolution of the Earth and Planets, Geophys. Monogr. Ser.*, Vol. 74, ed. by E. Takahashi et al., pp. 89-109, AGU, Washington, D. C., 1993.

Wysession, M. E., Large-scale structure at the core-mantle boundary from core-diffracted waves, *Nature*, 382, 244-248, 1996a.

Wysession, M. E., How well do we utilize global seismicity?, *Bull. Seismol. Soc. Am.*, 86, 1207-1219, 1996b.

Wysession, M. E., Imaging cold rock at the base of the mantle: The sometimes fate of Slabs?, in *Subduction: Top to Bottom*, edited by G. E. Bebout, D. Scholl, S. Kirby, and J. P. Platt, in press, AGU, Washington, D. C., pp. 369-384, 1996c.

Wysession, M. E., and E. A. Okal, Evidence for lateral heterogeneity at the core-mantle boundary from the slowness of diffracted *S* profiles, in *Structure and Dynamics of Earth's Deep Interior, Geophys. Monogr. Ser.*, Vol. 46, edited by D. E. Smylie and R. Hide, pp. 55-63, AGU, Washington, D. C., 1988.

Wysession, M. E. and E. A. Okal, Regional analysis of D″ velocities from the ray parameters of diffracted *P* profiles, *Geophys. Res. Lett.*, 16, 1417-1420, 1989.

Wysession, M. E., E. A. Okal and C. R. Bina, The structure of the core-mantle boundary from diffracted waves, *J. Geophys. Res.*, 97, 8749-8764, 1992.

Wysession, M. E., C. R. Bina and E. A. Okal, Constraints on the temperature and composition of the base of the mantle, in *Dynamics of the Earth's Deep Interior and Earth Rotation, Geophys. Monogr. Ser.*, edited by J.-L. LeMoüel et al., AGU, Washington, D.C., 181-190, 1993.

Wysession, M. E., R. W. Valenzuela, L. Bartkó, A.-N. Zhu, Investigating the base of the mantle using differential travel times, *Phys. Earth Planet. Int.*, 92, 67-84, 1995.

Wysession, M. E., K. M. Fischer, T. J. Clarke, G. I. Al-eqabi, M. J. Fouch, L. A. Salvati, P. J. Shore, R. W. Valenzuela, Slicing into the Earth: Seismic mapping with the Missouri-to-Massachusetts broadband deployment, *EOS*, 77, 477, 480-482, 1996.

Wysession, M. E., S. Robertson, G. I. Al-eqabi, A. Langenhorst, M. J. Fouch, K. M. Fischer, and T. J. Clarke, D″ lateral, vertical and anisotropic structure from MOMA core-diffracted waves, *Eos Trans. AGU*, 78(46), Fall Meeting suppl., F1, 1997.

Wysession, M. E., E. J. Garnero, T. Lay, J. Revenaugh, Q. Williams, R. Jeanloz, and L. Kellogg, The D″ discontinuity and its implications, in *Observational and Theoretical Constraints on he Core-Mantle Boundary Region*, ed. by M. Gurnis, M. Wysession, B. Buffett and E. Knittle, AGU, Washington, D. C., this volume, 1998.

Young, C. J., and T. Lay, Evidence for a shear velocity discontinuity in the lower mantle beneath India and the Indian Ocean, *Phys. Earth Planet. Inter.*, 49, 37-53, 1987.

Young, C. J., and T. Lay, Multiple phase analysis of the shear velocity structure in the D″ region beneath Alaska, *J. Geophys. Res.*, 95, 17,385-17,402, 1990.

M. E. Wysession, Department of Earth and Planetary Sciences, Washington University, St. Louis, Missouri, 63130.

R. W. Valenzuela, Instituto de Geofisica, Universidad Nacional Autonoma de Mexico, 04510. Mexico, D. F., Mexico.

[1]Now at Instituto de Geofisica, Universidad Nacional Autonoma de Mexico, Mexico City, Mexico

The Temperature Contrast across D"

Quentin Williams

Department of Earth Sciences, University of California, Santa Cruz, CA

The range of superadiabatic temperatures increases likely to be present within D" is constrained as a function of three primary parameters: the temperature at the inner core-outer core boundary (as constrained by the melting behavior of iron and its alloys); the superadiabatic temperature change within the Earth's transition zone; and the adiabataic temperature gradient within Earth's lower mantle. Much of the uncertainty in the magnitude of the temperature jump in D" is produced by variations in the inferred temperature of the inner core-outer core boundary. This is a consequence both of the controversial phase relations of iron, and uncertainties in the identity and high pressure/high temperature behavior of the lighter alloying component in the core. Nevertheless, a broad family of parameters produce superadiabatic temperature increases across D" of 1000-2000 K. Somewhat higher or lower values for the temperature jump are, however, not precluded: such values simply require more restrictive (but still possibly viable) choices of inner core-outer core temperatures and/or differing magnitudes of adiabatic or superadiabatic gradients in the remainder of the mantle.

INTRODUCTION

There is an extended history of inferring the temperature at the base of the mantle: most of the early estimates of this parameter were aliased by a lack of appreciation of the importance of convection in the Earth's deep mantle, and the resultant significance of the adiabatic gradient in determining the thermal regime of the deep Earth. Nevertheless, nearly half-a-century ago, *Verhoogen* [1953] utilized estimated adiabatic gradients extending from the approximate temperature at the base of the crust to the base of the mantle, coupled with extant quasi-theoretic estimates of the melting temperature of iron to infer that the temperature at the base of the mantle "is likely to lie between 1500 and 6000°C" [*Verhoogen*, 1953]. This article focuses on improvements in the understanding of the

thermal structure of the lowermost mantle which have been generated since Verhoogen's pioneering work. Despite rather extensive experimental and theoretical work on this topic, it is both a reflection of the tremendous uncertainties in inferring the temperature within the deep Earth and a modest embarassment in deep Earth geophysics that the major achievement since the early 1950's is a paring of Verhoogen's rather liberal inferred temperature range at the base of the mantle by about a factor of two. The role, placement and magnitude of superadiabatic gradients has become one of the primary uncertainties in inferring the thermal regime of the deep planet. Indeed, the dramatic contrast in material properties (particularly density) occurring between the mantle and core has long been speculated to produce a region characterized by superadiabatic thermal gradients: in short, a thermal boundary layer. The major recognition of relevance to D" over the last half century is that at least one zone of superadiabatic temperature gradients appears to be required in the mantle: if this occurs within D", it is likely that a temperature jump of at least 800 K (and possibly as much as 1500-2000 K) occurs near the base of the mantle.

The Core-Mantle Boundary Region
Geodynamics 28
Copyright 1998 by the American Geophysical Union.

The importance of a temperature jump in D" is multi-fold: the temperature distribution in this zone determines the rheology and thus the geodynamics of this layer, as well as modulates the heat flow out of Earth's core. However, both the expected thickness of this layer and the magnitude of the temperature contrast across the thermal boundary layer has remained uncertain. Several approaches have been utilized to provide constraints on each of these parameters: simple fluid dynamic boundary layer theory [e.g. *McKenzie et al.*, 1974; *Jeanloz and Richter*, 1979; *Doornbos et al.*, 1986] has been utilized to provide approximate constraints on the plausible depth extent of the thermal boundary layer; comparison of adiabatic back-extrapolations of the inferred temperature at the liquid-solid interface of the inner-core to the core surface coupled with a similar extrapolation from the transition zone in Earth's mantle (with or without a mid-mantle thermal boundary layer) have provided estimates of the temperature contrast across this layer [e.g., *Spiliopoulos and Stacey*, 1984; *Williams and Jeanloz*, 1990; *Boehler*, 1993]; and seismic velocity data over the depth of D" has been inverted using estimated temperature derivatives of elastic constants to derive constraints on temperature changes in the lowermost several hundred kilometers of the mantle [e.g., *Jones*, 1977; *Doornbos et al.*, 1986].

Nevertheless, the recognition that D" contains wide-spread layering at varying depths with markedly different seismic signatures [e.g. *Lay and Helmberger*, 1983; *Garnero and Helmberger*, 1996; *Williams and Garnero*, 1996] indicates that the thermal structure of this zone may be considerably more complex than that of a simple boundary layer: in a sense, although the material constraints on the likely thermal contrast across D" have improved, the complexity of this zone itself has been enhanced through improved seismic characterization of this layer. As such, given the notable lateral heterogeneity of structures within D" [*Garnero et al.*, this volume; *Wysession et al.*, this volume], it is difficult to utilize seismic data to construct a single average profile of the temperature contrast across this zone of the planet.

Here, I examine refinements on the original approach of *Verhoogen* [1953], comparing downward (and mostly adiabatic) extrapolations of temperature from the upper mantle with upward extrapolations of temperature from the inner core-outer core boundary, based on the inferred melting temperature of iron at this depth.

BASAL TEMPERATURES AT THE CMB

The inferred temperature at the top of the liquid outer core could, in concept, be nearly precisely constrained assuming that the core is well-mixed (no vertical stratification) and that the inner core-outer core boundary reflects equilibrium coexistence of solid and liquid. The primary information needed for such a determination are i) the rate of adiabatic temperature change in the core, ii) the chemistry of the solid inner core and the coexisting liquid outer core, and iii) the melting relations of the inner core material. The adiabatic temperature gradient is controlled by the thermodynamic identity

$$dT/dP = \gamma T/K_S \qquad (1)$$

where γ is the Grüneisen parameter and K_S is the isentropic bulk modulus. Fortunately, the Grüneisen parameter of liquid iron is reasonably well-constrained through a combination of low-pressure measurements under static conditions and modelling of high pressure shock data [*Hixson et al.*, 1990; *Anderson and Ahrens*, 1994]. At this juncture, (ii) is modestly uncertain: current estimates of the chemistry of the inner core often incorporate a depression in density of the inner core of 3-6% relative to pure iron, with the amount of lighter alloying component being dependent on the inferred temperature of the core [*Jephcoat and Olson*, 1987; *Anderson and Ahrens*, 1994]. Additionally, the amount of nickel incorporated within the inner core is uncertain: its abundance in the inner core will be controlled by its partitioning behavior between liquid and solid iron alloys under core conditions, which is almost completely unconstrained. Notably, nickel has only a small effect on the density of iron [*Mao et al.*, 1990], but it has been shown to have a moderate effect on the solid state phase relations of iron-nickel alloys [*Huang et al.*, 1992]. However, the effect of transition metal alloys, such as Cr, on the melting temperature of iron at ultra-high pressures may be small [*Bass et al.*, 1990].

To first order, the melting behavior of the core system may thus be modeled as mirroring that of iron. Representative values for the melting temperature of iron are shown in Figure 1, as derived from both static [*Williams et al.*, 1987; 1991; *Boehler*, 1993; *Saxena et al.*, 1994; *Yoo et al.*, 1995; *Jephcoat and Besedin*, 1996] and shock determinations of the melting curve [*Williams et al.*, 1987; *Bass et al.*, 1987; *Yoo et al.*, 1993]. The datum of *Ringwood and Hibberson* [1991] at 16 GPa is not shown, but is compatible (within error bars) with all other determinations; the single melting determination of *Chen and Ahrens* [1996] from shock temperature experiments is also not shown, as this experiment was both designated as preliminary and has no reported error bars. Within the error bars of each study, the differing melting curve determinations are compatible to a pressure of ~40-50 GPa, with a dramatic discrepancy between the measurements of *Saxena et al.* [1994] and *Boehler* [1993] with those of *Williams et al.* [1987; 1991] at higher pressures. The former measurements use largely identical apparatuses, and arrive at mutually compatible temperatures for the melting of iron to 150 GPa. Detailed discussions of the differing experimental apparatuses and approaches to measuring melting temperatures, including possible origins and assessments of systematic biases in such state-of-the-art measurements may be found elsewhere [*Jeanloz and*

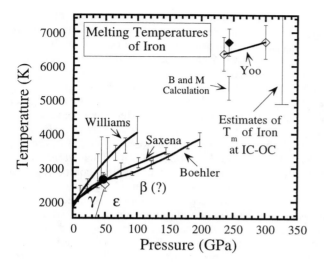

Figure 1. Experimental constraints on the melting curve of iron: representative error bars from the differing studies are shown. Curves labelled Boehler, Saxena, Williams and Yoo (and their error bars) are from *Boehler* [1993], *Saxena et al.* [1994], *Williams et al.* [1987; 1991] and *Yoo et al.* [1995], respectively. Open diamonds are from *Yoo et al.* [1993; 1995], the closed circle is the datum of *Jephcoat and Besedin* [1996], and the closed diamond is from *Bass et al.* [1987] and *Williams et al.* [1987]. The γ–ε phase boundary is from *Yoo et al.* [1995]. The error bar at 243 GPa represents a calculated temperature of the melting of iron from *Brown and McQueen* [1982]. The multiple temperature measurements conducted on shocked solid iron between pressures of 150 and 230 GPa and at temperatures in excess of 4000 K are not included as bounds in this plot [*Yoo et al.*, 1993; *Williams et al.*, 1987; *Bass et al.*, 1987].

Kavner, 1996; *Jephcoat and Besedin,* 1996; *Lazor and Saxena,* 1996]. The shock constraints on the melting temperature of iron at pressures between 240-300 GPa all clearly lie at notably higher temperatures than the extrapolations of the Boehler or Saxena et al. melting curves, but are compatible with an extrapolation of the *Williams et al.* [1987, 1991] melting curve. In explaining this discrepancy (which has been dismissed by some: *Lazor and Saxena,* 1996), possible explanations include extrinsic effects, such as systematic biases in either the shock data or the static data of Boehler or Saxena et al., or an intrinsic effect, such as a new solid phase of iron intercepting the melting curve at pressures in excess of 200 GPa. In terms of extrinsic effects, the possible role of optical effects and misalignments in biasing temperatures in the "narrow-slit" static experiments of Boehler and Saxena et al. has been discussed by *Jeanloz and Kavner* [1996], and the possible role of temperature-induced recrystallization in these experiments in producing severe biasing of apparent melting determinations towards low temperatures has been proposed by *Belonoshko and Dubrovinsky* [1997]. The

primary criticisms of the *Williams et al.* [1987; 1991] static measurements have been associated with the analyses of temperature gradients in these experiments and the size of their error bars [e.g. *Boehler,* 1996a; *Lazor and Saxena,* 1996]. The shock measurements have been criticized on two primary grounds: i) the dependence of shock temperatures on the thermal conductivity of the substrate material used in the shock experiments [*Boehler,* 1996a, *Lazor and Saxena,* 1996], and ii) the possibility of superheating of the solid phase of iron in the shock-loading process, producing an upward bias of the experiments [*Boehler,* 1996a]. It is notable that the shock temperature experiments have been conducted using three different substrates (LiF, Al_2O_3 [*Bass et al.,* 1987] and diamond [*Yoo et al.,* 1993]), with each substrate yielding mutually consistent temperature values. Preliminary reanalyses of portions of these data have been presented [*Gallagher and Ahrens,* 1994].

The suggestion that superheating of the solid phase may occur in shocked iron is based on observations of such superheating in temperature measurements in shock-loaded KBr and CsBr by *Boness and Brown* [1993]. *Boehler* [1996a] has attributed the discrepancy between his static measurements and the calculated temperatures of *Brown and McQueen* [1982; 1986] to such superheating. However, the temperatures reported by *Brown and McQueen* represent thermodynamic calculations of the temperature along the Hugoniot at which the sound speed of iron decreases to liquid values. Even if superheating is important in shock measurements on iron, the only bias associated with the Brown and McQueen calculation would be to overestimate the pressure at which melting occurs: no superheating-induced bias is expected in such thermochemically calculated temperatures. Moreover, the shock temperature measurements of *Yoo et al.* [1993], which represent the most accurate set of shock temperature measurements made to date on iron, show little evidence for a large decrease in temperature associated with melting of the superheated solid, as is expected if superheating is a major effect in shock-loaded iron [e.g., *Boness and Brown,* 1993]. Accordingly, little evidence exists that shock-induced superheating can be responsible for the large discrepancy between the various shock estimates [*Brown and McQueen,* 1982] and shock measurements [*Bass et al.,* 1987; *Yoo et al.,* 1993] for the temperature at which iron melts and the static measurements of *Boehler* [1993] and *Saxena et al.* [1994].

The key point here is that measurements of the phase diagram of iron at core conditions remain one of the most technically challenging aspects of high pressure research: conducting such experiments lies at the technological edge of what is currently feasible. In such a situation, even careful, well-documented experiments may produce results which are subject to modification: as a simple example, the apparently firmly located ε-γ-liquid triple point at about 100 GPa and about 2850 K [*Boehler,* 1993; 1996a] has recently been shown through in situ x-ray diffraction measurements to either 1) apparently occur at 50 GPa and 2500 K [*Yoo et al.*, 1995], or 2) not exist at all, with the appropriate triple point near 50 GPa on the liquidus of iron being ε–β-liquid, with the β phase likely being an orthorhombic polymorph of iron stable only at pressures greater than about 40 GPa and temperatures above 1500 K [*Andrault et al.*, 1997]. This ambiguity illustrates the controversy over whether a new high pressure, high temperature phase is on the liquidus of iron at pressures above about 50 GPa [*Boehler*, 1993; 1996a; *Saxena et al.*, 1992; *Andrault et al.*, 1997], as the ε–phase of iron has been observed at temperatures near the liquidus of iron to pressures of roughly 110 GPa [*Yoo et al.*, 1995; 1997]. Given this ongoing state of flux in our understanding of the iron phase diagram under extreme conditions, a number of different approaches can be used to assess the temperature of the outer core. Self-consistent data sets can be constructed using using either comparatively high [*Williams et al.*, 1987; *Bass et al.*, 1987; *Yoo et al.*, 1993] or low [*Boehler*, 1993; *Saxena et al.*, 1994] estimates of the iron melting curve under core conditions [*Williams et al.*, 1991; *Boehler*, 1996a; *Anderson and Duba*, 1997], yielding possible endmember core temperature estimates. Alternatively, all available measurements (and associated estimates of the magnitude of freezing point depressions due to lighter alloying components) can be incorporated into estimates of the possible range of core temperatures permitted by the current data set. While the uncertainties in core temperature produced by this all-inclusive approach are large, the combined roles of possible experimental biases in these state-of-the-art experiments, the ill-constrained high pressure polymorphism of iron, and ongoing alterations and refinements of the iron phase diagram render more parsimonious treatments of the extant data set prone to misadventure.

Therefore, it is likely premature to firmly state a precise temperature for the melting temperature of iron at the pressure of the inner core/outer core boundary. The present experimentally-derived estimates for the melting temperature of iron at this depth range from a lower limit of near 4850 "with an uncertainty of about ±200" K [*Boehler*, 1996a], and an upper estimate of 7600 (± 500) K [*Williams et al.*, 1987; *Bass et al.*, 1987]. Other experimentally based estimates typically fall between (but are not necessarily incompatible with) these two values, such as *Yoo et al.*'s [1993] estimate of 6830 (±500) K, *Saxena et al.*'s [1994] estimate of 6350 (±350) K for the temperature at the center of the planet, and *Brown and McQueen*'s [1982] older estimate of 6200 (± 500) K for the melting temperature of iron at the inner core-outer core boundary. It is notable that even this range of values for the melting temperature of iron at the inner core-outer core boundary represents a substantial enhancement of our understanding of the thermal regime of the deep Earth over that which existed prior to the considerable experimental

effort on determining the melting temperature of iron at very high pressures: comparatively low temperature geotherms, incorporating inner core-outer core boundary temperatures below 4000 K, are clearly incompatible with the available data.

To second order, values for the change in melting temperature of iron-rich alloys associated with the lighter alloying component(s) in the outer core have varied between about zero [*Knittle and Jeanloz*, 1991; *Boehler*, 1993; 1996a] to a depression of about 1000 K [*Stevenson*, 1981; *Williams and Jeanloz*, 1990]. This variation in estimates of the effect of lighter alloying components on the melting temperature is produced by whether solid solution or eutectic type behavior is assumed to exist between iron and lighter alloying components within the core. An experimental consensus has emerged that no freezing point depression is likely to take place if oxygen is the principal lighter alloying component in the core [*Knittle and Jeanloz*, 1991; *Boehler*, 1992]. The situation with sulfur is less clear: freezing point depression by sulfur clearly persists in the Fe-FeS system to 60 GPa in the experiments of *Boehler* [1996b] and to 90 GPa in those of *Williams and Jeanloz* [1990]. It appears that phase relations in this system may be complex at high pressures, with at least one phase with a new stoichiometry (Fe_3S_2) being stable at high pressures [*Fei et al.*, 1997]. Large volume press measurements of the eutectic in the Fe-FeS system yield a temperature of below 1000°C at 14 GPa [*Fei et al.*, 1997]; for comparison, *Boehler* [1996b] reports a eutectic temperature of near 1500°C in the iron-sulfur system at 14 GPa.

With respect to other possible alloying components, there is essentially no information on the melting behavior of the Fe-H system at pressures above 15 GPa [*Fukai*, 1992], and none on the Fe-C or Fe-Si systems at high pressures. Furthermore, information on ternary systems are scant: anecdotal information is presented by *Boehler* [1996a] that a mixture of 70% Fe: 20% FeO: 10% FeS (whether the percentages represent weight percents or mole percents is unclear) shows only a "few hundred degrees" freezing point depression at 50 GPa. The canonical method for calculating an upper estimate on the amount of freezing point depression of iron by lighter alloying constituents is through an application of the van Laar equation, in which

$$T/T_{m,Fe} = - ((\ln X_{Fe}/(\Delta S/R)) - 1)^{-1} \qquad (2)$$

[e.g., *Stevenson*, 1981; *Brown and McQueen*, 1982; *Williams and Jeanloz*, 1990]. Here, $T_{m,Fe}$ is the melting temperature of pure iron, X_{Fe} is the mole fraction of iron within the outer core, ΔS is the entropy of melting, and R is the gas constant. This equation implicitly assumes that ideal freezing point depression of iron by its lighter alloying constituent occurs: as such, it provides an upper estimate on the magnitude of freezing point depression. Non-ideality and/or solid solution behavior by any of the

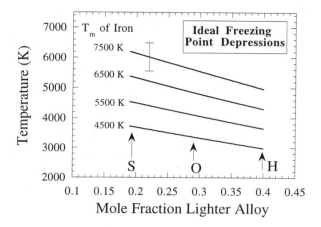

Figure 2. Estimated melting temperatures assuming ideal freezing point depression for iron with a range of concentrations of impurities: as such, these represent a likely upper bound on the temperature depression induced by lighter alloying constituents. The amounts of sulfur, oxygen and hydrogen needed to produce the density depression of the outer core relative to pure iron are shown with the arrows [*Ahrens and Jeanloz*, 1987; *Jeanloz and Ahrens*, 1980; *Badding et al.*, 1992; *Williams and Knittle*, 1997]. The mole fraction of iron is derived from the amount of elemental alloying component, rather than the amount of alloying molecule (for example, the mole fraction of H rather than of FeH is used to calculate freezing point depressions). The entropy of melting is assumed to equal the gas constant, R, with the error bar representing the variation in temperature induced by a variation of ±0.3 R in the entropy of melting.

lighter alloying constituents present in the outer core will lower the estimated magnitude of freezing point depression. While the entropy of melting of iron is known to closely approximate the gas constant at pressures to 6 GPa [*Strong et al.*, 1973], its value at higher pressures is ill-constrained, although it may converge on $0.693R$ $((ln2)R)$ for negligible volume changes on melting at high pressures [*Stishov*, 1988]. Accordingly, upper limits to the freezing point depression are calculated for differing mole fractions of iron in the outer core from Eqn. 2, and compared with estimated end-member stoichiometries for the outer core if a single light alloying element predominates in the outer core (Figure 2). Notably, if hydrogen is a major alloying element, the effects on core phase relations and the resultant inferred temperature at the inner core-outer core boundary could be profound: this is a simple consequence of the high molar percentage of hydrogen required to compensate for the density deficit of the outer core [*Badding et al.*, 1992].

In spite of the uncertainties implicit in Figure 2, the primary point of interest here is that the presently lowest experimentally-derived estimate for the melting temperature of iron at the inner core/outer core boundary [*Boehler,*

1993] is associated with the assertion that no freezing point depression by lighter alloying components exists [*Boehler,* 1996a]. At present, the experimental situation with regard to the role of lighter alloying components in alterring the melting temperature of iron continues to allow freezing point depressions (for alloying elements other than oxygen) which are on the order of 1000 K (or more, in the case of elements which could have large mole fractions in the core, such as hydrogen).

Figure 3 shows the adiabatic back-extrapolation of a suite of estimated melting temperatures of iron at the inner core/outer core boundary utilizing a standard density scaling of the Grüneisen parameter with volume in which γ is equal to $\gamma_0(\rho_o/\rho)^q$; parameters for this fit are derived from *Anderson and Ahrens* [1994]. Notably, both shock sound speed measurements and inversions of porous Hugoniot data yield values for the Grüneisen parameter of liquid iron at outer core pressures of near 1.4 [*Jeanloz*, 1979; *Brown and McQueen*, 1986; *Anderson and Ahrens*, 1994]: this value is largely independent of the precise formulation used to calculate the Grüneisen parameter under core conditions, and appears to be quite robust. As a result, although the precise value of the inner core-outer core boundary temperature has generally wide uncertainties, at least the Grüneisen parameter of the outer core appears to be well-constrained.

TEMPERATURES AT THE TOP OF D"

The inferred temperature at the top of D" is modulated by two primary effects: 1) the magnitude of superadiabatic temperature increases within the mantle; and 2) the rate of

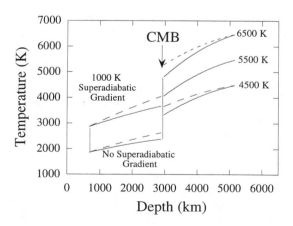

Figure 3. Upward adiabatic extrapolations of estimated temperatures at the inner core-outer core boundary compared with downward extrapolations of the estimated temperature at the 400 km discontinuity, utilizing an upper limit on the superadiabatic gradient through the transition zone of 1000 K and zero pressure Grüneisen parameters of the lower mantle of 1.3 (solid lines) and 1.8 (dotted lines).

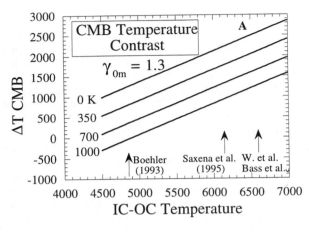 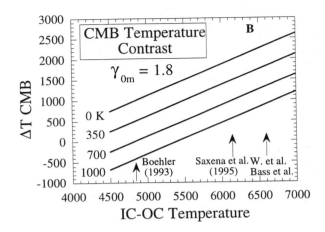

Figure 4a,b. Superadiabatic temperature increase across D" as a function of the inner core-outer core boundary temperature, as calculated from Eqn. 3. Contours represent differing amounts of superadiabatic temperature increases across the transition zone, and arrows denote different experimentally derived estimates of the temperature at the inner core-outer core boundary.

adiabatic temperature increase as a function of depth. The temperature at the 400 km discontinuity is well-known to be approximately 1850 K [*Ita and Stixrude*, 1994; *Akaogi et al.*, 1989], and downward extrapolation of this value hinges on both the value of the Grüneisen parameter of the lower mantle and on the magnitude of any superadiabaticity present between the 400 km discontinuity and the top of D". In this sense, the thermal environment of D" hinges crucially on whether the mantle convects in one layer or two. Canonical estimates for the magnitude of temperature increase across a thermal boundary layer with no material transport occurring across the boundary have typically been on the order of 1000 K [*Richter and McKenzie*, 1981]. Yet, models of mantle flow that incorporate the negative Clapeyron slopes of the perovskite forming reactions have produced superadiabatic temperature increases of in excess of 350 K within the transition zone over much of Earth history [*Peltier*, 1996]. Therefore, it appears that significant superadiabatic gradients may be maintained even if material exchange occurs between the upper and lower mantles [*Tackley et al.*, 1994; *Honda et al.*, 1993; *Peltier*, 1996]. Indeed, any impediments to whole-mantle convective circulation, whether thermodynamic or chemical in origin, are likely to produce a superadiabatic gradient within the mantle. Here, the implications for the temperature change within D" of superadiabatic temperature increases in the mantle of between 0 and 1000 K are examined.

There is additional uncertainty associated with the value of the Grüneisen parameter of the lower mantle, with two overall families of results having emerged for the ambient pressure Grüneisen parameter of the dominant mineral in the lower mantle, silicate perovskite: near 1.3 or 1.8-2.0 [e.g., *Anderson et al.*, 1996; *Hemley et al.*, 1992]. These two families of solutions are produced from differing

estimates of the thermal expansion of this material. Lower mantle geotherms are calculated using values of the Grüneisen parameter of both 1.3 and 1.8 (Figure 3) and a logarithmic dependence of the Grüneisen parameter on volume of 1. As the lower mantle is likely a mix of perovskite and magnesiowüstite and the zero pressure Grüneisen parameter of (Mg,Fe)O is near 1.5 [*Anderson et al.*, 1996; *Hemley et al.*, 1992], the role of added magnesiowüstite should be to moderate the difference between the two endmember sets of perovskite values. With the additional uncertainty in mantle adiabatic temperature gradients in mind, we are able to quantitatively evaluate the temperature change across the D" layer, which should equal

$$\Delta T_{CMB} = X(T_{IC\text{-}OC}) - Y(1850 + \Delta T_{Superadiabatic,\ TZ}) \pm \Delta T_{VE} \pm \Delta T_{Non\text{-}adiabatic,\ LM} \tag{3}$$

Here, X and Y are simply derived from assessing the temperature gradient from Eqn. 1, and are equal to about 0.75 and 1.29 or 1.42 (for zero pressure mantle Grüneisen parameters of 1.3 and 1.8, respectively). $\Delta T_{Superadiabatic,\ TZ}$ is the present superadiabatic temperature gradient across the transition zone. ΔT_{VE} is the magnitude of temperature change due to the latent heat of phase transitions (the Verhoogen effect): this is likely to be small (order 100 K maximum) [*Jeanloz and Morris*, 1986; *Ita and Stixrude*, 1992]. $\Delta T_{Non\text{-}adiabatic,\ LM}$ is included for completeness, as any non-adiabatic gradients in the bulk of the lower mantle will shift the inferred temperature change across D". The temperature change associated with such non-adiabatic gradients is presumed to be zero for the calculations of Fig. 4, as whether weakly superadiabatic or weakly subadiabatic gradients are present in the lower mantle is unclear [*Brown and Shankland*, 1981; *Jeanloz and Morris*, 1987]. The results from Eqn. 3 are shown in Fig. 4 for two different values of the mantle Grüneisen parameter as a function of inferred values of temperature at the inner core-outer core boundary. It is clear that a wide range of temperature contrasts are permitted across D", with estimates in the 1000-2000 K range being most compatible with inner core-outer core boundary temperatures in excess of 6000 K. For comparison, temperatures below 5000 K at this boundary are only compatible with a significant thermal boundary layer (O ~ 1000 K) if there is a small or negligible superadiabatic gradient across the mantle, and if the Grüneisen parameter of the mantle is relatively low. A temperature contrast across a chemically homogeneous D" of significantly above 1000 K would indicate that secondary convection would be induced in this layer [e.g., *Loper and Eltayeb*, 1986]. Yet, chemical stratification within D" would serve to stabilize this zone against such convective instabilities: a gradual increase in density of order 3-6% would likely stably stratify the base of the mantle, even for superadiabatic temperature anomalies of the order of 1000-2000 K within this layer [*Sleep*, 1988; *Williams and Jeanloz*, 1990; *Farnetani*, 1997]. That such chemical stratification may exist is indicated by observations of at least two separate, laterally varying seismic discontinuities, near 200-300 km above the CMB and between 5 and 40 km above the CMB [*Garnero et al.*, this volume; *Wysession et al.*, this volume]. The deeper of these discontinuities is likely associated with partial melting, but may also be associated with a difference in chemistry from the overlying mantle [*Williams and Garnero*, 1996; *Revenaugh and Meyer*, 1997]. The origin of the shallower discontinuity is unclear, but it is unlikely to be associated with an isochemical phase transition [*Wysession et al.*, this volume]. Therefore, the detailed dynamics of D" remain unclear, but possible chemical stratification and partial melting in this zone could indicate that the temperature increase across D" is significantly in excess of 1000 K.

CONCLUSION

The inferred temperature contrast across the D" region remains dependent on a range of difficult-to-assess parameters. In approximate order of importance, these are: 1) the identity (identities) of the light alloying component in the core and its (their) effect on the melting temperature of iron; 2) the melting temperature of iron at the inner core-outer core boundary; 3) the magnitude of superadiabatic (or subadiabatic) temperature increases within the mantle; and 4) the adiabatic temperature gradient within the lower mantle. The first two of these effects contribute net uncertainties exceeding 1000 K to estimates of the deep mantle thermal regime; the latter two effects likely produce uncertainties of less than 1000 K. The upper limits on the temperature contrast across this layer imposed by dynamical instabilities, and resultant removal of the

thermal boundary layer over time, are largely removed if there is significant density/chemical stratification within this layer. Similarly, the canonical upper limit for the temperature across this layer derived from the idea that the temperature in this zone must lie below the melting temperature of mantle materials is removed if the recently discovered basal ultra-low velocity zone is associated with large degrees of partial melting. A relatively large number of combinations of parameters yield temperature contrasts of the order of 1000-2000 K across this layer; lower values for the temperature contrast involve more restrictive assumptions about the amount of freezing point depression in the outer core, the melting temperature of iron at the inner core-outer core boundary of the planet, and/or either the superadiabatic or adiabatic temperature increase across the transition zone or lower mantle.

Acknowledgments. This work was supported by the NSF and the W.M. Keck Foundation. I thank T.J. Ahrens, M. Gurnis and E. Knittle for comments on the manuscript.

REFERENCES

Ahrens, T.J. and R. Jeanloz, Pyrite: Shock compression, isentropic release, and composition of the Earth's core, *J. Geophys. Res., 92*, 10363-10375, 1987.

Akaogi, M., E. Ito and A. Navrotsky, Olivine-modified spinel-spinel transitions in the system Mg_2SiO_4-Fe_2SiO_4: Calorimetric measurements, thermochemical calculation, and geophysical application, *J. Geophys. Res., 94*, 15671-15685, 1989.

Anderson, O.L. and A. Duba, Experimental melting curve of iron revisited, *J. Geophys. Res., 102*, 22659-22669, 1997.

Anderson, O.L., K. Masuda and D.G. Isaak, Limits on the value of δ_T and γ for $MgSiO_3$ perovskite, *Phys. Earth Planet. Inter,. 98*, 31-46, 1996.

Anderson, W.W. and T.J. Ahrens, An equation of state for liquid iron and implications for the Earth's core, *J. Geophys. Res., 99*, 4273-4284, 1994.

Anderson, W.W. and T.J. Ahrens, Shock temperature and melting in iron sulfides at core pressures, *J. Geophys. Res., 101*, 5637-5642, 1996.

Andrault, D., G. Fiquet, M. Kunz, F. Visocekas and D. Hausermann, The orthorhombic structure of iron: An in situ study at high-temperature and high-pressure, *Science, 278*, 831-834, 1997.

Badding, J.V., H.K. Mao and R.J. Hemley, High-pressure crystal structure and equation of state of iron hydride: Implications for the Earth's core, in *High Pressure Research: Application to Earth and Planetary Sciences*, Eds. Y. Syono and M.H. Manghnani, pp. 363-371, American Geophysical Union, Washington, D.C., 1992.

Bass, J.D., T.J. Ahrens, J.R. Abelson and T. Hua, Shock temperature measurements in metals: New results for an Fe alloy, *J. Geophys. Res., 95*, 21767-21776, 1990.

Bass, J.D., B. Svendsen and T.J. Ahrens, The temperature of shock compressed iron, in *High Pressure Research in Mineral Physics*, M.H. Manghnani and Y. Syono, Eds., pp. 393-402, American Geophysical Union, Washington, D.C., 1987.

Belonoshko, A.B. and L.S. Dubrovinsky, A simulation study of induced failure and recrystallization of a perfect MgO crystal under non-hydrostatic compression: application ot melting in the diamond-anvil cell, *Am. Mineral., 82,* 441-451, 1997.

Birch, F., Elasticity and constitution of the Earth's interior, *J. Geophys. Res., 57,* 227-286, 1952.

Boehler, R., Melting of the Fe-FeO and the Fe-FeS systems at high pressure: Constraints on core temperatures, *Earth Planet. Sci. Lett., 111,* 217-227, 1992.

Boehler, R., Temperatures in the Earth's core from melting point measurements of iron at high static pressures, *Nature, 363,* 534-536, 1993.

Boehler, R., Melting of mantle and core materials at very high pressures, *Phil. Trans. R. Soc. Lond. A, 354,* 1265-1278, 1996a.

Boehler, R., Fe-FeS eutectic temperatures to 620 kbar, *Phys. Earth Planet. Int., 96,* 181-186, 1996b.

Boness, D.A. and J.M. Brown, Bulk superheating of solid KBr and CsBr with shock waves, *Phys. Rev. Lett., 71,* 2931-2935, 1993.

Brown, J.M. and R.G. McQueen, The equation of state for iron and the Earth's core, in *High Pressure Research in Geophysics,* Eds. S. Akimoto and M.H. Manghnani, pp. 611-623, Center for Academic Publications, Tokyo, 1982.

Brown, J.M. and R.G. McQueen, Phase transitions, Grüneisen parameter and elasticity for shocked iron between 77 GPa and 400 GPa, *J. Geophys. Res., 91,* 7485-7494, 1986.

Brown, J.M. and T.J. Shankland, Thermodynamic parameters in the Earth as determined from seismic profiles, *Geophys J. Roy. Astr. Soc., 66,* 579-596, 1981.

Chen, G.Q., and T.J. Ahrens, High-pressure melting of iron: New experiments and calculations, *Phil. Trans. R. Soc. Lond. A, 354,* 1251-1263, 1996.

Doornbos, D.J., S. Spiliopoulos, and F.D. Stacey, Seismological properties of D" and the structure of a thermal boundary layer, *Phys. Earth Planet. Int., 41,* 225-239, 1986.

Farnetani, C.G., Excess temperature of mantle plumes: The role of chemical stratification across D", *Geophys. Res. Lett., 24,* 1583-1586, 1997.

Fei, Y, C.M. Bertka, L.W. Finger, High-pressure iron-sulfur compound, Fe_3S_2, and melting relations in the Fe-FeS system, *Science, 275,* 1621-1623, 1997.

Fukai, Y., Some properties of the Fe-H system at high pressures and temperatures, and their implications for the Earth's core, *High Pressure Research: Application to Earth and Planetary Sciences,* Eds. Y. Syono and M.H. Manghnani, pp. 373-385, American Geophysical Union, Washington, D.C., 1992.

Gallagher, K.G., and T.J. Ahrens, First measurements of thermal conductivity in griceite and corundum at ultra high pressure and the melting point of iron, *EOS, Trans. Am. Geophys. Un. 75,* 653, 1994.

Garnero, E.J., J. Revenaugh, Q. Williams, T. Lay and L. Kellogg, Ultralow velocity zone at the core-mantle boundary, *this volume.*

Hemley, R.J., L. Stixrude, Y. Fei and H.K. Mao, Constraints on lower mantle composition from P-V-T measurements of $(Fe,Mg)SiO_3$-perovskite and $(Fe,Mg)O$, *High Pressure Research: Application to Earth and Planetary Sciences,* Eds. Y. Syono and M.H. Manghnani, pp. 183-189,

American Geophysical Union, Washington, D.C., 1992.

Hixson, R.S., M.A. Winkler and M.L. Hodgdon, Sound speed and thermophysical properties of liquid iron and nickel, *Phys. Rev. B, 42,* 6485-6491, 1990.

Honda, S., D.A. Yuen, S. Blachandar and D. Reuteler, Three dimensional instabilities of mantle convection with multiple phase transitions, *Science, 259,* 1308-1311, 1993.

Huang, E., W.A. Bassett and M.S. Weathers, Phase diagram and elastic properties of Fe 30% Ni alloy by synchrotron radiation, *J. Geophys. Res., 97,* 4497-4502, 1992.

Ita, J. and L. Stixrude, Petrology, elasticity, and composition of the mantle transition zone, *J. Geophys. Res., 97,* 6849-6866, 1992.

Jeanloz, R., Properties of iron at high pressure and the state of the core, *J. Geophys. Res., 84,* 6059-6069, 1979.

Jeanloz, R. and T.J. Ahrens, Equations of state of FeO and CaO, *Geophys. J.R. Astr. Soc., 62,* 505-528, 1980.

Jeanloz, R. and A. Kavner, Melting criteria and imaging spectroradiometry in laser-heated diamond-cell experiments, *Phil. Trans. R. Soc. Lond. A, 354,* 1279-1305, 1996.

Jeanloz, R. and S. Morris, Temperature distribution in the crust and mantle, *Ann. Rev. Earth Planet. Sci., 14,* 377-415, 1986.

Jeanloz, R. and S. Morris, Is the mantle geotherm subadiabatic?, *Geophys. Res. Lett., 14,* 335-338, 1987.

Jeanloz, R. and F. Richter, Convection, composition, and the thermal state of the lower mantle, *J. Geophys. Res., 84,* 5497-5504, 1979.

Jephcoat, A.P. and S.P. Besedin, Temperature measurement and melting determination in the laser-heated diamond-anvil cell, *Phil. Trans. R. Soc. Lond. A, 354,* 1333-1360, 1996.

Jephcoat, A.P. and P. Olson, Is the inner core of the Earth pure iron?, *Nature, 325,* 332-335, 1987.

Jones, G.M., Thermal interaction of the core and the mantle and long-term behavior of the geomagnetic field, *J. Geophys. Res., 82,* 1703-1709, 1977.

Knittle, E. and R. Jeanloz, The high presssure phase diagram of $Fe_{0.94}O$, a possible constituent of the Earth's outer core, *J. Geophys. Res., 96,* 16169-16180, 1991.

Lay, T. and D.V. Helmberger, A lower mantle S-wave triplication and the velocity structure of D", *Geophys. J. Roy. Astr. Soc., 75,* 799-837, 1983.

Lazor, P. and S.K. Saxena, Discussion comment on melting criteria and imaging spectroradiometry in laser-heated diamond-cell experiments (by R. Jeanloz and A. Kavner), *Phil. Trans. R. Soc. Lond. A, 354,* 1307-1313, 1996.

Loper, D.E. and I.A. Eltayeb, On the stability of the D" layer, *Geophys. Astrophy. Fluid Dyn., 36,* 229-255, 1986.

Mao, H.K., Y. Wu, J.F. Shu and A.P. Jephcoat, Static compression of iron to 300 GPa and $Fe_{0.8}Ni_{0.2}$ alloy to 260 GPa: Implications for composition of the core, *J. Geophys. Res., 95,* 21737-21742, 1990.

McKenzie, D.P., J.M. Roberts and N.O. Weiss, Convection in the Earth's mantle: towards a numerical simulation, *J. Fluid Mech., 62,* 465-538, 1974.

Peltier, W.R., Phase-transition modulated mixing in the mantle of the Earth, *Phil. Trans. R. Soc. Lond. A, 354,* 1425-1447, 1996.

Revenaugh, J.S. and R. Meyer, Seismic evidence of partial melt within a possibly ubiquitous low velocity layer at the base of the mantle, *Science, 277,* 670-673, 1997.

Richter, F.M. and D.P. McKenzie, On some consequences and

possible causes of layered mantle convection, *J. Geophys. Res., 86,* 6133-6142, 1981.

Ringwood, A.E. and W. Hibberson, The system Fe-FeO revisited, *Phys. Chem. Minerals, 17,* 313-319, 1990.

Saxena, S.K., G. Shen and P. Lazor, Experimental evidence for a new iron phase and implications for Earth's core, *Science, 260,* 1312-1314, 1993.

Saxena, S.K., G. Shen, and P. Lazor, Temperature in Earth's core based on melting and phase transformation experiments on iron, *Science, 264,* 405-407, 1994.

Sleep, N.H., Gradual entrainment of a chemical layer at the base of the mantle by overlying convection, *Geophys. J., 95,* 437-447, 1988.

Spiliopoulos, S. and F.D. Stacey, The Earth's thermal profile: Is there a mid-mantle thermal boundary layer, *J. Geodyn., 1,* 61-77, 1984.

Stevenson, D.J., Models of the Earth's core, *Science, 214,* 611-619, 1981.

Stishov, S.M., Entropy, disorder, melting (in Russian), *Achiev. Phys. Sci., 154,* 1-122, 1988.

Strong, H.M., R.E. Tuft and R.E. Hanneman, The iron fusion curve and γ–ε-l triple point, *Metall. Trans., 4,* 2657-2661, 1973.

Tackley, P.J., D.J. Stevenson, G.A. Glatzmaier and G. Schubert, Effects of multiple phase transitions in a 3-dimensional spherical model of convection in Earth's mantle, *J. Geophys. Res., 99,* 15877-15901, 1994.

Verhoogen, J., Petrological evidence on temperature distribution in the mantle of the Earth, *Trans. AGU, 35,* 85-92, 1953.

Williams, Q. and E. Garnero, Seismic evidence for partial melt at the base of Earth's mantle, *Science, 273,* 1528-1530, 1996.

Williams, Q. and R. Jeanloz, Melting relations in the iron-sulfur system at ultra-high pressures: Implications for the thermal state of the Earth, *J. Geophys. Res., 95,* 19299-19310, 1990.

Williams, Q., R. Jeanloz, J. Bass, B. Svensen and T.J. Ahrens, The melting curve of iron to 250 GPa: A constraint on the temperature at Earth's center, *Science, 236,* 181-182, 1987.

Williams, Q., E. Knittle and R. Jeanloz, The melting curve of iron: A technical discussion, *J. Geophys. Res., 96,* 2171-2184, 1991.

Williams, Q. and E. Knittle, Constraints on core composition from the pressure dependence of the bulk modulus, *Phys. Earth Planet. Inter., 100,* 49-59, 1997.

Wysession, M.E., et al., The D" discontinuity and its implications, *this volume.*

Yoo, C.S., J. Akella, A.J. Campbell, H.K. Mao and R.J. Hemley, The phase diagram of iron by in situ x-ray diffraction: Implications for Earth's core, *Science, 270,* 1473-1475, 1995.

Yoo, C.S., J. Akella, A.J. Campbell, H.K. Mao and R.J. Hemley, Detecting phases of iron, *Science, 275,* 96, 1997.

Yoo, C.S., N.C. Holmes, M. Ross, D.J. Webb and C. Pike, Shock temperatures and melting of iron at Earth core conditions, *Phys. Rev. Lett., 70,* 3931-3934, 1993.

Q. Williams, Dept. of Earth Sciences, University of California, Santa Cruz, CA 95064

Elastic Constants and Anisotropy of MgSiO$_3$ Perovskite, Periclase, and SiO$_2$ at High Pressure

Lars Stixrude

Dept. of Geological Sciences, University of Michigan, Ann Arbor

We review recent results of first principles theory as applied to the elastic constants of important deep earth phases at high pressure, including MgSiO$_3$ perovskite, MgO periclase and the stishovite, CaCl$_2$ and columbite phases of SiO$_2$. The foundation and implementation of the theoretical technique, the plane-wave pseudopotential method, is briefly discussed as is the method for determining the structure and elastic constants at high pressure. The results show that predicted elastic constants are in good agreement with experiment: RMS deviation are 5 %, 7 % and 4 %, respectively for perovskite, periclase and stishovite. Much of the difference can be accounted for by the difference in temperature between the athermal calculations and the 300 K experiments. We discuss these results in terms of seismological observations paying particular attention to 1) the properties of aggregates and 2) anisotropy. Comparison of the seismic wave velocities of isotropic mono-phase aggregates with those of the lower mantle show that the P- and S-wave velocities of perovskite are consistent with those of the deep earth, supporting the view that this mineral is the most abundant constituent of the lower mantle. The properties of anisotropic mono-phase aggregates are estimated based on the theoretical results for single crystals and simple models of flow in D″ and of mineralogical deformation mechanisms. We find that the SH/SV anisotropy is large for aggregates of perovskite, periclase and silica (6-8 % in magnitude) compared with that observed in D″. While mono-phase aggregates of silica exhibit fast SH velocity at pressures corresponding to D″, perovskite and periclase have slow SH velocities. We discuss the implications of these results for the origin of anisotropy in D″.

1. INTRODUCTION

The elasticity of minerals determines the solid earth's response to a wide variety of loads that have time scales shorter than the Maxwell relaxation time and magnitudes less than the shear strength. Elasticity controls such diverse geologic phenomena as the form of sedimentary basins, moats and peripheral bulges surrounding oceanic volcanos, and forearc bulges. In the deep earth where the Maxwell relaxation time is much shorter than that in the lithosphere, elasticity is most important for understanding the propagation of seismic energy.

Seismological observations of the earth's interior in principle contain a wealth of information concerning its composition, thermal state, and dynamics. This information has proven difficult to extract, primarily because the elastic constants of mantle minerals under the relevant conditions are essentially unknown. While the elastic constants of most major mantle phases are known at ambient conditions, and, in some cases at ambient pressure and high temperature [*Bass*, 1995; *Anderson and Isaak*, 1995], little is known of the elastic

The Core-Mantle Boundary Region
Geodynamics 28
Copyright 1998 by the American Geophysical Union.

constants of any material at mantle pressures. This is a serious limitation as pressure is expected to have a much greater effect on elastic properties over the range of pressure and temperature relevant to the deep mantle.

Our lack of knowledge of high pressure elasticity has meant that we have been unable to make use of those aspects of mantle elasticity that are unique to the solid state: anisotropy, and the propagation of shear waves. This limitation likely accounts for the lack of consensus concerning the composition of the lower mantle [*Jeanloz and Knittle*, 1989; *Bukowinski and Wolf*, 1990; *Stixrude et al.*, 1992; *Wang et al.*, 1994; *Bina and Silver*, 1997] - of seismologically observable properties, it is precisely those of which we are ignorant (the shear modulus) which are most sensitive to composition, temperature and phase.

Recent experimental advances have led to the first measurements of the elastic constants of mantle minerals at the relevant pressures [*Zaug et al.*, 1993; *Duffy et al.*, 1995; *Chen et al.*, 1996]. A variety of techniques including light-scattering and ultrasonic methods hold tremendous promise for revealing the elasticity of mantle minerals at mantle pressure and temperature conditions. These methods have not yet been applied to the major minerals of the lower mantle.

Major advances in theory and computation now make it possible to calculate the full elastic constant tensor of mantle minerals from first principles [*Karki et al.*, 1997a; *Karki et al.*, 1997b; *Karki et al.*, 1997c; *da Silva et al.*, 1997]. These methods are parameter-free and are completely independent of experiment. They thus provide the ideal complement to the experimental approach, allowing for important cross-checks between laboratory observation and theory. First principles methods are unlimited in the range of pressure that they can explore and are equally applicable in principle to all mantle materials. They represent a potentially important tool for studying the wide range of pressure and composition encountered in the earth's interior.

We briefly discuss theoretical methods below, including techniques for calculating the full elastic constant tensor. We review some recent results on the most abundant constituents of the lower mantle [*Karki et al.*, 1997a; *Karki et al.*, 1997b; *Karki et al.*, 1997d; *Karki et al.*, 1997c]. Theory is compared with seismological observation with particular emphasis on the one-dimensional isotropic structure of the lower mantle, and observations of anisotropy in the deepest portions of this region. Careful attention is paid to the anisotropy of single crystals versus that of aggregates, and possible texturing mechanisms in the earth's interior.

2. THEORY

2.1. Density Functional Theory

The calculations are based on density functional theory [*Hohenberg and Kohn*, 1964; *Kohn and Sham*, 1965], in principle an exact theory of the ground state electronic total energy and band structure. We discuss the two essential approximations that are made in our calculations, 1) the local density approximation and in the following subsection, 2) the pseudopotential approximation. The methods for determining the structure and elastic constants are briefly described in the last subsection.

The essence of this theory is the proof that the ground state electronic properties of a material are a unique functional of the charge density $\rho(\vec{r})$. This central feature is appealing because the charge density is experimentally observable, for example, by x-ray diffraction. A variational principle leads to a set of single-particle, Schrödinger-like Kohn-Sham equations, with an effective potential

$$V_{KS} = V_{e-n}\left[\rho(\vec{r})\right] + V_{e-e}\left[\rho(\vec{r})\right] + V_{xc}\left[\rho(\vec{r})\right] \quad (1)$$

where the first two terms are the Coulomb potentials due to the nuclei and the other electrons, and the last, the exchange-correlation potential, subsumes the complex many-body interactions among the electrons. The power of density functional theory is that it allows one to calculate, in principle, the exact many-body total energy of a system from a set of single-particle equations. In practice, exact solutions are impossible at present because the exact form of V_{xc} is currently unknown.

Simple approximations to the exchange-correlation functional have proven to be very successful in studies of silicates and essentially all other classes of solids. In this study, we use the most widely studied approximation, the Local Density Approximation (LDA), which replaces V_{xc} at every point in the crystal by the precisely known value for a free electron gas with a density equal to the local charge density [*Lundqvist and March*, 1987]. The success of the LDA can be understood at a fundamental level in terms of the satisfaction of exact sum rules for the exchange-correlation hole [*Gunnarsson and Lundqvist*, 1976].

LDA calculations of silicates and oxides of geophysical interest agree very well with experimental data. The error due to the LDA approximation can be evaluated by comparing Linearized Augmented Plane Wave (LAPW) calculations, which make no further essential approximations beyond the LDA, with experiment. It has been found that errors in volumes are typically 1-4 % with theoretical volumes being uniformly smaller than experimental [*Mehl et al.*, 1988; *Cohen*, 1991; *Cohen*, 1992; *Stixrude and Cohen*, 1993]. Part of this small difference is due to the higher temperatures of experiments (300 K) compared with the athermal calculations [*Mehl et al.*, 1988]. This is a highly satisfactory level of agreement for a theory which is parameter free and independent of experiment.

2.2. *The Pseudopotential Method*

The basis of this method is the simple observation that only the valence electrons participate in the response of the crystal to most perturbations. Unless the perturbation is of very high energy (comparable to the binding energy of the core states), the tightly bound core states remain essentially unchanged. It is reasonable then to assume for most applications that the core electrons are frozen. The frozen core approximation is satisfied to a high degree of accuracy for many applications, for example in the case of finite strains of magnitudes typically encountered in the earth's interior.

The pseudopotential method replaces the nucleus and core electrons with a single object, the pseudopotential, which has the same scattering properties [*Pickett*, 1989]. The advantages of the pseudopotential method are 1) spatial variations in the pseudopotential are much less rapid than the bare Coulomb potential of the nucleus and 2) one need solve only for the valence electrons which show much less rapid spatial variation than the core electrons. This means that in the solution of the Kohn-Sham equations, potential and charge density can be represented by a particularly simple, complete and orthogonal set of basis functions (plane-waves) of manageable size (typically $10^3 - 10^5$ basis functions are required). With this basis set, evaluation of total energies, stresses, and forces acting on the atoms is particularly efficient.

The increased computational efficiency of the pseudopotential method comes at the cost of additional assumptions; the frozen core approximation, already mentioned, and, more seriously, the pseudopotential itself, which is an approximation to the potential that the valence electrons "see". The construction of the pseudopotential is a non-unique process, and several different methods have been developed [*Vanderbilt*, 1990; *Troullier and Martins*, 1991; *Kresse et al.*, 1992]. Care must be taken to demonstrate the transferability of the pseudopotentials generated by a particular method and to compare with all electron calculations where these are available. When these conditions are met, the error due to the pseudopotential is generally small (few percent in volume for earth materials) [*Kiefer et al.*, 1997].

In this study, pseudopotentials are generated by the Q_c-tuning method [*Lin et al.*, 1993], which yield results in excellent agreement with LAPW calculations where these are available [*Karki et al.*, 1997a; *Karki et al.*, 1997d; *Karki et al.*, 1997c], and with calculations [*Wentzcovitch et al.*, 1995] based on other pseudopotentials [*Troullier and Martins*, 1991]. Computational parameters were chosen so that calculated stresses were converged to within 0.02 GPa: the number of plane waves per atom ranges from 2,400 (MgO) to 22,000 ($MgSiO_3$) (maximum kinetic energy 62-66 Ry); the Brillouin zone is sampled on a 4x4x4 (oxides) or 4x4x2 ($MgSiO_3$) Monkhurst-Pack [*Monkhurst and Pack*, 1976] special k-point mesh.

2.3. *Structural Optimization and Determination of Elastic Constants*

Only recently has it become possible to determine from first principles the ground state atomic arrangement in a relatively complex structure such as $MgSiO_3$ perovskite [*Wentzcovitch et al.*, 1993]. The key innovation has been the development of a structural optimization strategy based on a pseudo-Lagrangian that treats the components of the strain tensor and the atomic positions as dynamical variables [*Wentzcovitch*, 1991]. The optimization is performed at constant pressure. At each step of the dynamical trajectory, the Hellman-Feynman forces and stresses [*Nielsen and Martin*, 1985] acting, respectively, on the nuclei and lattice parameters are evaluated and used to generate the next configuration. The optimization is complete when the forces on the nuclei vanish and the stress is hydrostatic and balances the applied pressure. This is an efficient procedure, requiring on the order of 10-20 steps even for a relatively complex structure such as orthorhombic perovskite with three lattice parameters and 7 internal degrees of freedom [*Wentzcovitch et al.*, 1995]. Even larger and more complex structures have been studied with this technique (e.g. olivine) [*Wentzcovitch and Stixrude*, 1997].

Once the equilibrium structure at a given pressure is determined, one can calculate the elastic constants. This is done in a straightforward way by applying a deviatoric strain to the lattice and calculating the resulting stress tensor. The elastic constant, c_{ijkl} is then given by the ratio of stress, σ_{ij} to strain ϵ_{ij}

$$\sigma_{ij} = c_{ijkl}\epsilon_{kl} \qquad (2)$$

To ensure that the calculated elastic constants refer to the appropriate linear (infinitesimal strain) regime, the ratio of stress to strain is calculated for several values of the strain magnitude (1-4 %) and then extrapolated to the limit of zero strain. Care must be taken to re-optimize the positions of the atoms in each strained configuration since vibrational modes typically couple with lattice strains in silicate structures.

2.4. Derived Quantities: Single Crystal Elastic Wave Velocities

Once the full elastic constant tensor has been determined, it is straightforward to calculate the elastic wave velocities of the single crystal for arbitrary propagation direction, $\vec{n}$ from the Cristoffel equation. The eigenvalues and eigenvectors of the matrix

$$c_{ijkl}n_j n_l \qquad (3)$$

yield the moduli (ρV^2) and polarizations, respectively of the three elastic waves, where ρ is the density and V is the speed of the wave. The three eigensolutions correspond to a quasi-longitudinal or P wave (velocity V_P) and two quasi-shear waves (velocities V_{S1} and V_{S2}). Examination of the eigenvectors shows that unless the anisotropy is very large (greater than 20 %), the modes are nearly pure (P-waves more than 90 % pure). Only in the vicinity of the stishovite to $CaCl_2$ phase transition have we found P-S mode mixing exceeding 10 %.

We summarize the dependence of elastic wave velocities on propagation direction in the single crystal by three measures of the anisotropy:

$$\Delta_{P,azimuthal} = \frac{max[V_P(\vec{n})] - min[V_P(\vec{n})]}{V_{Pagg}} \qquad (4)$$

$$\Delta_{S,azimuthal} = \frac{max[V_S(\vec{n})] - min[V_S(\vec{n})]}{V_{Sagg}} \qquad (5)$$

$$\Delta_{S,polarization} = \frac{max|V_{S1}(\vec{n}) - V_{S2}(\vec{n})|}{V_{Sagg}} \qquad (6)$$

which are, respectively, the maximum single-crystal azimuthal anisotropy in P- and S-waves and the maximum single-crystal polarization anisotropy of S-waves. The quantities V_{Pagg} and V_{Sagg} are the isotropic aggregates velocities.

3. PROPERTIES OF AGGREGATES

The earth's interior is a multi-phase aggregate, and any comparison of the properties of minerals with seismology must take this into account. For the case of geophysical interest, where the seismic wavelength is much larger than the size of the component crystals, the elastic wave velocities of an aggregate can be uniquely calculated from the single crystal elastic constants once the texture, that is the positions, shapes, and orientations of the constituent grains are known.

Because we have little knowledge of the texture of the deep mantle, we consider geophysically motivated approximations to the texture of aggregates in the earth's interior, for which the estimation of elastic wave velocities is relatively straightforward. The simplest case is the isotropic mono-phase aggregate, in which crystals of a single phase are oriented randomly. It is worth remembering that even for this simple case, the aggregate velocities are inherently uncertain since the shapes of the grains have not been specified [Watt et al., 1976]. However, it is possible to place strict bounds on the aggregate velocities, which, in most cases, differ by a few percent or less. Here, we have estimated the velocities of isotropic mono-phase aggregates by the average of the Hashin-Shtrikman bounds [Hashin and Shtrikman, 1962; Meister and Peselnick, 1966] in the case of cubic and tetragonal crystals, and by the Voigt-Reuss-Hill average [Hill, 1952] in the case of orthorhombic crystals.

While the bulk of the lower mantle is isotropic [Meade et al., 1995], anisotropic regions are of particular interest, because they occur within the D'' layer [Vinnik et al., 1995; Matzel et al., 1996; Kendall and Silver, 1996; Garnero and Lay, 1997], and because observations of anisotropy contain, in principle, information about mantle flow [Park and Yu, 1993]. In monophase aggregates, anisotropy arises through Lattice Preferred Orientation (LPO). LPO may be produced when the aggregate is subjected to deviatoric stresses, as in a convecting system.

In order to estimate the properties of anisotropic mono-phase aggregates in the deep earth, we follow Karato [1997] in making the following assumptions: 1) mantle flow in dynamical boundary layers is primarily horizontal; this is likely to be a good approximation in D'' 2) lower mantle minerals contain a dominant slip plane and slip direction which largely controls their rheology 3) the dominant slip planes and directions can be estimated on the basis of ambient pressure data and/or analog materials and 4) the texture is fully developed (perfect): slip planes and directions are parallel to the flow plane and direction, respectively. The latter assumption means that we will be examining upper limits to the anisotropy of aggregates. In general, alignment of all grains of the aggregate with the flow will not be perfect, and the anisotropy will be correspondingly reduced.

Table 1.   Dominant Slip Planes and Slip Directions

Phase	Slip Plane	Slip Direction
Perovskite[1]	(010)	[100]
Periclase[2]	{110}	$<111>$
Stishovite[3]	{101}	$<101>$
$CaCl_2$-type	(101)	[$\bar{1}$01]
Columbite-type	(101)	[$\bar{1}$01]

References: 1, [*Karato et al.*, 1995]; 2, [*Chin and Mammel*, 1973]; 3, [*Ashbee and Smallman*, 1963]. Deformation mechanisms in $MgSiO_3$ perovskite, and stishovite are taken from experiments on isostructural analogs: $CaTiO_3$ perovskite and TiO_2 rutile, respectively. Deformation of the silica phases is assumed to be controlled by the direction of the chains of edge-sharing octahedra which are parallel to the c-axis in all three phases.

There are two cases to consider: 1) azimuthal anisotropy and 2) transverse anisotropy. In the first case, the flow is coherent over a length scale that is large compared with the region sampled by the seismic wave. In this case, assuming perfect alignment, the elasticity of the aggregate is that of a single crystal. In the second case, the flow is still horizontal, but is azimuthally variable over the region sampled. This situation may occur in the vicinity of stationary points in the flow that will exist at centers of convergence (upwelling) or divergence (downwelling). In this case, slip planes will be horizontal, but slip directions may be approximately evenly distributed over the azimuthal direction. Transverse anisotropy may also have a rheological origin: if the deformation mechanism has a dominant slip plane, but no dominant slip direction, transverse anisotropy will result. The assumed dominant slip planes and slip directions for the phases considered here are summarized in Table 1.

The elastic constants of an azimuthally anisotropic aggregate are those of the single crystal, but rotated into the coordinate system of the flow. We define a crystallographic coordinate system (x) and a laboratory or mantle coordinate system (x') in which x'_3 defines the vertical and coincides with the normal to the flow plane and x'_1 coincides with the flow direction (Fig. 1). The rotation matrix relates the two coordinate systems:

$$a_{ij} = cos(\theta_{i'j}) \qquad (7)$$

where $\theta_{i'j}$ is the angle between x'_i and x_j. The elastic constant tensor of the azimuthally anisotropic aggregate

is then

$$c'_{ijkl} = a_{im}a_{jn}a_{ko}a_{lp}c_{mnop} \qquad (8)$$

where c_{mnop} is the elastic constant tensor in the crystallographic coordinate system [*Nye*, 1985].

In the case of transverse anisotropy, the aggregate is isotropic within the flow plane. We define a coordinate system x'' for which x''_3 coincides with x'_3, and x''_1 is related to x'_1 by the azimuthal angle ϕ (Fig. 1). The primed and double-primed coordinate systems are related by the rotation matrix:

$$\mathbf{b} = \begin{pmatrix} \cos(\phi) & \sin(\phi) & 0 \\ -\sin(\phi) & \cos(\phi) & 0 \\ 0 & 0 & 1 \end{pmatrix} \qquad (9)$$

The elastic constant tensor of the transversely isotropic aggregate is

$$c''_{ijkl} = \frac{1}{2\pi}\int_0^{2\pi} b_{im}b_{jn}b_{ko}b_{lp}c'_{mnop}\,d\phi \qquad (10)$$

The elastic constants in this case will always have hexagonal symmetry.

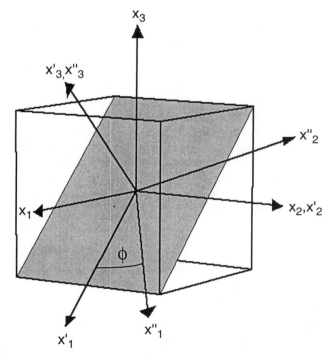

Figure 1.   Coordinate systems used in the analysis of anisotropic aggregates. The example shown corresponds to stishovite: (101) is the slip plane and [10$\bar{1}$] the slip direction.

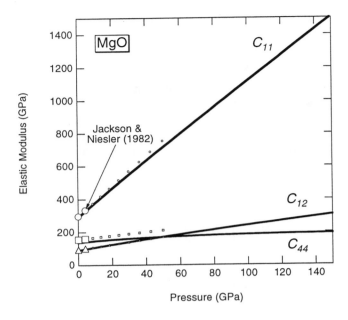

Figure 2. Elastic constants of periclase according to theory [*Karki et al.*, 1997a] (lines) and experiment [*Jackson and Niesler*, 1982] large symbols. Small symbols represent a linear extrapolation of the experimental results beyond the maximum pressure of the measurements (3 GPa).

To characterize the anisotropy of transversely anisotropic aggregates we consider the ratio PH/PV of the velocity of horizontally to vertically propagating P-waves, and the ratio SH/SV of the velocity of horizontally to vertically polarized shear waves propagating in the horizontal plane. For azimuthally anisotropic aggregates, we consider the shear wave splitting $S_\parallel/S_\perp$, the ratio of velocities of vertically propagating shear waves, one polarized in the flow direction, the other perpendicular to it. The velocities are calculated from an equivalent form of the Cristoffel Equation (3)

$$\rho V^2 = c_{ijkl} n_i w_j n_k w_l \qquad (11)$$

where $\vec{w}$ is the polarization direction.

4. RESULTS

4.1. Elastic Constants

First principles elastic constants of the the orthorhombic *Pbnm* structure of MgSiO$_3$ perovskite, the B1 structure of MgO (periclase), and stishovite are in excellent agreement with experimental data (Figs. 2-4, Tables 2,3). The agreement with observation substantially surpasses that achieved by previous theoretical approaches including semi-empirical ionic potentials [*Matsui et al.*,

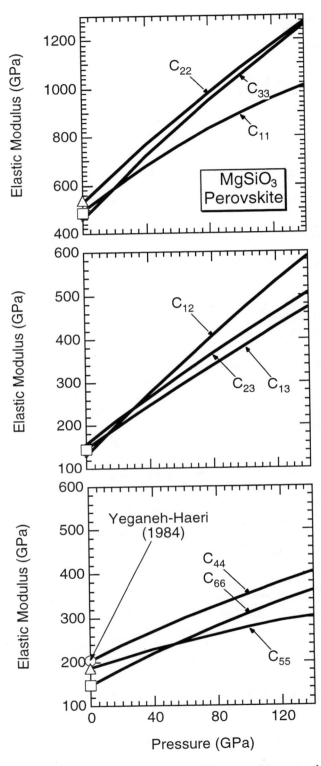

Figure 3. Elastic constants of MgSiO$_3$ perovskite according to theory [*Karki et al.*, 1997b] (lines) and experiment [*Yeganeh-Haeri*, 1994] symbols: c_{11}, c_{12}, c_{44} (○); c_{22}, c_{13}, c_{55} (△); c_{33}, c_{23}, c_{66} (□).

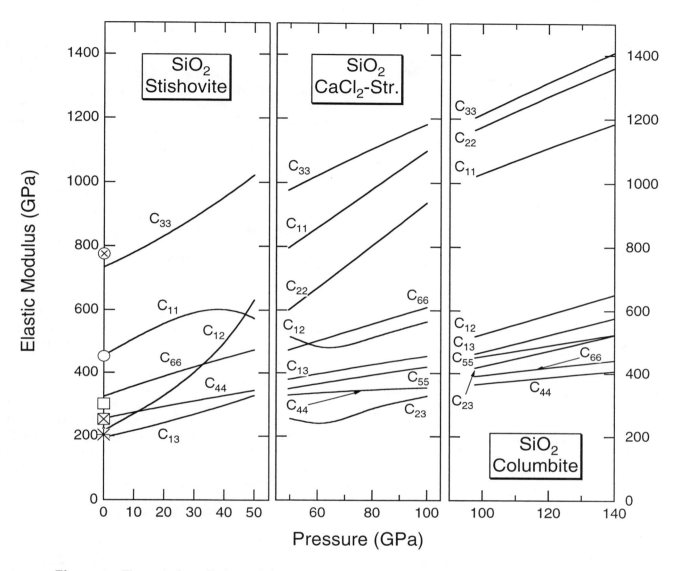

Figure 4. Theoretical predictions of the elastic constants of the three phases of SiO$_2$ [*Karki et al.*, 1997c] (lines) compared with experimental measurements of stishovite [*Weidner et al.*, 1982] symbols: c_{11} ($\circ$); c_{33} ($\otimes$); c_{12} ($\triangle$); c_{13} ($*$); c_{44} ($\times$); c_{66} ($\square$).

1987] or *ab initio* potential induced breathing methods [*Cohen*, 1987; *Isaak et al.*, 1990]. The better agreement is not surprising. Unlike previous approaches, the first principles calculations make no *a priori* assumptions regarding the nature of the electronic structure or bonding. The fact that the pseudopotential method predicts elastic constants in much better agreement with experiment than the modified electron gas model indicates that covalent or angle-dependent forces are important for understanding the elasticity of perovskite.

We attribute remaining differences between theory and experiment to 1) temperature; ambient temper-ature (300 K) is expected to reduce elastic constants by 2-6 % relative to the athermal (0 K, no zero point motion) values calculated theoretically [*Anderson and Isaak*, 1995] 2) the LDA; comparison between LAPW and experiment indicates that this error is of the order of 2 % in volume and 3 % in bulk modulus [*Mehl et al.*, 1988; *Karki et al.*, 1997a] and 3) the pseudopotential approximation. For the particular pseudopotentials used in this study, the latter two errors partially cancel one another. While the LDA tends to underestimate lattice parameters and overestimate elastic constants, the Q_c-tuning pseudopotentials tend to overexpand the lattice

Table 2. Moduli, M_0, and their first pressure derivatives M'_0 at zero pressure from first principles theory.

	MgSiO₃ Perovskite[1]		MgO Periclase[2]	
	M_0 (GPa)	M'_0	M_0 (GPa)	M'_0
c_{11}	493	5.15	291	9.00
c_{22}	523	6.56		
c_{33}	460	6.70		
c_{12}	135	3.33	90.0	1.91
c_{13}	145	2.55		
c_{23}	158	2.73		
c_{44}	201	1.98	135	1.30
c_{55}	183	1.44		
c_{66}	147	1.91		

References: 1, [Karki et al., 1997b]; 2, [Karki et al., 1997a]. Moduli and pressure derivatives are from third-order Eulerian finite strain fits to the first principles results.

relative to all electron LDA calculations by a similar amount. The maximum deviation between theory and experiment is 11 % in the case of c_{44} of periclase. The RMS deviations of the elastic constants between theory and experiments are 5 %, 7 % and 4 %, respectively for perovskite, periclase and stishovite.

Theory predicts that silica undergoes two phase transformations within the pressure regime of the lower mantle [Karki et al., 1997d]. Stishovite is predicted to transform to a CaCl₂ type structure at 47 GPa, in

essentially perfect agreement with all electron calculations and experiment [Cohen, 1992; Kingma et al., 1995]. This transition involves a shear-mode softening $(c_{11}-c_{12} \to 0)$, which has a major effect on the acoustic velocities of isotropic and anisotropic silica-containing aggregates. The phase of silica relevant for understanding the base of the lower mantle is isostructural with columbite (and α-PbO₂) to which the CaCl₂-type phase is predicted to transform at 98 GPa. This prediction is consistent with recent experimental data [Kingma et al., 1996].

The predicted elastic constants of MgSiO₃ perovskite, periclase and SiO₂ columbite at the pressure of the core-mantle boundary are reported in Table 4.

4.2. Velocities of Isotropic Aggregates

Theory predicts that the elastic wave velocities of mono-phase aggregates of perovskite lie in between those of its constituent oxides (Figs. 5,6). The P- and S-wave velocities (V_P and V_S, respectively) of the columbite structure exceed those of perovskite by 6 and 9 % respectively and those of periclase by 9 and 11 % respectively at pressures corresponding to the base of the mantle.

Silica is the seismically fastest material throughout the pressure regime of the lower mantle except near 47 GPa, where it is the slowest. This unusual behavior is caused by the phase transition from stishovite to the CaCl₂-type structure. This phase transition involves a shear-mode softening, and a rapid decrease

Table 3. Moduli, M_0, and their first and second pressure derivatives M'_0, M''_0 at zero pressure from first principles theory[1].

	SiO₂ Stishovite			SiO₂ CaCl₂-type	SiO₂ Columbite-type	
	M_0 (GPa)	M'_0	M''_0 (GPa⁻¹)	M (GPa)	M_0 (GPa)	M'_0
c_{11}	456	5.32	0.000	794	557	5.90
c_{22}				602	624	6.78
c_{33}	734	4.31	0.028	988	640	7.04
c_{12}	216	5.32	0.000	516	196	3.57
c_{13}	195	2.03	0.011	382	181	3.16
c_{23}				255	155	2.91
c_{44}	254	2.09	(-0.012)	332	230	1.88
c_{55}				350	245	2.59
c_{66}	325	3.30	(-0.017)	475	237	2.08

References: 1, [Karki et al., 1997c]. Moduli and pressure derivatives are from Eulerian finite strain fits to the theoretical results, except for CaCl₂ for which we tabulate values of the elastic constants at a pressure of 50 GPa. We used third-order fits for columbite and c_{44} and c_{66} of stishovite and fourth order for c_{33} and c_{13} of stishovite. Second derivatives in parentheses are from third order fits. The pressure dependence of c_{11} and c_{12} of stishovite is represented by $(c_{11} + c_{12})/2 = 336 + 5.32P$ and $(c_{11} - c_{12})/2 = 120[1 - (P/47)^{3.5}]$, where P is pressure in GPa.

Table 4. Elastic Constants at $P = 136$ GPa.

	MgSiO$_3$ Perovskite	Periclase	SiO$_2$ Columbite
c_{11}	996	1388	1171
c_{22}	1250		1341
c_{33}	1237		1391
c_{12}	577	286	639
c_{13}	463		566
c_{23}	495		513
c_{44}	398	192	403
c_{55}	300		518
c_{66}	355		437

in V_S and V_P over a narrow pressure interval. The lower bound on V_S vanishes at the transition while the Hashin-Shtrikman average value decreases by 50 % between 40 and 47 GPa. Once the transition is complete, velocities return to values similar to those far from the transition instantaneously, at a single pressure.

The P- and S-wave velocities of silica exhibit another discontinuous change at the transition from the CaCl$_2$- to the columbite-type phase at 98 GPa (2210 km depth). The P-wave velocity increases slightly at this transition (by 1 %) and the S-wave velocity decreases by a similar amount.

4.3. Single Crystal Anisotropy

The single crystal anisotropy of all three materials is predicted to depend strongly on pressure (Fig. 7). Not only do the magnitudes of P- and S-wave anisotropy change significantly over the pressure regime of the lower mantle, but the sense of anisotropy changes as well. In perovskite, the direction of slowest P-wave propagation is [001] at 0 GPa, but [100] at 140 GPa. This behavior can be understood in terms of the elastic constants and reflects the fact that of the longitudinal elastic moduli (c_{ii} for $i \leq 3$), c_{33} is the least at ambient pressure but c_{11} is the least at 140 GPa.

Periclase shows a larger change in anisotropy. The anisotropy is predicted to decrease with pressure initially, vanishing at 20 GPa before increasing again at high pressures. This behavior is not related to a structural or phase transition but results from the interchange of fast and slow directions as pressure increases. For example, the fast direction of P-wave propagation at zero pressure [111], becomes the slow direction above 20 GPa. Similar interchanges in the fast and slow directions of S-wave propagation and polarization occur at the same pressure. In a cubic material this behavior

can be understood in terms of a single combination of elastic constants $c_{11} - c_{12} - 2c_{44}$ which changes sign at 20 GPa and can be shown to determine P- and S-wave azimuthal and polarization anisotropy. This quantity is negative at ambient pressures but becomes positive at higher pressures because c_{11} depends on pressure much more strongly than c_{12} or c_{44}.

Silica shows the largest anisotropy and the largest change in anisotropy with pressure. The azimuthal S-wave anisotropy of stishovite changes by nearly a factor of 2 between 0 and 40 GPa, rising to a value of 190 % near the transition to the CaCl$_2$ structure. The anisotropy of the CaCl$_2$ structure is also large (60 % S-wave azimuthal anisotropy at 60 GPa). The anisotropy of the columbite structure is substantially smaller, but still greater than that of perovskite in the D'' layer.

4.4. Anisotropic Aggregates

Transverse and azimuthal anisotropy depend strongly on pressure so that measurements at ambient conditions provide little guidance as to the anisotropy at the base of the lower mantle (Fig. 8). For example, in periclase, the P- and S-wave anisotropies reverse sign at 20 GPa. This sign reversal and the vanishing of the single crystal anisotropy at the same pressure have the same origin; namely a change in sign of $c_{11} - c_{12} - 2c_{44}$. The magnitude of transverse P- and S-wave anisotropy in periclase

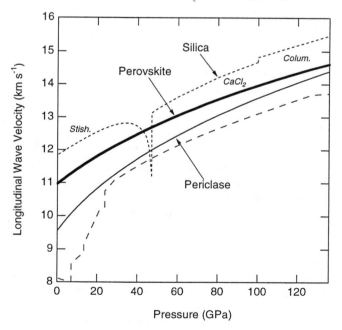

Figure 5. Longitudinal (P) wave velocity of isotropic mono-phase aggregates compared with the P-wave velocity of the lower mantle (long dashed line).

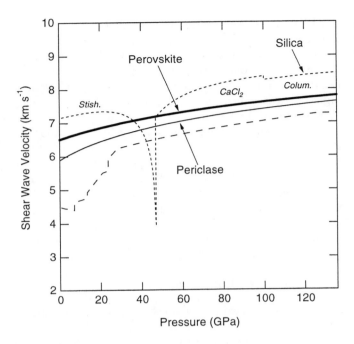

Figure 6. Shear (S) wave velocity of isotropic mono-phase aggregates compared with the S-wave velocity of the lower mantle (long dashed line).

at the base of the mantle are approximately 2 % and 6 %, respectively. The $S_\parallel/S_\perp$ anisotropy is substantially larger and reaches a value of 16 % at the base of the mantle.

The anisotropy of perovskite aggregates also depends strongly on pressure. The SH/SV anisotropy is essentially zero at ambient pressure but reaches -9 % at the base of the mantle. The slower SH velocity results from the much weaker dependence of c_{55} on pressure and its smaller value at high pressure compared with that of c_{44} and c_{66}. The $S_\parallel/S_\perp$ anisotropy reflects the ratio of c_{66} to c_{44} and varies from -15 % at ambient pressure to -5 % at the base of the mantle. Only the PH/PV anisotropy changes by a relatively small amount with pressure, from -4 % to -6 %. PH/PV anisotropy is largely controlled by the difference between c_{22}, corresponding to propagation along the b-axis, the vertical direction for the slip plane assumed here, and the average of c_{11} and c_{33}.

Silica aggregates generally show the largest variation with pressure. This is due to the two phase transitions, and to the anomalous behavior of stishovite and $CaCl_2$ in the vicinity of the phase transition at 47 GPa. Shear mode softening in the stishovite phase leads to the largest magnitudes of S-wave anisotropy of any phase studied here, up to -28 % in $S_\parallel/S_\perp$. Although large, this

anisotropy is substantially less than the single-crystal S-wave anisotropy. The reason for the difference is that the dominant slip plane does not coincide with the plane of shear-mode softening: the phase transition is driven by a rotation of the octahedra in the $a - b$ plane (001) while experiments on rutile indicate that {101} are the dominant slip planes.

5. DISCUSSION

5.1. The Lower Mantle

Our results have allowed us to compare, for the first time, the P- and S- wave velocities of major mantle phases with observed seismological properties of the lower mantle. Comparison with the isotropic homogeneous (radial) structure of the lower mantle shows that the P- and S-wave velocity profiles of $MgSiO_3$ perovskite are nearly parallel to those of the lower mantle [*Karki et al.*, 1997b]. Further, the velocities of this mineral are similar in magnitude to those of the lower mantle. Theoretical P- and S-wave velocities are higher by 6 and 8 %, respectively, than those observed seismologically. We attribute the differences between theoretical and seismological velocities to 1) the effect of Fe, which is expected to lower the velocities of perovskite by 1-2 % 2) the high temperatures of the lower mantle which are expected to reduce velocities by several percent relative to the athermal values determined theoretically and 3) the possible presence of other phases in the lower mantle including magnesiowüstite, $CaSiO_3$ perovskite, silica, and garnet.

This result represents an important test of the prevailing view that Mg-rich silicate perovskite is the most abundant mineral in the lower mantle. This hypothesis has been supported in the past by only a limited subset of the observed seismological properties of the lower mantle: the density and bulk modulus [*Jeanloz and Knittle*, 1989; *Bukowinski and Wolf*, 1990; *Stixrude et al.*, 1992; *Bina and Silver*, 1997]. Our results show for the first time that those properties which are most sensitively measured by seismic observations, V_P and V_S, are also consistent with a perovskite rich lower mantle. This is also significant because V_P and V_S are expected to be more sensitive to bulk composition and mineral structure than are density and bulk modulus.

The predicted velocities of periclase indicate that magnesiowüstite will have very low seismic velocities in the lower mantle. The effect of iron on the P- and S-wave velocities of magnesiowüstite is expected to be large both because of its large effect on the density and because magnesiowüstite is expected to be enriched in

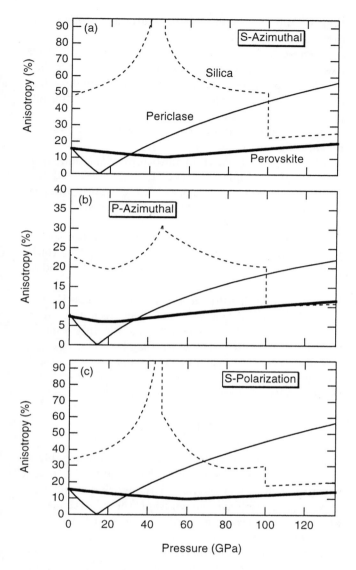

Figure 7. Single crystal anisotropy of MgSiO₃ perovskite (bold lines), periclase (light solid lines), and silica (dashed lines): a) azimuthal anisotropy of S-waves b) azimuthal anisotropy of P-waves c) maximum polarization anisotropy.

certainties in measured quantities such as the magnesiowüstite-perovskite Fe partition coefficient (K_{pv-mw}), and 4) uncertainties in the iron content of the lower mantle. Our estimates are based on an assumed bulk iron content of the lower mantle $X_{Fe} = \frac{Fe}{Mg+Fe} = 8-12$ %, and a range of values for the partition coefficient $K_{pv-mw} = 1/6 - 1/3$.

The very low seismic wave velocities of magnesiowüstite suggested by our results promise a resolution of the long-standing question of the major element bulk composition of the lower mantle. Despite the remaining

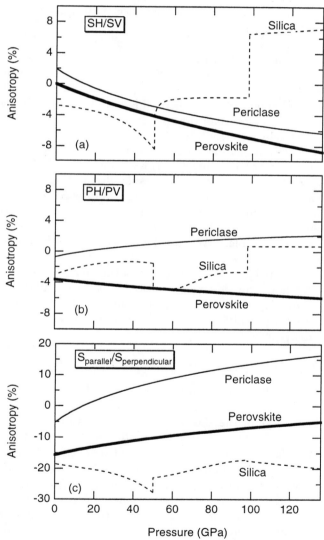

Figure 8. Anisotropy of mono-phase aggregates. a) $SH/SV - 1$ b) $PH/PV - 1$ and c) $S_\parallel/S\perp - 1$. Bold lines are for MgSiO₃ perovskite, light solid lines for periclase and dashed lines for silica.

iron compared with the bulk lower mantle. Existing data indicates that iron may lower the P- and S-wave velocities of magnesiowüstite by 6-12 and 8-16 %, respectively, relative to that of periclase. These estimates, primarily based on ambient pressure data are uncertain because 1) there are no experimental or theoretical results on the elasticity of Fe-bearing magnesiowüstites at high pressure 2) phase transitions in end-member FeO suggest that high pressure behavior may differ significantly from that at ambient conditions and 3) un-

uncertainties in the effect of Fe, it seems clear that the P- and S-wave velocities of magnesiowüstite will be substantially lower than that of perovskite at lower mantle conditions; the S-wave velocities may differ by 15 %. In contrast, the density of these two phases differ by only 2.5 % on average throughout the lower mantle. This small difference, combined with the fact that the density is less well determined seismologically than either V_P or V_S accounts for the difficulty of resolving the question of lower mantle composition on the basis of density alone. The relatively large difference in V_S and V_P between perovskite and magnesiowüstite reflects the sensitivity of the shear modulus to crystal structure and bulk composition.

The seismic wave velocities of silica remain significantly higher than those of the lower mantle throughout most of the relevant pressure regime (by as much as 20 % in V_S). However, for pressures near 50 GPa (1200 km depth), the athermal S-wave velocity of silica is 20 % smaller than that of the lower mantle. The unusual velocity structure associated with the stishovite to $CaCl_2$ transition - a rapid and large decrease in velocities with pressure (depth) followed by a sudden and equally large increase - is capable of reflecting seismic energy in the frequency band typically observed seismically. This presents the possibility that even small amounts of silica in the lower mantle may be seismically detectable. We have found that a lower mantle with as little as 2 volume percent silica would show significant reflectivity at the depth of the phase transition. This may provide an explanation of at least a subset of seismically reflective features within the lower mantle at depths of 710-785 km, 920 km, and 1180 km [*Karki et al.*, 1997c].

5.2. Anisotropy in D''

Hypotheses concerning the origin of anisotropy in D'' have been difficult to test because of our lack of knowledge of the elastic constants of possible D'' constituents. Our results allow us to test the hypothesis that the anisotropy originates in lattice preferred orientation. This hypothesis is motivated by the expected, nearly mono-mineralic composition of the lower mantle (75-100 volume % Mg-rich silicate perovskite) and by the expectation that deviatoric stress due to mantle flow are concentrated in the mantle's dynamical boundary layers.

We find that perovskite, periclase, and silica all have anisotropy of sufficient magnitude to explain seismological observations. Observations show that the anisotropy in D'' is typically less than 2 %, significantly smaller than the upper bounds on aggregate anisotropy found here (6-8 %, Fig. 8).

Our results show that lattice-preferred orientation is likely to produce anisotropy such that SV is faster than SH. Mono-phase aggregates of both perovskite and periclase show this sense of anisotropy. The columbite phase of silica has the opposite sense of anisotropy, but is unlikely to be a major constituent of D'' because of its very high velocities.

Seismological studies find anisotropy of the same sense ($SV > SH$) in some regions of D'' [*Pulliam and Sen*, 1998]. In these regions, theory indicates that lattice preferred orientation of a perovskite dominated aggregate is responsible for the observations.

In other regions of D'', SH is found to be faster than SV [*Kendall and Silver*, 1996; *Garnero and Lay*, 1997]. In these regions, theory supports explanations based on shape preferred orientation. This class of explanations is much more difficult to test at present because it relies on elastic constants of materials that are not yet measured or predicted. For example, one possible origin of lamellar structures in D'' which may be able to explain the observations are chemical reactions between mantle and core [*Knittle and Jeanloz*, 1991]. Aside from silica, the elasticity of the reaction products are unknown.

6. CONCLUSIONS

Modern first principles methods are now capable of realistic predictions of the elastic constants of complex silicates such as perovskite. Although independent of experiment, we have shown that these methods are able to reproduce even subtle features such as elastic anisotropy with good accuracy. We note that it is possible to extend these methods to high temperatures, although we have not done so here. Modern electronic structure theory thus promises an ideal complement to the experimental approach, and the ability to determine the elasticity and anisotropy of major constituents throughout the pressure and temperature regime of the earth's mantle.

Investigation of the deformation mechanisms of high pressure phases will be another important avenue of future research. This knowledge is critical for understanding the development of lattice preferred and in some cases, shape preferred orientation due to mantle flow. This area of investigation is now accessible to theory and experiment.

Acknowledgments. This work supported by the National Science Foundation under grant EAR-9628199.

REFERENCES

Anderson, O. L., and D. G. Isaak, Elastic constants of mantle minerals at high temperature. In T. J. Ahrens, editor, *Mineral Physics and Crystallography: A Handbook of Physical Constants*, pages 64–97. American Geophysical Union, Washington, DC, 1995.

Ashbee, K. H. G., and R. E. Smallman, The plastic deformation of titanium dioxide single crystals. *Proceedings of the Royal Society of London, Series A, 274*, 195–205, 1963.

Bass, J. D., Elasticity of minerals, glasses, and melts. In T. J. Ahrens, editor, *Mineral Physics and Crystallography: A Handbook of Physical Constants*, pages 45–63. American Geophysical Union, Washington, DC, 1995.

Bina, C. R., and P. G. Silver, Bulk sound travel times and implications for mantle composition and outer core heterogeneity. *Geophysical Research Letters, 24*, 499–502, 1997.

Bukowinski, M. S. T., and G. H. Wolf, Thermodynamically consistent decompression - implications for lower mantle composition. *Journal of Geophysical Research, 95*, 12583–12593, 1990.

Chen, G. L., B. S. Li, and R. C. Liebermann, Selected elastic moduli of single-crystal olivines from ultrasonic experiments to mantle pressures. *Science, 272*, 979–980, 1996.

Chin, G. Y., and W. L. Mammel, A theoretical examination of the plastic deformation of ionic crystalsii. analysis of uniqcial deformation and axisymmetric flow for slip on $(110)\langle100\rangle$ and $(100)\langle110\rangle$ systems. *Metallurgical Transactions, 4*, 335–340, 1973.

Cohen, R. E., Elasticity and equation of state of $MgSiO_3$ perovskite. *Geophysical Research Letters, 14*, 1053–1056, 1987.

Cohen, R. E., Bonding and elasticity of stishovite SiO_2 at high pressure: linearized augmented plane wave calculations. *American Mineralogist, 76*, 733–742, 1991.

Cohen, R. E., First-principles predictions of elasticity and phase transitions in high pressure SiO_2 and geophysical implications. In Y. Syono and M. H. Manghnani, editors, *High-Pressure Research: Applications to Earth and Planetary Sciences*, pages 425–431. TERRAPUB, Tokyo, 1992.

da Silva, C., L. Stixrude, and R. M. Wentzcovitch, Elastic constants and anisotropy of forsterite at high pressure. *Geophysical Research Letters, 24*, 1963–1966, 1997.

Duffy, T. S., C. S. Zha, R. T. Downs, and H. K. Mao et al., Elasticity of forsterite to 16 gpa and the composition of the upper mantle. *Nature, 378*, 170–173, 1995.

Garnero, E. J., and T. Lay, Lateral variations in lowermost mantle shear wave anisotropy beneath the north pacific and alaska. *Journal of Geophysical Research, 102*, 8121–8135, 1997.

Gunnarsson, O., and B. I. Lundqvist, Exchange and correlation in atoms, molecules, and solids by the spin-density-functional formalism. *Physical Review B, 13*, 4274–4298, 1976.

Hashin, Z., and S. Shtrikman, A variational approach to the theory of the elastic behavior of polycrystals. *Journal of the Mechanics and Physics of Solids, 10*, 343–352, 1962.

Hill, R., The elastic behavior of a crystalline aggregate. *Proceedings of the Physical Society, London, 65A*, 349–354, 1952.

Hohenberg, P., and W. Kohn, Inhomogeneous electron gas. *Physical Review, 136*, B864–B871, 1964.

Isaak, D. G., R. E. Cohen, and M. E. Mehl, Calculated elastic constants and thermal properties of MgO at high pressures and temperatures. *Journal of Geophysical Research, 95*, 7055–7067, 1990.

Jackson, I., and H. Niesler, The elasticity of periclase to 3 gpa and some geophysical implications. In S. Akimoto and M. H. Manghnani, editors, *High-Pressure Research in Geophysics*, pages 93–133. Center for Academic Publications, Tokyo, 1982.

Jeanloz, R., and E. Knittle, Density and composition of the lower mantle. *Philosophical Transactions of the Royal Society of London Series A, 328*, 377–389, 1989.

Karato, S., S. Zhang, and H.-R. Wenk, Superplasticity in earth's lower mantle: evidence from seismic anisotropy and rock physics. *Science, 270*, 458–461, 1995.

Karki, B. B., L. Stixrude, S. J. Clark, M. C. Warren, G. J. Ackland, and J. Crain, Structure and elasticity of MgO at high pressure. *American Mineralogist, 82*, 51–60, 1997a.

Karki, B. B., L. Stixrude, S. J. Clark, M. C. Warren, G. J. Ackland, and J. Crain, Elastic properties of orthorhombic $MgSiO_3$ perovskite at lower mantle pressures. *American Mineralogist, 82*, 635–638, 1997b.

Karki, B. B., L. Stixrude, M. C. Warren, G. J. Ackland, and J. Crain, *Ab initio* elasticity of three high-pressure polymorphs of silica. *Geophysical Research Letters, 24*, 3269–3272, 1997c.

Karki, B. B., M. C. Warren, L. Stixrude, G. J. Ackland, and J. Crain, *Ab initio* studies of high-pressure structural transformations in silica. *Physical Review B, 55*, 3465–3471, 1997d.

Kendall, J. M., and P. G. Silver, Constraints from seismic anisotropy on the nature of the lowermost mantle. *Nature, 381*, 409–412, 1996.

Kiefer, B., L. Stixrude, and R. M. Wentzcovitch, Elastic constants and anisotropy of Mg_2SiO_4 spinel at high pressure. *Geophysical Research Letters, 24*, 2841–2844, 1997.

Kingma, K. J., R. E. Cohen, R. J. Hemley, and H. K. Mao, Transformation of stishovite to denser phase at lower-mantle pressures. *Nature, 374*, 243–245, 1995.

Kingma, K., H. K. Mao, and R. J. Hemley, Synchrotron x-ray diffraction of SiO_2 to multimegabar pressures. *High Pressure Research, 14*, 363–374, 1996.

Knittle, E., and R. Jeanloz, Earth's core-mantle boundary - results of experiments at high pressures and temperatures. *Science, 251*, 1438–1443, 1991.

Kohn, W., and L. J. Sham, Self-consistent equations including exchange and correlation effects. *Physical Review, 140*, A1133–A1138, 1965.

Kresse, G., J. Hafner, and R. J. Needs, Optimized norm-conserving pseudopotentials. *Journal of Physics - Condensed Matter, 4*, 7451–7468, 1992.

Lin, J. S., Q. Qteish, M. C. Payne, and V. Heine, Optimised and transferable non-local separable ab-initio pseudopotentials. *Physical Review B, 47*, 4174–4180, 1993.

Lundqvist, S., and N. H. March, *Theory of the Inhomogeneous Electron Gas*. Plenum Press, London, 1987.

Matsui, M., M. Akaogi, and T. Matsumoto, Computational

model of the structural and elastic properties of the ilmenite and perovskite phases of MgSiO$_3$. *Physics and Chemistry of Minerals, 14*, 101–106, 1987.

Matzel, E., M. K. Sen, and S. P. Grand, Evidence for anisotropy in the deep mantle beneath alaska. *Geophysical Research Letters, 23*, 2417–2420, 1996.

Meade, C., P. G. Silver, and S. Kaneshima, Laboratory and seismological observations of lower mantle isotropy. *Geophysical Research Letters, 22*, 1293–1296, 1995.

Mehl, M. J., R. E. Cohen, and H. Krakauer, Linearized augmented plane wave electronic structure calculations for MgO and CaO. *Journal of Geophysical Research, 93*, 8009–8022, 1988.

Meister, R., and L. Peselnick, Variational method of determining effective moduli of polycrystals with tetragonal symmetry. *Journal of Applied Physics, 37*, 4121–4125, 1966.

Monkhurst, H. J., and J. D. Pack, Special points for brillouin-zone integrations. *Physical Review B, 13*, 5188–5192, 1976.

Nielsen, O. H., and R. Martin, Quantum mechanical theory of stress and force. *Physical Review B, 32*, 3780–3791, 1985.

Nye, J. F., *Physical Properties of Crystals: Their Representation by Tensors and Matrices*. Oxford, Oxford, UK, 2 edition, 1985.

Park, J., and Y. Yu, Seismic determination of elastic anisotropy and mantle flow. *Science, 261*, 1159–1162, 1993.

Pickett, W. E., Pseudopotentials in condensed matter systems. *Comp. Phys. Rep., 9*, 114–197, 1989.

Pulliam, J., and M. K. Sen, Seismic anisotropy in the core-mantle transition zone. *Geophysical Journal International*, in press, 1998.

Stixrude, L., and R. E. Cohen, Stability of orthorhombic MgSiO$_3$-perovskite in the earth's lower mantle. *Nature, 364*, 613–161, 1993.

Stixrude, L., R. J. Hemley, Y. Fei, and H. K. Mao, Thermoelasticity of silicate perovskite and magnesiowustite and stratification of the earth's mantle. *Science, 257*, 1099–1101, 1992.

Troullier, N., and J. L. Martins, Efficient pseudopotentials for plane-wave calculations. *Physical Review B, 43*, 1993–2003, 1991.

Vanderbilt, D., Soft self-consistent pseudopotentials in a generalized eigenvalue formalism. *Physical Review B, 41*, 7892–7895, 1990.

Vinnik, L., B. Romanowicz, Y. LeStunff, and L. Makeyeva, Seismic anisotropy in the d″ layer. *Geophysical Research Letters, 22*, 1657–1660, 1995.

Wang, Y. B., D. J. Weidner, R. C. Liebermann, and Y. S. Zhao, P-V-T equation of state of (Mg,Fe)SiO$_3$ perovskite - constraints on composition of the lower mantle. *Physics of the Earth and Planetary Interiors, 83*, 13–40, 1994.

Watt, J. P., G. F. Davies, and R. J. O'Connell, The elastic properties of composite materials. *Reviews of Geophysics and Space Physics, 14*, 541–563, 1976.

Weidner, D. J., J. D. Bass, A. E. Ringwood, and W. Sinclair, The single-crystal elastic moduli of stishovite. *Journal of Geophysical Research, 87*, 4740–4746, 1982.

Wentzcovitch, R. M., Invariant molecular dynamics approach to structural phase transitions. *Physical Review B, 44*, 2358–2361, 1991.

Wentzcovitch, R. M., and L. Stixrude, Crystal chemistry of forsterite: a first principles study. *American Mineralogist, 82*, 663–671, 1997.

Wentzcovitch, R. M., J. L. Martins, and G. D. Price, *Ab initio* molecular dynamics with variable cell shape: application to MgSiO$_3$ perovskite. *Physical Review Letters, 70*, 3947–3950, 1993.

Wentzcovitch, R. M., N. L. Ross, and G. D. Price, Ab initio study of MgSiO$_3$ and CaSiO$_3$ perovskites at lower-mantle pressures. *Physics of the Earth and Planetary Interiors, 90*, 101–112, 1995.

Yeganeh-Haeri, A., Synthesis and re-investigation of the elastic properties of single-crystal magnesium silica perovskite. *Physics of the Earth and Planetary Interiors, 87*, 111–121, 1994.

Zaug, J. M., E. H. Abramson, J. M. Brown, and L. J. Slutsky, Sound velocities in olivine at earth mantle pressures. *Science, 260*, 1487–1489, 1993.

L. Stixrude, Department of Geological Sciences, 425 E. University Av., University of Michigan, Ann Arbor, MI 48109-1063

Investigating Causes of D'' Anistropy

J–M. Kendall

Department of Earth Sciences, University of Leeds, Leeds, England.

P. G. Silver

Department of Terrestrial Magnetism, Carnegie Institution of Washington, Washington, D.C., USA

Evidence of seismic anisotropy in the D'' region holds tantalizing insights into the mineralogy and dynamics of this core–mantle boundary–layer. In this chapter we summarize the observations to date, characterize the type of anisotropy and finally consider the physical processes that may have produced this seismological signature. Although studies of D'' anisotropy are still a long way from achieving global coverage, it is clear that there are strong lateral variations in the degree of anisotropy, with large regions appearing to be isotropic. Reliable interpretations of this anisotropy require corrections for receiver–side upper–mantle anisotropy and complications due to source–side anisotropy can be mitigated using only deep (>500 km) events. To date, regions below the Caribbean and the north Pacific / Alaska show clear evidence for D'' anisotropy, while observations in regions beneath the central Pacific are less straightforward. The primary seismic constraints in the Caribbean and Alaskan regions are: (1) S/ScSH–phases arrive earlier than S/ScSV–phases (2) SKS does not appear to be sensitive to D'' anisotropy (3) these are regions of high seismic velocities and a D'' discontinuity. Transverse isotropy best describes the style of anisotropy in these regions. Two candidate anisotropy mechanisms are considered: (1) that due to lattice–preferred orientation (LPO) of constituent minerals and (2) that due to a shape–preferred orientation (SPO). It is difficult to explain the observations with the LPO mechanism. We use effective–medium modeling to investigate the more likely SPO–anisotropy, which is attributed to oriented inclusions within a matrix of contrasting seismic properties. The seismic constraints force us to conclude that the inclusions must be horizontally–aligned tabular bodies (disks or layers). It seems unlikely that the physical process responsible for this anisotropy is associated with infiltration of core material. Instead, there are a number of arguments which suggest that the anisotropy is associated with the hypothesis that, in places, D'' represents a graveyard for subducted material. High aggregate shear–velocities can be explained by the retained thermal anomaly of the slab and the anisotropy

1. INTRODUCTION

Recent observations of seismic anisotropy in the lowermost mantle layer, D'', constitute a new and intriguing feature of this enigmatic region. While these observations are very interesting from a seismological point

The Core-Mantle Boundary Region
Geodynamics 28
Copyright 1998 by the American Geophysical Union.

can be explained by contrasts in the material properties between what was formerly oceanic–crust and oceanic–mantle–lithosphere.

of view, they are equally if not more valuable as a geo-dynamic constraint on this region. In this chapter, we briefly summarize the observations to date, character-ize the type of anisotropy and finally consider the physi-cal processes that may have produced this seismological signature.

There are numerous regions of the Earth where seismic anisotropy is present. In the upper–mantle, anisotropy is attributed to the flow– or strain–induced lattice preferred orientation (LPO) of constituent min-erals, especially olivine. Another common form of anisotropy is that due to a shape preferred orienta-tion (SPO). In this case, oriented inclusions with one seismic velocity are embedded in a matrix character-ized by a second velocity. This generates an effectively anisotropic medium if the inclusions are small compared to a seismic wavelength. This is known to occur, for example, in the continental crust where it is is gener-ally attributed to the preferred alignment of fluid–filled cracks. Several effective–medium theories exist for esti-mating the elasticity of a medium given the inclusion or crack attributes [e.g., *Hudson*, 1980; *Tandon and Weng*, 1984]. The limiting case of large–aspect–ratio inclusions is periodic variations of thin layers of contrasting veloc-ities [*Backus*, 1962]. Regardless of which mechanism is operating, the presence of anisotropy suggests an or-dered medium, with a particular fabric or texture. This ordering in turn points to an underlying physical pro-cess, such as mantle flow. We may thus use anisotropy to understand this underlying process.

There are some important differences between wave propagation in weakly anisotropic media and in isotropic media. First, the wave velocity varies as a function of propagation direction. Second, the ray direction, parti-cle motion (for a P–wave) and wavefront normal are no longer coincident. Three, there are three distinct body waves, a P–wave and a slow and fast S–wave, rather than two (the familiar P– and S–wave). Indeed, it is the observation of two time–separated quasi–orthogonally polarized S–wave arrivals that is the most robust man-ifestation of wave propagation through an anisotropic zone. This phenomenon is commonly know as shear–wave splitting and we utilize this in the present study.

Initial suggestions of anisotropy in the D'' region were based primarily on splitting in core–diffracted shear–waves [*Vinnik et al.*, 1989a; 1995; *Lay and Young*, 1991 and the presence of anomalously large amplitudes for the vertically–polarized diffracted–waves (SVdiff) [*Vinnik et al.*, 1989a]. Interpretation of these effects

as due to anisotropy is not straightforward. For ex-ample, detailed modeling by *Maupin* [1995] showed that isotropic models could also explain the anomalous SVdiff–amplitudes. More recently, it has been possi-ble to examine this region with pre–diffracted arrivals. *Kendall and Silver* [1996a], for example, observed dra-matic splitting in shear–waves which turn in the low-ermost 250 km of the mantle beneath the Caribbean and Central America at pre–diffracted distances. The anisotropy was constrained to the D'' region through analyses of phases turning above and within the layer. *Matzel et al.* [1996] and *Garnero and Lay* [1997] ob-served splitting in pre–diffracted phases beneath the northern Pacific and Alaska. It is of historical inter-est to note that *Mitchell and Helmberger* [1973] noticed a travel–time separation in the arrival of the radial and transverse ScS–phases, although they did not attribute this to anisotropy (see *Lay et al.* [1998, this issue] for a historical perspective).

In the first two sections we review observations of seismic anisotropy in D''. While we are still a long way from a global picture of the anisotropy, it is already clear that its spatial distribution is non–uniform; it has been detected in some regions but not in others. This tells us that 1) anisotropy is not a general global feature of D'' and 2) that the actual distribution may be related to the underlying cause of the anisotropy. In the following sections we investigate the two most plausible causes of the anisotropy, mineral alignment or inclusion align-ment. Finally, we examine possible physical process for producing the anisotropy.

2. DATA PROCESSING AND CONSTRAINING THE ANISOTROPY TO THE LOWERMOST MANTLE

Observations of shear–wave splitting in phases which transit D'' can be diagnostic of anisotropy in this lower-most boundary–layer. Figure 1 shows a clear and large (> 5 secs) separation between the radial and transverse components of the shear–phases which turn within the D'' layer. At these epicentral distances the core reflec-tion, ScS, arrives only a few seconds behind S which makes it difficult to pick its onset. We therefore call this combined phase S/ScS. The SKS energy on the transverse component is evidence of mantle anisotropy as an isotropic 1D earth would not generate any SKSH energy, although it is clearly much less (<2s) than the S/ScS splitting. This earthquake is very deep (601 km)

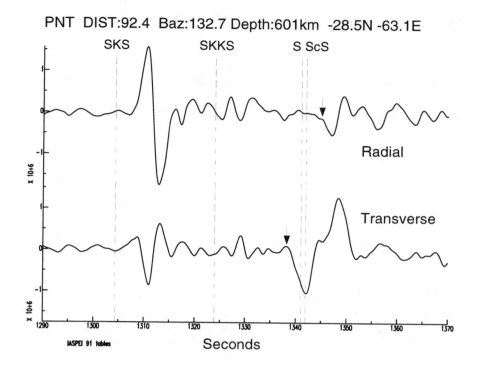

PNT DIST:92.4 Baz:132.7 Depth:601km -28.5N -63.1E

Figure 1. Example of clear S/ScS–splitting. The velocity records are for a 601 km deep earthquake beneath Argentina (1994–05–10) recorded 92.4° away at the station PNT in Canada. Significant SKS–energy on the transverse component is indicative of upper–mantle anisotropy at this station, but the degree of splitting us much less than that for the S/ScS–arrival.

and the S/ScS–phase and the SKS–phase have almost identical paths in the upper–mantle. Therefore, the cause of the additional splitting in the S/ScS–phase must be in the lower–mantle.

The effects of anisotropy in other regions of the Earth must be accounted for when interpreting anisotropy in D''. It is well known that the upper–mantle is seismically anisotropic and can generate over 2 seconds of splitting in SKS–phases [*Silver*, 1996]. Any interpretation of D'' anisotropy must correct for this upper–mantle anisotropy. There are some cases where even small degrees of upper–mantle induced splitting can seriously bias interpretations of lower–mantle anisotropy, as discussed below. A possible contribution of upper–mantle anisotropy near the source region is avoided by confining our analysis to events that are well below the olivine stability field (below 400 km). We thus have restricted our analyses to events which are deeper than 500 km.

Estimates of SKS–splitting from many events recorded at a given station can be used as a correction for upper–mantle anisotropy. What is now standard splitting–analysis can be applied to estimate the degree of SKS–

splitting and the polarization of the fast and slow shear–waves [e.g., *Vinnik et al.*, 1989b; *Silver and Chan*, 1991]. In order to get meaningful estimates of the splitting delay times in the S/ScS arrival it is crucial to make this upper–mantle anisotropy correction. Figure 2 illustrates how this correction can also help clarify the onset of each phase. It is important to note that this correction must be applied even if there is little transverse component SKS or SKKS energy, as the source–receiver back–azimuth may lie near the polarization azimuth of one of the shear–waves. In such a case there will be little SKS energy on the transverse component, but the S/ScS splitting times will be still affected by the upper–mantle anisotropy. Analyses of SKS splitting at some stations reveal azimuthal variations in upper–mantle anisotropy. The splitting parameters for the relevant azimuth should be used when making the upper–mantle correction. In practice, this is most easily done by performing the SKS analysis on the seismogram being analyzed.

Anisotropy in the upper–mantle can generate an apparent SVdiff arrival because this phase is normally very weak for arrivals beyond 110 degrees. Figure 3 shows

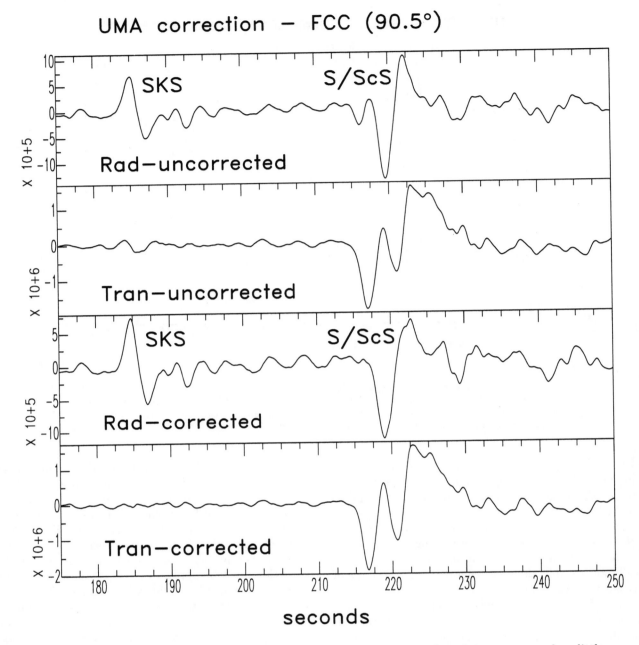

Figure 2. Example of the importance of correcting for upper–mantle anisotropy even when little transverse–component SKS–energy is visible. This event (1994–04–29) was 562 km beneath Argentina and recorded 90.5° away from the station FCC in Canada. The top 2 traces show the raw velocity data, while the lower 2 have been corrected for upper–mantle anisotropy. Note that the onset of the S/ScS phase on the radial component is much clearer after the correction.

how this apparent SVdiff vanishes when the upper–mantle correction is applied. In light of this sensitivity, we avoid phases which have diffracted substantially into the core shadow in our analyses. This also helps reduce possible effects due to core coupling that may be different for SVdiff and SHdiff. We note that the uncorrected SVdiff phase will apparently lag behind the SHdiff signal and appear higher in frequency content. This is due to the fact that the apparent SVdiff signal will look like the time derivative of the SHdiff signal.

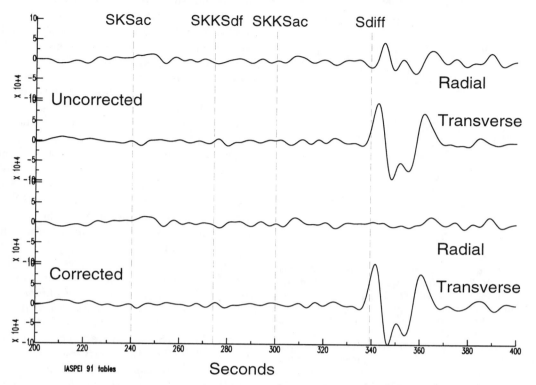

Figure 3. An example of diffracted–shear–wave splitting due to upper–mantle anisotropy. The event (1994–09–02) was 590 km beneath the Fiji Islands region and recorded at the Canadian station RES (109.6°). The top 2 traces show the original data and the lower 2 show the data after the correction for upper–mantle anisotropy. The traces have been band–pass filtered (0.001–0.1Hz). Note the apparent SVdiff signal in the original traces which is absent in the corrected traces. The correction is necessary despite the weak SKS or SKKS signals.

This is analogous to the case of SKS splitting where the SKSH signal looks like the time derivative of the SKSV signal [e.g., *Silver and Chan*, 1988].

The lower mantle above D'' is potentially another area where anisotropy could be present. This region appears, however, to be isotropic, as summarized by *Meade et al.* [1995] and *Silver* [1996]. In addition to this general conclusion, it is possible to utilize specific ray geometries to place tighter constraints on the possible contribution of regions above D''. In particular *Kendall and Silver* [1996a] utilized a source–receiver configuration which allowed investigation of phases which turn above and below the D'' seismic–discontinuity in this region. For example, in Figure 4 the early arriving S–phase at 77° turns above the D'' discontinuity which, in this region, lies 250 km above the core–mantle boundary (CMB) [*Kendall and Nangini*, 1996]. Splitting analysis on this phase shows 0.7 seconds of splitting and is in

very good agreement with SKS results for this station. In contrast there appears to be over 3 seconds of separation between the later arriving ScS–phase on the radial and transverse components. The only significant path difference between these two phases is the ScS segment through the D'' region.

3. REVIEW OF OBSERVATIONS

Studies of D'' anisotropy are, to date, limited in global coverage. Nevertheless, the observations suggest lateral variations in the magnitude and style of the anisotropy which reflect variations in the physical processes causing the anisotropy.

In *Kendall and Silver* [1996a] it was shown that the region beneath the Caribbean and Central America (Figure 5) had on average 1.8% anisotropy. Figure 6 shows the separation between the radial and transverse

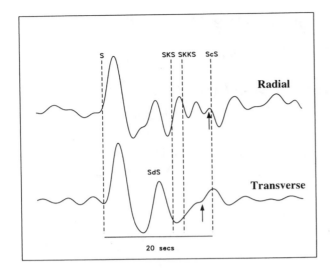

Figure 4. The radial and transverse components recorded at the station DRLN, 77.4° from a 562 km deep earthquake (1994–04–24) in South America. The S–phase turns in the lower mantle, but above the D'' region, and the SdS phase is a reflection from the top of the D'' discontinuity (250 km above the CMB in this region). Note that the transverse component ScS–phase arrives much earlier (3.4 secs) than the radial component ScS–phase indicating passage through an anisotropic region. In contrast, measurements of SKS splitting at this station for a wide range of azimuths and epicentral distances show only 0.7 seconds of separation. The radial component S–phase arrives slightly earlier than the transverse component and splitting analysis shows that this separation (0.7 seconds) can be entirely explained by upper–mantle anisotropy. The ScS raypath differs from the S path only in that it travels through D'' and reflects at the core–mantle boundary, thereby implying anisotropy in D''.

component recordings of the S/ScS arrival for a range of epicentral distances. The separations for the entire data range between 3 and 9 seconds with the transverse signal always leading the radial signal. They also show that the anisotropy exists throughout the D'' layer in this region which is bounded from above by a discontinuity which is anywhere from 250 km to 280 km above the CMB [*Kendall and Nangini*, 1996].

There are areas of inferred D'' anisotropy beneath Alaska and the northern Pacific (hereafter referred to as the Alaskan region). For example, *Lay and Young* [1991], *Matzel et al.* [1996] and *Garnero and Lay* [1997] observe 0–4 seconds of splitting in northwest Pacific events recorded on WWSSN stations and some more recent broadband stations in North America. The studies show that the anisotropy must be distributed throughout the D'' region beneath Alaska and the Aleutian Islands, but *Garnero and Lay* [1997] show that there is

lateral variability in the magnitude of the anisotropy. In a region south of this (Figure 5), *Kendall and Silver* [1996b] observe modest amounts of S/ScS splitting in CNSN (Canadian National Seismograph Network) recordings of a recent event beneath the Solomon Islands, suggesting that the anisotropy is weaker, but present in this region. In these studies the transverse component signal was observed to arrive earlier than the radial signal.

The region beneath the central Pacific is more complicated and a consensus on the form and magnitude of the anisotropy, if any, has not been reached. This region has been studied primarily using events from the Tonga–Fiji region recorded in North America. Vinnik and co–workers [*Vinnik et al.*, 1989a; 1995] have reported D'' anisotropy beneath the Pacific, based on observations of anomalously high-amplitude SVdiff arrivals at epicentral distances beyond 106°. *Vinnik et al.* [1995] note that SVdiff is delayed by roughly a quarter period and that it is higher in frequency content than the SHdiff arrival. In contrast, *Pulliam and Sen* [1996; 1998] have studied Fiji events recorded on stations in south-central USA. At the station HKT in Texas, they observe that S/ScS on the radial component arrives earlier than S/ScS on the transverse component. Figure 7 shows the S/ScS and Sdiff arrivals, after the upper–mantle–anisotropy correction, for a range of epicentral distances between Fiji events and various CNSN stations in Canada. These seismograms show little evidence of D'' anisotropy as the degree of separation between the radial and transverse components is on the order of 1 second, although in some seismograms the onset of the S/ScS phase is not sharp. *Kendall and Silver* [1996b] have studied Fiji events recorded by the BANJO array [*Beck et al.*, 1995] in Bolivia and Chile and see no evidence for splitting in D'' implying that the region beneath the southern Pacific is isotropic (see Figure 5).

4. ARGUMENTS FOR TRANSVERSE ISOTROPY

The form of the anisotropy in D'' beneath the Caribbean and Alaskan regions appears to be primarily that of transverse isotropy. Transverse isotropy is a special case of hexagonal symmetry where the symmetry axis is vertical. In such a medium, there is no azimuthal variation in wave velocity. (The Earth model PREM [*Dziewonski and Anderson*, 1981] is transversely isotropic in the top 200 km of the mantle). There are three main arguments in support of this conclusion. First, in these regions it is observed that the SH–phase

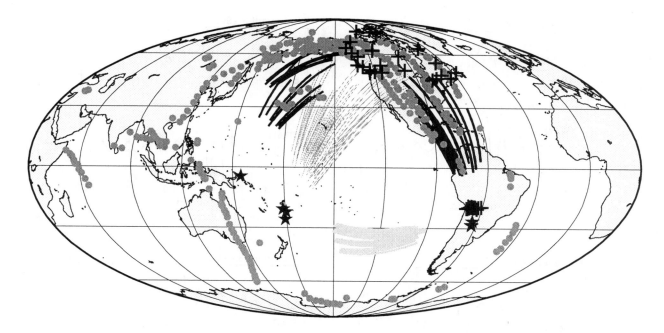

Figure 5. Regions where D'' anisotropy has been studied. The thick dark lines across the northern Pacific, Alaska and the Americas denote regions where there is evidence of D'' anisotropy with Vsh>Vsv in phases which propagate horizontally through the D'' region. The lines are surface projections of raypaths through the lowermost 250 km of the mantle. The lighter lines across southern Pacific show regions where there is no evidence for D'' anisotropy. The dashed, fine lines across the middle of the Pacific mark regions there isn't yet an agreement on the nature and magnitude of the anisotropy. The grey circles show the CMB sites of paleo–slabs as predicted by *Lithgow–Bertelloni and Richards* [1998]. The epicenters of events and stations in Canada (CNSN network) and South America (BANJO array) used in the studies of *Kendall and Silver* [1996a; 1996b] are marked by stars (events) and crosses (stations).

always leads the SV–phase indicating azimuthal independence. Second, SKS and SKKS phases do not appear to be affected by D'' anisotropy in these regions. As noted by *Silver* [1996] and *Meade et al.*, [1995], the overall delay–time contribution of D'' to these phases is at most 0.2s (the limit of resolution in measuring delay time), whereas the expected delay time through the D'' layer, assuming 2% azimuthal anisotropy, should be nearly 1 second. This is consistent with a medium with a vertical symmetry axis. It should be noted that even though SKS and SKKS will transit the D'' region at high angles from the vertical, they will not be split into a fast and slow shear–waves in transversely isotropic media. At a horizontal interface (the CMB in this case), the P–wave will only convert to an SV–wave. This is not true of more general forms of anisotropy or if the boundary in inclined from the horizontal. The final piece of evidence for transverse isotropy in these regions comes from waveform modeling. *Sen et al.* [1998] have shown good agreement between data and waveforms for a transversely isotropic model of the D'' region beneath Alaska. In the D'' region beneath the Americas,

Kendall and Silver [1996b] show that the shape of the observed waveforms agrees well with the synthetic waveforms for an isotropic model. They differ only in that the data possess a large time shift between the radial and transverse components that is not observed in the isotropic waveforms. Such simplicity in the waveforms suggest transverse isotropy as the SH and SV systems are decoupled for this special case of anisotropy [e.g., *Cormier*, 1986; *Maupin*, 1995].

The form of anisotropy, if any, for the region beneath the central Pacific is not clear, but the observations suggest a different underlying cause for the anisotropy from that in the Caribbean and Alaskan regions. The central Pacific region shows closely spaced patches where S/ScSH leads S/ScSV, patches which show the opposite and patches where there is no apparent D'' splitting. The anisotropy may still be locally transversely-isotropic, with either SH leading SV or SV leading SH, but there must be lateral changes in the anisotropy over short length–scales. If the region exhibits a more general form of anisotropy, the simplest being hexagonal symmetry with a horizontal symmetry axis, the

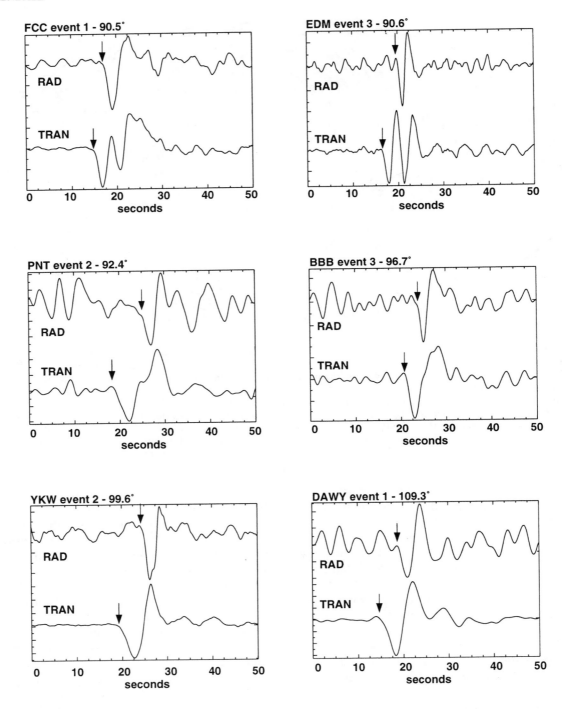

Figure 6. Evidence of D'' anisotropy in the region beneath the Caribbean and Central America (from *Kendall and Silver* [1996a]). The travel–time separations between the radial (upper) and transverse (lower) components at stations beyond 90° are shown. These seismograms (unfiltered velocity records) have been corrected for upper–mantle anisotropy and every time window is 50 seconds long. The nearer stations record an S/ScS arrival while the more distant station, DAWY, records the core diffracted phase, Sdiff. Note that the SH–arrival always leads the SV–arrival.

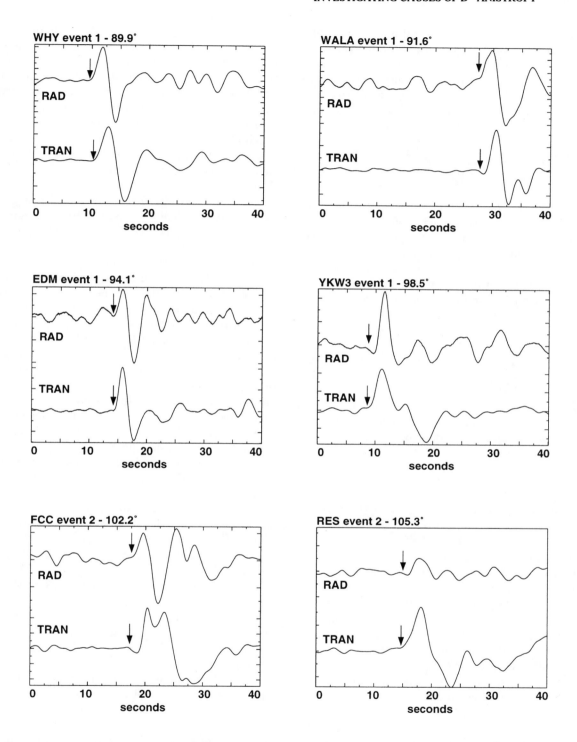

Figure 7. Evidence against D'' anisotropy in the region beneath the central Pacific Ocean. See Figure-caption 6 for details. The seismograms have been corrected for upper–mantle anisotropy and the time windows are 40 seconds in length. Separations between the radial and transverse components are on the order of $\sim \pm 1$ sec and can be therefore viewed as negligible.

radial and transverse components will not be decoupled from each other, as they are in the case of transverse isotropy. As a result, measuring the time separation between the radial and transverse components will be in general meaningless. It will also be difficult to rotate into the fast and slow shear-wave orientations as these waveforms can be quite different [*Maupin*, 1995].

5. ANISOTROPIC MECHANISMS: LPO OR SPO

As discussed in the introduction, seismic anisotropy may have two basic causes: lattice preferred orientation (LPO) of anisotropic minerals or shape preferred orientation (SPO) of inclusions within a matrix of differing elastic properties. A special case of SPO is layering of materials with contrasting velocities. Our first task is to determine which of these two mechanisms is more consistent with the seismic constraints. For the Caribbean region, these are transverse isotropy (vertical symmetry axis) and Vsh>Vsv by 1.8% (1.8% = 200 (Vsh - Vsv) / (Vsh + Vsv)) [*Kendall and Silver*, 1996a]. The constraints are similar in the Alaskan region except that the magnitude of the anisotropy appears to be more variable. As the anisotropy parameters for the region beneath the central Pacific are still being debated, we concentrate on finding a mechanism for the Caribbean and Alaskan regions.

5.1. Mineral Alignment

Crystals in an aggregate will develop LPO when subject to deformation if a dominant glide system exists. The best-documented case for LPO anisotropy is for olivine in upper-mantle peridotites. This is based on an extensive examination of mantle samples, laboratory studies of deformation, as well as the seismic observations of upper-mantle anisotropy that are well explained by this mechanism. In the upper mantle, anisotropy has been detected beneath the continents, and appears to be due to lithospheric deformation [*Silver*, 1996], although beneath the oceans, it appears to be more closely associated with the spreading process [*Hess*, 1964; *Blackman et al.*, 1996] and mantle flow in general.

In order to comprehensively evaluate mineral alignment as a viable mechanism for anisotropy, we would need to know the minerals present, their single-crystal elastic constants at CMB conditions, and the degree of LPO development induced by finite strain at these depths. This information for lower mantle minerals is presently incomplete, so we take two approaches. First,

we make the assumption that LPO is perfect, and then use the available single-crystal elastic constants to evaluate the anisotropy. This is equivalent to assuming that there is one dominant glide plane, and one dominant slip direction within this plane. While the most commonly expected lower mantle minerals (perovskite, MgO, and to a lesser extent stishovite) are strongly anisotropic, ranging from 6% to nearly 20% in splitting anisotropy (Figure 8, Figure 9, Figure 10, Figure 11) it is difficult to create the observed form of transverse isotropy (Vsh>Vsv). In the case of perovskite, it may exist in an orthorhombic or cubic form, each of which is anisotropic [*Cohen*, 1987]. The slowness- and wave-surfaces for both forms are shown in Figures 8 and 9. There is no orientation of wave-surfaces in Figures 8 and 9 that satisfy Vsh>Vsv for a horizontal propagation direction and a lack of splitting in the vertical direction. For magnesiowüstite (MgO), the slowness and wave surfaces for this cubic mineral are shown in Figure 10. As with perovskite, it is not possible to find an orientation of this crystal such that SH leads SV in a horizontal direction simultaneously with no splitting in the vertical direction. Finally, we consider stishovite, a mineral with tetragonal symmetry. The slowness- and wave-surfaces are shown in Figure 11. In this case, there are orientations where Vsh>Vsv for horizontal propagation (with no splitting in the vertical direction). However, there will also be azimuths where Vsv>Vsh. This has only been observed for a localized region beneath the central Pacific [*Pulliam and Sen*, 1996; 1998]. An added complication in the case of stishovite, is that there is expected to be, at CMB conditions, a transformation to a columbite structure with a lower symmetry (orthorhombic) [*Tsuchida and Yagi*, 1989; *Cohen*, 1992; *Kingma et al.*, 1995]. Again, there are orientations which predict no SKS splitting and S/ScSH leading S/ScSV, but there must be azimuths where S/ScSV leads S/ScSH (see Figure 12).

The second approach is to assume that LPO exists on one glide plane, but with arbitrary slip direction, and that the glide plane is parallel to the CMB. This would be appropriate for lateral flow along this boundary, and would generate a transversely-isotropic medium. This approach was used by *Stixrude* [1998, this issue], where the glide plane is taken to be that found for low-pressure analogs. In this case, it is found that perovskite and MgO show Vsh<Vsv. Only columbite supports the seismic data, where Vsh>Vsv by 8%.

While anisotropy due to aligned SiO_2 is conceivable, we regard it as unlikely for several reasons. First, SiO_2 is a minor constituent of the mantle, and anisotropy in the other more common phases, particularly perovskite

Perovskite - ortho

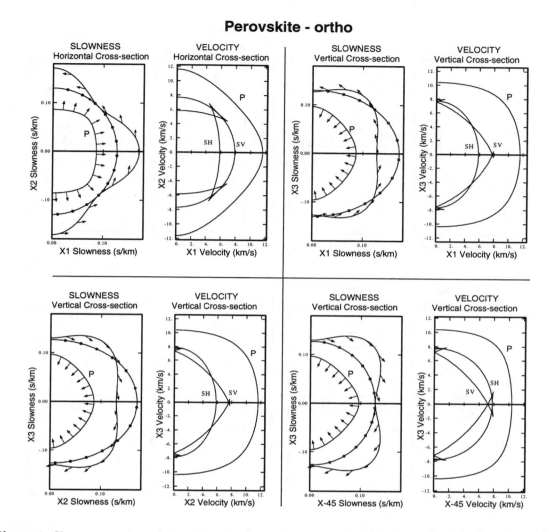

Figure 8. Slownwess surfaces (left–side) and wave surfaces (group velocity) (right–side) for orthorhombic perovskite from the elastic constants of *Cohen* [1987]. The wave surfaces can be thought of as snapshots of the P-wavefront (the outer surface) and the S-wavefronts (inner two surfaces) after 1 second in time. The P-wave is the inner line on the slowness diagrams. Arrows on the slowness surfaces show the direction of particle motion for each of the 3 wavetypes. The four panels show the slowness and velocity information in four planes: the X1–X2 plane (top left) and 3 planes through the X3 direction. The direction X-45 refers to the direction 45° from the X1 and X2 axes.

and MgO, would mask the effect of the anisotropy in SiO_2. At lower mantle conditions MgO will develop LPO [*Meade and Jeanloz*, 1988] while perovskite may not [*Meade et al.*, 1995]. Second, while the transverse isotropy could be explained by LPO of the columbite phase, this would require it to constitute a quarter of the material in D'' which is excessively high. We conclude that LPO in lower–mantle minerals is an unlikely cause for the anisotropy, at least in the Caribbean and Alaskan regions, although we must always allow for the possibility that there is an as–yet unknown mineralogy that dominates D'' [*Nataf and Houard*, 1993].

5.2. Shape Preferred Orientation and Layering

The other form of anisotropy that could be present in D'' is that due to a shape–preferred–orientation, SPO, in the form of aligned inclusions or periodic layering. This is often observed in crustal rocks that have been subjected to deformation. Indeed, it is observed that near–vertical shear waves passing through the continental crust generally exhibit shear–wave splitting, with delay times that average about 0.2s [*Silver*, 1996]. The splitting is most probably due to vertical, fluid (water)–filled cracks, whose orientation is determined by the

Perovskite - cubic

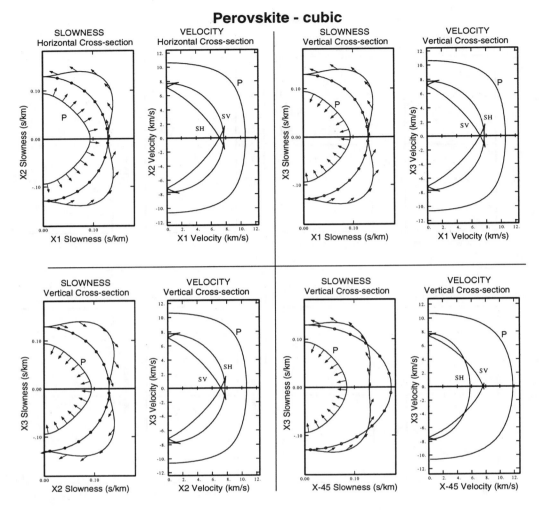

Figure 9. Slownwess and wave surfaces for cubic perovskite using the elastic constants of *Cohen* [1987]. See Figure–caption 8 for more detail.

ambient stress field. As another example, metamorphic segregation and deformation can produce *l*– and *s*–tectonites which exhibit highly aligned inclusions [*Spry*, 1969]. These tectonites have inclusions which may exist as rods (lineations or *l*–tectonites) or disks (shistostic or *s*–tectonites). At the time of formation the inclusions may be fluid–filled and related to the presence of melt. There is less consensus concerning the properties of such inclusions in the mantle [*Schmeling*, 1985]. They may exist in isolated melt packets, along grain edges in tubule–like inclusions, or along grain faces in disk–like inclusions. Disk–shaped melt inclusions with aspect ratios on the order of 0.05 have been observed in peridotites [*Faul et al.*, 1994]. Based on observed metamorphic textures [see *Spry*, 1969] it is thus reasonable to suppose that inclusions will be oriented parallel to the foliation or shear plane of finite strain. Extrapolating

to the CMB, then, lateral flow of mantle material along the CMB in the D'' region should align the inclusions sub–parallel to the CMB.

We model the seismic anisotropy due to SPO and layering using the effective medium theory of *Tandon and Weng* [1984]. The theory allows us to assess the effective elastic constants of a medium with preferentially–aligned spheroidal inclusions that may be of higher or lower velocity than that of the host matrix. It is assumed that the inclusions are much smaller in dimension than the dominant seismic wavelength. The inclusions may be oblate, resembling disks, or elongate, resembling tubules or cigars. We consider both possibilities.

Figure 13 shows the shear–wave anisotropy for both low– and high–velocity inclusions where the aspect ratio of the inclusions is 20 (the cigar is 20 times longer

MgO

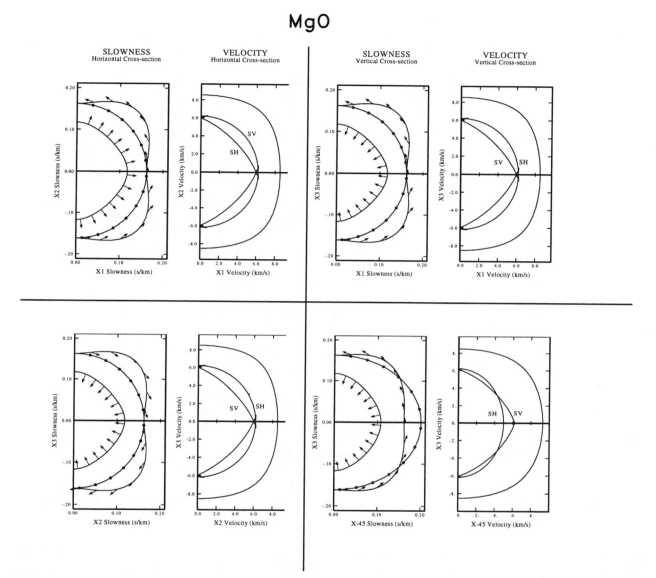

Figure 10. Slownwess and wave surfaces for magnesiowüstite using the elastic constants of *Meade and Jeanloz* [1988]. See Figure–caption 8 for more detail.

than it is wide). There is little variation in the properties beyond aspect ratios of 10. Smaller aspect ratios for fiber–like inclusion (aspect–ratios between 10 and 1) show less pronounced results. Dramatic velocity contrasts and high volume fractions are required in order to achieve 1.8% anisotropy. For example, if the inclusions have a higher velocity than the matrix, the velocity contrast ratio must be > 0.6. Given a matrix velocity of 7.5 km/s, the inclusion velocity must be > 12 km/s which would render exceptionally high aggregate–velocities for the medium as a whole. On the other hand, if the inclusions have a lower velocity than

the matrix, the required volume fractions will significantly reduce the aggregate–velocities for the medium as a whole.

Tubules oriented vertically produce transverse isotropy, but predict Vsv>Vsh for horizontal wave propagation. In order to explain the seismic observations for the Caribbean and Alaskan regions, these tubule–like inclusions must be oriented with the elongate axis parallel to the CMB. However, in this orientation, this form of SPO exhibits azimuthal anisotropy. Vsh > Vsv is only found for propagation directions roughly perpendicular to the orientation of the long axis of the tubules and

Stishovite

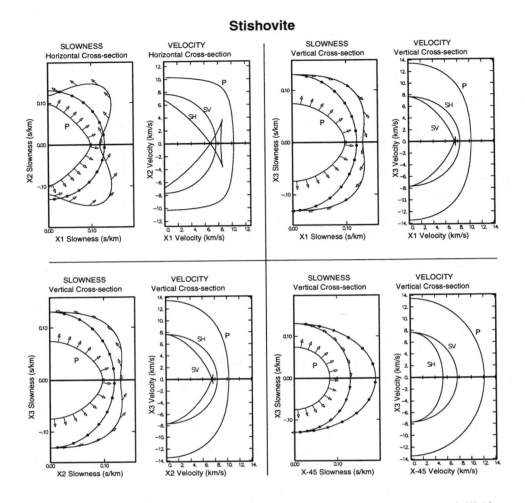

Figure 11. Slownwess and wave surfaces for stishovite using the elastic constants of *Weidner et al.* [1982]. See Figure–caption 8 for more detail.

there will be significant splitting in vertically propagating shear waves. In summary, tubule–like inclusions yield seismic properties which are inconsistent with observations.

We next consider disk–like inclusions where the aspect ratio is much less than 1. Figure 14 shows results for inclusions (aspect–ratio = 0.001) aligned with the normal to the disk–face oriented vertically. There is little difference between these results and those for a periodic thinly–layered medium [*Backus*, 1962]. We find that this form of SPO possesses the basic characteristics of the Caribbean and Alaskan data sets. In particular, it exhibits transverse isotropy with Vsh>Vsv. It is also clear that this geometry can produce levels of anisotropy which are inferred from the data with reasonable inclusion velocities and volume fractions, as long as aspect ratios are below a value of 0.1.

In principle, either high or low–velocity disk–like inclusions could explain the data. For example, inclusions with high velocities (>10 km/s, assuming the matrix velocity v_1=7.5 km/s) are required for volume fractions of 0.1 (10%) (Figure 14) to reach 1.8% anisotropy. Low–velocity inclusions with a volume fraction of 0.1 (10%) require a shear–velocity of 5.6 km/s (assuming v_1=7.5 km/s) in the inclusions. These correspond to differences of +35% and –25% respectively and illustrate that less contrast is need in the case of low–velocity inclusions. A particularly effective means of generating anisotropy is having inclusion velocities that are effectively zero. Indeed, a concentration of disk–like inclusions of a liquid or melt with a volume fraction of 0.0001 (0.01%) is sufficient to explain the observations. Such a small amount of melt will have negligible impact on the isotropic–aggregate velocities of the medium.

Columbite

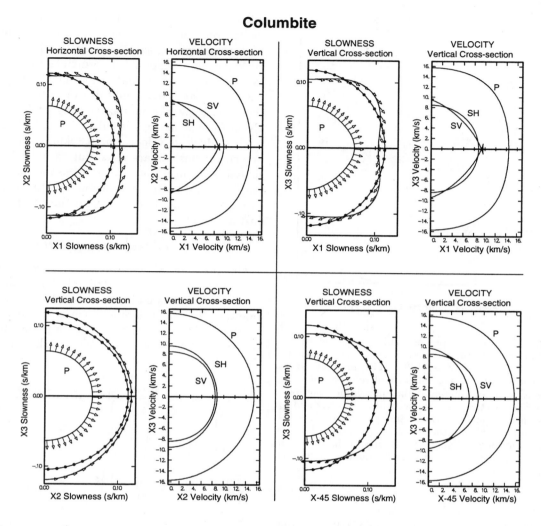

Figure 12. Slownwess and wave surfaces for columbite using the elastic constants of *Stixrude* [1998]. See Figure–caption 8 for more detail.

Weber [1994] has considered the isotropic wave propagation effects of such lamellae-like inclusions.

6. CANDIDATE PHYSICAL PROCESSES

The previous section leads to the conclusion that the most likely form of D'' anisotropy beneath the Caribbean and Alaskan regions is that due to SPO, consisting of either horizontal disk–like inclusions or horizontal layering. In order to further evaluate the characteristics of these regions, we consider two primary constraints. The first is the magnitude of the anisotropy. In the Caribbean region this is on average of 1.8% [*Kendall and Silver*, 1996a]. *Matzel et al.* [1996] use waveform modeling to develop a model which has

anisotropy grading between 3% at the top of D'' to 0% at the CMB in the region beneath Alaska. *Garnero and Lay* [1997] consider a broader area and find levels of anisotropy which vary between 0% and 1.5%. For the time being we adopt 1.8% as our constraint, recognizing that this may be an overestimation in parts of the Alaskan region. The second constraint is based on the observation that a seismic discontinuity has been detected at the top of the D'' layer in these regions. The velocity increase in Vsh is about 2.75%. Studies for the region beneath the Caribbean and Central America [*Lay and Helmberger*, 1983; *Kendall and Nangini*, 1996] have produced shear models (based on SH) with anomalously fast velocities and a discontinuity 250–280 km above the CMB. The region beneath Alaska and the

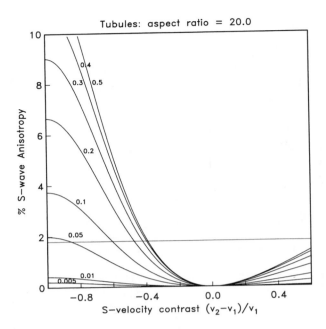

Figure 13. Modeling the percent anisotropy due to the preferential alignment of horizontal cigar–like inclusions using theory of *Tandon and Weng* [1984]. The different curves are the percent anisotropy as a function of velocity contrast, expressed as a ratio, for different inclusion volume–fractions. The term v_1 refers to the shear–velocity of the matrix and v_2 refers to the shear–velocity of the inclusion. The average observed anisotropy (1.8%) in the study of *Kendall and Silver* [1996a] is marked for reference. The density of the inclusion material is assumed to be 5700 kg/m^3 and the inclusions have an aspect ratio of 20.

northern Pacific also has high seismic velocities and a D'' discontinuity 243–278 km above the CMB [*Lay and Helmberger*, 1983; *Young and Lay*, 1990]. Furthermore, these are regions of high velocities in large–scale tomographic Earth models [*Su et al.*, 1994; *Masters et al.*, 1996]. The central and southern Pacific regions are sites of low velocities in these Earth models and evidence for a D'' discontinuity in this region has been equivocal.

The spatial association of D'' anisotropy in the Caribbean and Alaskan regions with high seismic velocities, and the presence of a D'' discontinuity, provides important constraints on the physical process that produces the anisotropy in these regions.

We first consider a model of D'' in which the matrix material has the same elastic properties as the region just above. To satisfy both constraints, the inclusion velocities must be faster than the matrix. Figure 15 shows the amount of anisotropy and the SH–velocity contrast as a function of inclusion velocity–contrast and concentration for horizontal disk–shaped inclusions. The only

allowable combinations (1.8% anisotropy and 2.75% velocity contrast) correspond to small volume–fractions with very high velocity contrasts (ratio of 0.8). While such a medium is possible in principle, there is no known lower mantle mineral that is this much faster than ambient mantle at CMB conditions (Vs=13.5 km/s for the inclusions, given a matrix velocity of Vs=7.5 km/s). Even if the anisotropy is only 1%, the inclusions would have to have a velocity of 9.8 km/s. If such high-velocity inclusions are ruled out as unlikely, then this model is inconsistent with the seismic data. We thus conclude that both the inclusions *and* the matrix consist of material that is elastically distinct from the mantle just above D''.

6.1. Core-Mantle Interaction Hypothesis

One possible cause for the distinct properties of D'' in these regions — SPO–related anisotropy and high Vsh — is the infiltration of core material into the lowermost mantle. This idea is based on experiments in which molten iron (core) reacts with mantle silicates at CMB pressures [*Knittle and Jeanloz*, 1991]. These authors

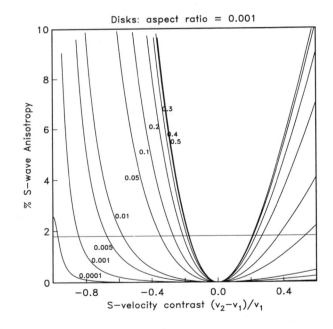

Figure 14. Modeling the percent anisotropy due to the preferential alignment of horizontal disk–like inclusions using theory of *Tandon and Weng* (1984). The density of the inclusion material is assumed to be 5700 kg/m3 and the aspect ratio is 0.001. See the caption for Figure 13 for more detail. Note that as the S–wave velocity of the inclusions tends to zero, very small inclusion volume–fractions will explain the observed anisotropy.

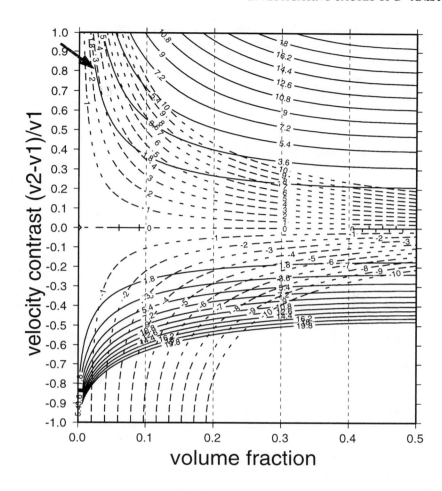

Figure 15. SH–velocity contrast (dashed contours) and percent–shear–wave anisotropy (solid lines) as a function of volume fraction and velocity contrast ratio for horizontally-aligned disk-shaped inclusions with an aspect ratio of 0.001. A matrix shear-velocity (v_1) of 7.5 km/s is assumed for calculating the SH velocity contrast. The seismic constraints from the D'' region beneath the Caribbean are 1.8% anisotropy and a SH-velocity increase of 2.75%. The point where this occurs is marked by the arrow and corresponds to a velocity contrast ratio of ~0.8. This implies an inclusion velocity of 13.5 km/s, given a matrix velocity of 7.5 km/s, which is unrealistically high.

found both silicate ($MgSiO_3$, SiO_2) and iron alloy (FeO, FeSi) reaction products which have high and low shear–velocities respectively. The velocity contrasts between these two groups are large, about 30–40%. In evaluating this model, we follow *Knittle and Jeanloz* [1991] and assume that the material just above D'' consists of $(Fe_{.1},Mg_{.9})SiO_3$, and that a variable amount of Fe is added to this. The reaction products are then assumed to separate into silicates (matrix) and iron alloys (inclusions). It is found that such a medium is capable of generating sufficient anisotropy to match observations. This assumes of course that the the inclusions are horizontally aligned and disk shaped, and that they persist

throughout the ~250 km thick D'' layer. However, the addition of iron will *reduce* Vsh from 1% to 6% , and cannot account for the observed increase in velocity. We are thus forced to reject the core–mantle interaction hypothesis on this basis, unless there is some as–yet unknown means of increasing the overall SH–velocity of D'' within the context of this model.

6.2. Slab Graveyard Hypothesis

We now consider the hypothesis that the properties of D'' beneath the Caribbean and Alaska are due to the presence of subducted slabs. There are several compelling reasons to entertain this hypothesis. There is increasingly strong evidence of slab penetration into the lower mantle [e.g., *van der Hilst et al.*, 1991; *Fukao et al.*, 1992; *Grand*, 1994; *van der Hilst*, 1995; *Grand et al.*, 1997]. Tomographic images of the mantle beneath the Americas show continuous high–velocity slab–like features extending down to the CMB. This is thought to be the remnant of the former Farallon plate and this is one of the regions where D'' anisotropy is clearly observed [*Kendall and Silver*, 1996a]. Geodynamic modeling also suggests slabs may descend to the CMB [*Christensen*, 1989; *Davies*, 1990; *Christensen and Hofmann*, 1994]. Finally, these observations of anisotropy correspond with the predicted locations of slabs in the lowermost mantle (Figure 5) based on paleo–subduction modeling [*Lithgow–Bertelloni and Richards*, 1998]. The question we ask is whether there is a means by which slab material can satisfy the two seismological constraints: the degree of anisotropy and the increase in SH-velocity.

Slab material is expected to be distinct in three ways. First, it is compositionally distinct as the basaltic and hartzburgitic components of the slab differ from the ambient mantle. Secondly, the slab will be thermally distinct, having retained a significant thermal anomaly acquired at the Earth's surface (see below). Finally, slabs denote regions of downwelling where stresses and strains will be high, especially as slab material spreads out along the CMB.

Within the context of SPO–anisotropy, we consider whether the anisotropy can be explained by appealing to the differing seismic properties of the formerly hartzburgitic lithosphere and the formerly basaltic oceanic–crust. Little is known of the state of oceanic crust and mantle–lithosphere at CMB conditions, so we propose of few scenarios which may explain the seismic observations.

One possible scenario is that the former–basalt layer produces substantial amounts of silica (25 wt %) [*Christensen and Hofmann*, 1994] which, as noted in section 5.1, has high seismic velocities. The layering of the high–velocity former oceanic–crust and former oceanic–mantle could give rise to the anisotropy. As the crust constitutes at most 10% of the subducted material, a shear–velocity contrast ratio of 0.4 would be needed to explain the observed anisotropy (Figure 14). This would require a shear–velocity of over 10.0 km/s (assuming $v_2 = 7.5$ km/s) in the crustal layer, which is unrealistically high. As discussed earlier (and shown in Figure 15) this also leads to SH-velocities which are much higher than those inferred from observations.

Alternatively, the former–basalt could represent the low–velocity component. This might occur if there is a low–melting–temperature component within the former–basalt, resulting from the relatively high concentrations of Fe, Al, and Ca. Recent melting experiments on probable basaltic components indeed suggest significant lowering of the melting point of perovskite due to the presence of Al and Ca [*Shen and Lazor*, 1995]. Compositionally, the difference between basalt and hartzburgite is that the basalt is enriched in Fe, Ca and Al (as well as Si). *Shen and Lazor* [1995] found that the effect of Ca and Al is to lower the melting point of perovskite, and that this effect will increase with increasing pressure. In addition, the melting temperature of diopside (mixture of Mg, Ca perovskite) had a lower melting temperature than either of the end members, suggesting the presence of eutectic melting. More recently, data on melting in basalt confirm that it indeed has a very low melting point. Diamond cell measurements of its melting point extrapolated to CMB conditions suggests that basalt melts at temperatures that are 3000°–4000o lower than magnesium perovskite or MgO, the probable matrix constituents [*Hirose et al.*, 1998]. Consequently, there may be a low–melting temperature component in former basalt. Figure 14 shows that very small amounts of melt (< 0.01%, a volume fraction of 0.0001) aligned in disk–like inclusions would explain the observed anisotropy. The overall decrease in Vsh velocity would be negligible, being reduced by the percent of the concentration (~0.01%). If the melt is confined to the former–basalt layer, which constitutes 10% of the subducted package, it would still only have to represent 0.1% of this layer. Of course, the amount of melt could be greater, since it could trade off against the aspect ratio of the inclusions. Another possibility is that the melt simply lowers the velocity of the former-basalt layer. Layering of this lower-velocity crustal component and the higher-velocity mantle–lithospheric component could produce the anisotropy if the velocity contrast is ~25% (a ratio of 0.25 in Figure 14).

This model must not only account for the anisotropy, but also explain the overall high shear–velocity of the region. High seismic velocities could be generated thermally if the slab retains its inherited thermal anomaly down to the CMB. This is likely to be the case, for the following reasons. First, slabs appear to thicken as they enter the lower mantle. This result derives from tomographic imaging [*Grand*, 1994] as well as modeling the slab contribution to the geoid [*Ricard et al.*, 1993]. It appears to be the result of a viscosity increase of about 30 from the upper to lower mantle. Both lines of reasoning suggest a thickening factor, T, of at least three. Assuming that descent transit time, τ_t, in the lower mantle scales as T and that thermal equilibration time,

τ_e, scales like T**2, then the ratio $\tau_e / \tau_t = kT$, where k is about equal to 1. Taking a typical upper mantle descent velocity of 6cm/yr for the upper mantle, then τ_t ~150my. The thermal equilibration time will be about 500my which is more than 3 times longer than the expected transit time to the CMB. We conclude that the slab should retain a significant portion of its original thermal anomaly. Lower–mantle tomography is consistent with this estimate. Regional, high resolution images of slabs in the lower mantle from portable seismic data [*VanDecar et al.*, 1997], suggest that the shear velocity anomaly of the slabs in the lower mantle is of order 3%, comparable in size to the velocity jump at the top of D''. As noted above, global tomographic images, at least beneath the Caribbean, are also consistent with this idea. Concerning the actual temperature of slab material at the CMB, *Wysession* [1996a] has estimated that as the slab reaches the CMB, its core will be a minimum of 600–800°C colder than the surrounding mantle. It thus appears that slabs provide a means of satisfying the two main constraints on D'' structure. The above model uses the thermal and compositional properties of the slab. One other possibility is shape orientation between the two main mantle constituents, perovskite and MgO. SPO-anisotropy may result from segregation due to significant shear in these regions if the velocity contrast between these minerals is sufficient. Figure 14 shows that in order to achieve 1.8% anisotropy the velocity contrast only needs to be ~17% (0.17) between the two phases if they are aligned in layers and constitute roughly equal portions volumetrically. The thermal anomaly is still required to give the overall 2.75% increase velocity. However, the results of *Stixrude* [1998, this issue] suggest contrasts much less than 15% between orthorhombic-perovskite and MgO at CMB conditions.

7. CONCLUSIONS

Shear-wave splitting in phases which transit the lowermost mantle are indicative of seismic anisotropy within D''. Studies of this anisotropy are limited in coverage, but it is clear that it is not a global feature. The observations for regions of the lowermost mantle beneath the Caribbean and Alaska strongly suggest transverse isotropy with SH–velocities faster than SV–velocities and no splitting for wave propagation in the vertical direction. It is important to be able to unambiguously constrain the anisotropy to the D'' region. For this reason, corrections for upper-mantle anisotropy are crucial, especially when dealing with shear–diffracted phases. It is also important, but not always feasible, to constrain the anisotropy to the lowermost mantle through comparisons of phases turning above and within the D'' layer.

It is hard to explain the anisotropy in these transversely isotropic regions with a preferred mineral alignment or LPO given our current knowledge of lowermost–mantle mineral physics. Instead we find it more likely that the anisotropy is due to the preferential alignment of inclusions within a background matrix. The seismic constraints on this SPO-anisotropy suggest that it cannot be due to aligned tubule–shaped inclusions, but is rather due to horizontally–aligned disk–shaped inclusions. Very low volume–fractions of melt–filled disk–shaped inclusions (<0.0001) explain the observations well. An important conclusion from the observations and modeling is that both the background matrix and inclusions must be distinct from the mantle just above the D'' layer.

We have considered candidate physical processes for producing the SPO-anisotropy in the Caribbean and Alaskan regions. It is hard to satisfy the seismic constraints with hypotheses which appeal to core–mantle interactions. There are, though, a number reasons to suspect the cause of the anisotropy is associated with subducting slabs descending to the CMB region. The anisotropy correlates with faster than average velocities in global tomographic models and observations of a seismic discontinuity atop D''. Tomographic images and geodynamic modeling suggest slabs descend to the lowermost mantle. More importantly, these regions are the predicted sites of paleo–slabs in the lowermost mantle. The high overall seismic velocities can be explained by slab thermal-anomalies and the anisotropy can be explained by contrasts in the material properties between what was formerly oceanic crust and oceanic mantle-lithosphere (Figure 16). The alignment of tabular melt inclusions in very low concentrations can explain the observed anisotropy and partial confirmation of this hypothesis has come from melting experiments on basalt.

We present what we feel is the most likely scenario given the seismic constraints. We are now in the exciting position of being able to test many of the hypotheses presented and they suggest investigations in both seismology and mineral physics. Studies of the seismological structure of D'' [e.g., *Weber*, 1993; *Kendall and Shearer*, 1994; *Garnero and Helmberger*, 1996; *Wysession*, 1996b] and the predictions of paleo–slab locations [e.g., *Engebretson et al.*, 1992; *Lithgow-Bertelloni and Richards*, 1998] will guide such future studies. Reliable observations of significant shear–wave splitting with SV leading SH are potentially important, as they would require a basically different mechanism, such as mineral

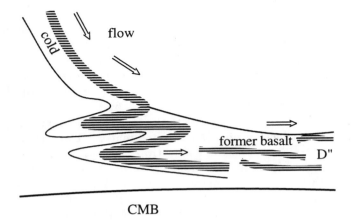

flow

cold

former basalt

D''

CMB

Figure 16. The scenario for D'' anisotropy which best fits our seismological observations. Subducted oceanic mantle–lithosphere sinks into the lower mantle and piles up at the core–mantle boundary (CMB). As the cold high–velocity material is swept along the CMB, former oceanic crust (basalt) becomes entrained as horizontal inclusions within the former peridotite. The contrasting velocities between the former–basalt and former–peridotite gives rise to an extrinsic anisotropy. Note that in reality the inclusions would be smaller and more abundant than what is depicted in the cartoon.

alignment (several minerals allow for this geometry), or at least a very different inclusion geometry. This may be central to understanding the emerging picture of D'' anisotropy in the central Pacific which is clearly different from that in the Caribbean and Alaskan regions. If slabs indeed strongly influence the thermal, compositional, and seismological properties of the CMB, they would then appear to play a major role in the mantle's two major thermal boundary–layers.

Acknowledgments. We would like to thank Carolina Lithgow-Bertelloni, Colin Sayers, Lars Stixrude and Raymond Jeanloz for helpful advice. We are greatful to CLB for providing models of the predicted sites of paleo–slabs at the CMB. Michael Wysession, Mrinal Sen and an anonymous reviewer are thanked for reviews which improved the manuscript.

REFERENCES

Backus, G. E. , Long–wave elastic anisotropy produced by horizontal layering, *J. Geophys. Res., 67,* 4427–4440, 1962.

Beck S., P. G. Silver, G. Zandt, T. Wallace, S. Myers, R. Kuehnel, S. Ruppert, L. Drake, E. Minaya, P. Soler, P. Baby, S. Barrientos, J. Telleria, M. Tinker and J. Swenson, Across the Andes and along the Altiplano: a passive seismic experiment, *Iris Newsletter, 8,* 1–2, 1994.

Blackman, D. K., J-M. Kendall, P. Dawson, H-R. Wenk, D. Boyce and J. Phipps Morgan, Teleseismic imaging of subaxial flow at mid-ocean ridges: travel-time effects of anisotropic mineral texture in the mantle, *Geophys. J. Int., 127,* 415-426, 1996.

Christensen, U. R., Models of mantle convection: one or several layers, *Philos. Trans. R. Soc. London, Ser. A, 328,* 417–424, 1989.

Christensen, U. R. and A. W. Hofmann, Segregation of subducted oceanic crust in the convecting mantle, *J. Geophys. Res., 99,* 19867–19884, 1994.

Cohen, R. E., Elasticity and equation of state of MgSiO3 perovskite, *Geophys. Res. Lett., 14,* 1053–1056, 1987.

Cohen, R. E., First–principles predictions of elasticity and phase transitions in high pressure SiO_2 and geophysical implications, in *High–pressure Research: Application to Earth and Planetary Sciences,* ed Y. Syono and M.H. Manghnani, pp 425–431, Terra Scientific , Washington D.C., 1992.

Cormier, V., Synthesis of body waves in transversely isotropic Earth models, *Bull. Seis. Soc. Amer., 76,* 231–240, 1986.

Davies, G. F., Mantle plumes, mantle stirring and hotspot chemistry, *Earth and Planet. Sci. Lett., 99,* 94–109, 1990.

Dziewonski, A. M. and D. L. Anderson, Preliminary reference earth model, *Phys. Earth Planet. Int., 25,* 297–356, 1981.

Engebretson, D. C., K. P. Kelley, H. J. Cashman and M. A. Richards, *Geol. Soc. Am. Today, 2,* 93–100, 1992.

Faul, U., D. R. Toomey and H. S. Waff, Intergranular basaltic melt is distributed in thin, elongated inclusions, *Geophys. Res. Lett., 21,* 29–32, 1994.

Fukao, Y., M. Obayashi, H. Inoue and M. Nenbai, Subducting slabs stagnant in the mantle transition zone, *J. Geophys. Res., 97,* 4809–4822, 1992.

Garnero, E. J. and D. V. Helmberger, Seismic detection of a thin laterally varying boundary layer at the base of the mantle beneath the central-Pacific, *Geophys. Res. Lett., 23,* 977–980, 1996.

Garnero, E. J. and T. Lay, Lateral variations in lowermost mantle shear wave anisotropy, *J. Geophys. Res., 102,* 8121–8135, 1997.

Grand, S. P., Mantle shear structure beneath the Americas and surrounding oceans, *J. Geophys. Res., 99,* 11591–11621, 1994.

Grand, S. P, R. D. van der Hilst and S. Widiyantoro, Global seismic tomography: A snapshot of convection in the Earth, *Geol. Soc. Am. Today, 7,* 1–7, 1997.

Hess, H. H., Seismic anisotropy in the uppermost mantle under oceans, *Nature, 203,* 629–631, 1964.

Hirose, K., Y. Ma, Y. Fei, and H. Mao, Fate of subducted basaltic crust in the lower mantle, submitted to *Nature,* 1998.

Hudson, J. A., Overall properties of a cracked solid, *Math. Proc. Camb. Phil. Soc., 88,* 371–384, 1980.

Kendall, J-M. and P. M. Shearer, Lateral variations in D'' thickness from long-period shear-wave data, *J. Geophys. Res., 99,* 11575–11590, 1994.

Kendall, J-M. and C. Nangini, Lateral variations in D'' below the Caribbean, *Geophys. Res. Lett., 20,* 379–383, 1996.

Kendall, J-M. and P. G. Silver, Constraints from seismic

anisotropy on the nature of the lowermost mantle, *Nature, 381*, 409–412, 1996a.

Kendall, J–M. and P. G. Silver, Evidence for and against anisotropy in the lowermost mantle, *EOS: Trans. Am. Geophys. Un., 77*, F708 1996b.

Kingma, K.J., R. E. Cohen, R. J. Hemley and H. K. Mao, Transformation of stishovite to denser phase at lower–mantle pressures, *Nature, 374*, 243–245, 1995.

Knittle, E. and R. Jeanloz, Earth's core–mantle boundary: results of experiments at high pressure and temperatures, *Science, 251*, 1438–1443, 1991.

Lay, T. and D. V. Helmberger, A lower mantle S–wave triplication and the shear velocity structure of D'', *Geophys. J. R. astro. Soc., 75*, 799–837, 1983.

Lay, T. and C. J. Young, Analysis of seismic SV waves in the core's penumbra, *Geophys. Res. Lett., 18*, 1373–1376, 1991.

Lay, T., E. J. Garnero, Q. Williams, B. Romanowicz, L. Kellogg and M. E. Wysession, Seismic wave anisotropy in the D'' region and its implications, this issue *Core–mantle boundary, AGU–monograph*, eds. Wysession, Gurnis, Knittle and Buffet, 1998.

Lithgow–Bertelloni, C. and Richards, The dynamics of Cenozoic and Mesozoic plate motions, *Rev. of Geophys., 36*, 27-78, 1998.

Masters, T. G., S. Johnson, G. Laske and H. F. Bolton, A shear–velocity model of the mantle, *Phil. Trans. R. Soc. Lond. A, 354*, 1385–1411, 1996.

Matzel, E., M. K. Sen and S. P. Grand, Evidence for anisotropy in the deep mantle beneath Alaska, *Geophys. Res. Lett., 23*, 2417–2420, 1996.

Maupin, V., On the possibility of anisotropy in the D'' layer as inferred from the polarization of diffracted S–waves, *Phys. Earth Planet. Int., 87*, 1–32, 1995.

Meade, C. P. and R. Jeanloz, Yield strength of MgO to 40 GPa, *J. Geophys. Res., 93*, 3261–3269, 1988.

Meade, C. P., G. Silver and S. Kaneshima, Laboratory and seismological observations of lower mantle isotropy, *Geophys. Res. Lett., 22*, 1293–1296, 1995.

Mitchell, B. J. and D. V. Helmberger, Shear velocities at the base of the mantle from observations of S and ScS, *J. Geophys. Res., 78*, 6009–6019, 1973.

Nataf, H–C. and S. Houard, Seismic discontinuity at the top of D'': A world–wide feature?, *Geophys. Res. Lett., 20*, 2371–2374, 1993.

Pulliam, J. and M. K. Sen, Evidence of azimuthal anisotropy in D'', *EOS: Trans. Am. Geophys. Un., 77*, F679, 1996.

Pulliam, J. and M. K. Sen, Anisotropy in the lowermost mantle beneath the Central Pacific, submitted to *Geophys. J. Int.*, 1998.

Ricard, Y., M. A. Richards, C. Lithgow–Bertelloni and Y. Le Stunff, A geodynamic model of mantle density heterogeneity, *J. Geophys. Res., 98*, 21895–21909, 1993.

Schmeling, H., Numerical models on the influence of partial melt on elastic, anelastic and electric properties of rocks. Part. 1: elasticity and anelasticity, *Phys. Earth. Planet. Int., 41*, 34–57, 1985.

Sen, M. K., E. Matzel and S. P. Grand, Seismograms for Earth models with transversely isotropic D'', in press *Geophys. J. Int.*, 1998.

Shen, G. and P. Lazor, Measurement of melting temperatures of some minerals under lower mantle pressures, *J.*

Geophys. Res., 100, 17699–17713, 1995.

Silver, P., Seismic anisotropy beneath the continents: Probing the depths of geology, *Annual Rev. Earth Space Sci., 24*, 385, 1996.

Silver, P. G. and W. W. Chan, Implications for continental structure and evolution from seismic anisotropy, *Nature, 335*, 34–39, 1988.

Silver, P. G. and W. W. Chan, Shear–wave splitting and subcontinental mantle deformation, *J. Geophys. Res., 96*, 16429–16454, 1991.

Spry, A., *Metamorphic Textures*, Pergamon Press, Oxford, U.K., 1969.

Stixrude, L., Elastic constants and anisotropy of $MgSiO_3$ perovskite, periclase, and SiO_2 at high pressure, this issue *Core–mantle boundary, AGU–monograph*, eds. Wysession, Gurnis, Knittle and Buffet, 1998.

Su, W. J., R. L. Woodward and A. M. Dziewonski, Degree–12 model of shear velocity heterogeneity in the mantle, *J. Geophys. Res., 99*, 6945–, 1994.

Tandon, G. P. and G. J. Weng, The effect of aspect ratio of inclusions on the elastic properties of unidirectionally aligned composites, *Polymer Composites, 5*, 327–333, 1984.

Tsuchida, Y. and T. Yagi, A new post–stishovite high–pressure polymorph of silica, *Nature, 340*, 217–220, 1989.

Van der Hilst, R. D., Complex morphology of subducted lithosphere in the mantle beneath the Tonga trench, *Nature, 374*, 154–157, 1995.

Van der Hilst, R. D., R. Engdahl, W. Spakman and G. Nolet, Tomographic imaging of subducted lithosphere below northwest Pacific island arcs, *Nature, 353*, 37–43, 1991.

VanDecar, J. C., D. James, P. Silver, M. Assumpcão, S. Myers, S. Beck, T. Wallace, G. Zandt, E. Minaya, G. Asch and R. Kuehnel, Mantle structure beneath central South America, *EOS: Trans. Am. Geophys. Un., 76*, F606, 1995.

Vinnik, L., F. Farra and B. Romanowicz, Observational evidence for diffracted SV in the shadow of the Earth's core, *Geophys. Res. Lett., 16*, 519–522, 1989a.

Vinnik, L. P., R. Kind, G. L. Kosarev and L. I. Makeyeva, Azimuthal anisotropy in the lithosphere from observations of long–period S–waves, *Geophys. J. Int., 99*, 549–559, 1989b.

Vinnik, L., B. Romanowicz, Y. LeStunff and L. Makeyeva, Seismic anisotropy in the D'' layer, *Geophys. Res. Lett., 22*, 1657–1660, 1995.

Weber, M., P– and S–wave reflections from anomalies in the lowermost mantle, *Geophys. J. Int., 115*, 183–210, 1993.

Weber, M., Lamellae in D''? An alternative model for lower mantle anomalies, *Geophys. Res. Lett., 23*, 2531–2534, 1994.

Weidner, D. J., J. D. Bass, A. E. Ringwood and W. Sinclair, The single–crystal elastic moduli of stishovite, *J. Geophys. Res., 87*, 4740–4747, 1982.

Wysession, M. E., Imaging cold rock at the base of the mantle: The sometimes fate of Slabs?, in *Subduction: Top to Bottom*, edited by G. E. Bebout, D. Scholl, S. Kirby, and J. P. Platt, AGU, Washington, D. C., pp. 369-384, 1996a.

Wysession, M. E., Large–scale structure of the core–mantle boundary from diffracted waves, *Nature, 382*, 244–248, 1996b.

Young, C. J. and T. Lay, Multiple phase analysis of the

shear velocity structure in the D'' region beneath Alaska, *J. Geophys. Res.*, *95*, 17385–17402, 1990.

J–M. Kendall, Department of Earth Sciences, University of Leeds, Leeds, LS2 9JT, England.   (e-mail: m.kendall@earth.leeds.ac.uk)

P. G. Silver, Department of Terrestrial Magnetism, Carnegie Institution of Washington, 5241 Broad Branch Road, N.W., Washington, D.C., 20015, USA. (e-mail: silver@dtm.ciw.edu)

The Solid/Liquid Partitioning of Major and Radiogenic Elements at Lower Mantle Pressures: Implications for the Core-Mantle Boundary Region

Elise Knittle

Department of Earth Sciences, University of California, Santa Cruz, CA

The liquid/solid partitioning of major (Mg, Fe, Si, Ca, Al) and minor (K, U) radiogenic components have been studied between $(Mg,Fe)SiO_3$ silicate perovskite and coexisting melts. Synthesis conditions for the perovskites are representative of those in the lower mantle: pressures of >50 GPa and temperatures of 2000-4500 K generated using the laser-heated diamond anvil cell. Major element partitioning between solid Mg-silicate perovskite and silicate liquid is similar to that observed at lower pressure conditions, with Fe, Al and Ca partitioning into the melt. The behavior of the radiogenic elements, U and K, are markedly different at lower mantle pressures with uranium remaining an incompatible element and potassium apparently becoming compatible within the solid perovskite-structured phase. The magnitude of partitioning of iron between solid and melt implies that melts in the lower mantle could be up to 3% denser than their coexisting solids, with the density difference being produced by a combination of small (or zero) volumes of fusion and iron-enrichment in the liquid. Any melting event occurring in the lower mantle could thus produce descent of melt into D'': such a process could both geochemically enrich D'' in incompatible elements, and produce pooling of melt in the lowermost mantle. Such melt pooling would be compatible with recent seismic observations of an ultra-low velocity zone at the base of the mantle, and the thickness of this zone could be an indicator of the degree of melting which has occurred in the overlying mantle.

INTRODUCTION

The partitioning of major and minor elements between the dominant mineral of the Earth's lower mantle, silicate perovskite $(Mg,Fe)(Al,Si)O_3$, and its coexisting melts has important implications for the chemical differentiation of the Earth's interior, the fractionation of radiogenic isotopes in the mantle and the production of the heat which drives mantle convection and ultimately plate tectonics [*Ringwood*, 1975; *Basaltic Volcanism Study Project*,

1981]. Moreover, the recent observation that laterally extensive regions of the lowermost mantle may be partially molten provides evidence that the detailed chemistry of solid/liquid partitioning at ultra-high pressures may be critical for understanding the chemical evolution of the core-mantle boundary region [*Williams and Garnero*, 1996; *Revenaugh and Meyer*, 1997]. Indeed, it has also been proposed based upon the high ratio of changes in shear velocity to changes in compressional wave velocity $(dlnVs/dlnVp)$ in low velocity regions of the lower mantle that small amounts of partial melt (0~1%) may be widespread within this region [*Duffy and Ahrens*, 1992]. In spite of the possible importance of melting processes in the deep mantle, little is known about the geochemistry of solid/melt partitioning at deep lower mantle pressures (> 25

The Core-Mantle Boundary Region
Geodynamics 28
Copyright 1998 by the American Geophysical Union.

GPa). Relevant work to date has focussed on element partitioning between perovskite and majorite mineral phases, the partitioning of elements between silicate liquids and silicate perovskites near the low-pressure boundary of the perovskite stability field and the partitioning of iron between silicates and core-forming liquids [*Kato et al.*, 1988a,b; *Ito and Takahashi*, 1987; *Ohtani et al.*, 1991; *Trønnes et al.*, 1992; *Drake et al.*, 1993; *McFarlane et al.*, 1994; *Gasparik and Drake*, 1995; *Kato et al.*, 1996].

The partitioning of major elements (Mg, Si, Fe, Ca and Al) between silicate perovskite and coexisting melts could have a profound effect on the differentiation of the lowermost mantle. In particular, a natural rationale for the presence of a partially molten zone at the base of D" may exist if melts formed near the base of the mantle are denser than their coexisting solids. Such density crossovers between melt and coexisting solid have been demonstrated to occur at transition zone conditions [*Rigden et al.*, 1984; 1988; *Agee and Walker*, 1988a,b], but whether melts which are denser than their coexisting solids are expected to exist at the base of the mantle remains uncertain [e.g. *Knittle and Jeanloz*, 1989, 1991]. In particular, if iron extensively partitions into melts under lower mantle conditions, then it would be expected that such melts would be denser than the solid mantle and might sink to the core-mantle boundary. There is evidence from relatively low pressure measurements (<25 GPa) [*Kato et al.*, 1988a,b; *Ito and Takahashi*, 1987; *Ohtani et al.*, 1991; *Trønnes et al.*, 1992; *Drake et al.*, 1993; *McFarlane et al.*, 1994] that iron may preferentially partition into melt relative to silicate perovskite, but no data exist at pressures which lie well within the perovskite stability field. Such iron enrichment in melts could also produce relatively high conductivity silicate liquids: if iron-enriched melt exists at the core-mantle boundary (with iron-enrichment produced by either core-mantle reactions [e.g., *Knittle and Jeanloz*, 1991] or simple equilibration with lower mantle assemblages, as examined here), the propagation of the geomagnetic field through melt-containing regions could be profoundly alterred.

Additionally, models of mantle convection are highly dependent on the location of heat sources in the deep Earth [*Turcotte and Schubert*, 1982; *Sleep*, 1979; *Davies*, 1980]. One endmember possibility is that the mantle is primarily heated from below, with secular cooling of the core providing the main heat source driving mantle convection. The second extreme possibility is that the primary heat source in the deep Earth is from radiogenic decay within the lower mantle itself, predominantly from K, Th and U [e.g., *Urey*, 1956; *Wasserburg et al.*, 1964]. Naturally, it is almost certain that mantle convection is driven by a combination of core-derived heat from below and internal radiogenic heating, with estimates of core heat flux often lying in the 10% range of total heat flux out of the Earth [e.g. *Sleep*, 1979]. However, an important aspect of understanding heat sources within the mantle is basic: how do heat producing elements partition between solid silicate perovskite and the liquid? That is, could radiogenic elements have been extracted from the lower mantle via melting processes through Earth history, and concentrated at the base of the mantle in a partially molten zone at the base of D" in much the same way that these elements have been progressively extracted from the upper mantle?

We present here experimental results on the partitioning of five major elements (Mg, Fe, Si, Ca and Al) and two trace elements (U and K) between solid and liquid silicate perovskite using the laser-heated diamond anvil cell. Typical experimental conditions access pressures corresponding to depths well within the Earth's lower mantle (~ 50 GPa or 1500 km). Our results are by neccessity semi-quantitative, since laser-heating involves large temperature gradients which render it difficult to extract precise partition coefficients from our data. However, as both disequilibrium effects and (on average) higher temperatures within melted material than in the surrounding solid material should each bias partition coefficients towards unity, we can readily derive (at least) the sign of the partition coefficient from these studies.

EXPERIMENTAL TECHNIQUE

The samples containing uranium were prepared by mixing oxides (SiO_2, MgO, $Fe_{0.94}O$ and UO_2) together to obtain pyroxene stoichiometry with approximate composition $Mg_{0.87}Fe_{0.13}U_{0.01}SiO_3$. The mixed oxides contained approximately 2 weight percent uranium: although this amount is much larger than that likely to be present in the planet, this enrichment is necessary in order that the samples contain a sufficient amount of uranium to determine partitioning, and is equivalent to the standard "percent level doping" technique used in low pressure experiments on element partitioning [e.g., *McKay*, 1989]. The mixed-oxide material was ground together with a natural Bamble orthopyroxene of composition $Mg_{0.87}Fe_{0.12}Ca_{0.01}SiO_3$ in a 1:1 ratio by weight of the oxides to the natural orthopyroxene under acetone for approximately one hour. Resultant grain sizes were ~ 1-3 μm. Visually, the sample material appeared quite uniform with opaque grains of FeO and UO_2 evenly distributed throughout the sample. The purpose of grinding together the mixed oxides with the natural sample was to facilitate the heating of the sample with a Nd:YAG laser; the addition of natural orthopyroxene produces a more uniform temperature distribution than that observed when granular wüstite is the only material in the sample which absorbs the 1.06 μm radiation of the laser.

The starting material for the samples containing potassium was an aluminous pyroxene glass synthesized in the laboratory of Prof. D. Burnett at Cal Tech. This sample material was ground together in a 10:1 ratio (by weight) with natural Bamble orthopyroxene for one hour under acetone. As with the uranium-doped material, this addition of natural orthopyroxene was necessary to laser-heat the sample, as the glass itself does not absorb the Nd:YAG laser radiation.

From these pyroxene stoichiometry starting materials, the high-pressure perovskite phase samples were synthesized using a Mao-Bell type diamond cell with Type I diamonds having 500 μm culets. Samples were contained between the diamonds in stainless steel gaskets and were typically 150-300 μm in diameter. Pressures were determined in separate calibration runs in which the pyroxene starting materials were compressed together with ruby and the pressures measured using the standard ruby fluorescence method [*Barnett et al.*, 1973; *Mao et al.*, 1978]. In addition to the ruby fluorescence, the spring length of the diamond cell was also measured. The variation of pressure with spring length was then used as a calibration from which to estimate the pressure in the diamond cell. This method to estimate pressure was utilized to avoid contamination of our samples by the ruby calibrant. The maximum estimated uncertainties in pressure caused by the combination of using spring length calibrations and by non-hydrostaticity are about ±20 %. The accuracy of this technique is more than sufficient to demonstrate that the pressure in the diamond cell has exceeded the pressure of the perovskite stability field (>25 GPa). Synthesis pressures for the uranium-doped samples ranged from 51±10 GPa to 58±10 GPa and for the potassium-doped samples from 52±10 GPa to 56 ±10 GPa.

After being compressed in the diamond cell to pressures of ~50 (±10) GPa, samples were heated in a sub-solidus temperature regime (~1800-2000 K) with a 25 W CW Nd:YAG laser tuned in the TEM_{00} mode (Gaussian power distribution) to convert the material to the high-pressure perovskite structure. The Nd:YAG was absorbed by both the iron in the orthopyroxene structure (as well as the FeO and/or UO_2 in the oxide mix), ensuring reasonably uniform heating conditions, and thus uniform transformation of the sample to the perovskite structure. The laser beam (focussed to a beam diameter of about 25 μm) was scanned across the samples to convert them to the perovskite structure. During synthesis, average temperatures in the laser-heated spot were measured using a spectroradiometric technique described previously [*Heinz and Jeanloz*, 1987a,b; *Williams et al.*, 1991] and ranged from 1875±110 K to 2315±105 K. Visual observations of the samples upon transformation were identical to that of natural Bamble orthopyroxene with no dopants added: the samples turned a fairly uniform brown color with some darker irregularities caused by grain boundaries, occasional minor amounts of contamination by the steel gasket material and in some cases, unreacted FeO or UO_2. For the uranium and potassium-bearing samples synthesized in this way, x-ray diffraction, electron microprobe analysis and scanning electron microscopy were carried out on two to three samples quenched from similar high pressure and temperature conditions in order to characterize the run products and determine their compositions.

Following conversion of two to three samples to the high pressure phase, the laser beam was held stationary over seven to ten individual regions of the sample for 10 to 100 seconds, and the laser power is increased until the sample is visually observed to melt. The areas chosen for melting were always in transparent and homogeneously colored parts of the sample. This spatial selection ensures that the experiments are uncontaminated by regions in the sample which have either unreacted FeO or UO_2 or gasket material. In addition to visual observations, melting was also detected by the presence of a quenched melt structure after the laser is turned off. Figure 1 illustrates a typical quenched melt structure in the perovskite samples. Notably, these laser-heated spots have large temperature gradients; therefore, only the central portion of the heated spot melts, the outer regions of the laser heated spot remain solid (but hot). However, these temperature gradients can be directly measured with a modification of the spectroradiometric technique used to measure average temperatures [*Williams et al.*, 1991; *Heinz and Jeanloz*, 1987b; *Jeanloz and Kavner*, 1996]. The temperature of every laser-melted spot and its associated temperature gradient was not measured; rather, an average temperature and temperature gradient measurement were made at a given laser power for one molten spot in each diamond cell sample. Average melting temperatures were all in excess of 3400 (±300) K. The remainder of the experiments in the same sample were conducted at the same laser power in regions of the sample with similar optical properties. Figure 2 illustrates a typical temperature profile across a laser-melted spot.

X-ray diffraction of quenched samples (at zero pressure and 293 K) was used for phase identification and lattice parameter determinations. A single diamond cell sample (~150-300 μm diameter, ~15-20 μm thick and ~ 1 μg in weight) was mounted in a 114.6 mm Debye-Scherrer x-ray camera, the sample was exposed to $CuK\alpha$ x-rays for 24 hours, and the pattern was collected on film. A diffraction pattern was collected from two different samples for both the U and K-doped perovskites to verify the reproducibility of the results.

The chemical composition of the two different silicate perovskites, the uniformity of the samples and the compositional variation across the laser-melted spots were examined using the electron microprobe and scanning electron microscope. As with the x-ray diffraction, all these analyses were carried out on single diamond cell samples quenched from high pressures and temperatures. Perovskite samples were removed from the diamond cell, mounted in epoxy and polished. During polishing, particular attention was paid to the removal of the top 2-3 μm of the sample. This is because the region of the sample next to the highly thermally conductive diamond (k = 9.9 $W\text{-}cm^{-1}\text{-}K^{-1}$) does not melt in laser-heated experiments, and

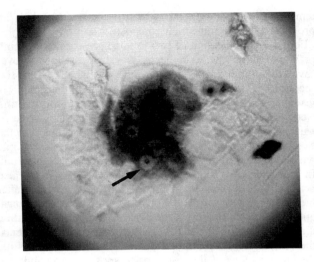

Figure 1. A photograph of a diamond cell sample of silicate perovskite quenched from high pressures and temperatures. The arrow indicates a quenched melt blob which is distinguished by its round structure reflecting the Nd:YAG laser beam shape. The dark inner region of the quenched spot was above the perovskite melting temperature while the lighter halo is the solid region of the heated spot. The dark color of the inner part of the quenched structure is the result of the concentration of primarily iron in what was the molten portion of the sample.

must be removed to expose the regions of quenched melt in the sample. The sample compositions are characterized using an 8-channel ARL electron microprobe operated at an accelerating voltage of 15 kV and are analyzed for Mg, Fe, Al, Si, O, Ca, U and K. Notably, the oxygen content of the samples was directly analyzed in these measurements and is not determined by difference.

The average compositions of the silicate perovskites, given in Table 1, were measured using at least two unmelted regions of samples and result from an average of ~100 individual compositional determinations for each type of perovskite. All samples studied were uniform with respect to each element analyzed on the 2-5 μm scale, with the exception of the iron content of a few samples which were non-uniform on the 5-10 μm scale. These latter compositional heterogeneities were likely produced by sample contamination by shreds of the stainless steel gasket which become embedded in the sample material at high pressures. Such iron-enriched regions of samples were avoided in the compositional analyses of the samples.

To characterize the compositional variations across a melted region, microprobe analyses were conducted across the melt blobs at 2 μm intervals. For the uranium-doped samples, satisfactory results were obtained on four regions

of quenched melt, while for the potassium-doped perovskite three melt regions yielded acceptable results. In each of these cases, the results at every 2 μm interval produce microprobe weight percent totals of 95% of greater. A large number of quenched melt blobs were impossible to analyze because the diamond cell samples are too thin (< 10 μm in sample thickness), yielding poor microprobe totals.

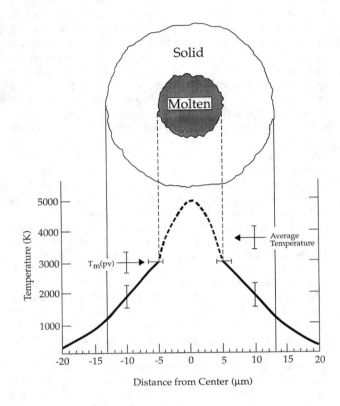

Figure 2. An illustration of a temperature profile in a laser-melted spot such as that shown in Figure 1. The extent of the molten region (~10 μm in diameter) is determined by the silicate perovskite melting temperature: here, at 55 GPa, the melting temperature is 3000 ± 300 K [*Knittle and Jeanloz*, 1989]. Therefore, the interface between liquid and solid provides provides a temperature "fixed point" in the sample. In both the outer, solid part of the laser-heated spot and the inner molten region, the temperature changes with an approximately Gaussian distribution; however, due to the different abilities of the solid and liquid to absorb the laser radiation, the molten zone exhibits a much steeper temperatuare gradient [cf. *Williams et al.*, 1991; *Godwal et al.*, 1990; *Jeanloz and Kavner*, 1996]. This rapid steepening of the temperature gradient in the liquid also contributes to the high average sample temperature during melting. Previous work on the measured temperature corrections necessary to convert average sample temperatures to peak temperatures, indicates that the peak temperature in this example is ~4500 K [*Heinz and Jeanloz*, 1987b; *Knittle and Jeanloz*, 1989; *Williams et al.*, 1991; *Jeanloz and Kavner*, 1996].

Table 1. Average Composition of Silicate Perovskite

Element	U-Doped Starting Material (Atomic %)	K-Doped Starting Material (Atomic %)
O	60.10 (±1.3)	59.52 (±1.1)
Mg	17.13 (±0.01)	15.73 (±0.35)
Fe	2.61 (±0.14)	0.44 (±0.27)
Si	19.90 (±0.15)	20.05 (±0.53)
Ca	0.13 (±0.01)	0.06 (±0.04)
U	0.06 (±0.04)	---
K	---	1.46 (±0.27)
Al	---	2.02 (±0.20)
Total	99.93 (±1.32)	99.28 (±1.34)

The same melt spots studied using the electron microprobe are also examined using scanning electron microscopy (SEM). An ISI WB-6 microscope with EDS analytical ability was used to obtain backscattered electron images and characteristic x-ray maps of the quenched melt regions. Although the SEM compositional analyses were not quantitative, the samples again appeared compositionally uniform (apart from both the iron-enriched regions and the laser-heated spots) down to at least the 400 nm scale.

RESULTS AND DISCUSSION

Phase Identification and Average Compositions

The x-ray diffraction pattern for the uranium-doped perovskite contains only diffraction lines which can be assigned to orthorhombic silicate perovskite: no x-ray lines are seen which can be attributed to minor contaminating phases. The lattice parameters of the perovskite phase are a = 479.8 (±0.3) pm, b = 493.6 (±0.2) pm and c = 691.0 (±0.3) pm with a resultant volume (163.65 (±0.46) x 10^6 pm^3). The average composition of the uranium-doped perovskite, as determined by electron microprobe analysis is given in Table 1, and is consistent with a perovskite stoichiometry of $(Mg_{0.861}Fe_{0.131}Ca_{0.006}U_{0.003})SiO_3$. Although the uranium is present in the starting material as UO_2, no separate UO_2 phase is observed in the high pressure samples either in the x-ray diffraction results or the microanalysis using the SEM and electron microprobe. In particular, the SEM analyses confirm that no UO_2 remains in the sample in particles larger than 400 nm. This result suggests that some or all of the uranium may have been incorporated into the perovskite structure upon synthesis. Alternatively, the UO_2 may be finely dispersed throughout the sample and the lack of a UO_2 x-ray pattern is due to the small amount of material in an individual diamond cell sample (about 1 μg total weight per sample).

There is also no evidence in these experiments for the minor amount of calcium present in the Bamble enstatite segregating into a separate $CaSiO_3$ perovskite phase [e.g. Mao et al., 1977; Liu, 1979; Tamai and Yagi, 1989; Kim et al., 1994]. Although we would not expect to see evidence of $CaSiO_3$ perovskite in the quenched samples because it amorphizes upon decompression [Liu and Ringwood, 1975), there is no evidence for the presence of this phase from the SEM analyses. Because of the small amount of calcium present, it may be that it is retained in solid solution in the orthorhombic structure.

The zero-pressure x-ray diffraction pattern of the potassium-doped perovskite gives lattice parameters for this material of a equals 479.1 (±0.2) pm, b equals 493.1 (±0.1) pm and c equals 690.3 (±0.2) pm with a volume of the K-doped phase (163.08 (±0.3) x 10^6 pm^3). Only diffraction lines of the perovskite structure are observed: in particular, no evidence for potassic hollandite is observed [Ringwood et al., 1967]. Table 1 lists the average composition of this perovskite as determined from the electron microprobe: $(Mg_{0.79}Fe_{0.02}K_{0.07}Al_{0.11}Ca_{0.01})SiO_3$. These samples are enriched in both K and Al and depleted in Fe relative to the uranium-doped samples. In this case, the nature of the potassium-bearing starting material, a pyroxene stoichiometry glass, and the x-ray and SEM data suggest that the potassium and aluminum are incorporated into the perovskite structure. Again, the x-ray diffraction, microprobe and SEM examination of the samples documents that the synthesized perovskite is single phase material down to the maximum resolution of these techniques (~ 400 nm with the SEM).

Partitioning of elements between solid and liquid silicate perovskite

The electron microprobe results for the behavior of Mg, Si, Fe, Ca, O and U across the melt-solid interface in these U-doped perovskite samples are plotted in Figures 3a-f. These data are averages of three successful microprobe traverses across two different laser-melted spots. On an atomic percent basis, silicon, oxygen and magnesium are depleted in the glass quenched from the central molten region of the laser-heated spot and uranium, calcium and iron are enriched. The microprobe results give an estimate of the stoichiometry of the solid and molten regions of the laser-heated spot (Table 2). For the solid portion, the silicate has a perovskite stoichiometry which is slightly magnesium enriched and iron- and calcium-depleted relative to the average uranium-free composition of the starting material: $Mg_{0.903}Fe_{0.086}Ca_{0.004}SiO_3$. Because of the limited amount of sample material in the laser-melted spot, the microprobe determinations are not as accurate as the results for the average silicate perovskite material. The results on the quenched glass are consistent with the following stoichiometry in the melt: $Mg_{0.43}Fe_{1.0}$

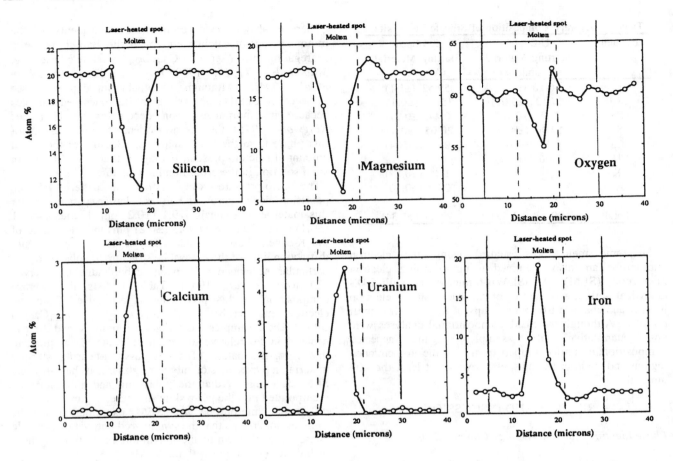

Figure 3a-g. The electron microprobe traverses for quenched uranium-doped silicate perovskite plotted as atomic % versus arbitrary distance across the laser-melted spot. The total diameter of the laser-melted spot is indicated by the solid lines and the molten region is delineated by the broken lines. These figures illustrate that uranium, iron and calcium are incompatible in lower mantle melts dominated by Mg-endmember perovskites.

Table 2. Composition of the Quenched Laser-Heated Spot for the U-Doped Perovskite

Element	Solid Region (Atomic %)	Molten Region (Atomic %)
O	60.64 (±2.1)	55.97 (±3.52)
Mg	18.22 (±0.05)	8.09 (±1.10)
Fe	1.73 (±0.25)	19.02 (±1.05)
Si	20.17 (±0.33)	11.60 (±0.56)
Ca	0.08 (±0.03)	2.43 (±0.41)
U	0.00	4.25 (±1.18)
Total	100.84 (±2.14)	101.36 (±4.07)

$Ca_{0.13}U_{0.23}Si_{0.62}O_3$. Mass balance of the liquid and solid portions of the laser-heated spot require that about 5% of the sample melted.

Figures 4a-g show the partitioning results for Si, Mg, Fe, O, Ca, Al and K in the potassium-doped sample. The example plotted in Figure 4 is a microprobe traverse across two adjacent quenched melt blobs, in which three separate microprobe scans across this region were averaged to obtain the atomic percent values. Again oxygen, magnesium, and silicon are depleted in the melt and iron is enriched. In addition, aluminum preferentially enters the melt while potassium surprising remains in the solid perovskite. Satisfactory microprobe results were not obtained for the chemical analysis of the solid portion of the laser heated spot for this composition. However, Table 3 gives the microprobe results on the composition of the molten region of the potassium-doped perovskite: the quenched glass composition corresponds to a stoichiometry of $Mg_{0.71}Fe_{0.08}Ca_{0.09}K_{0.02}Si_{0.82}Al_{0.56}O_3$.

Notably, the large temperature gradients present in the laser-heated spots cannot be a controlling factor in determining the observed element partitioning in these experiments. In fact, all the partitioning trends observed are opposite in sign to what would be expected if Soret diffusion played a major role in the chemical changes observed. In Soret diffusion, the heaviest elements migrate

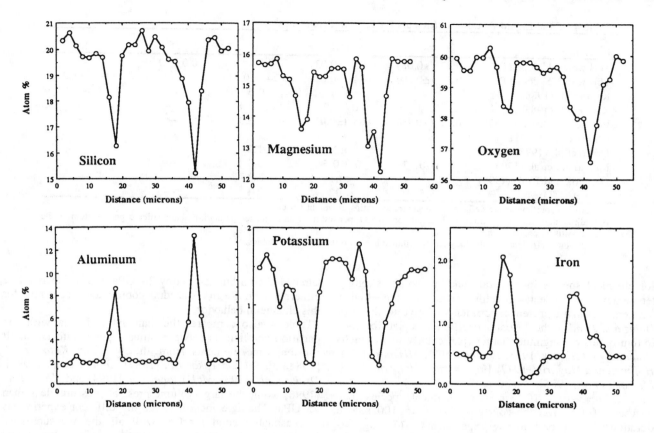

Figure 4a-f. The electron microprobe traverses for quenched potassium-doped silicate perovskite plotted as atomic % versus arbitrary distance across the laser-heated spot. Here, the microprobe intersects two quenched melt blobs at 15 and 40 μm distance. These figures illustrate that iron, calcium and aluminum are incompatible in lower mantle melts.

Table 3. Composition of the Quenched Laser-Heated Spot for the K-Doped Perovskite

Element	Molten Region (Atomic %)
O	57.91 (±0.59)
Mg	13.65 (±0.80)
Fe	1.53 (±0.29)
Si	15.74 (±0.53)
Ca	1.69 (±0.21)
K	0.44 (±0.24)
Al	10.90 (±2.30)
Total	101.90 (±2.60)

to the lowest temperature region of the sample: thus, in the case of laser-heated diamond cell samples, heavy elements would move towards the outermost edge of the laser-heated spot. This is in fact observed in diamond cell experiments where the laser is held on a given spot in the sample for long periods of time [~ 1 hour: see *Heinz and Jeanloz*, 1987a]. There is no evidence from the microprobe or SEM

examination of the samples to suggest that the heavier elements present in these experiments are migrating toward the edge of the laser-heated spot; in fact, the heavy elements, in general, are observed to partition into the central molten hottest part of the laser-heated spot .

Partition Coefficients

It is possible to estimate solid-liquid partition coefficients for silicate perovskite from these experiments despite the large temperature gradients in the samples. From the Soret diffusion experiments of *Heinz and Jeanloz* [1987a], it should take 45 (±15) minutes for most of the iron in a sub-solidus laser-heated sample (~2000 K) to migrate to the edge of a laser-heated spot. This result gives an approximate Soret diffusion coefficient (D) for iron in perovskite at this temperature of 10 (±5) x 10^{14} m^2/sec. In the partitioning experiments, the temperature in the bulk of the laser-heated spot during melting is higher than in the sub-solidus experiment by 1000-2000 K, which implies that this diffusion coefficient is a lower limit for our present experiment. This is because diffusion coefficients

Table 4. Solid/Liquid Partion Coefficients for Silicate Perovskites[a]

Reference	Fe	Ca	Al	U	K
This work: U-doped Pv	0.12 (±0.08)	0.05 (±0.03)	---	0.0 (±0.3)	---
This work: K-doped Pv	0.34 (±0.20)	0.04 (±0.02)	0.24 (±0.11)	---	3.92 (±1.21)
Kato et al. (1988a)[b]	---	0.2	0.5	---	---
Kato et al. (1988b)	0.21	0.07-0.21	0.9-1.8	---	---
Ito and Takahashi (1987)	0.31 (±0.26)	0.05 (±0.01)	0.97 (±0.1)	---	---
Ohtani et al. (1991)	0.1	---	---	---	---
Drake et al. (1993)	---	0.2	1.2	---	---
McFarlane et al. (1994)	0.45-0.57	0.2-0.5	1.0-1.3	---	---
Gasparik and Drake(1995)[c]	---	<0.1	3.1	---	---
Kato et al. (1996)	3.5	0.02	2.1	---	---

a. Partition coefficients are defined by concentration ratios in weight per cent.
b. Indirect derivation of partition coefficients for $MgSiO_3$ perovskite from majorite- garnet/liquid and silicate perovskite/majorite-garnet partition coefficients.
c. Partition coefficients in this study are determined in the presense of the volatiles.

for chemical species in minerals and melts are strongly temperature dependent; therefore, due to the higher temperatures in the present experiments relative to those of *Heinz and Jeanloz*, the diffusion coefficient is plausibly one to four orders of magnitude larger (particularly in the melt) [*Buening and Buseck*, 1973; *Foland*, 1974; *Misener*, 1974; *Hoffman and Magaritz*, 1977; *Magaritz and Hoffman*, 1978; *Morioka et al.*, 1985].

For a D = ~10 x 10^{-13} m^2/sec, chemical species may diffuse 3-10 μm over the course of the 10 to 100 second duration of these partitioning experiments. This suggests that local equilibrium may be obtained particularly near the hotter, central regions of the laser-heated spot, while not enough time has elapsed for Soret diffusion to dominate in the cooler, outer regions of the laser-heated spot. Examination of the highest quality data obtained from these experiments (that of Fe in the U-doped perovskite: Figure 3d) shows that at the edge of the laser-heated spot, a slight rise in iron content is obtained. That this increase in iron content in the colder region of the samples is small implies that the effect of Soret diffusion in minor in these experiments. Nevertheless, there may be a small contribution to the chemical gradient in these samples from Soret diffusion superimposed on the larger effects of chemical partitioning.

Table 4 lists the calculated partition coefficients for Fe, Ca, Al, K and U determined from this study. The behavior of the most abundant minor elements is essentially what would be expected from studies of solid-melt partitioning of lower pressure mineral assemblages: iron, calcium and aluminium preferentially partition into the melt relative to the coexisting solid. This result is comparable to the situation observed at much lower pressures in basalt petrogenesis [*Ringwood*, 1975; *BVSP*, 1981]. However, for these high pressure melts, Si is depleted in the melt: the opposite of the observed behavior of silicon in the formation of basalts which are enriched in Si relative to the peridotitic source rock. Presumably, the change in the nature of Si-O bonding between minerals with Si in

tetrahedral coordination below 24 GPa and and perovskite minerals with Si in octahedral coordination accounts for this change in behavior.

Table 4 also compares the diamond cell data with the partition coefficients between silicate perovskite and melt for the same elements from the work of *Kato et al.* [1988a,b], *Ito and Takahashi* [1987], *Ohtani et al.* [1991], *Drake et al.* [1993], *McFarlane et al.* [1994] and *Kato et al.* [1996] using the large-volume press at pressures less than 25 GPa. The agreement between the different experiments is reasonably good for Ca, with all the measurements indicating that this element partitions into the liquid relative to Mg-rich silicate perovskite. This partitioning behavior is consistent with limited Ca-solubility in Mg-rich perovksite, as this relatively large cation does not readily dissolve in the smaller, distorted magnesium silicate perovskite structure.

The results for Al are more problematic with a large range of partition coefficients observed between studies indicating both compatibility and incompatiblity of Al between silicate perovskite and liquid (sometimes in the same study: *Kato et al.*, 1988b). However, most of the previous work concludes that aluminum remains dissolved in solid $(Mg,Fe)SiO_3$ perovskite and does not partition into the melt in contrast to the present diamond cell results. We note that all the relevant previous experiments are conducted at maximum pressures 25 GPa, marginally within the perovskite stability field which may influence these results. Alternatively, as a trivalent ion with behavior analogous to that of Fe^{3+}, the solubility of Al^{3+} in silicate perovskite may be somewhat dependent on oxygen fugacity within the experimental charge. Diamond-cell experiments are likely to have a different oxygen environment that large volume press charges. In particular, diamond cell samples are likely to be more reduced than large volume pressure charges.

Notably and importantly for understanding the behavior of deep Earth melts, the diamond cell experiments indicate a strong enrichment of iron in liquids coexisting with silicate perovskite. As can be seen from Table 4, this behavior has also been noted in all the large volume press experiments except for the recent work of *Kato et al.* [1996]. Indeed, the present results at 50 GPa are consistent with even more robust partitioning of iron into the liquid than has been observed in the lower pressure experiments: this enhanced amount of iron in the liquid is anticipated to induce a marked increase in the density of silicate liquids under lower mantle pressure conditions.

In the Earth's upper mantle, uranium and potassium are incompatible elements which strongly partition into melt relative to solid and are thus strongly concentrated in the Earth's crust [*Seitz and Shimizu*, 1972; *Seitz*, 1973; *Benjamin et al.*, 1978; *Shimizu*, 1974; *McKay and Weill*, 1977; *Ringwood*, 1975, *BVSP* 1981]. From the experiments conducted at lower mantle pressures described here, uranium exhibits geochemical behavior similar to that at lower pressures, with strong partitioning into the melt phase (Table 4). However, the partitioning behavior of potassium is profoundly different at high pressures, possibly reflecting major changes in the electronic structure of K under pressure. The observations of compatible behavior of potassium within silicate perovskite is in marked contrast to the results of *Kato et al.* [1988a] who found, based on an indirect calculation of partition coefficients of silicate perovskite, that potassium is incompatible in the Mg-perovskite structure and strongly partitions into coexisting melt. However, these conclusions were based on samples that contained both calcium-silicate perovskite as well as majoritic garnet. Therefore, the two results are not directly analogous, in that the present experiments involve partitioning only between the $(Mg,Fe)SiO_3$ perovskite phase and liquid.

The partitioning of potassium is extremely different in the high-pressure experiments than would be expected from examining zero-pressure ionic radii. Specifically, the K^+ ion, which is extremely large at zero-pressure $((K^+-O^2)^{VIII-XII} = 294-306$ pm) and generally viewed as an incompatible ion within the upper mantle, remains in the perovskite structure relative to the melt at high pressures (Fig. 4). This transition of potassium from incompatible to compatible behavior with pressure may be viewed as due to a combination of two effects: first, the dodecahedral site in perovskite is larger than that in any upper mantle minerals; and second, the volume of the potassium ion is anticipated to markedly decrease under compression. For example, metallic K undergoes phase transitions at pressures of 11 and 20 GPa which involve an s-d electron transfer leading to a large decrease in the volume of the metal at high pressures ($V/V_0 = 0.45$ at 11 GPa: cf. *Bukowinski* [1976] and *Olijnyk and Holzapfel* [1983]). Such an alteration in the properties of potassium could also generate its ability to substitute into clinopyroxenes at pressures above about 8 GPa [*Harlow and Veblen*, 1991], a result which is in complete accord with the present observations of compatible behavior of the potassium ion within perovskite at very high pressures.

Implications for the Lower Mantle and Core-Mantle Boundary Region

Our observations of U and K partitioning provide insights into the possible location of these heat-producing elements in Earth's lowermost mantle. Melting events in the lower mantle will tend to partition uranium out of perovskite mineral and into the melt, while potassium may remain a compatible element in silicate perovskite at high pressures. Pervasive melting of the lowermost mantle is a likely consequence of giant impacts early in Earth's history [*Cameron*, 1986; *Melosh and Sonnett*, 1986; *Stevenson*, 1987]. Therefore, segregation of uranium from the bulk of the lower mantle could have been generated during solidification of an early magma ocean, if negatively buoyant melts percolated through lower mantle assemblages. Potassium, however, is likely to have been retained throughout the bulk of the lower mantle, providing a continuing internal heat source for mantle convection. Indeed, as the arguments for potassium being soluble in the iron alloy liquid of the Earth's outer core are weak on both theoretical and experimental grounds [*Sherman*, 1990; *Somerville and Ahrens*, 1980], the lower mantle could be the primary reservoir for potassium in the Earth, with potassium being a major internal heat source driving convection within the mantle. Uranium, however, could be segregated into the deep mantle and possibly remain trapped in the core-mantle boundary. Here, the enrichment of uranium could provide a long-term heat source for the core-mantle boudary region analogous to the role uranium plays as a heat source in the continental crust. Alternatively, uranium could be remixed into the bulk of the mantle during solid-state convection, or possibly dissolve into the core [*Murrell and Burnett*, 1986].

One implication of the partitioning of iron into the melt relative to the solid is to shift the density contrast between the solid and liquid. At upper mantle pressures, silicate liquids appear to become denser than their coexisting solids because of the relatively high compressibility of silicate liquid [*Stolper et al.*, 1981; *Rigden et al.*, 1984, 1988; *Agee and Walker*, 1988b; *Williams and Jeanloz*, 1988]. From the partition coefficients determined in this study, the effect of element partitioning on the densities of coexisting solid perovskite and liquid may be evaluated. We note that silicate perovskite is likely to be the solidus phase for lower mantle mineral assemblages which would include $(Mg,Fe)O$ magnesiowustite as an accessory phase [*Williams*, 1990].

The precise melting curve of silicate perovskite at high pressures is controversial, with two different families of results: those which produce a nearly constant melting temperature of ~3000 K at pressures between 30 and 60 GPa [*Heinz and Jeanloz*, 1987b; *Knittle and Jeanloz*, 1989; *Heinz et al.*, 1994], and those which when extrapolated produce a melting temperature of near 7500 K at the core-mantle boundary [*Zerr and Boehler*, 1993]. The possible origins of this discrepancy have been discussed [*Zerr and*

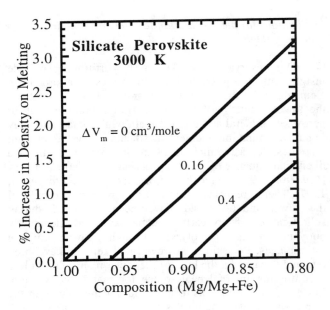

Figure 5. The density increase on melting of silicate perovskite plotted as a function of the perovskite Mg/(Mg+Fe) ratio. The calculations are all done at 3400 K, the melting temperature of perovskite at a mid-mantle pressure of ~80 GPa [*Knittle and Jeanloz*, 1989]. The assumed volume change on fusion in cm³/mole is indicated on each line. Here, the increase in density of the liquid is completely due to the iron partitionng into the melt relative to the solid ($D_{solid/liquid}$= 0.23 ± 0.11 for Fe) at deep mantle conditions. For a total mantle bulk composition of Mg/(Mg+Fe) = 0.88, the maximum Mg-content of the perovskite phase would range from Mg/(Mg+Fe)~0.94 to 0.90 depending on the perovskite/magnesiowustite ratio of the lower mantle [*Mao et al.*, 1997].

Boehler, 1993; *Heinz et al.*, 1994; *Jeanloz and Kavner*, 1996]. The results of *Knittle and Jeanloz* (1989) indicate a volume change on melting of 0 (±0.2) cm³/mole between 30 and 60 GPa with the volume change on melting increasing to 0.16 (±0.03) cm³/mole above 60 GPa. Zerr and Boehler's results imply a positive volume change on melting, but do not constrain the absolute value of this change. Regardless of the precise magnitude of the volume change on melting, the density contrast between solid and liquid could be profoundly affected by the relative iron contents of the two phases.

Figure 5 illustrates the density change on melting for different perovskite iron contents for volume changes ranging between 0 cm³/mole and 0.4 cm³/mole The molar volumes for the liquid and solid are taken from *Knittle and Jeanloz* [1989], and the densities are corrected to account for the observed iron partitioning coefficient of 0.23 (±0.11) between solid and liquid. Notably, this analysis does not treat effects of Ca, Al or volatile components on melt densities. Clearly, for small volume changes of fusion, the

effect of iron partitioning will be to produce a melt which is 1-3 % denser than its coexisting solid depending on the iron content of the silicate perovskite. Our results also constrain the magnitude of the volume change on melting which would result in a densification of the liquid relative to the solid due to iron partitioning. For likely iron contents of silicate perovskites in realistic mantle mineral assemblages (perovskite compositions of $Mg_{0.94-0.90}Fe_{0.06-0.1}SiO_3$ for a total mantle Mg/(Mg+Fe) ratio of 0.88: see *Mao et al.* [1997]), volume changes on melting of ~0.3 cm³/mole or less are required.

Therefore, negatively buoyant melts may be a natural consequence of small volume changes of fusion coupled with iron partitioning into the melt under lower mantle conditions. The implications of this phenomenon are rather profound: any melt which occurs in the deep mantle may drain into D", forming a melt- (and iron-) enriched layer above the core-mantle boundary: a result in reasonable accord with recent seismic observations of an ultra-low velocity layer in the lowermost 5-40 km of portions of the mantle [*Garnero and Helmberger*, 1995; *Garnero et al.*, 1996; *Williams and Garnero*, 1996; *Revenaugh and Meyer*, 1997]. Notably, the amount of partial melting of the lower mantle required to generate the ultra-low velocity zone is small. Utilizing maximum estimates of near 40 km for the thickness of this layer, 30% partial melting present within it, and assuming a (clearly overestimated) global lateral extent of the layer [*Williams and Garnero*, 1996; *Garnero and Helmberger*, 1995] indicates that exposing the lower mantle to less than 0.5% partial melting could generate the amount of melt inferred to be present in this layer: this is similar in magnitude to the amount of partial melt proposed by *Duffy and Ahrens* [1992] to be present in low velocity regions of the lowermost mantle. Accordingly, there may exist a causal relationship between the high values of dlnVs/dlnVp in the lower mantle and the ultra-low velocity zone at its base.

Acknowledgments. This work was supported by the NSF, the A.P. Sloan Foundation and the W.M. Keck Foundation. I thank D.L. Burnett for providing the K-doped pyroxene starting material; J. Donovan for technical assistance at the U.C. Berkeley microprobe; J. Krupp for technical assistance at the U.C. Santa Cruz electron microscopy lab; and the reviewer, Q. Williams, R. Jeanloz, C. Agee, M.L. Drake and D. Walker for comments on this work.

REFERENCES

Agee, C. and D. Walker, Mass balance and phase density constraints on early differentiation of chondritic mantle,*Earth Planet. Sci. Lett.*, 90, 144-156, 1988a.

Agee, C. and D. Walker, Static compresion and olivine flotation in ultrabasic silicate liquid, *J. Geophys. Res.*, 93, 3437-3449, 1988b.

Barnett, J., S. Block and G. Piermarini, An optical fluorescence system for quantitative pressure measurement in the diamond anvil cell, *Rev. Sci. Instrum.*, 44, 1-9, 1973.

Basaltic Volcanism Study Project, *Basaltic Volcanism on the Terrestrial Planets*, 1286 pp., Pergamon Press, New York, 1981.

Benjamin, T.M., R. Heuser and D.S. Burnett, Solar system actinide abundances. I. Laboratory partitioning between whitlockite, diopsidic clinopyroxene and anhydrous melt, *Proc. Lunar Planet. Sci. Conf. IX*, 70-77 (1978).

Buening, D.K. and P.R. Buseck, Fe-Mg lattice diffusion in olivine, *J. Geophys. Res.*, 78, 6852-6862, 1973.

Bukowinski, M.S.T., The effect of pressure on the physics and chemistry of potassium, *Geophys. Res. Lett.*, 3, 491-494, 1976.

Cameron, A.G.W., in *Origin of the Moon*, W.K. Hartmann, R.J. Phillips and G.J. Taylor, Eds., pp. 609-616, Lunar and Planetary Institute, Houston (1986).

Davies, G.F., Thermal histories of convective Earth models and constraints on radiogenic heat production in the Earth, *J. Geophys. Res.*, 85, 2517-2530, 1980.

Drake, M.J., E.A. McFarlane, T. Gasparik and D.C. Rubie, Mg-perovskite/silicate melt and majorite garnet/silicate melt partition coefficients in the system CaO-MgO-SiO_2 at high temperatures and pressures, *J. Geophys. Res.*, 98, 5427-5431, 1993.

Duffy, T. and T.J. Ahrens, Lateral variation in lower mantle seismic velocity, in *High Pressure Research: Application to Earth and Planetary Sciences*, Y. Syono and M.H.Manghnani, Eds., pp. 197-205, Terra Scientific, Tokyo, 1992.

Foland, K.A., Alkali diffusion in orthoclase, in *Geochemical Transport and Kinetics, Publ. 634*, A. Hofmann, B. Gilletti, H. Yoder and R. Yund, Eds., pp. 77-98, Carnegie Inst. Washington, Washington, D.C. (1974).

Garnero, E.J. and D.V. Helmberger, A very slow basal layer underlying large-scale low-velocity anomalies in the lower mantle beneath the Pacific: Evidence from core phases, *Phys. Earth Planet. Inter.*, 91, 161-176, 1995.

Gasparik, T. and M.J. Drake, Partitioning of elements among two silicate perovskites, superphase B, and volatile-bearing melt at 23 GPa and 1500-1600°C, *Earth Planet. Sci. Lett.*, 134, 307-318, 1995.

Godwal, B.K., C. Meade, R. Jeanloz, A. Garcia, A.Y. Liu and M.L. Cohen, Ultrahigh pressure melting of lead: A multidisciplinary study, *Science*, 248, 462-464, 1990.

Harlow, G.E. and D.R. Veblen, Potassium in clinopyroxene inclusions from diamonds, *Science*, 251, 652-655, 1991.

Heinz, D.L. and R. Jeanloz, Measurement of the melting curve of $Mg_{0.9}Fe_{0.1}SiO_3$ perovskite under lower mantle conditions and its geophysical implications, *J. Geophys. Res.*, 92, 11437-11444, 1987a.

Heinz, D.L. and R. Jeanloz, Temperature measurement in the laser-heated diamond cell, in *High Pressure Research in Mineral Physics*, M.H. Manghnani and Y. Syono, Eds.,pp. 113-127, American Geophysical Union, Washington, D.C., 1987b.

Heinz, D.L., E. Knittle, J.S. Sweeney, Q. Williams and R. Jeanloz, High pressure melting of $(Mg,Fe)SiO_3$-perovskite, *Science*, 264, 279-280, 1994.

Hofmann, A.W. and M. Margaritz, Diffusion of Ca, Sr, Ba and Co in basaltic melt: implications for the geochemistry of the mantle, *J. Geophys. Res.*, 82, 5432-5440, 1977.

Ito, E. and E. Takahashi, Melting of peridotite at uppermost lower mantle conditions, *Nature*, 328, 514-517, 1987.

Jeanloz, R. and A. Kavner, Melting criteria and imaging spectroradiometry in laser-heated diamond-cell experiments, *Phil. Trans. R. Soc. Lond. A*, 354, 1279-1305, 1996.

Kato, T., E. Ohtani, Y. Ito and K. Onuma, Element partitioning between silicate perovskites and calcic ultrabasic melt, *Phys. Earth Planet. Inter.*, 96, 201-207, 1996.

Kato, T., A.E. Ringwood and T. Irifune, Experimental determination of element partitioning between silicate perovskites, garnets and liquids: Constraints on early differentiation of the mantle, *Earth Planet. Sci. Lett.*, 89, 123-145, 1988a.

Kato, T., A.E. Ringwood and T. Irifune, Constraints on element partition coefficients between $MgSiO_3$ perovskite and liquid determined by direct measurements, *Earth Planet. Sci. Lett.*, 90, 65-68, 1988b.

Kim, Y.H., L.C. Ming and M.H. Manghnani, High pressure phase transformations in a natural crystalline diopside and a synthetic $CaMgSi_2O_6$ glass, *Phys. Earth Planet. Inter.*, 83, 67-79, 1994.

Knittle, E. and R. Jeanloz, Melting curve of $(Mg,Fe)SiO_3$ perovskite to 96 GPa: Evidence for a structural transition in lower mantle melts, *Geophys. Res. Lett.*, 16, 421-424, 1989.

Knittle, E. and R. Jeanloz, The Earth's core-mantle boundary: results of experiments at high pressures and temperatures, *Science*, 251, 1438-1443, 1991.

Liu, L.G., The high pressure phase transformations of monticellite and implications for upper mantle mineralogy, *Phys. Earth Planet. Inter.*, 20, 25-29, 1979.

Mao, H.K., P.M. Bell, J. Shaner and R. Steinberg, Specific volume measurements of Cu, Mo, Pd, and Ag and calibration of the ruby R1 fluorescence pressure gauge from 0.06 to 1 Mbar, *J. Appl. Phys.*, 49, 3276-3283, 1976.

Mao, H.K., T. Yagi and P.M. Bell, Mineralogy of the Earth's deep mantle: Quenching experiments on mineral compositions at high pressures and temperature, *Year Book Carnegie Inst. Wash.*, 76, 502-504, 1977.

Mao, H.K., G. Shen and R.J. Hemley, Multivariable dependence of Fe-Mg partitioning in the lower mantle, *Science*, 278, 2098-2100, 1997.

Margaritz, M. and A.W. Hofmann, Diffusion of Sr, Ba and Na in obsidian, *Geochim. Cosmochim. Acta*, 42, 595-605, 1978.

McFarlane, E.A., M.J. Drake and D.C. Rubie, Element partitioning between Mg-perovskite, magnesiowüstite, and silicate melt at conditions of the Earth's mantle, *Geochim. Cosmochim. Acta*, 58, 5161-5172, 1994.

McKay, G.A., Partitioning of rare earth elements between major silicate minerals and basaltic melts, *Rev. Mineral.*, 21, 45-77, 1989.

McKay, G.A. and D.F. Weill, KREEP petrogenesis revisited, *Proc. Lunar Planet. Sci. XIII*, pp. 2338-2355, Lunar and Planetary Institute, Houston, 1977.

Melosh, H.J. and C.P. Sonnett, in *Origin of the Moon*, W.K. Hartmann, R.J. Phillips and G.J. Taylor, Eds., pp. 609-616, Lunar and Planetary Institute, Houston, 1986.

Misener, D.J., Cationic diffusion in olivine to 1400°C and 35 kbar, in *Geochemical Transport and Kinetics, Publ. 634*, A. Hofmann, B. Gilletti, H. Yoder and R. Yund, Eds., pp. 117-129, Carnegie Inst. of Washington, Washington, D.C., 1974.

Morioka, M., K. Suzuki and H. Nagasawa, Trace element diffusion in olivine: Mechanism and possible implication to natural silicate systems, in *Point Defects in Minerals*, R.N. Schock, Ed., pp. 116-121, American Geophysical Union, Washington, D.C., 1985.

Murrell, M.T. and D.S. Burnett, Partitioning of K, U and Th between sulfide and silicate liquids: Implications for radioactive heating of planetary cores, *J. Geophys. Res., 91*, 8126-8136, 1986.

Ohtani, E., T. Kato and E. Ito, Transition metal partitioning between lower mantle and core materials at 27 GPa, *Geophys. Res. Lett., 18*, 85-88, 1991.

Olijnyk, H. and W.B. Holzapfel, Phase transitions in K and Rb under pressure, *Phys. Lett., 99A*, 381-383, 1983.

Revenaugh, J.S. and R. Meyer, Seismic evidence of partial melt within a possibly ubiquitous low velocity layer at the base of the mantle, *Science, 277*, 670-673, 1997.

Rigden, S.M., E. Stolper and T.J. Ahrens, Densities of liquid silicates at high pressure, *Science, 226*, 1071-1074, 1984.

Rigden, S.M., T.J. Ahrens and E. Stolper, Shock compression of a molten silicate: Results for a model basaltic composition, *J. Geophys. Res., 93*, 367-382, 1988.

Ringwood, A.E., *Composition and Petrology of the Earth's Mantle*, McGraw Hill, New York, 1975.

Ringwood, A.E., A.F. Reid and A.D. Wadsley, High-pressure $KAlSi_3O_8$, an aluminosilicate with six-fold coordination, *Acta Cryst., 23*, 1093-1095, 1967.

Seitz, M.G., Uranium and thorium partitioning in diopside-melt and whitlockite-melt systems, *Yearbook Carn. Inst. Wash., 72*, 581-586, 1973.

Seitz, M.G. and N. Shimizu, Partitioning of uranium in the Di-An system and spinel lherzolite system by fission track mapping, *Yearbook Carn. Inst. Wash., 71*, 548-553, 1972.

Sherman, D.M., Chemical bonding and the incorporation of potassium into the Earth's core, *Geophys. Res. Lett., 17*, 693-696, 1990.

Shimizu, N., An experimental study of the partitioning of K, Rb, Cs, Sr and Ba between clinopyroxene and liquids at high pressure, *Geochim. Cosmochim. Acta, 38*, 1789-1798, 1974.

Sleep, N.H., Thermal history and degassing of the Earth: Some simple calculations, *J. Geology, 87*, 671-686, 1979.

Somerville, M. and T.J. Ahrens, Shock compression of $KFeS_2$ and the question of potassium in the core, *J. Geophys. Res., 85*, 7016-7024, 1980.

Stevenson, D.J., Origin of the moon -- The collision hypothesis, *Ann. Rev. Earth Planet. Sci., 15*, 271-315, 1987.

Stolper, E.M., D. Walker, B.H. Hager and J.F. Hays, Melt segregation from partially molten source regions: the importance of melt density and source region size, *J. Geophys. Res., 86*, 6161-6171, 1981.

Tamai, H. and T. Yagi, High-pressure and high-temperature phase relations in $CaSiO_3$ and $CaMgSi_2O_6$ and elasticity of perovskite-type $CaSiO_3$, *Phys. Earth Planet. Inter., 54*, 370-377, 1989.

Trønnes, R.G., D. Canil and K. Wei, Element partitioning between silicate minerals and coexisting melts at pressures of 1-27 GPa, and implications for mantle evolution, *Earth Planet. Sci. Lett., 111*, 241-255, 1992.

Turcotte, D. and G. Schubert, *Geodynamics*, J. Wiley, New York, 1982.

Urey, H.C., The cosmic abundances of potassium, uranium and thorium and the heat balances of the Earth, moon and Mars, *Proc. Nat. Acad. Sci., 42*, 889-891, 1956.

Walker, D. and C. Agee, Partitioning "equilibrium", temperature gradients and constraints on Earth differentiation, *Earth Planet. Sci. Lett., 96*, 49-60, 1989.

Wasserburg, G.J., G.J.F. Macdonald, F. Hoyle and W.A. Fowler, Relative contributions of uranium, thorium and potassium to heat production in the Earth, *Science, 143*, 464-467, 1964.

Williams, Q., Molten $(Mg_{0.88}Fe_{0.12})_2SiO_4$ at lower mantle conditions: Melting products and structure of quenched glasses, Geophys. Res. Lett., 17, 635-638, 1990.

Williams, Q. and E.J. Garnero, Seismic evidence for partial melt at the base of Earth's mantle, *Science, 273*, 1528-1530, 1996.

Williams, Q. and R. Jeanloz, Spectroscopic evidence for pressure-induced coordination changes in silicate glasses and melts, *Science, 239*, 902-905, 1988.

Williams, Q., E. Knittle and R. Jeanloz, The melting curve of iron at high pressures: A technical discussion, *J. Geophys. Res., 96*, 2171-2184, 1991.

Zerr, A. and R. Boehler, Melting of $(Mg,Fe)SiO_3$-perovskite to 625 kilobars: Indication of a high melting temperature in the lower mantle. *Science, 262*, 553-555, 1993

Elise Knittle, Dept. of Earth Sciences, University of California, Santa Cruz, CA 95064

Is There a Thin Electrically Conducting Layer
at the Base of the Mantle?

J. P Poirier, V. Malavergne, J. L. Le Mouël

Institut de Physique du Globe de Paris

There is seismological evidence for an ultra-low velocity zone (ULVZ), tens of kilometers thick, at the base of the mantle, close to the outer core. On the other hand, it has been suggested that a layer with high electrical conductivity could exist in the same region. We first examine whether partial melting of silicates, suggested as a phenomenon that could account for the lowering of seismic velocities, could also impart a high electrical conductivity to this layer. We find that the increase in conductivity due to partial melting would then be negligible. We then investigate the claim that infiltration of liquid iron from the core into the base of the mantle could produce a high conductivity layer. We review the experimental evidence for reaction of molten iron with solid silicates, and present new data showing that the reaction is possible in the absence of water. We estimate the maximum thickness of the layer and find it would be much too thin to be identified with the ULVZ. We therefore think it very unlikely that a high conductivity layer, more than a few tens of meters thick, can exist at the base of the mantle and we discuss a few geophysical consequences.

1. INTRODUCTION

Ohmic dissipation in a thin electrically conducting layer at the base of the mantle has been proposed as the source of a small residual out-of-phase component in the retrograde annual nutation of the Earth, once the effects of ocean tides and mantle anelasticity have been taken into account [Buffett, 1992].

The idea that a thin conducting layer might be responsible for the observed effect had been suggested by experiments in laser-heated diamond-anvil cell [Knittle and Jeanloz, 1989; Jeanloz, 1990], which had shown that molten iron reacted with constituent minerals of the lower mantle: silicate perovskite (Mg,Fe)SiO_3 and magnesio-wüstite (Mg, Fe)O. It had therefore been thought possible that the liquid iron alloy of the core could infiltrate the

bottom of the lower mantle by a surface tension mechanism, thus creating a high conductivity layer. [Buffett, 1992] found that such a layer about 200 m thick, with a conductivity equal to that of the fluid core, was needed to account for the out-of-phase component of the nutation.

More experimental evidence of chemical reaction between iron and mantle oxides was provided cell [Knittle and Jeanloz, 1991]. [Goarant et al. 1992], using analytical transmission electron microscopy, confirmed these results and directly observed that molten iron wetted the oxide grain boundaries.

[Jeanloz 1990] and [Knittle and Jeanloz, 1991] suggested that infiltration of the core fluid into the mantle could be responsible for the presence of a heterogeneous layer of varying thickness, detected through seismic velocities anomalies at the base of the mantle (D" layer); they speculated that metallic heterogeneities in the layer could pin, or deflect, magnetic field lines, influencing the magnetic field observed at the Earth's surface.

[Poirier and Le Mouël, 1992] gave a quantitative estimate of the height to which liquid iron, corroding grain

The Core-Mantle Boundary Region
Geodynamics 28
Copyright 1998 by the American Geophysical Union.

boundaries, could rise into the mantle by capillarity. They found it unlikely that the fluid could rise more than a few tens of meters into the mantle, and contended that no local heterogeneity of conductivity could possibly affect the magnetic field observed at the surface. If, as assumed by [*Jeanloz*, 1990], the infiltrated fluid were swept up in the whole D" layer, the resulting electrical conductivity would not be significantly different from that of the silicate lower mantle [*Poirier and Le Mouël*, 1992; *Poirier*, 1993; *Le Mouël et al.*, 1997].

However, despite these challenges to the existence and possible effects of a zone of high-conductivity, it has almost become a received opinion that such a layer is present at the base of the lower mantle. The opinion was apparently buttressed by recent seismological studies [*Garnero and Helmberger*, 1995, 1996] which reported the presence of an ultra-low velocity zone (ULVZ), close to the core-mantle boundary, where the velocity of P-waves is reduced by 10%. This layer of laterally varying thickness, up to 40 km, lies below a zone of large-scale low velocities (2 to 4% reduction).

Its nature is still a matter of debate, as the observed low velocity could be ascribed to the presence of metallic material resulting from core fluid *infiltration* [*Manga and Jeanloz*, 1996] or to some degree of partial melting of the lower mantle material [*Williams and Garnero*, 1996; *Holland and Ahrens*, 1997]. In both these eventualities, the layer might be assumed to possess a high electrical conduc-tivity [*Williams and Garnero*, 1996], possibly high enough to couple the mantle and the core electromagnetically. It has also been proposed that heterogeneities in velocity might be correlated with conductivity heterogeneities that could cause the virtual geomagnetic pole to follow preferred paths, during magnetic polarity reversals [*Aurnou et al.*, 1996].

Now, to what extent can the ULVZ be identified with the supposed high-conductivity layer? This is what we intend to examine critically in the present paper.

We will first try to put bounds on the electrical conductivity that can be achieved by partial melting, assuming this to be the cause of the velocity drop in the ULVZ.

We will then review the evidence for core-mantle chemical reactions, in the light of recent experimental observations in the laser-heated diamond-anvil cell, which contradict the view that the reactions cannot occur between dry materials [*Boehler et al.*, 1995]. The extent of infiltration and the resulting electrical conductivity will then be evaluated.

Finally, we will discuss the consequences with regard to core-mantle coupling.

2. ELECTRICAL CONDUCTIVITY OF A PARTIALLY MOLTEN LAYER

Whereas most rocks are poor electrical conductors, molten rocks are electrolytes and their electrical conductivity is usually higher than that of the solid. It is therefore to be expected that the electrical conductivity of rocks increases with the degree of partial melting.

The connectivity of the melt has a considerable influence on the effective electrical conductivity of the partially molten rock. For the same melt fraction, the conductivity is obviously higher if the melt wets the grain boundaries in a totally interconnected network, than if there are isolated melt pockets.

The conductivity σ of a partially molten rock is usually related to the melt fraction f by Archie's law (e.g. [*Shankland and Waff*, 1974]):

$$\sigma/\sigma_m = C \, \phi^n \qquad (1)$$

where σ_m is the conductivity of the melt, C is a constant, and $1 < n$. [*Hermance*, 1979] proposed a modification of Archie's law for distributions of melt intermediate between melt pockets and a completely interconnected network of melt, in the case where the conductivity of the solid cannot be neglected:

$$\sigma = \sigma_s + (\sigma_m - \sigma_s) \, \phi^2 \qquad (2)$$

where σ_s is the conductivity of the solid.

We can now try to estimate the electrical conductivity of the ULVZ, assuming the drop in seismic velocities is due to partial melting. The amount of partial melt necessary to generate a reduction of 10% in P-wave velocity depends on the geometry of the melt. [*Williams and Garnero*, 1996] calculate that 25 to 30 % melt fraction is needed if the melt is in the shape of spherical inclusions, whereas a melt fraction as low as 5% would suffice for films of melt with an aspect ratio of 0.01.

We will use those extremal values in equation (2), but we need first to estimate the conductivity of the solid σ_s and that of the melt σ_m. Recent experiments have shown that the high-pressure lower mantle assemblage can undergo eutectic melting [*Holland and Ahrens*, 1997]. It seems therefore reasonable to identify the ULVZ with a partially molten zone in the lower mantle and to assume that the conductivity of the solid matrix σ_s is equal to that of the lower mantle assemblage, perovskite and magnesiowüstite. [*Poirier and Peyronneau*, 1992, *Shankland et al* 1993] and [*Poirier et al.*, 1996] have measured the electrical conductivity of this assemblage and, extrapolating their results to the pressure and temperature conditions at the base of the mantle, they find a conductivity between 1 and 10 S/m. We will therefore adopt $1 < \sigma_s < 10$ S/m.

There are no data on the conductivity of the melt. However, experiments by [*Manghnani and Rai*, 1978] on partial melting of spinel lherzolite and garnet peridotite at ambient pressure yield values of the conductivity of the liquid of the order of 30 S/m. Now, the melt is an ionic conductor and it is not unreasonable to assume that conductivity, like diffusion and viscosity, scales with melting temperature [*Poirier*, 1988]. The conductivity of the melt, near melting temperature, would then be roughly the same at ambient pressure or at the core-mantle boundary pressure. We will therefore take σ_m 30 S/m.

The use of Hermance's modification of Archie's law is justified by the fact that the conductivity of the ionic melt is

almost of the same order of magnitude as the conductivity of the solid.

One can surmise that a melt fraction of 30%, implying a high degree of partial melting may not be realistic, and that it is more likely that a smaller melt fraction, as thin films of low aspect ratio, not necessarily totally interconnected, at grain boundaries, may be responsible for the drop in seismic velocity.

We will therefore consider the case where the melt fraction is equal to f = 0.05. For σ_s 1 S/m, and σ_m 30 S/m, we find a value of the effective conductivity $\sigma = 1.07$; for $\sigma_s \approx 10$ S/m, and σ_m 30 S/m, we find $\sigma = 10.05$. For σ_s 1 S/m, and the extreme value $\sigma_m \approx 100$ S/m, we find σ 1.25 S/m. In all those cases, the effective conductivity is practically undistinguishable from that of the solid.

In the less realistic case of a melt fraction equal to 30%, the values of the effective conductivity would still be of the same order of magnitude as the conductivity of the solid: e.g., $\sigma = 3.6$ for σ_s 1 S/m, and σ_m 30 S/m.

The possibility that the electrical conductivity of the partial melt is increased by preferential partitioning of iron into the melt, must be considered. The few experimental data existing suggest that this is not the case. [*McFarlane et al.*, 1994] find that the iron content of the melt in equilibrium with perovskite and magnesiowüstite at 24.5 GPa is equal to that of the starting material.

It seems therefore that, if the ULVZ is a partially molten zone, it is not much more electrically conducting than the solid mantle.

3. CHEMICAL REACTIONS BETWEEN MOLTEN IRON AND THE LOWER MANTLE MATERIAL

[*Knittle and Jeanloz*, 1989] first "simulated" the core-mantle boundary in the diamond anvil-cell, by putting in contact molten iron and $(Mg,Fe)SiO_3$ perovskite at pressures up to 70 GPa and temperatures above 3700 K. By visual inspection and X-ray diffraction study of the recovered quenched samples, they showed that liquid iron and silicate perovskite had chemically reacted producing iron oxide FeO, iron-silicon FeSi and SiO_2 stishovite as well as depleting the perovskite in iron. Ulterior investigations [*Knittle and Jeanloz*, 1991], using electron microprobe, confirmed these findings.

[*Goarant et al.*, 1992] further studied the reactions between molten iron and lower mantle material, at high pressure and temperature. They observed the recovered diamond-anvil cell samples by means of analytical transmission electron microscopy, which allows the identification and analysis of the reaction products with a spatial resolution better than 0,1 mm. They performed experiments at 70 GPa, using as starting material natural enstatite and olivine (transformed into perovskite and an assemblage of perovskite and magnesiowüstite, respectively), on the one hand, and iron or a mixture of iron and iron sulfide FeS, on the other hand. They found that the liquid iron reacted with the silicate perovskite, depleting it in iron, while dissolving oxygen. When in contact with the assemblage transformed from olivine, liquid iron preferentially depleted magnesiowüstite in iron. Furthermore, it was directly observed that molten iron wetted the grain boundaries. The reaction rate was increased, and wetting of the grain boundaries was enhanced when iron-sulfur alloys were used. One experiment at the core-mantle boundary pressure (130 GPa) confirmed the results obtained at 70 GPa.

[*Song and Ahrens*, 1994] calculated the Gibbs free-energies for the reactants and products of the reaction between liquid iron and perovskite silicate at core-mantle boundary conditions and concluded that the experimentally observed reactions were thermodynamically possible and could take place in the Earth.

However, on the basis of visual observation and preliminary microprobe analysis of samples that had been dried overnight at 100°C before being pressurized and heated in the diamond anvil cell, [*Boehler et al.*, 1995] and [*Hillgren and Boehler*, 1996] claimed that no sign of reaction was visible. Despite the obvious fact that a sample left overnight at 100°C cannot be considered really free of adsorbed water [*Poirier and Peyronneau*, 1992], they contended that reactions between molten iron and lower mantle silicates or oxides could not take place in the absence of water.

[*Malavergne*, 1996] dried mixtures of San Carlos olivine (Fe/(Fe+Mg) = 0.16) and metallic powders (Fe, FeS) at 1200°C, during 12 to 17 hours, in a furnace under controlled atmosphere. The oxygen partial pressure was fixed by a flowing mixture of CO and CO_2 (log PO_2 = -13 to -9). The dry samples were then loaded in a diamond-anvil cell, pressurized to 40 or 70 GPa, and laser heated until the metal melted. Observation of the quenched samples by analytical transmission electron microscopy confirmed the results of [*Goarant et al*, 1992]: liquid iron (or iron-sulfur alloy) reacts with perovskite and magnesiowüstite, depletes them in FeO and wets the grain boundaries, while dissolving oxygen. There seems to be no doubt that the reactions between molten iron and solid oxides and silicates occur in thoroughly dry systems.

At the pressure and temperature of the core-mantle boundary, it is therefore conceivable that the fluid iron alloy of the core can penetrate upward into the lower mantle, by dissolving the mantle material along the grain boundaries. Let us emphasize here that fluid iron cannot infiltrate the mantle by just ascending between the grains by capillarity, as water would in a sandbag. Grain boundaries are defects in the solid mantle material, not vacant spaces between grains, a misconception probably due to the fact that crustal rocks are pervaded by intergranular cracks, as a result from decompression and cooling. At the pressure of the core-mantle boundary, fluid iron cannot infiltrate the mantle without first dissolving its way through the material at the grain boundaries. Reaction between iron and silicates is absolutely necessary for infiltration to occur.

4. ELECTRICAL CONDUCTIVITY OF A CORE-INFILTRATED LAYER

Let us now consider the case where the core fluid corrodes and infiltrates the grain boundaries of the lower mantle oxides. We will investigate whether the infiltrated layer can be identified with the ULVZ and we will estimate its electrical conductivity.

[Poirier and Le Mouël, 1992] calculated the maximum height, h_{cap}, to which the fluid can ascend into the lower mantle by capillarity. Assuming perfect wetting, they found a relation between h_{cap} and the width δ (in meters) of the channel invaded by the fluid at the grain boundaries:

$$h_{cap} = 1.6 \cdot 10^{-5}/\delta$$

The time t_{cap} (in seconds) taken by the fluid to rise to that height, by corrosion of the grain boundaries, was found to be:

$$t_{cap} = 2.3 \cdot 10^6 \, h^2$$

For grain sizes between 1 mm and 1 cm, values of δ may reasonably range between 1 and 50 μm, which yields values of h_{cap} between 16 m and 30 cm, and values of t_{cap} between about 19 years and 2 days, respectively.

It is therefore likely that the infiltration of the core into the mantle can take place very rapidly, but over a very small thickness.

The infiltrated zone, thus formed, would then consist of a network of corroded grain boundaries filled with liquid iron enriched with reaction products (oxygen, silicon, etc.). The network, which may have dead-end branches, cannot extend higher than the maximum height of capillary rise. Also, there must always be a free path in the liquid, along the grain boundaries, from any part of the network to the core.

An important consequence of this geometry is that it should be impossible to entrain any liquid upward by solid-state convection of the lower mantle. The residence time of the infiltrated layer in contact with the core is $\tau \approx L/v$, where $L \approx 1000$ km is a typical length scale of the core-mantle boundary , and $v \approx 1$ to 10 cm/year is a typical velocity of the convective flow; we therefore find that $10^7 < t < 10^8$ years. The residence time is very much longer than the characteristic time for filling the layer. Now, let us assume that the infiltrated layer is uplifted in a convective upwelling. The bottom of the layer, originally in contact with the core, will move up a few centimeters per year, at a rate slow enough for the new material to be infiltrated. The infiltrated network will therefore remain connected to the core. The dense fluid in its upper parts, brought above the maximum height of capillary rise, will rapidly adjust back to that height, and the excess fluid will flow back into the core. The infiltrated layer will thus be stabilized to a thickness limited to the maximum height of capillary rise above the core-mantle boundary.

Several models have been proposed, whereby at least some dense fluid can be entrained into the convective lower mantle and cause heterogeneities of seismic velocities, and electrical or thermal conductivity in a layer tens or hundreds of kilometers thick. These models, however, rest on asssumptions that the distribution of the dense fluid is continuous, e.g., that the growth of the dense reaction product can be modeled as a diffusion process [Kellogg and King, 1993], or that viscous stresses are balanced by buoyancy on finite size blobs of dense matter [Manga and Jeanloz, 1996]. They do not take into account the geometry of the distribution, nor the kinetics of the reaction, which make it impossible for the infiltrated layer to grow thicker than 20 m at most [Poirier and Le Mouël, 1992], or for the iron fluid to be entrained upward to form heterogeneities at the scale of tens of kilometers.

We have so far accepted that the products of the reaction between iron and silicate are fluid. However, [Manga and Jeanloz, 1996] consider that the reaction between iron and silicates or oxides of the lower mantle forms a solid layer of FeO. On the basis of our observations by transmission electron microscopy, we think this is very unlikely. We do indeed find evidence of iron oxides in the intercrystalline films of our quenched samples, by electron diffraction patterns, but we never could see crystals of FeO [Goarant et al., 1992; Malavergne, 1996]. It is very probable, in our opinion, that liquid iron readily dissolves oxygen at high pressure and temperature, but exsolves microcrystals of iron oxides, less than one nanometer in size, during the quench.

Besides, one can imagine that, if the reaction products were solid, they would rapidly clog the channels and stop the reaction and infiltration.

Finally, we could envisage the possibility of iron being present in a layer at the base of the mantle, without coming from the core. This could for instance be the case if the core formation process were still going on, or if the perovskite or magnesiowüstite of the lower mantle were out of equilibrium, in reducing conditions. However, the thickness and electrical conductivity of a layer enriched in iron by such hypothetic processes would be a matter of pure speculation.

Let us then go back to an ultra-thin fluid-infiltrated layer, and estimate its electrical conductivity. We will then use the traditional form (1) of Archie's law, since the conductivity of the iron fluid is considerably higher than that of the lower mantle rock. We will assume the conductivity of the fluid is equal to that of the core, σ_m $7 \cdot 10^5$ S/m [*Poirier*, 1994].

An upper bound on the conductivity of the layer would be obtained by assuming each grain, of size d, to be surrounded by a film of fluid of thickness d/2, which corresponds to a fluid fraction $\phi = 3d/d$. For a grain size d=1mm and a thickness of film δ=1 μm, we have ϕ=$3 \cdot 10^{-3}$, and for a grain size d=1cm and a thickness of film δ=50 μm, we have ϕ=$1.5 \cdot 10^{-2}$.

The assumption that the grains are entirely surrounded by fluid would imply n= 1 in Archie's law, since the network would then be entirely connected. This upper bound is unrealistic, since the oxide or silicate grains would then be entirely separated by films of fluid. The layer would then be weaker in shear than the lower mantle material, and might not be effective for core-mantle coupling.

For the much more realistic case where n=2 in Archie's law and for fluid fractions ϕ=$3 \cdot 10^{-3}$, and ϕ=$1.5 \cdot 10^{-2}$, we would have σ 6 S/m, and σ 150 S/m, respectively.

It seems difficult to identify the ULVZ with an infiltrated layer, which cannot reasonably be thicker than 20 m at most. This very thin infiltrated layer would have an effective electrical conductivity of about 100 S/m, considerably less than demanded, for instance, by [*Buffett*, 1992].

5. DISCUSSION AND CONCLUSIONS

Let us summarize the discussions of the previous sections.

The conductivity of the lower mantle is basically the conductivity of the solid silicates and oxides, except possibly in a very thin infiltrated layer, a few meters thick, in close contact with the core.The results of recent experimental studies, extrapolated to the conditions of the lower mantle, down to the CMB, lead to the conclusion that the conductivity is not higher than a few Sm^{-1} [*Shankland et al.*, 1993; *Poirier et al.*, 1996]. As to the conductivity of the ultrathin layer, infiltrated with iron from the core, it could reach 100 Sm^{-1}.

In other words, when dealing with most geophysical problems, the electrical conductivity of the deep mantle can be considered as small (<10 Sm^{-1}) and spherically symmetrical. The ULVZ of the seismologists, with its horizontally varying thickness of the order of a few tens of kilometers, has no electrical signature.

A few consequences on the geomagnetic field and its secular variation directly follow. There is no inhomogeneity important enough to pin or deflect the magnetic field lines and influence the magnetic field observed at the Earth's surface, as proposed by [*Jeanloz*, 1990]. Neither can a thin, heterogeneous, highly conductive layer play a part in the dynamo process by generating poloïdal components of the magnetic field from

toroïdal components; a much too high conductance should indeed be required [*Busse and Wicht*, 1992].

A third —positive— consequence arises from the mantle conductivity profile being governed by the conductivity of the solid silicates (from ≈ 1 Sm^{-1} at a depth of 600 km to about 10 Sm^{-1} at the CMB). It has indeed been shown that the time variation of the core-generated magnetic field exhibits rapid changes, completed in one year, or so [e.g., *Courtillot et al.*, 1978; *Alexandrescu et al.*, 1995]. But, there is no reason for the spectrum of the secular variation—understood as the time variation of the field of internal origin— to be limited to 12 months. If the conductance of the mantle is only of the order of 10^7 Ω^{-1}, as proposed here, the electromagnetic time constant of the mantle could be as small as a few 10^6 s, i.e., a month. It is, then, worthwhile to look for these short components of the secular variation (the difficulty being, of course, in separating them from the external field variations).

Electromagnetic torques have been considered for a long time as possible agents of core-mantle coupling, which could account for for the decadal variation of the length of day (e.g., [*Stix and Roberts*, 1972]). However, a conductance as high as 10^9 Ω^{-1} would be requested for the electromagnetic torque to be tight enough to account for the l.o.d. data (e.g., [*Jault and Le Mouël*, 1991]). With the above estimates, the conductance of the mantle is two orders of magnitude too small. Electromagnetic core-mantle coupling has also been invoked to account for the decade component of the motion of the pole [*Greff-Lefftz and Legros*, 1995], but the required conductance would be much too high. Topographic coupling is a good candidate for axial coupling and l.o.d. decade variations [*Jault and Le Mouël*, 1989], but the problem of equatorial coupling (or pole motion) appears more difficult to solve [*Hulot et al.*, 1996; *Le Mouël et al.*, 1997].

The existence of a thin, highly conductive layer at the base of the mantle has also been proposed to account for a discrepancy between the observed value of the out-of-phase component of the retrograde annual nutation and the theoretically calculated value [*Buffett*, 1992]. A layer 200 m thick, with a conductivity as high as $5 \cdot 10^5$ Sm^{-1}, would allow ohmic dissipation to take place and reconcile observed and computed values. However, if our conclusions are adopted,the extra dissipative source is to be found elsewhere. In the same way, there is no hope to locate close to the CMB an ohmic dissipation higher than previously estimated [*MacDonald and Ness*, 1961; *Rochester*, 1976; *Toomre*, 1966], which could explain the strong decrease of the Earth's obliquity, that some geological observations appear to require [*Williams*, 1993].

Another use of a heterogeneous, highly conductive layer close to the CMB has been made, after [*Laj et al.*, 1991] published their paper about preferred geomagnetic virtual pole (VGP) paths during reversals. They found that the preferred paths are concentrated over two longitude bands, 180∞ apart, centered on the Americas and their antipodes.

Although [*Valet et al.*, 1992] and [*Courtillot et al*, 1992] argued that this bimodal distribution of VGP paths was a spurious effect of the distribution of the sampling sites, which would disappear for a more uniform sampling, it was proposed to interpret the non-uniform distribution of VGP paths as due to a horizontally heterogeneous higher conductivity layer at the base of the lower mantle [*Runcorn*, 1992 ; *Aurnou et al.*, 1996]. Indeed, such a lack of symmetry in the behavior of a magnetic field generated in a spherical ball of fluid could legitimately be attributed to interactions with a heterogeneous mantle. [*Runcorn*, 1992] made the hypothesis that there was a conducting layer at the core-mantle boundary under the Pacific ocean, accounting at the same time for the subdued secular variation observed over the Pacific hemisphere, which, incidently, may be just a transient feature, present only since a few centuries [*Hulot and Le Mouël*, 1994] ; [*Yukutake*, 1993]. More recently, [*Aurnou et al.*, 1996] proposed a similar mechanism to explain the distribution of VGP paths. However, for this mechanism to work, the electrical conductivity in the D" layer must be comparable to the core conductivity. This is quite unrealistic in the framework of our hypothesis.

Finally, [*Manga and Jeanloz*, 1996] assumed that infiltrated iron, or metallic iron oxides, infiltrated in a laterally heterogeneous layer at the base of the mantle, could impart to it a high electrical conductivity, hence a high thermal conductivity (since thermal and electrical conductivities are proportional in metals). Momentous consequences as to the dynamics of the mantle and patterns of convection were derived from the temperature distribution arising from variations in thickness of the layer by as much as 300 km. Such an effect simply could not exist if, as proposed above, the thicknes of the infiltrated layer does not exceed 20 m.

In conclusion, we have considered two mechanisms potentially capable of increasing the electrical conductivity of a layer at the base of the mantle. Partial melting of the mantle silicates might account for the ULVZ seen by seismologists, but the electrical conductivity of the partially molten layer would be hardly higher than that of the solid mantle. The possibility of infiltration of the highly con-ducting iron of the core into the lower mantle has been experimentally demonstrated, but we find that the high-conductivity infiltrated layer cannot be thicker than a few tens of meters, and cannot be assimilated to the ULVZ. There might perhaps be a way of reconciling a thick layer of high electrical conductivity with the ULVZ, but we have not found it.

6. REFERENCES

Alexandrescu, M., D. Gibert, G. Hulot, J.-L. Le Mouël, and G. Saracco, Detection of geomagnetic jerks using wavelet analysis, *J. Geophys. Res,100*, 12557-12572, 1995.

Aurnou, J.M., J.L., Buttles, G.A., Neumann, and P.L. Olson, Electromagnetic core-mantle coupling and paleomagnetic reversal paths, *Geophys. Res. Lett.*, 23, 2705-2708, 1996.

Boehler, R., A. Chopelas, and A. Zerr, Temperature and chemistry of the core-mantle boundary, *Chemical Geology, 120*, 199-205, 1995.

Buffett, B.A., Constraints on magnetic energy and mantle conductivity from the forced nutations of the Earth, *J. Geophys. Res*, 97, 19581-19597, 1992.

Busse, F. H., and J. Wicht, A simple dynamo caused by conductivity variations, *Geophys.astrophys.Fluid Dyn.*, 64, 135-144, 1992.

Courtillot, V., J. Ducruix, and J.-L. Le Mouël, Sur une accélération récente de la variation séculaire du champ magnétique terrestre, *C.R. Acad. sci. Paris,287*, 1095-1098, 1978.

Courtillot, V., J.-P. Valet, G. Hulot, and J.-L. Le Mouël, The Earth's magnetic field: which geometry?, *Eos Trans. AGU*, 73, 337 and 340-342, 1992.

Garnero, E.J., and Helmberger, D.V., Seismic detection of a thin laterally varying boundary layer at the base of the mantle beneath the central-Pacific, *Geophys. Res. Lett.*, 23, 977-980, 1996.

Goarant, F., F. Guyot, J. Peyronneau, and J.P. Poirier, High-pressure and high-temperature reactions between silicates and liquid iron alloys, in the diamond anvil cell, studied by analytical electron microscopy, *J.Geophys.Res.,97*, 4477-4487, 1992.

Greff-Lefftz, M., and H. Legros, Core-mantle coupling and polar motion, *Phys. Earth Planet. Interiors*, 91, 273-283, 1995.

Hermance, J.F., The electrical conductivity of materials containing partial melt: a simple model from Archie's law, *Geophys. Res. Lett.*, 6, 613-616, 1979.

Hillgren, V.J., and Boehler, R., High pressure reactions between Fe-metal and mantle silicates, *Eos Trans. AGU*, 77, 673, 1996.

Holland, K.G., and T.J. Ahrens, Melting of $(Mg,Fe)_2SiO_4$ at the core-mantle boundary of the Earth, *Science*, 275, 1623-1625, 1997.

Hulot, G., and J.-L. Le Mouël, A statistical approach to the Earth's main magnetic field, *Phys. Earth Planet. Interiors*, 82, 167-183, 1994.

Hulot, G., M. Le Huy, and J.-L. Le Mouël, Influence of core flows on the decade variations of the polar motion, *Geophys.astrophys. Fluid Dyn.*, 82, 35-67, 1996.

Jault, D., and J.-L. Le Mouël, Exchange of angular momentum between the core and the mantle, *J. Geomag. Geoelectr.*, 43, 111-129, 1991.

Jault, D., and J.-L. Le Mouël, The topographic torque associated with a tangentially geostrophic motion at the core surface and inferences on the flow inside the core, *Geophys.astrophys. Fluid Dyn.*, 48, 273-296, 1989.

Jeanloz, R., The nature of the Earth's core, *Annu. Rev. Earth Planet. Sci.*, 18, 357-386, 1990.

Kellogg, L.H., and S.D. King, Effect of mantle plumes on the growth of D" by reaction between the core and the mantle. *Geophys. Res. Lett.*, 20, 379-382, 1993.

Knittle, E., and R. Jeanloz, Earth's core-mantle boundary: Results of experiments at high pressures and high temperatures, *Science*, 251, 1438-1443, 1991.

Knittle, E., and R. Jeanloz, Simulting the core-mantle boundary: an experimental study of high-pressure reactions between silicates and liquid iron, *Geophys. Res. Lett., 16,* 609-612, 1989.

Laj, C., A. Mazaud, R. Weeks, M. Fuller, and E. Herrero-Bervera, Geomagnetic reversal paths, *Nature, 351,* 447, 1991.

Le Mouel, J.L., G. Hulot, and J.P. Poirier, Core-mantle interactions, *Earth's deep interior,* D.J. Crossley ed., Gordon and Breach, 197-221, 1997.

MacDonald, G.J.F., and N.F. Ness, A study of the free oscillations of the Earth, *J. Geophys. Res, 66,* 1865-1911, 1961.

Malavergne, V., Etude exprimentale à haute pression et haute température du partage des éléments de transition entre les phases du manteau et du noyau de la Terre, Doctorate thesis, University Paris 7, 1996

Manga, and R. Jeanloz, Implications of a metal-bearing chemical boundary layer in D" for mantle dynamics, *Geophys. Res. Lett., 23,* 3091-3094, 1996.

Manghnani, M.H., and C.S. Rai, Electrical conductivity of a spinel lherzolite and a garnet peridotite to 1550°C: relevance to the effects of partial melting, *Bull. Volcanol., 41,* 328-332, 1978.

McFarlane, E.A., M.J. Drake, and D.C. Rubie, Element partitioning between Mg-perovskite, magnesiowüstite, and silicate melt at conditions of the Earth's mantle, *Geochimica Cosmochimica Acta, 58,* 5161-5172, 1994.

Poirier J.P., and J. Peyronneau, Experimental determination of the electrical conductivity of the material of the Earth's lower mantle, in: *"High pressure research in mineral physics: Applications to Earth and planetary science",* edited by Y. Syono and M. H. Manghnani, Geophysical Monograph 67, 77 - 87, AGU, 1992.

Poirier, J.P., A. Goddat, and J. Peyronneau, Ferric iron dependence of the electrical conductivity of the Earth's lower mantle material, *Phil. Trans. R.Soc. London, 354,* 1361-1369, 1996.

Poirier, J.P., and J.L. Le Mouël, Does infiltration of core material into the lower mantle affect the observed geomagnetic field? *Phys.Earth Planet. Interiors, 73,* 29-37, 1992.

Poirier, J.P., Core-infiltrated mantle and the nature of D" layer, *J. Geomag. Geoelectr., 45,* 1221-1227, 1993.

Poirier, J.P., Physical properties of the Earth's core, *CR Acad. Sci. Paris, 318 II,* 341-350, 1994.

Poirier, J.P., Transport properties of liquid metals and viscosity of the Earth's core. *Geophys.J., R. astron. Soc. , 92,* 99-105, 1988.

Rochester, M.G., The secular decrease in obliquity due to dissipative core-mantle coupling, *Geophys. J. R. astron. Soc, 10,* 289-315, 1976.

Runcorn, S.K., Polar path in geomagnetic reversals, *Nature, 356,* 654-656, 1992.

Shankland, T.J., a d H.S. Waff, Conductivity in fluid-bearing rocks, *J. Geophys. Res, 79,* 4863-4868, 1974.

Shankland, T.J., J. Peyronneau, and J.P. Poirier, Electrical conductivity of the Earth's lower mantle, *Nature, 366,* 453-455, 1993.

Song, X., and T.J. Ahrens, Pressure-temperature range of reactions between liquid iron in the outer core and mantle silicates, *Geophys. Res. Lett., 21,* 153-156, 1994.

Stix, M., and P.H. Roberts, Electromagnetic core-mantle coupling, *J. Geomag. Geoelectr., 24,* 231-259, 1972.

Toomre, A., On the coupling of the Earth's core and mantle during the 26000 year precession, in *The Earth-Moon system,* edited by B.G. Marsden and A.G.W. Cameron, Plenum, New York, 1966.

Valet, J.-P., P. Tucholka, V. Courtillot, and L. Meynadier, Paleomagnetic constraints on the geometry of the geomagnetic field during reversal, *Nature, 356,* 400, 1992.

Williams, G.E., History of the Earth's obliquity, *Earth-Science Reviews, 34,* 1-45, 1993.

Williams, Q., and Garnero, E.J., Seismic evidence for partial melt at the base of the Earth's mantle, *Science, 273,* 1528-1530, 1996.

Yukutake, T., The geomagnetic non-dipole field in the Pacific, *J. Geomag. Geoelectr., 45,* 1441-1453, 1993.

J. L. Le Mouël, Département de Géomagnétisme, Institut de Physique du Globe de Paris, 4 place Jussieu, 75252 Paris cedex 05, France.

V. Malavergne, Département des Géomatériaux, Institut de Physique du Globe de Paris, 4 place Jussieu, 75252 Paris cedex 05, France.

J. P. Poirier, Département des Géomatériaux, Institut de Physique du Globe de Paris, 4 place Jussieu, 75252 Paris cedex 05, France.

Electromagnetic Core-Mantle Coupling II: Probing Deep Mantle Conductance

Richard Holme

Department of Geology and Geophysics, University of Edinburgh, Scotland.

Observed changes in the Earth's length of day on a decadal time scale are usually attributed to the exchange of angular momentum between the solid mantle and fluid core. The change in core angular momentum may be accommodated by motions within the core which are constant on cylindrical annuli. Electromagnetic core-mantle coupling is a possible mechanism for the angular momentum exchange, dependent on the fluid flow at the top of the core. Models of mantle structure allow for a highly conducting layer of material at the CMB: if the conductance of this layer is 10^8S or greater, then flows exist which explain the observed magnetic secular variation, can account for changes in Earth rotation by EM coupling only, and also accommodate the required transfer of angular momentum. If the conductance is lower, it is unlikely that EM coupling could be the sole mechanism for angular momentum exchange.

INTRODUCTION

The rate of rotation of the Earth is not constant, but varies slightly over a wide range of time scales [*Hide and Dickey*, 1991]. It is generally accepted that the largest of these variations, of a few milliseconds variation over decadal time scales, arises from the exchange of angular momentum between the solid mantle and fluid core. Four possible coupling mechanisms have been suggested to effect the angular momentum exchange. Viscous coupling, arising from the drag of flow at the top of the core on the core-mantle boundary (CMB) is probably too small to be significant [*Rochester*, 1984], but gravitational coupling between density heterogeneities in the core and mantle [*Jault and Le Mouël*, 1989; *Buffett*, 1996b], topographic coupling from fluid pressure against an irregular CMB [*Hide*, 1969; *Jault and*

Le Mouël, 1991; *Hide et al.*, 1993; *Hide*, 1995] [although see also *Kuang and Bloxham*, 1997b], or electromagnetic (EM) coupling [*Stix and Roberts*, 1984; *Stewart et al.*, 1995], each have their proponents. In a recent paper [*Holme*, 1998, henceforth paper 1], we demonstrated that EM coupling in isolation could fully explain the observed variations in length of day (ΔLOD). Here, we assume that EM coupling is indeed the dominant cause of decadal variations in LOD, and explore what information can then be derived about the electrical conductivity of the deep mantle, a procedure first suggested by *Runcorn* [1955] over forty years ago.

The magnetic field of the Earth is generated by magnetohydrodynamic dynamo action in the highly electrically conducting fluid iron core. Electric currents will also flow in the more weakly conducting mantle, both induced by the time-variation of the magnetic field (the magnetic secular variation), and due to diffusion of current from the core. The magnitudes of these currents, and hence also of the EM torque which they generate, are crucially dependent on the conductivity of the mantle, particularly the very deepest mantle. Recent

The Core-Mantle Boundary Region
Geodynamics 28
Copyright 1998 by the American Geophysical Union.

experimental results [*Poirier et al.*, 1996] suggest that the conductivity of the bulk lower mantle is probably only of order $1 Sm^{-1}$, much less than core conductivity (which we take to be $5 \times 10^5 Sm^{-1}$), and too small for significant EM coupling. However, it has been suggested that the conductivity may be significantly higher close to the CMB within the D″ layer, a region of great seismic heterogeneity and anisotropy. This deep mantle conductivity is optimal for efficient EM coupling. *Knittle and Jeanloz* [1989] have suggested that D″ is a reaction zone between core and mantle material [see also *Jeanloz*, 1990, 1993; *Poirier*, 1993]. Alternatively, numerical calculations by *Cohen et al.* [1997] of atomic structure suggest that at D″ pressures, the molecular structure of magnesiowüstite might undergo "magnetic collapse", changing bonding from ionic to metallic. Either of these mechanisms could lead to significantly enhanced electrical conductivity. Additionally, solid state or capillary diffusion could lead to direct infiltration of core fluid into the mantle. *Buffett* [1992] has suggested that features in the forced nutation of the Earth can be explained by a layer of core conductivity 200m thick, giving a conductance of 10^8S. However, other explanations for D″ (for example, a graveyard for subducted slabs) do not obviously lead to enhanced electrical conductivity [for a comprehensive review, see *Loper and Lay*, 1995].

We begin by outlining the method by which the EM torque can be calculated. This calculation requires a model of the fluid flow at the top of the core, determined from magnetic secular variation: unfortunately, such models are non-unique, and it is precisely the undetermined part of the flow that is most significant in generating the EM torque [*Jault and Le Mouël*, 1991; *Holme*, 1998]. By adopting an inverse method, we demonstrate that if deep mantle conductance exceeds 10^8S, then simple flows exist which can explain both secular variation and ΔLOD through EM coupling. A conductance of 3×10^7S requires flows that seem unreasonably complex, while for 10^7S, no suitable flows can be obtained. *Jault et al.* [1988] and *Jackson* [1989] have suggested that changes in the angular momentum of the core may be taken up by motions uniform on cylindrical annuli (perhaps torsional oscillations): our initial flows do not fit this model, but by applying a further constraint on the inversion, flows can be obtained which do. We conclude that for a deep mantle conductance of 10^8S or greater, EM coupling is capable of fully explaining ΔLOD, but this hypothesis is implausible if the conductance is lower.

THE ELECTROMAGNETIC CORE-MANTLE TORQUE

We begin with a brief outline of the formalism by which we calculate the EM torque on the mantle due to the core. Considerably more detail is given in paper 1. The torque is generated by the Lorentz force within the mantle

$$\mathbf{\Gamma} = \frac{1}{\mu_0} \int_{\text{Mantle}} \mathbf{r} \wedge ((\nabla \wedge \mathbf{B}) \wedge \mathbf{B}) \, dV \quad (1)$$

of which only the z-component contributes to changes in Earth rotation. The integral is taken over the electrically conducting volume of the mantle, $\mathbf{B}$ is the magnetic field and μ_0 is the magnetic permeability of free space. Applying the divergence theorem, equation (1) may be converted to a surface integral over the core-mantle boundary, radius r_c,

$$\Gamma_z = -\frac{r_c}{\mu_0} \oint_{\text{CMB}} B_\phi B_r \sin\theta dS \quad (2)$$

We use a poloidal-toroidal decomposition for magnetic field in the mantle, writing

$$\mathbf{B} = \nabla \wedge T\mathbf{r} + \nabla \wedge \nabla \wedge S\mathbf{r} \quad (3)$$

The radial field is poloidal only, and by downward continuation of a model of the magnetic field at the Earth's surface, radius r_e, is estimated as

$$\begin{aligned} B_r &= \sum_{l=1}^{\infty} (l+1) \left(\frac{r_e}{r_c}\right)^{l+2} \sum_{m=0}^{l} P_l^m(\cos\theta) \times \\ &\quad (g_l^m \cos(m\phi) + h_l^m \sin(m\phi)) \end{aligned} \quad (4)$$

where g_l^m and h_l^m are the Gauss coefficients calculated from observations of the downward-continued potential magnetic field at the Earth's surface [e.g., *Langel*, 1987], and the real spherical harmonics are Schmidt normalised, so that

$$\int_0^{2\pi} \int_0^{\pi} (P_l^m(\cos\theta))^2 \cos^2 m\phi \sin\theta d\theta d\phi = \frac{4\pi}{2l+1} \quad (5)$$

At the CMB, the estimated value of B_ϕ is dominated by the potential field observed at the Earth's surface, but that field generates no currents and so no torque. The remainder of the poloidal field and the toroidal field are zero outside the conducting part of the mantle, and therefore not observable at the Earth's surface. Instead,

B_ϕ must be calculated by solving the diffusion equation in the (weakly) conducting mantle. Assuming that mantle conductivity σ is a function of radius only, the torque can be divided into poloidal and toroidal parts corresponding to the poloidal and toroidal parts of B_ϕ.

If mantle conductivity is known, the poloidal torque can be calculated directly from a time-dependent model of the geomagnetic field [Stix and Roberts, 1984], giving

$$\Gamma_z = 4\pi \sum_{l=1}^{\infty} \frac{(l+1)}{l(2l+1)} r_e^{2l+4} \int \sigma r^{-2l} dr \times$$

$$\sum_{m=1}^{l} m \left(g_l^m \dot{h}_l^m - h_l^m \dot{g}_l^m \right) \qquad (6)$$

where the overdot indicates a time derivative. To determine the induced toroidal torque (the "advective torque"), we require a model of surface fluid flow $\mathbf{u}$ at the top of the core. We assume that over the time period modelled, advection of magnetic field dominates diffusion [the "frozen-flux approximation" of Roberts and Scott, 1965], and so use the radial component of the diffusionless induction equation

$$\frac{\partial B_r}{\partial t} + \nabla_H . (B_r \mathbf{u}) = 0 \qquad (7)$$

to invert for the flow from observed secular variation. ∇_H is the non-radial part of the divergence. For a comprehensive review of this procedure, see Bloxham and Jackson [1991]. Motivated by the possibility of high conductivity in D″, and following Stewart et al. [1995], we assume that the conductivity of the mantle is confined to a thin layer, thickness δ, above the core-mantle boundary. The advective torque is then given by

$$\Gamma_z = -r_c^2 G \oint_{\text{CMB}} \frac{\partial}{\partial \theta} \left(\text{L}^{-2} [\hat{\mathbf{r}}.\nabla_H \wedge B_r \mathbf{u}] \right) B_r \sin\theta dS \qquad (8)$$

[Holme, 1998], where L^{-2} is the inverse of the angular momentum operator [Edmonds, 1960; Gubbins and Roberts, 1987]. The torque depends only on the overall conductance of the thin layer

$$G = \int_{r_c}^{r_c+\delta} \sigma dr \qquad (9)$$

The determination of core surface flow is highly nonunique [Backus, 1968], essentially because equation (7) provides only one constraining equation with which to

obtain the two orthogonal components of surface velocity. Unfortunately, the part of the flow required to calculate the advective torque is precisely that part which is unconstrained by measurements of the secular variation [Jault and Le Mouël, 1991; Holme, 1998]. The induced torque associated with the well-determined part of the flow is included in the poloidal torque: if frozen-flux holds exactly, then the poloidal torque is precisely this induced torque. Additional assumptions can reduce flow non-uniqueness: this is useful for determining possible models of flow at the CMB, but less satisfactory for investigating EM coupling. While it may be a good approximation that the flow at the core surface is (for example) steady, or tangentially geostrophic (assuming that the Lorentz force at the CMB is much weaker than the Coriolis force), neither condition can hold exactly, and small departures from steadiness or geostrophy can significantly affect the torque. Further, for secular variation data with finite uncertainties, there are an infinite number of possible flow models which fit the data to within a chosen tolerance. Just because one such model does not produce a torque that correlates with the observed torque, does not mean that another viable flow will not.

A further contribution to the toroidal torque arises from possible diffusion of toroidal field within the core (in other words, a breakdown of the frozen-flux approximation for the toroidal field). This component of the torque (the "leakage torque") cannot be determined, because we are unable to observe the radial gradient of toroidal field at the top of the core. However, Jault [1990] [also Jault and Le Mouël, 1991] argued that because of its high conductivity, the diffusive response of the core would be too slow to account for rapid changes in LOD. Nevertheless, it could give rise to a slowly varying offset between the calculated induced torque (poloidal plus advective) and the torque required by observations [Stix and Roberts, 1984].

In paper 1, we demonstrated existence: for a lower mantle conductance of $G = 3 \times 10^8 \text{S}$, we obtained almost-steady, geostrophic flows that are both consistent with the observed secular variation and also explain the decadal variation in length of day by EM coupling alone. However, this value of conductance may be too high. In this paper we investigate what conductance range allows EM coupling to explain ΔLOD: by assuming the dominance of EM coupling, we use the observed ΔLOD to constrain the electrical conductivity structure at the base of the mantle.

METHOD

We briefly outline the numerical approach used to determine the core flow. For more details, see paper 1, and the references given therein. We use a poloidal-toroidal decomposition for the flow

$$\mathbf{u} = \nabla_H r\mathcal{S} + \nabla_H \wedge \mathcal{T}\mathbf{r} \qquad (10)$$

and then expand the scalars $\mathcal{S}$ and $\mathcal{T}$ in Schmidt-normalised real spherical harmonics in space and cubic B-splines in time. We obtain flow models by the construction and solution of a finite linear system of equations. We truncate the flow expansions at harmonic degree 14, and adopt a knot spacing of 5 years, a sufficiently detailed basis that our results are insensitive to either of these choices. For the period 1900 to 1980, we then seek flows that explain both the torque and the secular variation. We use two forms of secular variation "data". As in paper 1, we seek flows that can explain a global model of the secular variation calculated from the time-dependent magnetic field model ufm of *Bloxham and Jackson* [1992]. Following *Le Mouël et al.* [1985], we minimise the misfit to secular variation over the Earth's surface

$$\mathcal{M}_0 = \int \int_{r_e} \left(\dot{\mathbf{B}}(\mathbf{u}) - \dot{\mathbf{B}}_{\text{ufm}} \right)^2 dS dt \qquad (11)$$

where $\dot{\mathbf{B}}$ is related to the flow $\mathbf{u}$ through upward continuation of equation (7). In this paper, we also fit magnetic secular variation data directly. Magnetic observatories publish annual means of the magnetic field at a particular location through time: the difference of the values between successive years gives a first-difference approximation to the magnetic secular variation. We minimise

$$\mathcal{M}_1 = \sum_i (\dot{B}_i(\mathbf{u}) - \dot{B}_{i\,\text{obs}})^2/\sigma_i^2 \qquad (12)$$

where the sum is over all available data for the period 1900–1980 for the three components (X, Y, Z) of the magnetic field. These data were used in constructing the ufm model and there is thus some duplication of input as the secular variation of ufm is most sensitive to these data (J. Bloxham, personal communication). However, fitting raw data gives us far more confidence in our results than fitting only the model. In effect, we can regard our inversion as being to fit the secular variation data from the observatories, with additional data at lower weight provided by all other sources of magnetic field data, included through the ufm model. For numerical tractability, we assume that errors on the data are all independent and uncorrelated.

The observed torque is a smoothed form of data derived from *McCarthy and Babcock* [1986] by A. Jackson (personal communication), with short term variations filtered out, and an estimate of the torque due to tidal effects subtracted (corresponding to a ΔLOD of 1.4ms/century [*Jackson et al.*, 1993]). In this study, we fit the torque in two ways. First, as in paper 1, we seek flows for which the advective torque can fully explain ΔLOD, minimising

$$\mathcal{M}_2 = \int (\Gamma_z(\mathbf{u}) - \Gamma_{z\,\text{obs}})^2 dt \qquad (13)$$

Second, we investigate the hypothesis that the advective torque is responsible for most or all of the *variation* in torque, other torques (including the leakage torque) giving an offset that is assumed slowly varying. We seek flows which match the time derivative of the observed torque, minimising

$$\mathcal{M}_3 = \int (\dot{\Gamma}_z(\mathbf{u}) - \dot{\Gamma}_{z\,\text{obs}})^2 dt \qquad (14)$$

Some readers may worry that we are overstretching the data: change in LOD is calculated from the first derivative of the observed siderial angle, the second derivative gives the torque, so with the torque derivative we fit the third derivative of the observations. However, we do not require our fit to be smooth: we seek only a good visual fit to the trends in the time-variation of the torque, quite sufficient for the existence problem considered here.

Both ufm and the torque could be fit as continuous functions, but for simplicity, we fit discreet values calculated yearly. As in paper 1, we adopt a lag time for propagation of the secular variation due to mantle conductivity [*Backus*, 1983; *Benton and Whaler*, 1983] of 1 year.

We regularise our inversion [*Shure et al.*, 1982; *Parker*, 1994] by seeking flows which are smooth over the CMB. To ensure spatial smoothness, and numerical convergence by the truncation degree, we follow *Bloxham* [1988b] by minimising a norm of the second derivatives of the two components of velocity

$$\mathcal{M}_4 = \frac{r_c^2}{4\pi} \int \int_{\text{CMB}} \left[(\nabla_H^2 u_\theta)^2 + (\nabla_H^2 u_\phi)^2 \right] dS dt \qquad (15)$$

To reduce temporal variability, we seek a flow as close as possible to a steady flow. We choose to minimise

$$\mathcal{M}_5 = \frac{1}{4\pi r_c^2} \int \int_{\text{CMB}} \left(\frac{\partial \mathbf{u}}{\partial t} \right)^2 dS dt \qquad (16)$$

For a tangentially geostrophic flow

$$\nabla_H.(\mathbf{u}\cos\theta) = 0 \qquad (17)$$

[Le Mouël, 1984]. Therefore, in seeking flows that are close to tangentially geostrophic, we minimise the quantity

$$\mathcal{M}_6 = \int\int_{\text{CMB}} (\nabla_H.(\mathbf{u}\cos\theta))^2 \, dSdt \qquad (18)$$

The flow is then calculated by minimising

$$\mathcal{M}_0 + \lambda_1\mathcal{M}_1 + \lambda_2\mathcal{M}_2 + \lambda_3\mathcal{M}_3 + \lambda_4\mathcal{M}_4 + \lambda_5\mathcal{M}_5 + \lambda_6\mathcal{M}_6 \qquad (19)$$

where $\{\lambda_i\}$ are damping parameters (formally Lagrange multipliers) chosen to ensure adequate smoothness and fit to the data. All $\{\mathcal{M}_i\}$ are quadratic in $\mathbf{u}$, and thus the least squares problem posed here is linear. Ideally, we should solve the nonlinear flow problem, using field measurements rather than the secular variation derived from them [Voorhies, 1986; Bloxham, 1988a, 1989]. However, for this study, which seeks only to determine whether flows exist which match both the observed secular variation and the torque, a linear approach was deemed sufficient. Further, as the damping parameters give six degrees of freedom to the inversion, a linear approach is crucial to allow a reasonable exploration of parameter space.

RESULTS

As with many damped least-squares problems, one difficulty is to decide what constitutes a good fit to the data, or a model that is sufficiently smooth. Here, we shall be conservative (or at least what we regard as conservative!) We aim to fit ufm to a rms error of 5nT/year (the rms secular variation predicted by ufm is between 70nT/year and 84nT/year for 1900–1980), and the observatory annual mean first differences to 1.5 nominal standard deviations. The ufm model fits the observatory data to 1.32σ; Bloxham and Jackson [1992] note that this might overfit the observatory data, as a significant part of the secular variation can be attributed to external sources of field [e.g., Langel et al., 1996]. The rms flow speed is generally adjusted so as to match the westward drift rate of $0.2°$ per year ($\equiv 11$ km/year). We reject any flow with rms velocity greater than 30km/year. We define a characteristic time for the flow variation by

$$\tau = \sqrt{\int\int_{\text{CMB}} \mathbf{u}^2 dSdt \Big/ \int\int_{\text{CMB}} \left(\frac{\partial\mathbf{u}}{\partial t}\right)^2 dSdt} \qquad (20)$$

We reject flows that vary on a time scale shorter than the secular variation and torque that we are trying to explain, requiring $\tau \geq 10$. For the torque, we define a diagnostic

$$\Delta_\Gamma = \sqrt{\left(\int(\Gamma_z - \Gamma_{z\,\text{obs}})^2\,dt \Big/ \int(\Gamma_{z\,\text{obs}})^2\,dt\right)} \qquad (21)$$

and for its time-derivative

$$\Delta_{\dot\Gamma} = \sqrt{\left(\int\left(\dot\Gamma_z - \dot\Gamma_{z\,\text{obs}}\right)^2\,dt \Big/ \int\left(\dot\Gamma_{z\,\text{obs}}\right)^2\,dt\right)} \qquad (22)$$

where the limits $\Delta_\Gamma \leq 0.05$ and $\Delta_{\dot\Gamma} \leq 0.25$ are chosen heuristically to ensure a good qualitative fit.

We consider mantle conductances of $3\times10^8\text{S}$, 10^8S, $3\times10^7\text{S}$ and 10^7S. In table 1, we consider possible combinations of layer thicknesses and conductivities which could give rise to these conductances, either through a weakly conducting D'' an infiltration layer of core conductivity. Our results are summarised in table 2, where for some characteristic flow solutions we list the conductance, the form of the torque constraint ($\lambda_3 = 0$ for fitting the torque, and $\lambda_2 = 0$ for fitting the derivative), the rms flow speed, the characteristic time of the flow variation, the misfit from the secular variation model (in nT/year), the misfit from the observatory data (in nominal standard deviations as estimated by Bloxham and Jackson [1992]), the scaled misfit to the torque or its derivative, and the fraction of the flow power in ageostrophic modes (calculated using the projection matrix method of Bloxham [1989]). In each case, we repeat the inversion with the same values of the damping parameters $(\lambda_1, \lambda_4, \lambda_5, \lambda_6)$ (see equation 19), but with no constraint on the torque ($\lambda_2 = \lambda_3 = 0$). This is helpful in determining whether a particular solution is reasonable: if eliminating the torque constraint vastly alters the flow structure, then we regard the fit to the torque as contrived, and the EM coupling predicted as unlikely to be physically realisable. To provide a better feel for the meaning of the various tabulated diagnostics, we focus on flow (iv). We plot snapshots of this flow in figure 1: it is similar in appearance to many previously determined flows [Bloxham and Jackson, 1991], and is particularly reminiscent of the time-dependent geostrophic flow of Jackson et al. [1993]. Figure 2 shows the fit to the annual mean first differences for four representative magnetic observatories. The general trends in secular variation are well explained, although there is some evidence of systematic error (for example, the predictions for AL2X are consistently low, which over

Table 1. Possible values for deep mantle conductance.

Conductance (S)	Thickness	Conductivity (Sm^{-1})
3×10^8	150 km / 600 m	2000 / 5×10^5
1×10^8	200 km / 200 m	500 / 5×10^5
3×10^7	150 km / 60 m	200 / 5×10^5
1×10^7	200 km / 20 m	50 / 5×10^5

time would give a significant deviation from the observed field.) The same offset is also present if we do not attempt to fit the torque: it seems to arise from requiring the flow to be tangentially geostrophic, and has been observed previously (*Jackson et al.* [1993] and A. Jackson, in *Whaler and Davis* [1997]). It might be eliminated by performing a nonlinear inversion, fitting values of the main field rather than the secular variation [*Voorhies*, 1986; *Bloxham*, 1988a, 1989].

For mantle conductance of 3×10^8S or 10^8S, we find that we can fit the torque easily, with little additional complexity required in flows that are fully tangentially geostrophic. There is little difference between fitting the torque and its derivative: in both cases the fit is good, as is shown for flows (iii) and (iv) in figures 3 and 4. As seen from table 2, the principal requirement for

fitting the torque is an increase in flow time-dependence (reduction in τ), consistent with the results of paper 1 that the steady part of the flow explains most of the secular variation and the time-varying part most of the torque.

This situation changes significantly for $G = 3 \times 10^7$S. We list two sets of results: (vi), (vii) and (viii) with a slightly weakened geostrophic constraint applied, and (ix), (x) and (xi) with a substantially weakened constraint. To obtain a fit, we require either significantly stronger flows, or a departure from geostrophy. Formally, *Backus and Le Mouël* [1986] have shown that the assumption of tangential geostrophy fully resolves flow nonuniqueness in certain regions, and partially elsewhere. Numerically, for a harmonic truncation level L in both field and flow coefficients, a flow has $2L(L + 2)$ degrees of freedom, but the observed secular variation equation (7) provides only $L(L + 2)$ constraints. Geostrophy provides another $L(L+4)$ constraints [*Le Mouël et al.*, 1985], in principal resulting in a slighly overdetermined linear system, although certain linear combinations of flow coefficients may be poorly constrained. By relaxing this assumption, the flow is provided with many additional degrees of freedom with which to fit the torque. From a physical perspective, ageostrophy requires significant non-radial gravity or Lorentz forces at the core-mantle boundary. The former would generate significant gravitational coupling, which

Table 2. Selected results to test the required mantle conductance for EM coupling to explain ΔLOD. Columns are respectively mantle conductance, torque constraint (fitting the observed torque, its time derivative, or no constraint), the root mean square flow speed averaged over the CMB (nominal maximum 30 km/yr), the characteristic time of flow variation (nominal minimum 10 years), mean misfit to ufm at the Earth's surface (nominal value 5nT/yr), misfit to the observatory first differences (nominal value 1.5σ), misfit to the torque ($\lambda_2 \neq 0$) (nominal value 0.05), misfit to the torque derivative (λ_2 or $\lambda_3 \neq 0$) (nominal value 0.25), and fraction of the flow power in ageostrophic modes. The last flow of each section is obtained with the same inversion parameters, but no torque constraint.

	G (S)	Constraint	v_{rms} (km/yr)	τ (yr)	Δ_{ufm} (nT/yr)	$\Delta_{\mathrm{obs}}(\sigma)$	Δ_Γ	$\Delta_{\dot{\Gamma}}$	v^2_{ageo}/v^2
(i)	3×10^8	Torque	18.8	25.6	4.80	1.47	0.046	0.20	9.1×10^{-6}
(ii)	3×10^8	Deriv.	17.6	25.2	4.73	1.45		0.21	1.1×10^{-5}
(iii)	1×10^8	Torque	19.2	19.0	5.04	1.48	0.050	0.22	1.2×10^{-5}
(iv)	1×10^8	Deriv.	17.9	19.1	4.93	1.47		0.22	1.4×10^{-5}
(v)			17.5	27.6	4.70	1.45			1.1×10^{-5}
(vi)	3×10^7	Torque	29.9	11.4	4.92	1.50	0.049	0.22	3.2×10^{-4}
(vii)	3×10^7	Deriv	29.2	12.7	4.71	1.46		0.22	2.7×10^{-4}
(viii)			24.3	43.9	2.27	1.31			5.6×10^{-5}
(ix)	3×10^7	Torque	21.8	10.2	4.97	1.54	0.050	0.22	0.061
(x)	3×10^7	Deriv	21.2	11.3	4.68	1.48		0.22	0.055
(xi)			13.3	63.9	1.37	1.27			0.062
(xii)	1×10^7	Torque	50.6	10.0	7.00	1.56	0.060	0.25	0.17
(xiii)	1×10^7	Deriv.	46.3	10.2	6.39	1.51		0.27	0.16
(xiv)			10.7	43.1	1.09	1.23			0.30

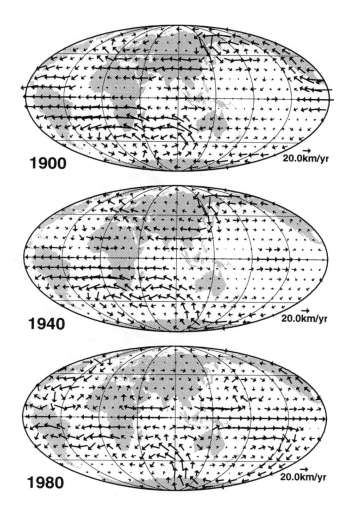

Figure 1. Plot of three snapshots of flow (iv)

we are assuming is negligible, while the latter would imply strong toroidal fields, most likely to be supported if mantle conductance is high. However, the flow is still dominantly geostrophic, so this objection might be ignored. More damningly, however, fitting the torque requires significant additional complexity in the flow. For example, comparing (ix) and (x) with (xi), the mean velocity is almost doubled, the characteristic variation time τ is reduced by a factor of 6, and the fit to ufm worsened by a factor of 3. That including only one additional equation of constraint should require so much additional complexity in the flow seems unrealistic. The situation is even worse for $G = 10^7 \mathrm{S}$, for which by our criteria we are unable to obtain any flows which fit the torque ((xii) and (xiii) give the closest "near misses").

We therefore conclude that for EM coupling to explain all of ΔLOD, we require mantle conductance of

at least $10^8 \mathrm{S}$. It could be argued that we have required an unreasonably good fit to the torque: however, allowing a weaker fit does not significantly alter our conclusions. An alternative possibility would be to allow significantly greater flow speeds, as indicated by flows (vi), (vii), (viii). Consideration of equation (8) suggests that a reduction in conductance G could be balanced by an increase in overall flow speed **u**. *Gubbins and Kelly* [1996] argue that advection by a steady flow will lead to a magnetic field that is not altered by the flow, and therefore not usable to obtain that flow. With finite errors in the data, there are an infinite number of possible flows that fit the data, and by seeking smooth models we will preferentially obtain those with the smallest magnitude. Hence, the flow orthogonal to the secular variation, and so contributing to the torque, may be stronger than has previously been assumed. However, it seems wise to be cautious in postulating a similar effect for time-dependent flows.

CONSERVATION OF ANGULAR MOMENTUM

Measurement of changes in LOD allows calculation of the change in angular momentum of the mantle, and consequently the torque required to generate that change. However, conservation of angular momentum requires an equal and opposite change in the angular momentum of the core (CAM). Unfortunately, this cannot be determined directly, as it would require knowledge of flow at all depths within the core, and we possess only partial knowledge of the flow at the core surface. However, *Jault et al.* [1988] have argued that for decadal changes in LOD, ΔCAM is likely to be absorbed by motions uniform on cylindrical annuli. Then changes in flow at the core surface may also reflect such changes at depth. Ignoring the inner core (and its recently inferred superrotation, [*Song and Richards*, 1996; *Su et al.*, 1996]), and approximating the density of the core as uniform, this part of the CAM is given by

$$L = r_c^4 \rho \int u_\phi \cos^2 \theta \sin \theta d\Omega \qquad (23)$$

[*Jault and Le Mouël*, 1989; *Jackson et al.*, 1993] where u_ϕ is the calculated ϕ-velocity at the surface of the core, $d\Omega$ is an element of solid angle, and ρ is the mean density. Only two flow modes contribute to the CAM, both zonal toroidal, so that

$$L = \frac{8\pi r_c^4 \rho}{15} \left(t_1^0 + \frac{12}{7} t_3^0 \right) \qquad (24)$$

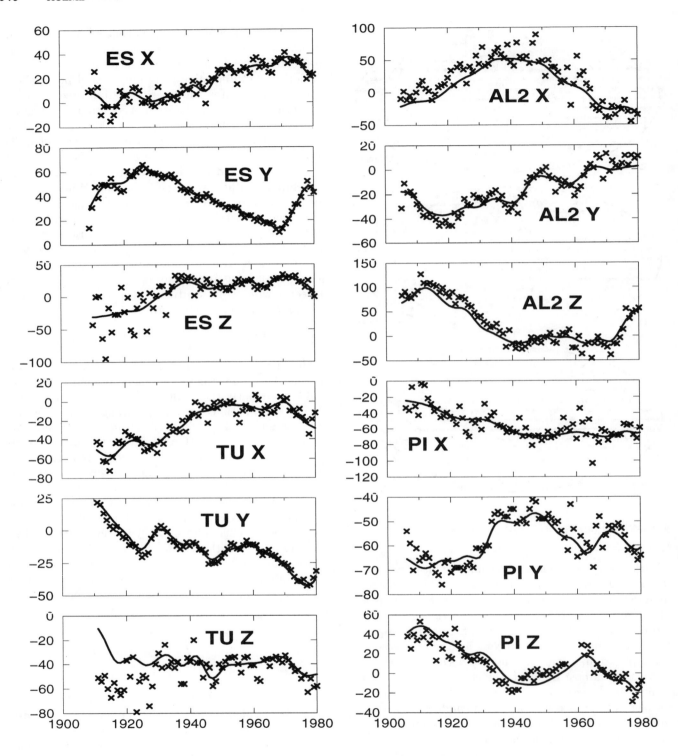

Figure 2. Fit to the annual mean first differences at four observatories for flow (iv). These are ES: Eskdalemuir, Scotland (55.3°N,3.2°W), TU: Tuscon, Arizona, USA (32.3°N,70.8°W), AL2: Alibag, India (18.6°N,72.9°E), and PI: Pilar, Argentina (31.7°S,63.9°W). Vertical axis in nT/year, horizontal axis is year, crosses are measured values, curve is secular variation predicted from flow.

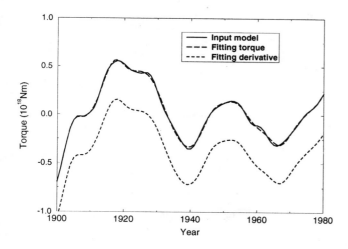

Figure 3. Fit to the observed torque for flows (iii) and (iv), table 2.

In figure 5, we perform a similar calculation to *Jault et al.* [1988] and *Jackson et al.* [1993], and compare the fit of the angular momentum of a geostrophic flow with the observed LOD. The offset between the two curves is arbitrary. The similarity between the two profiles is striking, although we note that we have also determined many quite reasonable tangentially geostrophic flows for which the comparison is much less impressive. We also include the LOD variations predicted by flow (iv) from table 2, which fits the torque. The fit to the LOD variation is poor: the predicted ΔLOD of this flow less reminiscent of the LOD variations than of the torque, their time derivative (compare with figure 3). This arises because the torque is strongly dependent on the zonal toroidal flow modes which contribute to the angular momentum. If we approximate the field at the CMB by the axial dipole only (its dominant component), then from equation (8), the advective torque is given by

$$\Gamma_z = 4r_c^4 G \left(\frac{r_e}{r_c}\right)^6 (g_1^0)^2 \times$$

$$\iint \frac{\partial}{\partial\theta} \left(\mathrm{L}^{-2}\nabla_H\cdot[\hat{\mathbf{r}}\wedge\cos\theta\mathbf{u}]\right)\cos\theta\sin^2\theta d\theta d\phi$$

$$= 4r_c^3 G \left(\frac{r_e}{r_c}\right)^6 (g_1^0)^2 \int u_\phi \cos^2\theta \sin\theta d\Omega \quad (25)$$

where we have successively integrated by parts with respect to θ, transferred the L^{-2} operator (see Appendix A, paper 1), and applied the surface form of Gauss'

theorem [e.g., *Backus et al.*, 1996, § 7.4.6]. Then substituting from equation (23),

$$\Gamma_z = L\frac{4Gr_e^6 g_1^{0^2}}{r_c^7\rho} \equiv L/T \quad (26)$$

where $T \approx 30$ years for $G = 3 \times 10^8$. Thus we might expect the strong correlation between the torque and estimated angular momentum seen in figure 5. *Zatman* [1997] suggests that this result underlies the claim of *Stewart et al.* [1995] of a strong correlation between EM torque and rate of change of length of day with a five year lag: ΔCAM is approximately harmonic, and so with suitable choice of phase lag will be well correlated with its own derivative, the torque. The dynamical implications of equation (26) are significant. The torque on the core is equal and opposite to that on the mantle, so the equation of motion for the core can be written

$$\dot{L} = -L/T + f(t) \quad (27)$$

where $f(t)$ is the rest of the torque (electromagnetic or non-electromagnetic). If $f(t)$ is a harmonic forcing, the response will lead slightly that expected from an undamped system. Equation (27) is an expression of Lenz's law: the core-mantle system responds to try to limit the differential velocity, so as to decrease the electromagnetic induction in the conducting mantle [*Bullard et al.*, 1950; *Gubbins and Roberts*, 1987, §3.4]. *Jault and Le Mouël* [1991] also noted the importance of the solid body rotation flow term (t_1^0) on the torque, and as a consequence expressed doubt as to whether

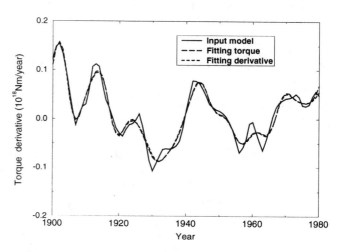

Figure 4. Fit to the observed derivative of the torque for flows (iii) and (iv), table 2.

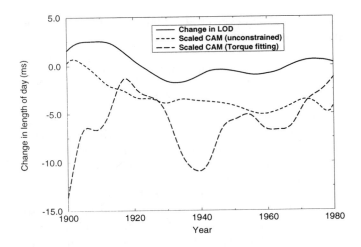

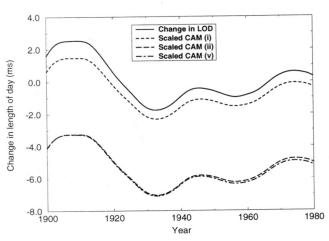

Figure 5. Comparison of LOD observations with predictions from an unconstrained geostrophic flow and flow (iv), table 2.

Figure 6. Match to LOD observations of CAM for flows $(i)_L$, $(ii)_L$, $(v)_L$, table 3.

the torque and angular momentum constraints could be satisfied simultaneously.

Hence, we test here whether it is possible to simultaneously fit the required torque, and also explain the variation in CAM by torsional modes uniform on cylinders. We add another constraint to the flow inversion: namely that the rate of change of CAM defined by equation (24) should be equal and opposite to the torque on the mantle. We minimise

$$\mathcal{M}_7 = \int \left(\dot{L} + \Gamma_{z\,\text{obs}}\right)^2 dt \qquad (28)$$

We define a diagnostic for the angular momentum similar to equations (21-22)

$$\Delta_L = \sqrt{\left(\int \left(\dot{L} + \Gamma_{z\,\text{obs}}\right)^2 dt \middle/ \int \left(\Gamma_{z\,\text{obs}}\right)^2 dt\right)} \approx 0.05 \qquad (29)$$

It is easiest to demonstrate our general conclusions by simply adding this constraint to flows (i)–(v) from ta-

ble 2 above, with results presented in table 3. Angular momentum profiles from $(i)_L$, $(ii)_L$ and $(v)_L$ are given in figure 6 (the profiles from $(iii)_L$ and $(iv)_L$ are almost indistinguishable from flows $(i)_L$ and $(ii)_L$ respectively). As can be seen comparing table 2 and table 3, the additional constraint has little effect on the general characteristics of the flow. However, it does alter which flow modes generate the torque. Here, the t_1^0 and t_3^0 modes must match the angular momentum, but otherwise these modes contribute strongly to generating the required torque.

The action of the poloidal torque can also be considered in the context of Lenz's law. The evolution of each multipole pair of real spherical harmonics can be written in terms of amplitude and phase, viz.,

$$g_l^m(t) = k_l^m(t)\cos\left(m\phi_l^m(t)\right)$$
$$h_l^m(t) = k_l^m(t)\sin\left(m\phi_l^m(t)\right)$$

From equation (6), the contribution to the poloidal torque on the core from the harmonics of degree l, order

Table 3. Selected results to test whether torque and angular momentum constraints can be fit simultaneously. Columns as for table 2, with addition of rms misfit to angular momentum, nominal value 0.05.

	v_{rms} (km/yr)	τ (yr)	Δ_{ufm} (nT/yr)	$\Delta_{\text{obs}}(\sigma)$	Δ_Γ	$\Delta_{\dot{\Gamma}}$	Δ_L	v_{ageo}^2/v^2
$(i)_L$	18.8	25.0	4.82	1.47	0.046	0.21	0.049	9.7×10^{-6}
$(ii)_L$	17.6	24.5	4.75	1.46		0.22	0.053	1.2×10^{-5}
$(iii)_L$	19.3	17.5	5.18	1.49	0.050	0.22	0.042	1.7×10^{-5}
$(iv)_L$	18.0	17.5	5.05	1.47		0.23	0.067	1.9×10^{-5}
$(v)_L$	17.6	27.5	4.69	1.45			0.041	1.1×10^{-5}

m is

$$\Gamma_z(l, m) \propto -\left[k_l^m(t)\right]^2 m \frac{d\phi_l^m}{dt}$$

In particular, the westward drift of the equatorial dipole produces an eastward torque on the core to try to reduce the drift. To maintain a steady westward drift, another torque must balance this. *Stix and Roberts* [1984] suggest that the leakage torque could fulfill this role. While such a suggestion is not observationally testable, some insight might be gained from recently developed computer simulations of the geodynamo. For example, the model of *Kuang and Bloxham* [1997a] includes only EM coupling between core and mantle, but produces a strong, predominantly steady, westward drift. The leakage torque is the only possibility which allows this to be maintained.

CONCLUSIONS

We have shown that by assuming EM coupling is the principal mechanism effecting exchange of angular momentum between the core and mantle, it is possible to constrain the conductivity structure of the lowermost mantle. A conductance of 10^8S or greater allows effective production of the required torque. This estimate is in agreement with the value required by *Buffett* [1992] to explain features in the Earth's forced nutations. Such a conductance also permits the required change in core angular momentum to be matched by motions constant on cylindrical annuli throughout the core. It is still possible to explain the torque with a conductance of 3×10^7S, although this requires a considerable increase in flow complexity over that required to explain the magnetic secular variation. A conductance of 10^7S would require a significantly stronger flow than is usually assumed present, although this is not impossible.

Two important physical effects are missing from this study. First, we have ignored diffusion, both in our very limited consideration of the leakage torque, and in the assumption of poloidal field frozen-flux used to determine the flow. The importance of diffusion is difficult to determine for the Earth. However, numerical output from recently developed computational simulations of the geodynamo [e.g. *Glatzmaier and Roberts*, 1995, 1996; *Kuang and Bloxham*, 1997a] will enable the study of model systems which may provide insight into the role of magnetic diffusion in the real Earth. Second, we have assumed no lateral variations in electrical conductivity. D″ is seismically heterogeneous and anisotropic, and it is therefore highly possible that the electrical conductivity may be similarly non-uniform. *Buffett*

[1996a] has recently considered the influence of small-scale heterogeneity on EM coupling, while *Aurnou et al.* [1996] have begun to examine large-scale heterogeneity by modelling D″ as having uniform conductivity but variable thickness. It is likely that non-uniform conductivity will increase the effectiveness of EM coupling, allowing input from flow modes only weakly coupled with a uniformly conducting layer. This issue is currently under study.

Acknowledgments. I am supported by a NERC postdoctoral fellowship. I thank Stephen Zatman for helpful discussions.

REFERENCES

Aurnou, J. M., J. L. Buttles, G. A. Neumann, and P. L. Olson, Electromagnetic core-mantle coupling and paleomagnetic reversal paths, *J. Geophys. Res., 23*, 2705–2708, 1996.

Backus, G., R. Parker, and C. Constable, *Foundations of Geomagnetism.* Cambridge University Press, 1996.

Backus, G. E., Kinematics of geomagnetic secular variation in a perfectly conducting core, *Philos. Trans. R. Soc. London A, 263*, 239–266, 1968.

Backus, G. E., Application of mantle filter theory to the magnetic jerk of 1969, *Geophys. J. Roy. Astron. Soc., 74*, 713–746, 1983.

Backus, G. E., and J.-L. Le Mouël, The region on the core-mantle boundary where a geostrophic velocity field can be determined from frozen flux magnetic data, *Geophys. J. Roy. Astron. Soc., 85*, 617–628, 1986.

Benton, E. R., and K. A. Whaler, Rapid diffusion of the poloidal geomagnetic field through the weakly conducting mantle: a perturbation solution, *Geophys. J. Roy. Astron. Soc., 75*, 77–100, 1983.

Bloxham, J., The determination of fluid flow at the core surface from geomagnetic observations, in *Mathematical Geophysics, A Survey of Recent Developments in Seismology and Geodynamics*, edited by N. J. Vlaar, G. Nolet, M. J. R. Wortel, and S. A. P. L. Cloetingh. Reidel, Dordrecht, 1988a.

Bloxham, J., The dynamical regime of fluid flow at the core surface, *Geophys. Res. Lett., 15*, 585–588, 1988b.

Bloxham, J., Simple models of fluid flow at the core surface derived from geomagnetic field models, *Geophys. J. Int., 99*, 173–182, 1989.

Bloxham, J., and A. Jackson, Fluid flow near the surface of the Earth's outer core, *Rev. Geophys., 29*, 97–120, 1991.

Bloxham, J., and A. Jackson, Time-dependent mapping of the magnetic field at the core-mantle boundary, *J. Geophys. Res., 97*, 19537–19563, 1992.

Buffett, B. A., Constraints on magnetic energy and mantle conductivity from the forced nutations of the Earth, *J. Geophys. Res., 97*, 19581–19597, 1992.

Buffett, B. A., Effects of a heterogeneous mantle on the velocity and magnetic fields at the top of the core, *Geophys. J. Int., 125*, 303–317, 1996a.

Buffett, B. A., A mechanism for decade fluctuations in the length of day, *Geophys. Res. Lett., 23*, 3803–3806, 1996b.

Bullard, E. C., C. Freeman, H. Gellman, and J. Nixon, The westward drift of the Earth's magnetic field, *Philos. Trans. Roy. Soc. London A, 243,* 61–92, 1950.

Cohen, R. E., I. I. Mazin, and D. G. Isaak, Magnetic collapse in transition metal oxides at high pressure: implications for the Earth, *Science, 275,* 654–657, 1997.

Edmonds, A. R., *Angular momentum in quantum mechanics.* Princeton Univertiy Press, 1960.

Glatzmaier, G. A., and P. H. Roberts, A three-dimensional self-consistent computer simulation of a geomagnetic field reversal, *Nature, 377,* 203–209, 1995.

Glatzmaier, G. A., and P. H. Roberts, On the magnetic sounding of planetary interiors, *Phys. Earth Planet. Int, 98,* 207–220, 1996.

Gubbins, D., and P. Kelly, A difficulty with using the frozen flux hypothesis to find steady core motions, *Geophys. Res. Lett., 23,* 1825–1828, 1996.

Gubbins, D., and P. H. Roberts, Magnetohydrodynamics of the Earth's core, in *Geomagnetism,* edited by J. A. Jacobs, vol. 2, chap. 1. Academic, San Diego, Calif., 1987.

Hide, R., Interaction between the Earth's liquid core and solid mantle, *Nature, 222,* 1055–1056, 1969.

Hide, R., Reply to comment on "The topographic torque on a bounding suface of a rotating gravitating fluid and the excitation by core motions of decadal fluctuations in the Earth's rotation", *Geophys. Res. Lett., 22,* 3563–3565, 1995.

Hide, R., and J. O. Dickey, Earth's variable rotation, *Science, 253,* 629–637, 1991.

Hide, R., , R. W. Clayton, B. H. Hager, M. A. Spieth, and C. V. Voorhies, Topographic core-mantle coupling and fluctuations in the Earth's rotation, in *Relating Geophysical structures and processes: The Jeffreys volume,* edited by K. Aki, and R. Dmowska, pp. 121–127. AGU/IUGG, 1993.

Holme, R., Electromagnetic core-mantle coupling I: Explaining decadal variations in the Earth's length of day, *Geophys. J. Int., 132,* 167–180, 1998.

Jackson, A., The Earth's magnetic field at the core-mantle boundary., Ph.D. thesis, Cambridge University, 1989.

Jackson, A., J. Bloxham, and D. Gubbins, Time-dependent flow at the core surface and conservation of angular momentum in the coupled core-mantle system, in *Dynamics of the Earth's deep interior and Earth rotation,* edited by J.-L. Le Mouël, D. E. Smylie, and T. Herring, pp. 97–107. AGU/IUGG, 1993.

Jault, D., Variation séculaire du champ geomagnetique et fluctuations de la longeur du jour, Ph.D. thesis, University of Paris VII, 1990.

Jault, D., and J. L. Le Mouël, The topographic torque assiciated with a tangentially geostrophic motion at the core surface and inferences on flow inside the core, *Geophys. & Astrophys. Fluid Dyn., 48,* 273–296, 1989.

Jault, D., and J. L. Le Mouël, Exchange of angular momentum between the core and the mantle, *J. Geomag. Geoelect., 43,* 111–129, 1991.

Jault, D., C. Gire, and J. L. Le Mouël, Westward drift, core motions and exchanges of angular momentum between core and mantle, *Nature, 333,* 353–356, 1988.

Jeanloz, R., The nature of the Earth's core, *Ann. Rev. Earth Planet. Sci, 18,* 357–386, 1990.

Jeanloz, R., Chemical reactions at the earth's core-mantle boundary: summary of evidence and geomagnetic implications, in *Relating Geophysical structures and processes: The Jeffreys volume,* edited by K. Aki, and R. Dmowska, pp. 121–127. AGU/IUGG, 1993.

Knittle, E., and R. Jeanloz, Simulating the core-mantle boundary: an experimental study of high-pressure reactions between silicate and liquid iron, *Geophys. Res. Lett., 13,* 1541–1544, 1989.

Kuang, W., and J. Bloxham, An Earth-like numerical dynamo model, *Nature, 389,* 371–374, 1997a.

Kuang, W., and J. Bloxham, On the dynamics of topographical core-mantle coupling, *Phys. Earth Planet. Int., 99,* 289–294, 1997b.

Langel, R. A., The main field, in *Geomagnetism,* edited by J. A. Jacobs, vol. 1, chap. 4. Academic, San Diego, Calif., 1987.

Langel, R. A., T. J. Sabaka, R. T. Baldwin, and J. A. Conrad, The near-Earth magnetic field from magnetospheric and quiet-day ionospheric sources and how it is modeled, *Phys. Earth Planet. Int., 98,* 235–267, 1996.

Le Mouël, J.-L., Outer core geostrophic flow and secular variation of Earth's magnetic field, *Nature, 311,* 734–735, 1984.

Le Mouël, J.-L., C. Gire, and T. Madden, Motions at the core surface in the geostrophic approximation, *Phys. Earth Planet. Int., 39,* 270–287, 1985.

Loper, D. E., and T. Lay, The core-mantle boundary region, *J. Geophys. Res., 100,* 6397–6420, 1995.

McCarthy, D. D., and A. K. Babcock, The length of day since 1656, *Phys. Earth Planet. Inter., 44,* 281–292, 1986.

Parker, R. L., *Geophysical Inverse Theory.* Princeton University Press, Princeton, NJ., 1994.

Poirier, J.-P., Core-infiltrated mantle and the nature of the D'' layer, *J. Geomag. Geoelec., 45,* 1221–1227, 1993.

Poirier, J.-P., A. Goddat, and J. Peyronneau, Ferric iron dependence of the electrical conductivity of the Earth's lower mantle material, *Phil. Trans. R. Soc. Lond. A, 354,* 1361–1369, 1996.

Roberts, P. H., and S. Scott, On the analysis of secular variation, 1, A hydromagnetic constraint: Theory, *J. Geomag. Geoelectr., 17,* 137–151, 1965.

Rochester, M. G., Causes of fluctuations in the rotation of the Earth, *Philos. Trans. R. Soc. London A, 313,* 95–105, 1984.

Runcorn, S. K., The electrical conductivity of the Earth's mantle, *Trans. Am. Geophys. Union, 36,* 191–198, 1955.

Shure, L., R. L. Parker, and G. E. Backus, Harmonic splines for geomagnetic modelling, *Phys. Earth Planet. Inter., 28,* 215–229, 1982.

Song, X. D., and P. G. Richards, Seismological evidence for differential rotation of the Earth's inner core, *Nature, 382,* 221–224, 1996.

Stewart, D. N., F. H. Busse, K. A. Whaler, and D. Gubbins, Geomagnetism, Earth rotation and the electrical conductivity of the lower mantle, *Phys. Earth Planet. Inter., 92,* 199–214, 1995.

Stix, M., and P. H. Roberts, Time-dependent electromagnetic core-mantle coupling, *Phys. Earth Plan. Int., 36,* 49–60, 1984.

Su, W.-J., A. M. Dziewonski, and R. Jeanloz, Planet within

a planet – rotation of the inner-core of Earth, *Science*, *274*, 1883–1887, 1996.

Voorhies, C. V., Steady surficial core motions: an alternate method, *Geophys. Res. Lett.*, *13*, 1537–1540, 1986.

Whaler, K. A., and R. G. Davis, Probing the Earth's core with geomagnetism, in *Earth's Deep Interior*, edited by D. J. Crossley, pp. 114–166. Gordon and Breach, Amsterdam, 1997.

Zatman, S., Angular momentum and the core of the Earth, Ph.D. thesis, Harvard University, 1997.

R. Holme, Department of Geology and Geophysics, University of Edinburgh, Grant Institute, West Mains Road, Edinburgh EH9 3JW, Scotland. (email: holme@mail.glg.ed.ac.uk)

Free Oscillations in the Length of Day: Inferences on Physical Properties near the Core-mantle Boundary

Bruce A. Buffett

Department of Earth and Ocean Sciences, University of British Columbia, Vancouver, Canada

Small variations in the rotation rate of the Earth over periods of several decades are generally attributed to exchanges of angular momentum between the core and the mantle. Since the mechanisms that permit this exchange of angular momentum depend on physical conditions near the boundaries of the core, observations of the Earth's rotation can provide insights into the properties of the boundary region. The relationship between boundary properties and the observed fluctuations is established by calculating the periods of free oscillations in the length of day. Motions in the fluid outer core are described in terms of torsional oscillations, while angular momentum is transferred between the core and the mantle by electromagnetic, topographic, and gravitational torques. Each of these coupling mechanisms may be assessed and compared by examining the predicted modes of oscillations. The most viable coupling mechanism is due to the gravitational torques, although plausible situations can be devised where both electromagnetic and topographic torques at the core-mantle boundary are also significant. Since each coupling mechanism has a distinct influence on the oscillations, it is possible to invert relevant geodetic and geomagnetic measurements for a variety of physical properties at the core-mantle boundary.

1. INTRODUCTION

Measurements of fluctuations in the Earth's rotation vector contain valuable information about the core-mantle boundary (CMB) region. The utility of these measurements arises from the mobility of the fluid core, which is free to rotate separately from the mantle. Differences in rotation do occur between the core and the mantle, both in rate and direction. These differences cause torques on the mantle (and opposite torques on the core) by a variety of mechanisms. Torques on the mantle alter the rotation of the Earth's surface, which may be detected in precise geodetic measurements [e.g.,

McCarthy and Babcock, 1986; *Herring et al.*, 1991]. By comparing observations with theoretical predictions for the effects of known coupling mechanisms, we can assess the importance of the various coupling mechanisms and the physical properties on which these mechanisms depend.

An instructive example involves the variations in the direction of the Earth's polar axis (nutations) due to tidal forces [*Mathews and Shapiro*, 1992]. Since the tidal forces are well-known, measurements of the nutations can be used to investigate the Earth's response to an imposed external torque [*Sasao et al.*, 1980; *Wahr*, 1981; *Dehant*, 1990; *Mathews et al.*, 1991]. Responses at certain tidal frequencies are particularly sensitive to relative rotations between the core and the mantle. At these frequencies, comparisons between the measurements and theoretical predictions have proven especially useful for investigating the CMB region. Observed discrepancies have been reconciled by altering the shape

The Core-Mantle Boundary Region
Geodynamics 28
Copyright 1998 by the American Geophysical Union.

of the CMB [*Gwinn et al.*, 1986] and by introducing a thin, electrically conducting layer at the base of the mantle [*Buffett*, 1992].

Observations of longer period changes in the rotation rate can be used in a similar way to study the CMB region. Changes in rotation rate are typically described in terms of the corresponding changes in the length of day. The measured fluctuations in the length of day over periods of several decades appear to be well explained by exchanges of angular momentum between the core and mantle [*Jault et al.*, 1988; *Jackson et al.*, 1993]. Since there are no known external torques at these periods, the oscillations must arise solely from internal interactions between the core and mantle. Less is known about the excitation of these fluctuations, but more is to be gained from an understanding of the physical processes involved. For example, the observed periodicities are probably dependent on the structure of the magnetic field inside the core. This suggests that a quantitative theory for decade fluctuations could be used to extract information about the internal field (see, for example, *Zatman and Bloxham*, this issue). Additional inferences can be drawn about the relevant coupling mechanisms and the source of excitation that sustains these oscillations. At present, however, there is no comprehensive theoretical framework for interpreting the length-of-day measurements. This situation has hampered progress in understanding the decade fluctuations. By contrast, studies of the nutations have been more successful because it has been possible to compare the observations with theoretical predictions. Specific predictions for the decade fluctuations will ultimately depend on details of the excitation source, but we can make substantial progress by constructing models that allow plausible excitation mechanisms to be tested.

The purpose of this paper is to develop a quantitative model for decade fluctuations in the length of day. Low-frequency motions in the fluid core are characterized by torsional oscillations [*Braginsky*, 1970], while rigid-body rotations of the inner core and the mantle are described by separate angular momentum equations. Interactions between the fluid core and the rest of the Earth are due to the effects of electromagnetic coupling at the boundaries of the fluid core [*Bullard et al.*, 1950; *Rochester*, 1960] and topographic coupling at the CMB [*Hide*, 1969; 1989]. Gravitational interactions between the solid inner core and the mantle are also incorporated [*Buffett*, 1996a], including the possibility of viscous deformation of the inner core. Free oscillations in the length of day are calculated for the coupled motions of the mantle, fluid outer core and solid inner core. The periods of oscillation are readily adjusted to

match the observed periodicities, providing useful constraints on the strength of the magnetic field inside the fluid core. Quantitative predictions also show that each of the coupling mechanisms has a distinct influence on the oscillations. Thus it is possible, in principal, to distinguish between the various types of coupling and to assess a variety of physical properties in the CMB region.

2. TORSIONAL OSCILLATIONS

Several observational constraints guide the form of a suitable theory for decade fluctuations in the length of day. First, the measured variations require torques on the mantle with peak amplitudes of 10^{18} Nm (see Fig. 1). Fluctuations in the torque typically occur over periods of several decades, yet the mean torque over the last century is less than 10^{16} Nm [*Kuang and Bloxham*, 1993]. This suggests that the measured variations represent an oscillation about a quasi-static equilibrium between the core and the mantle. Changes in the magnetic field and the convective flow in the core should cause this equilibrium to evolve, but the time scale for change must be slow compared with decade periods. It is reasonable under these circumstances to neglect the slow convective motions and assume that the oscillations occur about a static initial state.

A second observational constraint emerges from the flow inferred at the top of the core from geomagnetic variations [*Bloxham and Jackson*, 1991]. The strong influence of rotation in the core causes low-frequency fluid motions to be largely two-dimensional, with vanishingly small gradients in the direction of the rotation axis. Axisymmmetric motions, which carry angular momentum in the core, are characterized by rotations of fluid cylinders that extend along the rotation axis. Velocities at the ends of the cylinders can be inferred from the axisymmetric part of the flow at the top of the core, so it is possible to estimate the motion of these cylinders and the corresponding angular momentum in the core. Fluctuations in the flow at the top of the core imply changes in angular momentum, which correlate with the observed fluctuations in the length of day [*Jault et al.*, 1988; *Jackson et al.*, 1993]. The conclusions drawn from these studies are two-fold. First, they confirm that decade fluctuations arise from exchanges of angular momentum between the core and the mantle. Second, they suggest that the fluid motions are adequately represented by rigidly rotating fluid cylinders.

A suitable dynamical description of the fluid core is provided by the theory of torsional oscillations [*Braginsky*, 1970]. Motions are characterized by the angular velocity of fluid cylinders as a function of distance s from

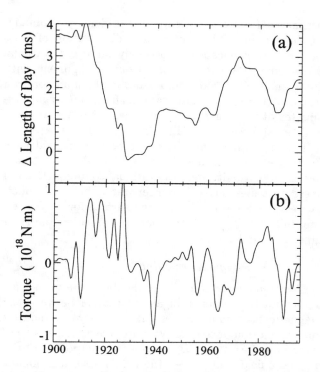

Figure 1. (a) Observed long-period changes in the length of day based on *McCarthy and Babcock* [1986] and more recent VLBI data from *Herring* (personal communication); (b) the associated torque on the mantle is smoothed using a 3-point Hanning filter.

the rotation axis. (We denote the *angular* velocity by $u(s)$). Rotations of adjacent cylinders are coupled when their surfaces are threaded with a magnetic field. The magnetic field provides the restoring force which causes the fluid cylinders to oscillate. Torsional oscillation interact with the inner core and mantle through surfaces forces at the ends of the fluid cylinders. The original study of *Braginsky* [1970] considered the influence of magnetic stresses at the fluid-solid interfaces, while the present study adds the influences of topographic coupling and gravitational interactions between the inner core and the mantle. These extensions permit a direct comparison of the commonly proposed coupling mechanisms.

The spherical surfaces that approximate the boundaries of the fluid core are defined in cylindrical coordinates (s, φ, z) by $z_i(s) = (r_i^2 - s^2)^{1/2}$ and $z_f(s) = (r_f^2 - s^2)^{1/2}$, where r_i and r_f are the spherical radius of the inner-core and core-mantle boundaries (see Fig. 2). (We let $z_i = 0$ when $s > r_i$). The introduction of small-amplitude topography on these boundaries has an interesting influence on the oscillations in the core, but we defer this discussion to the section on topographic

coupling. In the absence of boundary topography the oscillation period depends primarily on the magnetic restoring force associated with the radial component B_s of the magnetic field. It is convenient to express this restoring force in terms of a magnetic tension,

$$\tau(s) = 4\pi s^3 (z_f - z_i)\{B_s^2\}/\mu_0 \,, \qquad (1)$$

where μ_0 is the magnetic permeability and $\{\cdot\}$ denotes the average over a cylindrical surface.

The governing equation for the angular velocity $u(s)$ is obtained from an angular momentum balance for the fluid cylinders. When $u(s)$ is proportional to $\exp(-i\omega t)$, the angular momentum equation for the fluid can be written as [*Braginsky*, 1970]

$$-\omega^2 m(s)u(s) = \frac{d}{ds}\left(\tau(s)\frac{du(s)}{ds}\right) + i\omega f(s)\,, \qquad (2)$$

where $-f(s)$ is the torque on the ends of the cylinders due to surface forces at the fluid-solid boundaries and

$$m(s) = 4\pi \rho s^3 (z_f - z_i) \qquad (3)$$

is the moment density of the cylinders. It follows from the definition of $m(s)$ that the moment of inertia of the entire fluid core is

$$C_f = \int_0^{r_f} m(s)\,ds\,. \qquad (4)$$

while the moments of the inner core and the mantle are denoted by C_i and C_m.

The explicit form of $f(s)$ is established by the nature of the forces that couple the fluid core to the rest of the Earth. These coupling mechanisms typically depend on the relative motion across the fluid-solid boundaries. Thus the coupling $f_m(s)$ at the CMB depends on the angular velocity u_m of the mantle, while the coupling $f_i(s)$ at the inner-core boundary (ICB) involves the angular velocity u_i of the solid inner core. These two contributions to $f(s)$ may be expressed in the form

$$f_m(s) = \mathcal{F}_m(s)[u(s) - u_m]\,, \qquad (5)$$

$$f_i(s) = \mathcal{F}_i(s)[u(s) - u_i]\,, \qquad (6)$$

where amplitudes $\mathcal{F}_m$ and $\mathcal{F}_i$ are determined by the type of coupling mechanisms and the relevant physical properties near the fluid-solid boundaries.

In order to evaluate $f(s)$ we require additional equations to determine the angular velocities of the mantle u_m and the inner core u_i. Changes in u_m and u_i are caused by the surface forces $f_m(s)$ and $f_i(s)$, and we assume that a mutual gravitational torque acts on the mantle and the inner core [*Buffett*, 1996a]. The gravitational torque on the inner core Γ_i takes the form

$$\Gamma_i = \tilde{\Gamma}(\varphi_m - \varphi_i), \qquad (7)$$

where φ_i and φ_m are small angular displacements of the inner core and the mantle, which are related to the angular velocities by $u_i = d\varphi_i/dt$ and $u_m = d\varphi_m/dt$. An equal and opposite gravitational torque acts on the mantle. Consequently, the angular momentum equation for the mantle can be written as

$$-\omega^2 C_m u_m = -\tilde{\Gamma}(u_m - u_i) - i\omega \int_0^{r_f} \mathcal{F}_m(s)\,[u(s) - u_m]\,ds\,, \qquad (8)$$

while the angular momentum equation for the inner core is

$$-\omega^2 C_i u_i = \tilde{\Gamma}(u_m - u_i) - i\omega \int_0^{r_i} \mathcal{F}_i(s)\,[u(s) - u_i]\,ds\,. \qquad (9)$$

Expressions for $\mathcal{F}_m(s)$, $\mathcal{F}_s(s)$, and $\tilde{\Gamma}$ are given in the next section.

3. COUPLING MECHANISMS

There are a number of mechanisms that can transfer angular momentum between the core and the mantle. Most mechanisms involve forces at or near the CMB. For example, electromagnetic coupling involves Lorentz forces near the base of the mantle [*Bullard et al.*, 1950; *Rochester*, 1960], while topographic coupling is due to the effects of fluid pressure in the core acting on aspherical features in the CMB [*Hide*, 1969]. Coupling forces also arise from viscous stresses at the base of the mantle, although these effects are thought to be small [e.g., *Kuang and Bloxham*, 1993].

A different type of coupling mechanism involves gravitational interactions between the core and the mantle [*Jault and LeMouël*, 1989]. In this case angular momentum is transferred by gravitational forces, which are distributed over the volumes of the core and the mantle. Quantitative estimates of this coupling mechanism require information about the mass anomalies in both the core and the mantle. Global seismic models can provide an indirect estimate of the large-scale density anomalies in the mantle [e.g. *Su et al.*, 1994], but the corresponding anomalies in the core are too small to detect seismically [*Stevenson*, 1987]. However, an alternative means of determining density anomalies in the core can permit quantitative estimates of the gravitational torque [*Buffett*, 1996a]. Perturbations in the gravitational potential due to density anomalies in the mantle produce density variations inside the core by deflecting surfaces of constant potential. Since surfaces of constant potential and constant density coincide in a hydrostatic fluid, the density structure of the core may be estimated from a knowledge of the density anomalies in the mantle. This hydrostatic distribution of mass in the core can cause surprisingly large gravitational forces when the core is rotated out of alignment with the mantle. Convective anomalies in the fluid core can also contribute to the gravitational force, but typical estimates of the convective anomalies suggest that these additional forces are small.

The mechanisms usually proposed to explain decade fluctuations in the length of day include electromagnetic, topographic, and gravitational coupling. Each of these processes is considered separately below. Explicit expressions are obtained for the surface forces $f(s)$ on the fluid core and the associated torques on the solid inner core and the mantle.

3.1. Electromagnetic coupling

Electromagnetic coupling of the fluid core to the rest of the Earth results from an interaction between the magnetic field that crosses the fluid core boundaries and the motion of conductive materials on either side of these boundaries. Let $\bar{B}_r$ denote the component of the magnetic field that crosses the nearly spherical boundaries of the outer core and $\tilde{B}_\phi$ be the azimuthal perturbation induced by tangential motion across these boundaries. *Braginsky* [1970] has shown that the surface forces due to electromagnetic interactions are given by

$$f_m(s) = -4\pi s^2 \left(\frac{r_f}{z_f}\right)\left(\frac{\bar{B}_r \tilde{B}_\phi}{\mu_0}\right) \qquad (10)$$

at the CMB and

$$f_i(s) = 4\pi s^2 \left(\frac{r_i}{z_i}\right)\left(\frac{\bar{B}_r \tilde{B}_\phi}{\mu_0}\right) \qquad (11)$$

at the ICB. The electrical conductivity near the boundaries of the fluid core enters (10) and (11) through the magnetic perturbation $\tilde{B}_\phi$. It is convenient to assume that the conductivity in the mantle is confined to the lowermost region in the form $\sigma(r) = \sigma_m \exp[(r_f - r)/\Delta]$, where Δ is the effective thickness of the layer and σ_m is the value on the mantle side of the CMB. The conductivity σ_f in the fluid outer core is assumed to be constant and equal to the value in the solid inner core.

Solutions for $\tilde{B}_\phi$ at both the CMB and ICB were given by *Buffett* [1992] for the proposed distribution of conductivity. The magnetic perturbation at the CMB can be approximated by

$$\tilde{B}_\phi = -\bar{B}_r\,\mu_0 \sigma_m \Delta\,s\,[u(s) - u_m] \qquad (12)$$

when the layer thickness Δ is much less than the skin depth [e.g., *Gubbins and Roberts*, 1988, p. 38]

$$\delta_m = \left(\frac{2}{\omega\mu_0\sigma_m}\right)^{1/2} \qquad (13)$$

Since the skin depth δ_m at $\omega \approx 10^{-10}$ s^{-1} is comparable to the depth of the mantle for plausible conductivities, the approximation invoked in (12) is valid when the conductivity is confined to the lowermost mantle. The solution for the magnetic perturbation at the ICB is given by

$$\tilde{B}_\phi = \bar{B}_r(1+i)\,\mu_0\sigma_f\,\delta_f\,s\,[u(s)-u_i]/4,, \qquad (14)$$

where the skin depth δ_f in the fluid is nominally 30 km. Substituting these solutions for $\tilde{B}_\phi$ into the expressions for $f(s)$, we obtain the amplitudes

$$\mathcal{F}_m^{(b)}(s) = 4\pi s^3 B_r^2 \sigma_m\,\Delta\left(\frac{r_f}{z_f}\right) \qquad (15)$$

$$\mathcal{F}_i^{(b)}(s) = \pi s^3(1+i)B_r^2\sigma_f\,\delta_f\left(\frac{r_i}{z_i}\right) \qquad (16)$$

where superscript (b) signifies electromagnetic coupling. Equation (15) for amplitude of electromagnetic coupling at the CMB indicates that rotational variations of the mantle can provide information about the strength of the radial field B_r and the conductance $\sigma_m\,\Delta$ at the base of the mantle.

3.2. Topographic coupling

Small departures in the hydrostatic shape of the CMB disturb the motion of the underlying fluid and cause variations in the fluid pressure P. The distribution of P over the aspherical boundary can produce a net torque [e.g., Roberts, 1988],

$$\Gamma_m = \int_S \mathbf{r} \times P\hat{\mathbf{n}}\,dS\,, \qquad (17)$$

where the integral is take over the surface of the CMB and the unit normal $\hat{\mathbf{n}}$ is directed into the mantle. Quantitative estimates of Γ_m are obtained by first calculating the pressure perturbation caused by a prescribed boundary topography. The resulting distribution of pressure is then used to evaluate Γ_m. It is important to emphasize that the pressure perturbation which acts on the boundary topography is caused by the effect of boundary topography on the torsional oscillations. This does not rule out the possibility that pressure perturbations due to other physical processes can correlate with the topography to yield a net torque. However, if we assume that the length-of-day fluctuations are associated with torsional oscillations about a quasi-static equilibrium state, then it follows that time variations in the pressure perturbation are solely a consequence of

topographic effects on the torsional oscillations. This is significant because it means that the torque on the mantle can be calculated once the boundary topography is prescribed.

The influence of topography on zonal flows with the symmetry associated with torsional oscillations was previously investigated by *Anufriyev and Braginsky* [1977]. In that study the surface of the CMB was described by

$$z_\pm(s,\phi) = \pm\left[z_f(s) - h_\pm(s,\phi)\right]\,, \qquad (18)$$

where $h_\pm(s,\phi)$ represent the boundary deviations in the northern $(+)$ and southern $(-)$ hemispheres. The quantity $z_f(s)$ is always taken to be positive and the topography h is defined to be positive in the direction of the fluid core. Noting that

$$\hat{\mathbf{z}}\cdot(\mathbf{r}\times\hat{\mathbf{n}}_\pm) = \frac{z_f}{r_f}\partial_\phi h_\pm\,, \qquad (19)$$

the polar component of the torque on the mantle becomes

$$\hat{\mathbf{z}}\cdot\Gamma_m = \int_{S_+}\frac{z_f}{r_f}(\partial_\phi h_+)\,P_+\,dS + \int_{S_-}\frac{z_f}{r_f}(\partial_\phi h_-)\,P_-\,dS\,, \qquad (20)$$

where

$$dS = \frac{r_f}{z_f}\,s\,d\phi\,ds\,. \qquad (21)$$

Symmetries in the linearized momentum and induction equations for the disturbance are exploited by *Anufriyev and Braginsky* [1977] to decompose the solution for the velocity and pressure perturbations into two parts. These two solutions are associated with topography which is symmetric and antisymmetric about the equator (see Fig. 3). The undisturbed zonal flow is assumed to be quasi-geostrophic, consistent with the character of the torsional oscillations, while the initial magnetic field $\bar{B}_\phi(s,z)\hat{\phi}$ is purely zonal and antisymmetric (see *Kuang and Bloxham* [1997b], for a representative distribution of $\bar{B}_\phi$ inside the core). When the zonal flow encounters a deviation in the boundary, the disturbance must evolve so that line elements aligned with the rotation axis are unstrained. This condition is a consequence of the geostrophic nature of the motions. If the boundary topography is symmetric about the equator, then the velocity disturbance occurs mainly in the radial component v_s, while antisymmetric topography produces a disturbance in the vertical component v_z.

Simple kinematic arguments for the velocity disturbance caused by antisymmetric topopgraphy at the CMB (e.g., $h_+ = -h_-$) show that

$$v_z(s,\phi) = -[u(s) - u_m]\partial_\phi h_+\,. \qquad (22)$$

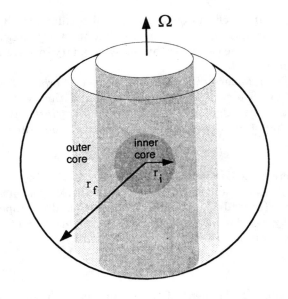

Figure 2. Fluid cylinders in the outer core as a function of distance s from the rotation axis Ω. The solid inner core is shown with radius r_i and the radius of the outer core is denoted by r_f.

Substituting this velocity field into the vertical component of the momentum equation yields [*Anufriyev and Braginsky*, 1977]

$$\frac{dP}{dz} = \left(\rho \partial_t^2 - \frac{\bar{B}_\phi^2(s,z)}{s^2 \mu_0} \partial_\phi^2 \right) [\varphi(s) - \varphi_m] \, \partial_\phi h_+(s,\phi), \tag{23}$$

where $\varphi(s)$ and φ_m are angular displacements related to $u(s)$ and u_m by $\partial_t \varphi = u$, and ∂_t denotes differentiation with respect to time. The first term on the right side of (23) represents the inertia associated with the disturbance, while the second term is the principal part of the Lorentz force due to the effect of v_z on the zonal main field $\bar{B}_\phi$. Integrating (23) for P yields $P_+(z_f) = -P_-(-z_f)$ since the sole dependence of dP/dz on z appears in B_ϕ^2, which is symmetric about the equator. It follows from (20) that the northern and southern hemispheres make equal contributions to the torque because the topography is antisymmetric (e.g. $h_+ = -h_-$). Thus the topographic torque for antisymmetric h reduces to an integral over the northern hemisphere only

$$\hat{z} \cdot \Gamma_m = 2 \int_0^{2\pi} d\phi \int_0^{r_f} ds (s \, P_+ \partial_\phi h_+). \tag{24}$$

A similar result can be obtained for the torque when topography is symmetric, but the pressure perturbations on the leading and trailing sides of a bump on the

boundary are identical, and the torque vanishes. As a result, the topographic torque exerted on the mantle by torsional oscillations is due to antisymmetric topography. When a general pattern of topography $h_\pm(s,\phi)$ is prescribed, we use only the antisymmetric part $h^a = (h_+ - h_-)/2$ in (24).

Integral (24) for the torque on the mantle is cast in the form needed in (8) by recalling that all motions are periodic with frequency ω. Letting $h^a(s,\phi) = |h(s)| \cos m\phi$, we obtain

$$\mathcal{F}_m^{(t)}(s) = -2\pi m^2 |h(s)|^2 z_f(s) \left[i\rho \omega s - i \frac{\{\bar{B}_\phi^2\}}{\mu_0} \frac{m^2}{\omega s} \right] \tag{25}$$

so that the surface force on the ends of the fluid cylinders at the CMB is given by

$$f_m(s) = \mathcal{F}_m^{(t)}(s) [u(s) - u_m]. \tag{26}$$

For typical values of the parameters in (25) the second term involving the average azimuthal field $\{\bar{B}_\phi^2\}$ in the core is the largest contribution to $\mathcal{F}_m^{(t)}$. This contribution is due to a pressure perturbation inside the core that arises when Lorentz forces oppose the vertical fluid motions (e.g. the Lorentz forces act to resist the distortion of the azimuthal main field). This pressure disturbance is transmitted parallel to the rotation axis toward the CMB to produce a torque on the mantle. *Anufriyev and Braginsky* [1977] characterize the resulting torque as an "elastic" response since the

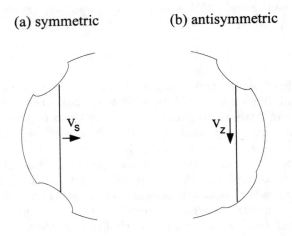

Figure 3. Fluid disturbance associated with (a) symmetric and (b) antisymmetric topography. Symmetric topography causes a velocity disturbance mainly in the radial component v_s, while antisymmetric topography produces a disturbance in the vertical component v_z.

contribution in (25) is proportional to the angular displacement (e.g. $(u - u_m)/i\omega$). It should be noted that the pressure torque does not depend on the conductivity of the mantle nor the strength of the azimuthal field at the CMB. Indeed, the largest contribution to $\{\bar{B}_\phi^2\}$ should occur inside the core where the azimuthal field is expected to be strongest. This dependence on the azimuthal main field is potentially important because it provides a connection to the observations of the Earth's rotation. At present, there no other methods available to assess this component of the field inside the core. Finally, we note that the main contribution to the topographic torque does not explicitly depend on accelerations in the fluid core. While a topographic torque on the core produces fluid accelerations in order to alter the angular momentum, these accelerations are best viewed as a consequence rather than a cause of the topographic torque. The topographic torque can be related to accelerations in the fluid by global conservation of angular momentum [*Kuang and Bloxham*, 1997a], but this relationship does not discern the mechanism that causes the pressure torque.

A topographic torque on the inner core could also be obtained in an analogous manner, but the probable magnitude of this torque is negligible compared with the electromagnetic torque because of the high conductivity on either side of the ICB. The inner core should also influence the torque on the mantle because its presence alters the motion of fluid cylinders inside the tangent cylinder (e.g. $s < r_i$). Since the expression for $\mathcal{F}_m^{(t)}$ does not account for the inner core, we must assume, for consistency, that $|h(s)|$ vanishes inside the tangent cylinder (e.g. $s < r_i$). Such an approximation has little influence on the total torque because the main contribution occurs outside the tangent cylinder. From (25), we infer that this torque depends on the amplitude $|h|$ and wavelength $\lambda = 2\pi s/m$ of the topography, as well as the average strength of the toroidal field $\bar{B}_\phi$ inside the core.

3.3. Gravitational coupling

Of the possible distributions of mass in the core below a heterogeneous mantle, the hydrostatic distribution minimizes the gravitational potential energy of the whole Earth. When the density field of the core rotates out of its equilibrium alignment with the mantle, the gravitational potential energy increases and a restoring force on the core and the mantle is produced [*Buffett*, 1996a]. The main interaction occurs between the inner core and the mantle because the density of the nearly adiabatic fluid in the outer core adjusts as it slowly moves through lateral variations in the gravitational potential. Consequently, the density field in the fluid core remains nearly fixed relative to the mantle. A similar adjustment of the inner-core shape must occur by either melting and solidification at the inner-core boundary or by viscous relaxation. Both of these processes are expected to be slow enough to permit significant misalignment of the density fields in the inner core and the mantle.

If the inner core behaves elastically over periods of several decades then the amplitude $\tilde{\Gamma}$ of the torque on the inner core in (7) is a constant. We denote this constant by $\tilde{\Gamma}(\infty)$ and adopt the value $\tilde{\Gamma}(\infty) = 10^{21}$ Nm from Model 2 of *Buffett* [1996a]. A reduction in the amplitude of the gravitational torque occurs if the shape of the inner core relaxes significantly over the period of an oscillation. We can accommodated this reduction in the gravitational torque by making the amplitude $\tilde{\Gamma}$ frequency dependent. Suppose that a relative rotation $\varphi_m - \varphi_i$ is introduced between the mantle and the inner core. Relaxation of the inner-core shape reduces the misalignment of the density fields from $\varphi_m - \varphi_i$ to, say, $\varphi_m^{(r)} - \varphi_i^{(r)}$. Assuming that the relaxation rate is linearly dependent on the misalignment $\varphi_m^{(r)} - \varphi_i^{(r)}$, we can express the rate of change of this misalignment as a superposition of differential rotation and relaxation by

$$\partial_t(\varphi_m^{(r)} - \varphi_i^{(r)}) = \partial_t(\varphi_m - \varphi_i) - (\varphi_m^{(r)} - \varphi_i^{(r)})/\tau \quad (27)$$

where τ is the characteristic time for relaxation of the inner-core shape. Since the torque on the inner core is most naturally written in terms of the misalignment of the density fields as

$$\Gamma_i = \tilde{\Gamma}(\infty)(\varphi_m^{(r)} - \varphi_i^{(r)}), \quad (28)$$

the equivalent torque given in (7) in terms of $\varphi_m - \varphi_i$ requires

$$\Gamma_i = \tilde{\Gamma}(\omega)(\varphi_m - \varphi_i) \quad (29)$$

where

$$\tilde{\Gamma}(\omega) = \tilde{\Gamma}(\infty)\left[1 + \frac{i}{\omega\tau}\right]^{-1}. \quad (30)$$

The complex, frequency dependent amplitude $\tilde{\Gamma}(\omega)$ is used in (8) and (9) to account for the influence of a single relaxation process on the gravitational torque. Relaxation of the inner-core shape reduces the strength of the gravitational torque and damps the oscillations. In fact, the mechanical work expended in deforming the inner core can be an important source of loss for these oscillations.

4. OSCILLATIONS IN THE LENGTH OF DAY

Equations (2), (8), and (9) for $u(s)$, u_i, and u_m represent a set of coupled angular momentum equations for the oscillations in the length of day. The coupling forces $f(s)$ in (2) depend linearly on the angular velocities, while the "friction" coefficients $\mathcal{F}$ for electromagnetic and topographic coupling are given in (15), (16) and (25). The amplitude of the gravitational torque in (8) and (9) is given by (29) and (30). Solving these homogeneous equations for $u(s)$, u_i, and u_m yields an eigenvalue problem for the complex frequency ω and the spatial structure of the normal modes. Angular velocities u_i and u_m are complex constants, whereas $u(s)$ is determined by integrating (2), subject to regularity conditions at the singular points $s = 0$, r_i, and r_f. The regularity condition at $s = 0$ requires

$$\left(\frac{du}{ds} \right) = 0 \tag{31}$$

so the angular velocity in the fluid near $s = 0$ is constant. This solution for $u(s)$ near the origin is integrated towards the singularity at $s = r_i$. In practice, the solution is integrated numerically to $s = r_i - \epsilon$ and continued analytically across a small interval ϵ to $s = r_i$. We obtain the incremental change in the solution across the interval ϵ by analytically integrating (2) over the singularity in $f(s)$ at $s = r_i$. To leading order in ϵ we find that $u(s)$ is continuous and that the increment in du/ds is proportional to $\epsilon^{1/2}$. For practical purposes we make ϵ sufficiently small that continuity of both $u(s)$ and du/ds can be imposed at $s = r_i$. This solution at $s = r_i$ is then integrated numerically towards a second singularity in $f(s)$ at $s = r_f$. Integrating (2) across this singularity assuming that $u(s)$ is continuous yields the final boundary condition at $s = r_f - \epsilon$,

$$\left(\frac{du}{ds} \right) = i\omega \mu_0 \sigma_m \Delta (u - u_m) \tag{32}$$

where we make use of the fact that $\bar{B}_r = \{\bar{B}_s\}$ at $s = r_f$.

The three constants $u(0)$, u_i, and u_m may be viewed as integration constants for the numerical solution of $u(s)$. Three linearly independent solutions for $u(s)$ are obtained by successively setting one of these constants to unity and the remaining constants to zero. A linear combination of these three independent solution must satisfy the boundary condition (32) and the two angular momentum conditions (8) and (9) for the mantle and inner core. The real and imaginary parts of the oscillation frequency ω are iteratively adjusted and the integration for $u(s)$ repeated until these three (complex) conditions are simultaneously satisfied.

Specific results for the oscillations depend on the structure of the magnetic field inside the core, as well as a number physical properties near the boundaries of

Table 1. Physical parameters used in calculations

Parameters		Value
radius of CMB	r_f	3.48×10^6 m
radius of ICB	r_i	1.22×10^6 m
density of core	ρ	10^4 kg m^{-3}
moment of inner core	C_i	5.87×10^{34} kgm^2
moment of mantle	C_m	7.12×10^{37} kgm^2
RMS field at CMB	$\bar{B}_r(r_f)$	5 G
RMS field at ICB	$\bar{B}_r(r_i)$	5 G
mantle conductance	$\sigma_m \Delta$	10^8 S
core conductivity	σ_f	5×10^5 Sm^{-1}
CMB topography	$|h(s)|$	5×10^3 m
angular order	m	10
gravitational torque	$\Gamma(\infty)$	10^{21} Nm

the core (see Table 1). The oscillation periods depend primarily on the cylindrically averaged radial field $\{\bar{B}_s^2\}$ through the core. Figure 4 shows the cylindrical averages $\{\bar{B}_s^2\}^{1/2}$ and $\{\bar{B}_\phi^2\}^{1/2}$ from a recent dynamo calculation of Kuang and Bloxham [1997b]. The value of the radial field is typically less than 1 G, yielding an oscillation period of several hundred years for the lowest frequency (fundamental) mode. This period is longer than the usual decade periods attributed to length-of-day variations, which suggests that the estimate of $\{\bar{B}_s^2\}^{1/2}$ is too small. While higher order harmonics can yield shorter period oscillations, the fluid motions for these modes are increasingly confined to the equatorial region of the core, which is inconsistent with the fluid motions inferred from geomagnetic variations [Jackson et al., 1993]. Consequently, we use the form of the magnetic fields given in Figure 4, but scale the amplitudes by a factor of four to reproduce the observed oscillation periods. While such a scaling overestimates the strength of the predicted dipole field at the surface by roughly a factor of 2, our motivation is to compare coupling mechanisms using realistic estimates of the fluid velocities in the outer core. Thus the peak values of $\{\bar{B}_s^2\}^{1/2}$ and $\{\bar{B}_\phi^2\}^{1/2}$ are approximately 4 and 8 G, respectively. With this choice of magnetic field, the fundamental mode has a typical period of 60 years.

4.1. Influence of electromagnetic coupling

Figure 5 shows the structure of the oscillations in the fluid core when the coupling to the rest of the Earth is due solely to electromagnetic friction. The normal component $\bar{B}_r$ of the magnetic field at both the ICB and CMB is assumed to be dipolar for convenience and both fields have root-mean-square (RMS) amplitudes of 5 G.

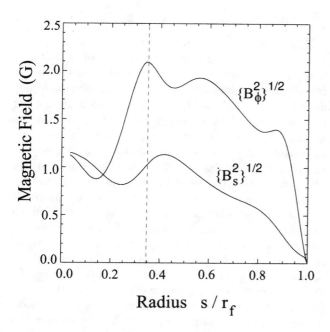

Figure 4. Cylindrical averages $\{\bar{B}_s^2\}^{1/2}$ and $\{\bar{B}_\phi^2\}^{1/2}$ through the core, based on recent dynamo calculations of *Kuang and Bloxham* [1997b]. The amplitudes of both fields are increased 4× in all calculations to reproduce the observed periodicities in the length of day.

The fluid velocity $su(s)$ for both the fundamental and first harmonic are shown as a function of radius s/r_f. The values of radius inside the tangent cylinder (e.g., $s < r_i$) are shaded for emphasis. Since the velocity is proportional to the displacement of the fluid cylinders from equilibrium, figure 5 provides a view of the disturbance from the north pole looking down on the tops of the fluid cylinders. The largest motions occur near the equator at the CMB where $\{\bar{B}_s^2\}$ is weakest, while the motions inside the tangent cylinder are nearly rigid-rotations. Fluid cylinders that terminate on the surface of the inner core are strongly coupled to the solid by electromagnetic forces. As a result, the fluid cylinders in the region $s < r_i$ tend to follow the motion of the solid inner core, regardless of the fluid motions outside the tangent cylinder. When the eigenfunctions are scaled to make $u_i = 1$, we find very little difference between the fundamental mode and the first harmonic in the region $s < r_i$.

Oscillations in the fluid outside the tangent cylinder are only weakly influenced by electromagnetic coupling at the CMB, but the angular velocity of the mantle u_m is directly dependent on this coupling. To explain the observed fluctuations in the length of day, u_m must reach peak values of 3×10^{-12} s^{-1}, corresponding to

a torque on the mantle of roughly 10^{18} Nm. Scaling the predicted eigenfunctions to yield the required u_m provides an estimate of the corresponding velocities in the fluid core. Alternatively, we can use estimates of fluid velocities inferred from geomagnetic variations to constrain the plausible amplitude of u_m for a given combination of coupling mechanisms. By limiting the RMS fluid velocity in core to 0.4 mm s^{-1}, we estimate u_m from the eigenfunctions and infer the corresponding torque on the mantle.

Torques on the mantle due to electromagnetic coupling at the CMB are given in Table 2 for the fundamental mode and the first two harmonics. We note that the oscillations are very heavily damped by the effects of electromagnetic friction on the mantle. The damping time for each mode is nominally 29 years, whereas the oscillation periods can be much longer. Indeed, the fundamental mode is so heavily damped that it is hardly an oscillation at all. Although these damping times can be reduced by decreasing the electromagnetic friction on the mantle, this makes electromagnetic coupling less viable as a coupling mechanism. Even the heavily damped oscillations in Table 2 yield electromagnetic torques on the mantle that are too small

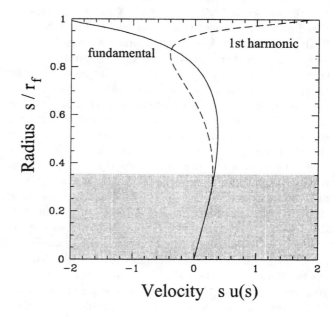

Figure 5. Fluid velocity $s\,u(s)$ in the outer core for the fundamental mode and the first harmonic. The fluid oscillations are coupled to the rest of the Earth solely by the effects of electromagnetic friction at the CMB and ICB. Shading below the radius $s/r_f = 0.35$ indicates the region inside the tangent cylinder, where strong electromagnetic coupling causes the fluid to follow the motion of the inner core.

Table 2. Influence of CMB coupling on free oscillations

Coupling	Harmonic	Period	Damping	Torque
		(year)	(year)	(10^{15} N m)
e.m.	0	63.9	29.6	1.9
	1	39.1	29.5	4.2
	2	28.3	29.3	6.1
topo.	0	60.3	4×10^4	3.4
	1	38.2	5×10^3	2.1
	2	27.9	2×10^3	1.0

Real ω_r and imaginary ω_i parts of the frequency are used to calculate the period $2\pi\omega_r^{-1}$ and the damping time ω_i^{-1}. Harmonic 0 denotes the fundamental mode.

by several orders of magnitude to account for the observed variations. However, electromagnetic coupling warrants closer scrutiny before being dismissed. Note that the torques on the mantle increase with higher harmonics. This behaviour can be understood by considering the form of the fluid oscillations in Figure 5 and the structure of the magnetic field $\bar{B}_r$ at the CMB. The largest fluid velocities occur in the equatorial region, but the associated frictional stresses are small because the dipole field is weak in this region. The magnetic field is stronger in the polar region, but the fluid velocities are weaker and directed in the opposite direction. Thus there is significant cancellation in the net torque on the mantle. Fluid motions associated with the higher harmonics are increasingly confined to the equatorial regions, so that less cancellation occurs and the net torque on the mantle increases. Increases in the electromagnetic torque are also possible if $\bar{B}_r$ includes nondipole components. Hypothetically, if $\bar{B}_r$ is predominantly short-wavelength, then we might reasonably approximate $\bar{B}_r^2(s)$ with a constant mean-square value over the surface of the CMB. Assuming that the RMS value of $\bar{B}_r$ remains 5 G and the conductivity of the mantle is unchanged, then the resulting electromagnetic torque on the mantle increases to 5.7×10^{17} Nm for the fundamental mode. (The damping time decreases to 20.4 years, but the period remains nearly unchanged at 64.3 years.) Since this torque is comparable to the value needed to explain the length-of-day variations, we conclude that electromagnetic torques can provide a viable coupling mechanism. Evidence for this coupling mechanism might be sought in geophysical observations by looking for the consequences of heavy damping on the oscillations in the fluid core.

4.2. Influence of topographic coupling

The introduction of small-amplitude topography on the CMB provides an alternative means of coupling the torsional oscillations to the mantle. In order to isolate the effects of topography, we first eliminate electromagnetic stresses at the CMB by making the mantle a perfect insulator. The effects of electromagnetic stresses at the inner-core boundary are retained in these calculations, but the effects of gravitational coupling are eliminated by setting $\Gamma(\infty) = 0$. (Gravitational coupling was also eliminated in the preceding section on electromagnetic coupling.) The structure of the resulting torsional oscillations in the core are nearly identical to those shown in Figure 5, which indicates that the spatial form of the fluid oscillation is relatively insensitive to the nature of the coupling at the CMB. However, the period and the damping of the oscillations are significantly different. Table 2 summarizes the results for the fundamental and the first two harmonics. The periods for topographic coupling are systematically shorter than those given for electromagnetic coupling at the CMB. This decrease in the period of oscillation can be attributed to the influence of the azimuthal magnetic field $\bar{B}_\phi$, which gives the fluid an effective rigidity. When fluid cylinders are rotated out of their equilibrium positions, the associated fluid disturbance due to topography is resisted by $\bar{B}_\phi$, which contributes to the restoring force. The damping times are substantially longer than those due to electromagnetic coupling because topographic effects do not directly dissipate mechanical energy. The sole source of dissipation in these calculations is due to ohmic losses at the ICB.

The topographic torques on the mantle for all three modes are well below the value needed to explain the length-of-day observations. The largest torque is achieved by the fundamental mode because the main contribution to the torque originates at mid-latitudes, where the fundamental mode has a slightly larger amplitude. The amplitude of the torque is sensitive to the ratio of the boundary topography h^a to the longitudinal wavelength λ, which means that larger torques are possible if h^a/λ could be increased. Significant increases in h/λ are unlikely at the longest wavelengths, but increases may be possible at short wavelengths. Larger torques can also be obtained by increasing the amplitude of $\bar{B}_\phi$. For example, if $\{\bar{B}_\phi^2\}^{1/2}$ has the form shown in Figure 4, but a peak amplitude of 80 G, then the fundamental mode can yield a topographic torque of 6.8×10^{17} Nm. (The period of the fundamental mode decreases to 49.8 years in this case.) Although the parameter values chosen in Table 1 suggest that topographic and electromagnetic

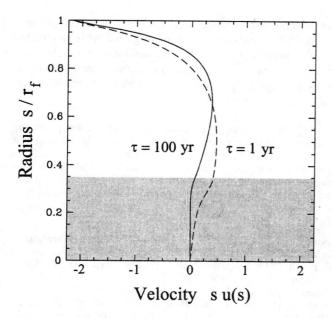

Figure 6. The influence of gravitational coupling on the fundamental oscillation in the outer core. Electromagentic stresses at the ICB are included, but there are no electromagnetic or topographic torques at the CMB. Two different values of relaxation time τ are chosen for the gravitational torque. One values is slightly longer than the period of the oscillation, while the other value is significantly shorter.

torques at the CMB are too weak to explain the observed variations in the length of day, it is possible to devise physical situations where both coupling mechanisms become significant. Even if both mechanisms are viable, it should be possible to distinguish between them on the basis of the damping of the associated oscillations.

4.3. Influence of gravitational coupling

The preceding calculations for topographic and electromagnetic coupling at the CMB do not include the influences of gravitational coupling between the inner core and the mantle. We now consider the role of gravitational coupling when there is no topographic or electromagnetic coupling at the CMB. The torsional oscillations in the fluid core are coupled to the inner core by the effects of magnetic friction at the ICB, while the inner core is coupled to the mantle by gravitational forces. The calculations presented in this section are similar to those given previously by *Buffett* [1996b], but the present study now allows for relaxation of the inner-core shape with a characteristic timescale τ.

Figure 6 shows the fundamental mode calculated using two different relaxation times. One value of τ is

chosen to be longer than the nominal 55-year period of the oscillation, while the other value is chosen to be substantially shorter. Remarkably, both oscillations yield torques on the mantle which are roughly 10^{18} Nm. This constancy of the torque can be explained by changes in the structure of the oscillations as the relaxation time decreases. For example, when the relaxation time is long the fluid motions inside the tangent cylinder are small. These fluid motions reflect the small amplitude of the inner core rotation, which is constrained by gravitational forces to remain nearly aligned with the mantle. Since the moment of inertia of the mantle is much larger than that of the fluid core, the mantle undergoes small-amplitude oscillations and the inner core follows.

On the other hand, large amplitude motions occur inside the tangent cylinder when the relaxation time is short. Electromagnetic stresses rotate the inner, but rapid relaxation keeps the misalignment of density fields small. Indeed, this misalignment must be roughly the same for both calculations shown in Figure 6 because the (gravitational) torque on the mantle is nearly the same. In order to maintain this misalignment when the relaxation time is short, it is necessary to continually rotate the inner core to compensate for the effects of relaxation. It follows that large amplitude rotations of the inner core are needed for a small, but significant misalignment of the density fields to persist when the relaxation time is much shorter than the period of the oscillations. The mode shown in Figure 6 with $\tau = 1$ yr still produces a large torque on the mantle because the large rotation amplitude compensates for the short relaxation time.

There are practical limits on the rotation amplitude of the inner core, so there must be very small values of τ where gravitational coupling ceases to be effective. The preceding calculations demonstrate that a relaxation time of $\tau = 1$ yr is long enough to make gravitational coupling viable. Specifically, the torque on the mantle in this particular case is 1.8×10^{18} Nm. By comparison, the torque on the mantle decrease to 3.5×10^{17} Nm when the relaxation time decreases to 0.1 yr. Shorter relaxation times are inconsistent with seismic estimates of shear attenuation in the inner core [*Widmer et al.*, 1991; *Durek and Ekstrom*, 1996], so we expect the gravitational torque to equal or exceed 3.5×10^{17} Nm. A second constraint on the value of τ is provided by recent evidence for differential rotation of the inner core [*Song and Richards*, 1996; *Su et al.*, 1996]. In order to reconcile the inferred rotation of the inner core with strong gravitational coupling to the mantle, it is necessary to allow the inner core shape to adjust with timescale $\tau = 0.6$ yr [*Buffett*, 1997]. This value would permit the inner core to rotate relative to the mantle at a rate of 1 deg. yr^{-1}, but still allow a gravitational torque on the mantle in excess of 10^{18} Nm. In this case, gravitational coupling should be the principal coupling mechanism for the decade fluctuations in the length of day.

5. CONCLUSIONS

A theoretical model for free oscillations in the length of day is used to establish a connection between the observed fluctuations and physical conditions inside the Earth. The model shows that the predicted oscillations depend on the structure of the magnetic field inside the core, as well as a number of physical properties near the boundaries of the core. The adopted structure of the magnetic field inside the core is based on recent dynamo calculations by *Kuang and Bloxham* [1997b], although a modest increase in the overall amplitude of this field is needed to match the observed periodicities. Angular momentum is transferred between the core and the mantle by electromagnetic and topographic coupling at the CMB, and by gravitational coupling between the inner core and the mantle. Gravitational coupling is the mechanism that most readily reproduces the amplitude of the observed oscillations, even when the influences of viscous deformation of the inner core are included in the calculations. A gravitational torque in excess of 10^{18} Nm is obtained when the viscous relaxation time is as short as six months.

Representative calculations for the effects of electromagnetic and topographic torques at the CMB suggest that these effects are small, although plausible situations can be devised where both torques become significant. Specifically, electromagnetic coupling becomes a viable coupling mechanism if the mantle conductance is 10^8 S and significant levels of non-dipole field are present at the CMB. Such a coupling mechanism could be identified by the heavy damping of the resulting oscillations. Topographic coupling can also become more important if the CMB is rough at short wavelengths (e.g. large values of h/λ). However, topography of several kilometers over wavelengths of several thousand kilometers, corresponding to $h/\lambda \approx 10^{-3}$, is insufficient to produce observable torques unless the average azimuthal field in the core is very large. Indeed, the average azimuthal magnetic field $\{\bar{B}_\phi^2\}^{1/2}$ needed to produce a significant torque is 80 G, which is about $40\times$ larger than the average field predicted by *Kuang and Bloxham* [1997b]. The resulting topographic torque on the mantle could be distinguished from other coupling mechanisms by the near absence of damping in the oscillations. Since each coupling mechanism has a distinct influence on the damping and structure of the oscillations in the fluid core, it should be possible to invert relevant geodetic and geomagnetic measurements for the physical properties on which these various coupling mechanism depend.

Acknowledgments. I thank Weijia Kuang for providing estimates of the magnetic field from his recent dynamo calculations with J. Bloxham. Richard Holme made many helpful comments that improved the paper. This work was supported by Research Grant OGP0138213 from the Natural Sciences and Engineering Research Council of Canada.

REFERENCES

Anufriyev, A.P., and S.I. Braginsky, Influence of irregularities of the boundary of the Earth's core on fluid velocity and the magnetic field. III, *Geomagn. Aeron.*, Engl. Transl., **17**, 492–496, 1977b.

Bloxham, J., and A. Jackson, Fluid flow near the surface of the Earth's outer core, *Rev. Geophys.*, **29**, 97–120, 1991.

Bloxham, J., and D. Gubbins, Thermal core-mantle interactions, *Nature*, *325*, 511–513, 1987.

Braginsky, S.I., Torsional magnetohydrodynamic vibrations in the Earth's core and variations in day length, *Geomag. Aeron.*, Engl. Transl., *10*, 1–8, 1970.

Buffett, B.A., 1992. Constraints on magnetic energy and mantle conductivity from the forced nutations of the Earth, *J. Geophys. Res.*, *97*, 19,581–19,597.

Buffett, B.A., Gravitational oscillations in the length of day, *Geophys. Res. Lett.*, *23*, 2279–2282, 1996a.

Buffett, B.A., A mechanism for decade fluctuations in the length of day, *Geophys. Res. Lett.*, *23*, 3803–3806, 1996b.

Buffett, B.A., Geodynamic estimates of the viscosity of the Earth's inner core, *Nature*, *388*, 571–573, 1997.

Bullard, E.C., C. Freedman, H. Gellman, and J. Nixon, Westward drift of the Earth's magnetic field, *Phil. Trans. R. Soc. Lond.*, A243, 67–92, 1950.

Dehant, V., On the nutations of a more realistic Earth model, *Geophys. J. Int.*, *100*, 477–483, 1990.

Durek, J.J. and G. Ekstrom, A radial model of anelasticity consistent with long-period surface wave attenuation, *Bull. Seis. Soc. Am.*, *86*, 144–158, 1996.

Gubbins, D., & Roberts, P.H., 1987. Magnetohydrodynamics of the Earth's Core, in *Geomagnetism*, vol. 2, edited by J.A. Jacobs, Academic Press, London.

Gwinn, C.R., T.A. Herring, and I.I. Shapiro, Geodesy by radio interferometry: Studies of the forced nutations of the Earth, 2, Interpretation, *J. Geophys. Res.*, *91*, 4755–4765, 1986.

Herring, T.A., B.A. Buffett, P.M. Mathews, and I.I. Shapiro, Forced nutations of the Earth: Influence of inner core dynamics, 1. Theory, *J. Geophys. Res.*, *96*, 8219–8242, 1991.

Hide, R., Interactions between the Earth's liquid core and mantle, *Nature*, *222*, 1055–1056, 1969.

Hide, R., Fluctuations in the Earth's rotation and the topography of the core-mantle interface, *Phil Trans. R. Soc. Lond. A*, *328*, 351–363, 1989.

Jackson, A., J. Bloxham, and D. Gubbins, Time-dependent

flow at the core surface and conservation of angular momentum in the coupled core-mantle system, in *Dynamics of Earth's Deep Interior and Earth Rotation*, AGU Monograph 72, Eds. J-L. LeMouël, D.E. Smylie, and T.A. Herring, 97–107, 1993.

Jault, D., C. Gire, and J.-L. Le Mouël, Westward drift, core motions and exchanges of angular momentum between core and mantle, *Nature*, *333*, 353–356, 1988.

Jault, D., & Le Mouël, J.-L., The topographic torque associated with a tangentially geostrophic motion at the core surface and inferences on the flow inside the core, *Geophys. Astrophys. Fluid Dyn.*, *48*, 273–296, 1989.

Kuang, W., & J. Bloxham, The effect of boundary topography on motions in the Earth's core, *Geophys. Astrophys. Fluid Dyn.*, *72*, 161–195, 1993.

Kuang, W., & J. Bloxham, On the dynamics of topographical core-mantle coupling, *Phys. Earth Planet. Inter.*, *99*, 289–294, 1997a.

Kuang, W., & J. Bloxham, An Earth-like numerical dynamo model, *Nature*, *389*, 371–374, 1997b.

McCarthy, D.D., and A.K. Babcock, The length of day since 1656, *Phys. Earth Planet. Inter.*, *44*, 281–292, 1986.

Mathews, P.M., and I.I Shapiro, Nutations of the Earth, *Ann. Rev. of Earth Planet. Sci.*, *20*, 469–500, 1992.

Mathews, P.M., B.A. Buffett, T.A. Herring, and I.I. Shapiro, Forced nutations of the Earth: Influence of inner core dynamics, 1. Theory, *J. Geophys. Res.*, *96*, 8219–8242, 1991.

Roberts, P.H., On topographic core-mantle coupling, *Geophys. Astrophys. Fluid Dynamics*, *44*, 181–187, 1988.

Rochester, M.G., Geomagnetic westward drift and irregularities in the Earth's rotation, *Phil. Trans. R. Soc. A*, *252*, 531–555, 1960.

Sasao, T., S. Okubo, and M. Saito, A simple theory on the dynamical effects of a stratified fluid core upon the nutational motions of the Earth, *Proc. IAU Sym. 78, Nutation and Earth Rotation*, 1980.

Song, X. and P. Richards, Seismic evidence for the rotation of the inner core, *Nature*, *382*, 221–224, 1996.

Su, W.-J., Woodward, R. L. & Dziewonski, A. M., Degree-12 model of shear velocity heterogeneity in the mantle, *J. Geophys. Res.*, *99*, 6945–6980, 1994.

Su, W.-J, A.M. Dziewonski, and R. Jeanloz, Planet within a planet: Rotation of the inner core of the Earth, *Science*, *274*, 1883–1887, 1996.

Wahr, J. M., The forced nutations of an elliptical, rotating, elastic and oceanless Earth, *Geophys. J. R. Astr. Soc.*, *64*, 705–727, 1981.

B. Buffett, Department of Earth and Ocean Sciences, University of British Columbia, Vancouver, Canada (email: buffett@eos.ubc.ca)

Interpreting the Paleomagnetic Field

David Gubbins

School of Earth Sciences, University of Leeds, UK

We will only understand the geodynamo in detail by combining theory with an accurate paleomagnetic record. Theory is not sufficiently well developed to predict paleomagnetic field behaviour, but some general results can guide data interpretation. Symmetry considerations suggest that (a) polarity reversals involve complete sign reversal of the field, (b) field morphology show a preference for reflection symmetry about the equator, and (c) there should be no preferred longitude. The inner core has a stabilising effect on the dynamo because its magnetic diffusion time (2–5 kyr) is probably the longest imposed time scale in the system. The solid mantle can produce lateral and temporal variations in the boundary conditions that break the symmetries and add a very long time scale (10–100 Myr).

Paleomagnetic measurements are derived from lavas and sediments during both stable polarity and transition. They give evidence of long-term (50–100 Myr) changes in dipole moment and reversal frequency, and departures from axial symmetry in both the time-averaged stable magnetic field and during polarity transition. The departures from axial symmetry can be explained if core convection concentrates magnetic flux on specific longitudes during stable and transitional times.

Many of these results are controversial, but both data and theory are improving rapidly. This paper reviews the available data and the theoretical framework available for its interpretation.

1. INTRODUCTION

The Earth's magnetic field is generated by dynamo action in the liquid iron core, fuelled by a combination of thermal and gravitational energy as the Earth cools and heavier constituents settle towards the centre. The geomagnetic field is very old, at least 3.5 Ga, and has left a detailed record of its evolution. The present field has been mapped globally with a spatial resolution of a few tens of kilometers, but our knowledge of that part of the field that originates in the core is limited to wavelengths of several hundred kilometers, shorter wavelengths being masked by the field of permanently magnetised crustal rocks.

The modern period of direct magnetic observation stretches back 400 years or so, and thanks to the magnetic field's importance for navigation the early records are quite good. Four centuries is a short time compared to the time for the main magnetic field to change significantly and we must rely on indirect paleomagnetic measurements to extend the data interval. Paleomagnetic measurements are less accurate than direct measurements and sparse in both space and time. To interpret the paleomagnetic record we must use the present as a guide to the past by comparison with the modern geomagnetic field and resort to studies of gross properties of the field such as the time average, rather than detailed changes.

The paleomagnetic record has revealed a complex time evolution, with fluctuations of 30–50% in both intensity and direction during periods of stable polarity

The Core-Mantle Boundary Region
Geodynamics 28
Copyright 1998 by the American Geophysical Union.

being interrupted by complete polarity reversal accompanied by a fall by a factor 5–10 in intensity every few hundred thousand years. This fits with a highly nonlinear, chaotic, dynamo with the dominant Coriolis force producing the organising effect necessary for dynamo action and a dominant axial dipole. The dynamo equations are invariant under change of sign of the magnetic field, thus allowing large secular variation fluctuations to occasionally develop into a complete magnetic reversal by change of sign.

Unfortunately, this scenario holds little hope for using observations to shed light on details of core dynamics because even simple nonlinear dynamical systems can generate very complex solutions. However, lateral variations in the mantle can, through the boundary conditions, stabilise the dynamo and promote simpler solutions such as stationary flows and a fixed longitude. This stabilisation could simplify both theory and data interpretation by promoting simpler solutions and providing comparisons with other datasets such as the seismic structure of the lowermost mantle. Paradoxically, adding the apparent complication of core-mantle interaction to an otherwise spherically symmetric core can simplify the problem.

Paleomagnetic evidence for the influence of the mantle on core dynamics comes from changes on very long time scales and persistent longitude-dependent effects. The frequency of polarity reversals has changed with time from its present rate of a few per million years to a long "superchron" from 118–83 Ma when the field was in the normal (present) polarity and there were no reversals. Nonlinear systems like the dynamo can persist in one state for long periods before changing suddenly to another, and "blocking" phenomena are known in the atmosphere where structures can persist for long times. The overturn time for core fluid is about 1 kyr, but the Cretaceous normal superchron lasted for tens of thousands of overturn times, and the reversing pattern that set in afterwards has lasted another hundred thousand overturn times. This is very unlikely to arise spontaneously but fits nicely with the time scale of mantle convection, suggesting changes in the lower mantle bring about changes in core dynamics through the boundary conditions. Changes in the geomagnetic field during the historical period of direct recording give a hint that some regions behave differently than others, and studies of the field at the core surface show clearly that the Pacific hemisphere is changing very little, with no westward drift. The duration of the historical record is too short (half an overturn time or less) to prove core-mantle interaction. The paleomagnetic record is long enough but much less reliable.

2. THEORETICAL CONSIDERATIONS

2.1. The Dynamo

The simplest governing equations are those of induction, a combination of Maxwell's equations that determines the evolution of the magnetic field with time

$$\frac{\partial \mathbf{B}}{\partial t} = \nabla \times (\mathbf{v} \times \mathbf{B}) + \eta \nabla^2 \mathbf{B}, \qquad (1)$$

where η is the magnetic diffusivity, $\mathbf{B}$ the magnetic field, and $\mathbf{v}$ the fluid velocity; motion in the Boussinesq approximation:

$$\frac{D\mathbf{v}}{Dt} + 2\boldsymbol{\Omega} \times \mathbf{v} = \mathbf{g}\frac{\delta\rho}{\rho_0} - \nabla p + \frac{1}{\mu_0\rho}(\nabla \times \mathbf{B}) \times \mathbf{B} + \nu\nabla^2\mathbf{v}, \qquad (2)$$

where ν is the kinematic viscosity, $\boldsymbol{\Omega}$ the Earth's angular velocity, μ_0 the permeability of free space, ρ the density of the undisturbed state, and $\delta\rho$ departures from that state; and continuity:

$$\nabla \cdot \mathbf{v} = \nabla \cdot \mathbf{B} = 0. \qquad (3)$$

A source of buoyancy is essential to drive radial flow in at least part of the core because magnetic field cannot be generated by purely toroidal motion. The usual choice is thermal convection, which adds a final equation:

$$\frac{DT}{Dt} = \kappa \nabla^2 T + Q \qquad (4)$$

where κ is the thermal diffusivity, Q the imposed heat source, and T the departures in temperature from the basic state. Density and temperature are related through the coefficient of thermal expansion:

$$\rho = \rho_0 \left(1 + \alpha T\right) \qquad (5)$$

Many authors have considered embellishments of these basic equations, including replacing the Boussinesq approximation with the anelastic approximation, thermal with compositional buoyancy, and adding electrical heating to the thermal equation (4). The dependent variables are $\mathbf{v}, \mathbf{B}$, and T. The equations are completed with suitable initial and boundary conditions at the core-mantle boundary: either zero stress or no slip for $\mathbf{v}$, constant temperature or heat flux for T, and, most importantly for any offer of hope of comparison with observation, matching $\mathbf{B}$ to a potential field outside the core. The solid inner core is included realistically as a solid conductor with the same electrical and thermal properties as the outer core, freely allowed to rotate about the spin axis.

Recently, full equations similar to these have been integrated numerically in time, with very considerable effort, by *Glatzmaier and Roberts* [1995] and Kuang and Bloxham (this volume). These calculations confirm the viability of the dynamo theory and provide important new understanding of the dynamical processes involved. However, attempts to relate them to detailed observations of the geomagnetic field are premature because it is impossible to represent the viscosity and inertial effects correctly. These shortcomings are serious because most of our observations are of rapid time variations that are strongly influenced by the boundary layer at the top of the core.

The governing equations have symmetry properties that are useful in classifying different types of solution. In the paleomagnetic literature these have been incorrectly attributed to special approximations such as the kinematic dynamo [equation (1) only] or the mean-field approximation. These approximations are unnecessary: the symmetry properties are fundamental and apply to the full set of nonlinear equations, even when additional complexities such as electrical heating are introduced. They take the form of invariance under transformation of the dependent or independent variables. They have been discussed in detail by *Gubbins and Zhang* [1993] and are listed here:

1. *Reversal (or inversion)*, $i : \mathbf{B} \to -\mathbf{B}$. Equation (1) is linear in $\mathbf{B}$ while the others are quadratic. Therefore change of sign of $\mathbf{B}$ leaves the equations the same; if we find one solution generating a field $\mathbf{B}$ we can construct a second solution generating $-\mathbf{B}$. Note that $\mathbf{v}$ and T remain unchanged. This symmetry is important when discussing reversals.

2. *Equatorial reflection*, $E : \theta \to \pi - \theta$. This transformation also leaves the equations unchanged, so that we can generate a second solution by looking at the reflection in a mirror with its plane normal to the spin axis. Each variable may be separated into a part that changes sign on reflection (E^A) and one that remains the same (E^S), and these form separate solutions of the equations. The axial dipole is antisymmetric about the equator (it actually remains invariant under reflection because magnetic induction is a pseudovector that itself changes sign under reflection, but this is usually ignored) while the axial quadrupole is symmetric. The possible separation of the geomagnetic field by this symmetry property has stimulated a number of geomagnetic and paleomagnetic studies. It is important to recognize that the solutions are separable but not independent: they

can exist alone but cannot be combined without destroying the symmetry.

3. *Polar rotation*, $P : \phi \to \phi + \psi$. Neither the geometry nor the equations contain any longitude reference, and we can therefore generate a different solution by changing the longitude reference. Some solutions are invariant under this transformation: axisymmetric solutions are possible, as are those invariant under rotation by π. Note that we do not guarantee dynamo action: a "solution" in this case is any non-trivial combination of variables satisfying the equations, including the case $\mathbf{B} = 0$ (and $\mathbf{v}, T \neq 0$). For example, Cowling's theorem states that axisymmetric fields (invariant under all polar rotations) cannot be generated by dynamo action, whereas *Kumar and Roberts* [1975] found a dynamo that generated a magnetic field that is invariant under polar rotation through an angle π. Without a longitude reference the generated magnetic field will not exhibit any longitude preference. This is important when considering observational evidence for persistent longitude-dependent features.

4. *Time translation*, $T : t = t + \tau$. The solution depends on the initial conditions but not on the choice of origin time. Solutions therefore exist that are independent under this transformation; they may be steady (invariant) or oscillatory (combined with inversion, they change sign under the translation but are invariant under a combination of the two). This symmetry may be relevant to understanding the dynamics of the reversal process.

It is important to understand the time scales suggested by the theory if we are to understand the historical and paleomagnetic record. The imposed natural physical time scales are those of diffusion and rotation. Thermal and viscous diffusion times are so long as to be effectively infinite, at least when based on molecular values, and the rotation period (day) is so short as to be effectively zero. The magnetic diffusion time c^2/η, where c is the core radius, is 10–100 kyr for accepted values of the electrical conductivity of the core, but the magnetic decay time of smaller observed features in the geomagnetic field is much less, perhaps a century or two (for example, *Gubbins and Roberts* [1987]). The magnetic diffusion time of the solid inner core, r_i^2/η where r_i is the inner core radius, is about 50 kyr, giving the decay time for a dipolar field of about 5 kyr. This will limit the speed of changes in the magnetic field in the inner core: it imposes a "magnetic inertia". Furthermore, because rotational dynamics resists variations along the spin axis (because of the Proudman-Taylor theorem) the fluid in the tangent cylinder above and below the inner core may also be stabilized *Hollerbach and Jones* [1993].

The natural intrinsic time scales are the overturn time c/v and period of Alfvén waves $c\sqrt{\mu_0\rho}/B$. Secular variation estimates give an overturn time of about a century and the Alfvén period only 1–25 yr, depending on the chosen field strength in the core. In a rotating system planetary magnetohydrodynamic (or MAC) waves are more important than Alfvén waves, and these have longer time scales of decades to centuries, again depending on the field strength. A configuration in which rotation is unimportant arises when field lines are parallel to the spin axis; in this case the magnetic forces resist twisting of cylinders of fluid on which the normally-dominant Coriolis force is zero. These *torsional oscillations* can have periods as long as a century. The dynamo effect itself can give rise to periodic solutions in the form of dynamo waves, as studied by *Parker* [1955], when magnetic field is generated in the opposite direction to the existing field. These form the basis of a theory of sunspots and the solar cycle, in which the magnetic field reverses every 11 years. The geomagnetic field does not oscillate, but something like dynamo waves could exist in local regions and be related to reversal behavior. Their time scale depends on details of the dynamo process.

2.2. Core Mantle Interactions

The basic dynamo equations are independent of the origin of longitude, yet some geomagnetic and paleomagnetic observations indicate a longitude dependence of the geomagnetic field, which requires some addition to the theory. Lateral variations in the mantle appear in the mathematics through inhomogeneous boundary conditions that depend on longitude. As always, the mathematical problem can be reformulated to homogeneous boundary conditions by addition of an inhomogeneous source term in the differential equations. For example in the case of laterally varying temperature boundary conditions the transformation leads to an additional heat source in equation (4) and buoyancy term in (2) [*Gubbins and Zhang*, 1993].

Three physical mechanisms have been proposed for long-term mantle control of core dynamics through lateral variations. All are implicit in the early paper by *Hide* [1967]:

1. *Topography.* This idea, developed by *Hide and Malin* [1970] with a correlation between the low-order geoid and rotated non-dipole field, draws on a meteorological analogy: that mountains produce winds and weather. This specific correlation is not now thought to be significant because more detailed information about the geomagnetic field at the CMB shows that the features do not rotate with a westward drift as was once thought. The physical mechanism remains viable. Topography of the CMB has been studied by seismology and is discussed elsewhere in this volume. If it exists at all the bumps are small, but small bumps of large lateral extent may be all that is required. The fluid mechanics of a rotating sphere with a bumpy boundary has been studied recently [*Bassom and Soward*, 1996; *Bell and Soward*, 1996]. When the bumps are sufficiently large the flow is dominated by a steady convective pattern locked to the bumps; as the bump size is decreased the preferred mode propagates eastward, modulated on a long azimuthal length scale.

2. *Electrical conductivity.* An electrically conducting mantle will allow current to leak out of the core and may produce an identifiable magnetic anomaly. A large variation in conductivity is required, which could be provided by dissolved iron from the core. Constraints are placed on the extent of the conducting region by the secular variation because it would be screened by high conductivity through a significant depth of mantle. Weak secular variation in the Pacific region has led some, notably *Runcorn* [1992], to propose a hemisphere of highly conducting lower mantle centered on the Pacific that would influence core flow. We return to this theory when discussing reversal transition fields. Current thinking gives the lower mantle a very low conductivity, however: *Shankland et al.* [1993] obtain the very low values of 3–10 Sm^{-1} from extrapolations of laboratory measurements on materials representative of the lower mantle. A chemically distinct layer at the base of the mantle, possibly D$''$ and possibly resulting from a high iron content, could provide the high conductivity required, but again *Poirier and LeMouël* [1992] argue against liquid core iron penetrating any significant distance into the mantle. Lateral variations in lower mantle conductivity can assist dynamo action [*Busse and Wicht*, 1992], although the effect is too small to account for the whole geodynamo effect.

3. *Heat flow.* A convecting mantle whose bottom temperature is held uniform by rapidly flowing core iron must inevitably have lateral variations

in heat flow. These can be estimated, in principle, from seismic velocity estimates in the mantle boundary layer, discussed elsewhere in this volume. The core flow must react to this demand for heat, and flow rates inferred from secular variation are of the right order of magnitude to supply the necessary variation [*Bloxham and Gubbins*, 1987]. *Zhang and Gubbins* [1993a] studied the thermal wind response in a rotating, non-magnetic sphere and, in a separate study, the effect of imposed boundary temperature on convection [*Zhang and Gubbins*, 1993b]. In the first case a simple thermal wind blows across the temperature anomalies. The steady wind becomes unstable for larger boundary variations. In the second case, convection that normally drifts in longitude is locked to the boundary for sufficiently strong variations. The equatorial symmetry of the boundary conditions is important: E^S temperatures are far more effective at locking convection and producing resonances with the underlying convection, while the E^A part is relatively ineffectual. The boundary conditions will contain both symmetries, but the E^S part may have the strongest influence on the core. *Olson and Glatzmaier* [1996] studied the problem with mixed-symmetry boundary conditions, an imposed magnetic field, extreme variations in boundary heat flux (ten times the radial flux), and just two combinations of parameters. They failed to find locking or a simple relation with the thermal wind.

The three physical effects will be difficult to distinguish from data. For example, bumps may produce a thermal effect because a depression into the core acts like a cold finger; in a dynamic mantle the effect may be as large as lateral variations in heat flow [*Bloxham and Gubbins*, 1987]. Again, bumps may produce a conductivity effect: stationary fluid in a hollow at the CMB will have the same effect as a high conductivity mantle anomaly. The simple solutions studied so far are all stable, but the corresponding effect in the core is likely to be unstable. For example, the stable regime of the thermal wind of *Zhang and Gubbins* [1993a] is small and the flow becomes time dependent for modest buoyancy numbers. Even allowing for a turbulent thermal diffusivity, the buoyancy number in the core could be very large. Similarly, topography has only been studied for a limited parameter range and boundary morphology.

2.3. Data Interpretation and Inversion

First consider how modern magnetic data are analyzed. We are only interested in that part of the field which originates in the core. This provides a powerful constraint: upward continuation through the insulating mantle attenuates the shorter wavelengths, leaving a larger scale field. Mathematically the field is the gradient of a scalar potential, which is conveniently represented by a spherical harmonic series. To produce an IGRF, for example, the spherical harmonic series is truncated at some suitable degree and order and a least squares fit performed to determine the geomagnetic coefficients. Origin in the core gives a geometric convergence condition for the geomagnetic coefficients which can be used to replace truncation of the series. A modern field model typically uses tens of thousands of observations of the field for a single epoch with truncation at degree 12 or so. With good data, core field smoothing produces a very similar set of coefficients and a similar number of degrees of freedom, but if we require error interpretation of the field at the core surface the results are heavily dependent on the smoothing assumption. This problem lies at the heart of inference from a finite number of inaccurate observations. The IGRF is a mapping tool and the only considerations are consistency, accuracy in representing the geomagnetic field at the Earth's surface, and ease of use; the geomagnetic coefficients of a truncated series satisfy these criteria adequately. On the other hand, core field inversions are done for inference, to test some specific theory. For the test to be meaningful the theory must be posed in a way that allows us to use the data to discriminate between candidate models. Thus additional information from the physics of the system must constrain those coefficients left undetermined by the data. For a more detailed discussion see *Backus* [1988] and *Bloxham et al.* [1989].

Making inferences from paleomagnetic data is no different in principle from using modern measurements, but there are great differences in practice because paleomagnetic data is so sparse. The dual temptations to oversimplify the underlying field and overinterpret the data have rarely been resisted. Paleomagnetism has an equivalent "IGRF": the geocentric axial dipole (GAD). It has just one term in the spherical harmonic series, is the assumed field for all geological work using paleomagnetism, and has served very well indeed. The geomagnetic field has been dominated by a GAD throughout most of its history, the only tricky issue being how measurement errors map into the magnetic axis, for which there are well-established statistical procedures. However, when we wish to go further and investigate the nondipole components of the stable field, secular variation, and reversal transitions, theory gives no simple choices among the infinity of geomagnetic coefficients. Unfortunately, the paleomagnetic literature has often failed to discriminate between two separate activities: mapping (finding the most convenient representation for a dataset) and inference (testing theories such as whether the time-averaged field is axisymmetric or the transition field dipolar). How many measurements (and of what type) are needed to establish the position of the geomagnetic pole? The answer depends on how complicated you allow the field to be. If you assume the field is a dipole the answer is just two directions at one site, which is how paleorotations are determined; otherwise the answer depends on the size of the non-dipolar field, since you need enough measurements to characterize it properly.

3. DATABASES

3.1. Paleomagnetic Observations

Paleomagnetic measurements are made on sediments and lavas, often with the primary aim of estimating tectonic movement. A *rock unit* (sedimentary sequence or igneous cooling units) is sampled at a wide range of *sites*, and each *sample* provides several *specimens* for measurement. Structural corrections must be made if the orientation of the bedding is not horizontal. This procedure checks for deformation and variations in magnetic mineralogy across the unit. A wide range of samples with different ages is used to average out the secular variation; an ideal averaging time is 10^5 yr but usually no confirmation is possible because the samples are not accurately dated [*Butler*, 1992].

Provided we can correct independently for plate movement, and there is no within-plate deformation of the rock unit, the magnetic direction at each site can be mapped into a virtual geomagnetic pole (VGP) by assuming a dipolar field. Data from the last 5 Myr or so need little or no plate motion correction. The average over the rock unit gives a paleomagnetic pole, which usually lies closer to the geographical pole than the VGPs because the secular variation averages out the non-dipole field to a large extent. The paleomagnetic pole is used for geological reconstruction, but where the rock unit has remained in place the data provide an estimate of the average magnetic field and its variation for the duration of formation of the rock unit.

The mean directions have been used to map the time-average of the field, a very demanding task since departures of the time-average from the GAD are only a few degrees. The scatter in values for each rock unit gives an error on the mean, which can be used for weighting in an inversion. Another check on consistency can be made by comparing neighboring rock units when the intervening distance is small enough for differences in the geomagnetic field to be negligible—typically 100–1,000 km. *Kelly and Gubbins* [1997] find this regional scatter largely consistent with the in-site scatter for igneous rock units, but 2–4 times larger for marine sediments, probably because of small systematic orientation errors in the hole.

McElhinny et al. [1996] argue that declinations are unreliable because of unknown geological rotations; inclinations are more reliable because any tilt would have been seen in the geology and corrected for. Even data from recent times (<5 Ma) could be affected in active tectonic regions. However, there is no direct evidence for this, and most rock units selected for this work are not in regions likely to have undergone significant rotations.

A further problem is overprinting of the original magnetisation by subsequent changes to the magnetisation of the rocks, including a viscous acquisition of an additional magnetic moment, sometimes by as much as 50%. Paleomagnetic techniques can separate the overprint, but the danger of a small residual overprint remains, particularly for older data. Overprinting by the modern field could reduce the magnetisation of a rock with an original magnetisation from a reversed field could give false evidence of a difference between the normal and reversed states of the geodynamo. A lack of N-R asymmetry indicates weak overprinting. Overprint of a sample formed in a weak field, such as during polarity transition, could be more serious.

3.2. Lavas

Lavas are good magnetic recorders but are erupted spasmodically in time and therefore give poor time coverage. Databases for the direction of the stable field have been assembled by *Lee* [1983], *McFadden et al.* [1991], and more recently by *Quidelleur et al.* [1994] and *Johnson and Constable* [1996]. There are fewer intensity measurements because they are more difficult to make and are less reliable. An intensity database from lavas has been published by *Tanaka et al.* [1995], including directional data from the same sites where this is available. *Prévot et al.* [1990] discuss the long-term variation in intensity. Intensities appear to have a log-normal distribution [*Kelly and Gubbins*, 1997; *McFadden and McElhinny*, 1982; *Tanaka et al.*, 1995].

Accurate dating is not needed to establish the time average, only a representative range of ages. However, it is often difficult or impossible to establish the age range of a suite of lava flows. This has led to two controversies: whether secular variation has been low in Hawaii [*Jacobs*, 1994] and whether transition fields show persistent directions: see *Prévot and Camps* [1993]; *Love*

[1998, *J. Geophys. Res.*, in press]. The in-site scatter can, in principle, be used to estimate statistical properties of paleosecular variation (PSV), but no detail can be recovered unless the site contains a suite of lava flows with known chronology. Recently, long cores from containing several hundred lava flows dating back over 400 kyr have given new insight into both eruption rates and PSV [*Garnier et al.*, 1996; *Holt et al.*, 1996].

Some lava sequences have captured reversals and left a record of the transition field. Often the reversal is not identified because the dating is not sufficiently accurate, but there are several studies from the last 5 Myr of individual, identifiable reversals. The most spectacular lava sequence is on Steens mountain in Oregon, which captured an unidentified reversal aged about 15 Ma [*Prévot et al.*, 1985]. This is probably the most detailed record of a transition, but problems still remain with measurements from some flows that appear to show rapid variations in direction, too rapid to be explained by a core source [*Camps et al.*, 1995]. *Prévot and Camps* [1993] give a compilation of transition fields for reversals and excursions in the last 12 Myr. They count flows with similar directions from a rock unit only once because they may be erupted within a short space of time. *Love* [1998, *J. Geophys. Res.*, in press] gives a more recent compilation using similar criteria but retaining results with similar directions unless the original authors specifically observed them to be from the same stratigraphic sequence.

3.3. Sediments

There are now some excellent sediment records of both relative intensity and direction [*Guyodo and Valet*, 1996; *Tauxe and Hartl*, 1997]. *Schneider and Kent* [1990] have produced a database of inclination measurements from marine sediments at 176 sites covering the last 2.5 Myr. Each site gives something approaching a continuous record and can therefore be used to study both the time averaged field and PSV. *Athanassopoulos* [1993] has produced a large database of reversal transition records which includes many records from sediments, and the compilation by *Love and Mazaud* [1997] for the Matuyama-Brunhes transition has both lava and sediment records.

Sediments are relatively poor magnetic recorders but are laid down continuously in time and therefore have potentially good time coverage; marine sediments also have the potential of good geographical coverage. Intensity is not well recorded because of variations in the magnetic mineralogy and oxidation state; usually only relative intensities can be recovered. Declination is often lost because the cores are not orientated. Com-

paction of the sediment after lock-in can reduce the inclination. Inclination shallowing moves the VGP away from the site, a tendency that has been sought and claimed for some transitions.

A further problem arises with the time a sediment takes to acquire its magnetisation. With lavas the issue is quite straightforward: magnetization locks in as the lava cools below the Curie temperature, which takes a few hours or days from eruption. In sediments the remanent magnetization may be acquired during compaction or after other diagenetic processes. The result may be a sediment that acquires its magnetization gradually over a thousand years or more and a depth of several meters, thereby providing us with a record of the geomagnetic field that has been smoothed in time, perhaps by convolution with a filter whose characteristics depend on the lock-in length and details of the compaction and other diagenetic processes. This may be what is required to account for the differences with lava records, which are more erratic: the erratic behavior may be averaged out by the sediments [*Langereis et al.*, 1992; *Rochette*, 1990].

A worldwide "sawtooth" pattern in magnetic intensity [*Meynadier and Valet*, 1996; *Meynadier et al.*, 1994] suggests that reversals are associated with a long-term decay in intensity followed by a rapid recovery. However, this pattern can also be explained if the sediments partially remagnetise throughout their history [*Kok and Tauxe*, 1996] or the grains continue to reorientate throughout the upper part of the sedimentary column [*Mazaud*, 1996].

Despite these problems with sediment magnetisation, improving techniques and understanding of diagenesis promise more consistent results for the future. Only sediments offer the hope of good geographical and temporal coverage of the geomagnetic field.

4. THE STABLE GEOMAGNETIC FIELD

4.1. The Time Average

Paleomagnetism cannot give us a snapshot of the geomagnetic field at one instant of time because it is impossible to date rocks at different sites with sufficient accuracy; at best we can hope for an average picture over a "long" period of time. The averaging time must eliminate the secular variation, but unfortunately the historical record is not long enough to tell us exactly how long this should be. Theory suggests it should be many times the overturn time, or possibly the diffusion time for the inner core if that limits the speed of geomagnetic changes, giving 10–50 kyr. Paleomagnetists have always averaged magnetic directions throughout a rock unit to obtain a mean magnetic direction. The geocentric axial dipole (GAD) hypothesis then allows the rock unit to be reoriented back to its position at the time of its formation. This procedure has been enormously successful, and the individual VGPs have been shown to cluster about a mean paleomagnetic pole that is close to the geographic axis.

We now ask the question: can we find departures from the GAD in the time average? This question was first answered by *Wilson* [1970], who found good evidence that inclinations were systematically low and that paleomagnetic poles tended to plot on the far side of the geographic pole for each site. He showed that this "far-sidedness" could be explained if the GAD were replaced with an axial dipole displaced a hundred km or so from the Earth's centre along the spin axis. He also found evidence for "right-handedness", that the paleomagnetic poles tended to plot to the right of the geographical pole. For any field configuration dominated by a GAD this implies either a current flowing in the mantle or a strong non-axisymmetric component [*Gubbins and Kelly*, 1995], which is unlikely.

Another fundamental question we can ask is whether there are any differences between the normal and reversed states. According to theory, any dynamo generating a normal field can also generate the reversed field with all components reversed. This does not necessarily mean that a reversed state that exists in the Earth must be exactly minus the normal state: it means dynamo solutions come in pairs. If the reversed state (**R**) is just slightly different from minus the normal state (-**N**), then **R** would have a counterpart, -**R**, slightly different from the Earth's normal state **N**. The observed non-GAD field appears to reverse along with the dipole (several authors claim that the quadrupole (g_2^0) and/or other low-order harmonics reverse with the dipole, but these claims are false because no effort is made to distinguish these harmonics from other plausible terms that cannot be resolved by the data—see "Data interpretation and inversion" above). Some studies have advocated differences, mostly based on inclination anomaly differences at the same site for normal and reversed fields (e.g. *Schneider and Kent* [1990]).

The differences claimed between states **R** and -**N** are some ten times smaller than the secular variation. We therefore have to explain how the Earth's field could vary about the state -**N** without ever entering the very close state **R**. *Hide* [1981] offered one explanation: that only the visible dipole field reversed, leaving the toroidal field with the same polarity. The two states are then not close, and the objection disappears. However, this scenario is difficult to justify on symmetry grounds because the toroidal and poloidal fields are linked through the dynamics and Coriolis force and should satisfy a consistent equatorial symmetry. Numerical dynamo models may shed some light on this question; meanwhile we should wait for much stronger paleomagnetic evidence for **N-R** asymmetry before invoking exotic new theories.

Attempts to map the time-averaged field have produced two schools of thought: the traditional view, that the data should be represented by the smallest number of low degree harmonics, almost invariably the axisymmetric, zero-order terms as advocated most recently by *McElhinny et al.* [1996]; and a newer approach in which the data are fitted to their estimated errors by a simple core field [*Gubbins and Kelly*, 1993; *Johnson and Constable*, 1995; *Kelly and Gubbins*, 1997]. There is an enormous apparent statistical difference between the two approaches. While the first uses a model with 2 or 3 coefficients (degrees of freedom), the second requires 20 or 30 to represent the same data!

A controversy has arisen over the interpretation of these mathematical models of the time-averaged field, the crucial difference involving non-axisymmetric and high-degree terms in the spherical harmonic series. I shall attempt to discuss the debate in two separate parts: first the ideas (or prejudices) derived from theory, then the requirements of the data. These two have been confused in the past. One theory (rarely if ever stated explicitly) asserts that, as we see mainly westward drift of the geomagnetic field at the Earth's surface during historical times, non-axisymmetric parts of the field will be averaged out over time to leave a purely axisymmetric time average. This makes a great deal of sense because convective systems in rotating spheres have no frame of reference, are known to drift in the laboratory, and have been shown to exhibit drifting solutions both analytically and numerically. The appropriate model is therefore one of an axisymmetric field, describable by relatively few terms (although convection experiments and the present geomagnetic field morphology suggest going significantly above degree 3, ideally to include the effect of the inner core cylinder—say degree 6 or higher). Any disagreement with data would then demand a revision of the theory or reassessment of the data quality.

Another theory draws on more recent analyses of historical secular variation that show westward drift confined to one part of the core surface, no tendency to drift through a full rotation, near-zero secular variation at the core surface beneath much of the Pacific, and other evidence that the core is influenced by the solid mantle. It asserts that core-mantle interaction can produce a non-axisymmetric time average that might appear as a time-smoothed version of the present field. This theory is not as simple as the first but it is perfectly plausible and receives support from a wide range of geophysical observations. Departures from axial symmetry releases a large number of spherical harmonic coefficients or degrees of freedom in the model and there is no problem in fitting the datasets provided they are consistent with a potential field.

In principle the data can discriminate between these two theories by detecting the non-axisymmetric part of the field and by producing a time average containing recognizable features, such as the inner core cylinder. *McElhinny et al.* [1996] argue against the use of non-axisymmetric terms because they depend mainly on declinations, which are prone to error from unidentified geological rotations and core mis-alignment (although they continue to use declinations to constrain their axisymmetric model, which seems inconsistent). *Kelly and Gubbins* [1997] compare the histograms for inclination and declination residuals and find no statistical differences provided they are weighted with total and horizontal intensity respectively, in the usual way, showing that there is no evidence in the databases to support McElhinny et al.'s suggestion. *Gubbins and Kelly* [1993] and *Johnson and Constable* [1995] find that non-zonal fields are required to fit the data to within the required tolerances, and further discussion should centre around whether the weights assigned to the data are correct and if systematic errors are present, including declination errors arising from tectonic rotations.

The main success of the non-axisymmetric studies is the recovery of a time-averaged field with some features in common with the modern field (Figure 1). The flux concentrations at the core surface in the northern hemisphere are in the same place as in the modern field, while their absence in the southern hemisphere could be due to lack of data or more vigorous secular variation in that region, as seen in the present day. This could only happen if the mantle were to lock core convection, or influence it in some more complex way that led to this average. This has prompted a number of studies of core-mantle interaction, and even incorporation into large-scale dynamo calculations.

Very long-term changes in the geomagnetic field, over tens of millions of years or more, properly belong in the discussion of the time-averaged field because the time scale is so long. Data from older times are sparse, but *Lee and Lilley* [1986] comment on changes in the apparent time average over 100 Myr. *Prévot et al.* [1990] gives evidence for low intensity during the Mesozoic. This long-term variation in intensity is now widely accepted, although it is not clear whether a relatively few

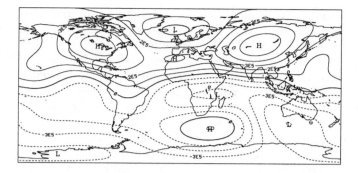

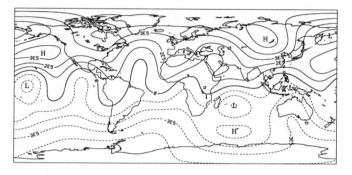

Figure 1. Global models of the geomagnetic field at the core surface based on a time average using paleomagnetic data from lavas from the last 5 Myr (lower plot) and the modern field using data degraded to similar quality (but better geographical distribution) (upper plot). Similarities in the non-axisymmetric parts give evidence of mantle influence over the geodynamo. From *Kelly and Gubbins* [1997].

measurements over an enormous length of time justify the broader conclusions.

4.2. Paleosecular Variation

Ideally, given a geomagnetic field as a function of time, we would like to estimate both the time average of the field:

$$\bar{\mathbf{B}}\left(\mathbf{r}\right) = \int_0^T \mathbf{B}\left(\mathbf{r}, t\right) dt \qquad (6)$$

where T is a suitable averaging time, and a time average of the secular variation such as

$$\mathcal{V}\left(\mathbf{r}\right) = \int_0^T \left(\mathbf{B} - \bar{\mathbf{B}}\right)^2 dt \qquad (7)$$

It would be interesting to examine this variance function for geographical variations or latitude dependence, because physical effects that produce persistent patterns in the time average may also produce patterns in the secular variation. The historical field shows quite distinct differences between the Pacific hemisphere, where directional changes are limited to 1° or so, and more active regions such as South America, Africa, and the South Atlantic, which have suffered changes up to 12°.

The same datasets that were analyzed for the time average can be used, in principle, to estimate secular variation by considering the scatter of directions and intensities about the mean at each site. *Kelly and Gubbins* [1997] plot the SV variance from the historical field and point estimates from the paleomagnetic database, and find no systematic geographical effects. Unfortunately, the databases are biased against high SV because data are often rejected if the samples at a single site show high variability (e.g. *Johnson and Constable* [1995], *Quidelleur et al.* [1994]). There may also be some bias against regions with exceptionally low SV because sequences of lava flows are rejected if they are clustered too tightly. This negative result is not surprising in view of the success achieved by *Constable and Parker* [1988] with a model consisting of a time average with g_1^0, g_2^0 and SV with coefficients drawn from a giant Gaussian process with zero mean. Presumably, changing the time average to a non-axisymmetric field would change the required statistics for the SV model, but this has not been considered. *Hulot and LeMouël* [1994] allow also for temporal covariance in their statistical model.

A more empirical approach to the data, pioneered by Cox [1970], suggests it does contain some systematic information. These studies use directly the variation in direction at each site, estimated from the angular distance Δ of the VGP from the geographic pole:

$$S^2 = \frac{1}{N} \sum_{i=1}^{N} \Delta_i^2 \qquad (8)$$

where i numbers the samples at a single rock unit. Some authors use $N - 1$ in the denominator and others use a slightly different definition of Δ. S^2 is called the *VGP scatter function*; it is a simple way to represent the variation of the geomagnetic field direction about the mean. Results from different sites are then averaged in longitude bands to give the mean S^2 as a function of latitude only. The variation with latitude appears to be significant.

McFadden et al. [1988] used a complicated argument to produce a quadratic dependence of S^2 on colatitude. The field is split into parts symmetric and antisymmetric about the equator, and by analogy with the modern field the symmetric part is supposed to give a contribution to S^2 that is independent of latitude while the antisymmetric part gives one that increases linearly with latitude. This model fits the paleomagnetic data quite

well. *Gubbins and Kelly* [1995] showed that this form for S^2 arises from the modern field from harmonics with $l = m = 1$ (the equatorial dipole) for the E^S part and $l = 2, m = 1$ for the E^A part, not the axisymmetric coefficients g_2^0 and g_3^0 as one might at first expect. Independently, *Kono and Tanaka* [1995] made a thorough study of the VGP scatter predicted by the random PSV model of *Constable and Parker* [1988] and found that it predicts a flat VGP scatter function, not the one observed. It is perhaps encouraging that the Constable and Parker model should fail, as a flat spectrum in both latitude and longitude has infinite energy and would not be expected from core dynamics. *Kono and Tanaka* [1995] also find that S^2 can be attributed mainly to harmonics with $l = m = 1$ and $l = 2, m = 1$, and *Hulot and Gallet* [1996] argue that higher degree harmonics of order 1 are also significant. *Kono and Hiroi* [1996] used paleointensities to investigate the variation of the non-dipole field.

5. THE GEOMAGNETIC FIELD IN TRANSITION

5.1. Reversals, Excursions, and Events

Reversals have occurred throughout geological time. Intervals of constant polarity, with duration about 1 Myr, are called *chrons* and the shorter intervals within these chrons are called *events* or, in the context of the timescale, *subchrons*, lasting an order of magnitude less time. Very long periods of no reversals, such as the one during the Cretaceous, are called *superchrons*. In addition to full reversals there are *excursions* when the reversed polarity is never quite reached. There is a gradation from secular variation through excursions to "full" polarity reversal, the latter sometimes becoming well established and sometimes switching polarity again in a relatively short time.

These empirical definitions are useful descriptions of the observations and essential in establishing a magnetic stratigraphy, but are without theoretical foundation. The only theoretical distinction one can make between an excursion and a reversal is whether the reversed polarity becomes completely established, perhaps defined as minus the original field, and has effectively no memory of the previous state. This working hypothesis can be tested against both paleomagnetic data and numerical dynamo experiments.

The recent (last 5 Myr) reversal time scale shown in Figure 2 exhibits a random pattern of reversals and excursions, suggestive of a chaotic nonlinear system. However, *Langereis et al.* [1997] identify a number of excursions within the Brunhes chron at several sites

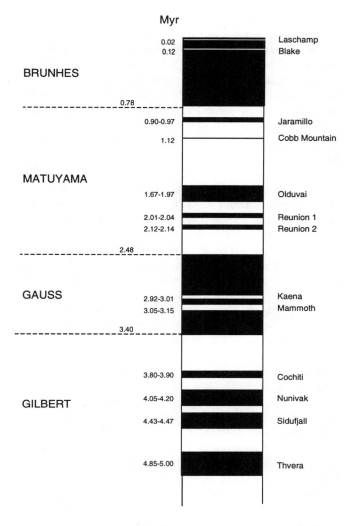

Figure 2. Reversal time scale for the last 5 Myr

an insulating mantle by the factor $(c/a)^l$, where c is the core radius, a the Earth's radius, and l the spherical harmonic degree, which produces a factor of 10 difference between the dipole (degree 1) and degree 4 harmonics. The intensity change takes longer than the directional change—perhaps as long as 10 kyr—while directional changes can be quite rapid. Indeed, there is no reason to restrict the speed of a directional change when the intensity is weak, whereas intensity change is restricted by the shortest natural time scales in the system, perhaps the overturn time or some shorter timescale of magnetohydrodynamic waves in the core. The geomagnetic field as a whole is limited by the longer time scale of diffusion in the inner core. There appears to be no asymmetry between R→N and N→R transitions, nor any time asymmetry between the early and late part of the transition, in the analysis of *Camps and Prévot* [1996].

The superchrons are fairly strong evidence of mantle influence on the core because core dynamics has no natural time scales that long. This provides a suitable theoretical definition of a superchron: a period when the extrinsic parameters define a non-reversing dynamo.

BRUNHES

Reversal excursions

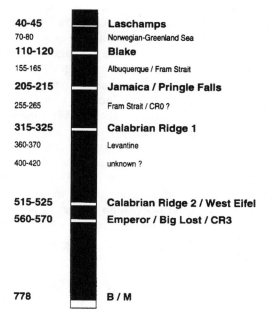

Figure 3. Excursions during the last 1 Myr from *Langereis et al.* [1997].

worldwide that all last a similar 5–10 kyr (Figure 3). This suggests a single dominant time scale. There is no similar evidence for a uniform time scale in older excursions, but perhaps this is because of poorer magnetic recording and dating. The time scale corresponds to the diffusion time of the inner core, so perhaps an excursion represents a change in the magnetic field throughout the fluid outer core that does not persist for long enough to reverse the field in the inner core.

Camps and Prévot [1996] have studied the relationship between intensity and direction during transitions from the last 20 Myr. The VDM appears to fall to about 20% of its normal value during the direction change (Figure 4). This intensity change could mean either a fall in core field strength or a cascade of magnetic energy to smaller length scales. The core field is attenuated by

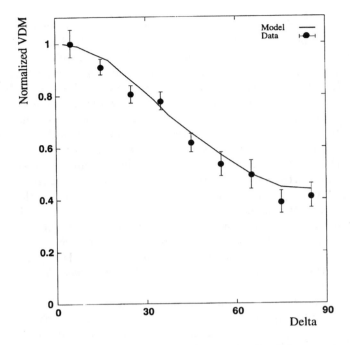

Figure 4. Normalised virtual dipole moment of the transition field plotted against VGP latitude (Delta), based on a compilation of data from many reversals and excursions by *Camps and Prévot* [1996].

There is further evidence that the frequency of reversals has increased since the last superchron [*Lowrie*, 1982] and that rapid reversal frequency is associated with stronger intensities in the Oligocene [*Tauxe and Hartl*, 1997]. Two different theories have been proposed: in one the entire throughput of heat changes [*Larson and Olson*, 1991; *Merrill and McElhinny*, 1983] while in the other the lateral distribution of anomalies (in heat flux, topography, or electrical conductivity) at the base of the mantle changes, both on the 50 Myr time scale of mantle convection. The former theory has been linked to dramatic changes at the Earth's surface associated with the higher heat flow through the mantle, while the second involves only slight changes with little or no associated surface effects. These ideas can go no further until we understand the dynamo better: at present it is not known what controls the dynamo regime, whether reversals are favored by high or low heat throughput, or the influence of lateral variations.

5.2. VGP Paths

Observations of the transitional directions have produced the following suggestions in turn: the idea that the VGPs pass through the site, indicative of an axisymmetric transition field; 90° away from the site, which

can only be due to a paleomagnetic bias or error; and on a great circle passing through the Americas and Asia, which requires control by the solid mantle. The current debate centers on whether VGPs show a longitude preference. This is important, since any longitude bias is indicative of mantle control. Initial claims were based on sediments [*Clement*, 1991; *Laj et al.*, 1991] and contested by *Valet et al.* [1992]. Lavas showed no preference in the study by *Prévot and Camps* [1993] but do in the later study by *Love* [1998 in press].

Langereis et al. [1992] have argued that the VGP paths may be an artefact of smoothing the field between non-antipodal N and R fields. The recorded transition field then reflects an average of the stable field directions before and after the transition, and not the true direction during the reversal. This result is often quoted as evidence against longitude confinement, but the argument is spurious. If a reversal is defined as between two well established stable fields then, as argued above, non-antipodal directions means N-R asymmetry, for which there is little or no evidence. Alternatively, the antipodal directions may not represent stable fields, in which case the reversal had not quite "finished", and the end points should properly be considered part of the transition field.

Hoffman [1992] claims that persistent intermediate directions are recorded by lavas in Hawaii and French Polynesia. Sediments might average such intermediate directions into a smooth longitude-confined VGP path. The substantive issue is whether there are persistent, longitude-dependent features in the transition field, not whether VGP paths lie along a single longitude, because both provide direct evidence of core-mantle interaction.

Most data compilations of lavas combine excursions and events with major reversals because there are insufficient data from lavas to examine full reversals or a single transition. This raises the interesting question of whether excursions and events behave differently from other reversals. *Gubbins and Coe* [1993] pointed out that R–N symmetry required any VGP path recorded at a site for a transition of one sense (e.g. R-N) should be accompanied by an antipodal VGP path for an N-R transition recorded at the same site. Any theory based on the magnetohydrodynamic equations that produces one VGP path for one complete transition type must inevitably produce the same path *for the opposite magnetic pole* for the other transition type. Since VGPs are always the south magnetic pole, the path will map to the opposite point on the Earth's surface. This assumes the transition starts and ends with a stable field; if this is not the case, as in an excursion or perhaps the boundary of a subchron, the geomagnetic field carries a memory of the original field direction and we should not expect such symmetry.

The preferred VGP longitudes are the same as those where flux is concentrated during stable polarity. This very significant observation suggests the same physical mechanism is operating throughout. The preferred longitudes also lie close to the Pacific rim, where subduction is thought to have cooled the lower mantle to produce a high seismic velocity anomaly [*Dziewonski and Woodhouse*, 1987], providing the connection with mantle control.

The observations are difficult to interpret because the transition field is unlikely to be dipolar: indeed, the presence of two VGP longitudes, not one, suggests a site-dependence which requires a non-dipolar field. *Gubbins and Sarson* [1994] found a dynamo-generated core field configuration that exhibits two VGP paths. The field has reflection symmetry about the equator, rotational symmetry by 180° about the polar axis, and flux concentrated near two longitudes 180° apart. The reversal proceeds by flux migrating towards the poles or equator. With this field configuration, sites in two quadrants record one VGP path while sites in the other quadrants record the other path. The flux concentrations lie in the centers of the 4 quadrants. Sites on the nodal lines separating the quadrants record rapid and erratic VGP paths. *Gubbins and Love* [1998, *Geophys. Res. Lett.*, under review] tested this model against the Matuyama-Brunhes database of *Love and Mazaud* [1997] and found general consistency with poleward flux migration.

Flux concentration can therefore explain both the time-averaged field during stable polarity and preferred transitional directions, but a dynamical mechanism is still required. Heat flow anomalies would tend to concentrate flux in the same regions as during stable polarity, with the reversal setting in by a dynamo wave instability as proposed by *Gubbins and Sarson* [1994]. Topography could act in a similar way. Mantle conductivity anomalies might also exert forces on the liquid core [*Aurnou et al.*, 1996] and rotate it into the preferred direction. None of these mechanisms have been examined in detail.

6. SUMMARY

The traditional view of the geodynamo is one of a complex nonlinear system exhibiting chaotic behaviour, with large fluctuations in dipole moment, random fluctuations in the non-dipole field, and no longitude reference. Variations on the very long time scale (> 10 Myr) are controlled by changes in conditions of the solid mantle. Evidence is now appearing of a short time scale in

the reversal record, that of excursions and perhaps the magnetic diffusion time of the inner core, the longest imposed time scale.

Theory strongly suggests the reversed field is minus the normal field, and there is little paleomagnetic evidence to the contrary. R-N asymmetry has wider implications than simple opposite signs of the two polarities of the stable field: on dependence of transitional VGP paths on the sense of the transition, and the definition of an excursion as opposed to a reversal.

An excursion is best defined as a large secular variation event in which negative field fails to be established throughout the core; a superchron as when the externally imposed parameters (such as the Rayleigh number) are such as to promote a non-reversing dynamo. Reversal frequency also depends on the imposed parameters. Chrons and subchrons appear to be non-linear behaviour of the dynamo, with no evidence of a dominant time scale.

A critical question is whether lateral variations in the lower mantle control the dynamo. Many authors have argued that an axisymmetric time average satisfies the paleomagnetic data from the last few million years, that secular variation appears random, and that reversal transitions involve directions with no longitude preference. All these aspects of the paleomagnetic record are consistent with the traditional view, which we can regard as the simplest scenario.

Now consider the effects of lateral variations in mantle topography, heat flow, or electrical conductivity. The theory has considerable appeal because these effects promote large scale, slowly-varying or stable solutions that are therefore simpler to compute. Recent geodynamo simulations have been marked by their rapid time variations and short interval between reversals, despite parameter choices that should produce models that are simpler than the real Earth, hinting that the theory needs some stabilising influence from the solid boundary.

Another advantage is that of any testable theory, it produces observable effects: a longitude-dependent time average, areas of persistently low or high secular variation, and transition fields with persistent directions (or VGP paths). Lateral variations can also account for changes in reversal frequency rather easily by mantle convection changing the pattern of lateral variations at the base of the mantle: no dramatic changes in overall heat throughput is needed. While lateral variations will certainly be present and have some effect, do they exert a large enough influence to produce observable effects? Preliminary theoretical studies (e.g. *Bell and Soward* [1996], *Zhang and Gubbins* [1993b]) suggest that they will

The evidence for lateral variations in the time-averaged field depends rather critically on a good assessment of the random data errors, absence of small systematic errors, and potentially troublesome declination measurements. However, there is no evidence in the databases of any of these problems, and therefore nothing inherent to make us prefer the axisymmetric models. There is weak, indirect evidence of systematic geographical variations in PSV, including low SV in the Pacific as seen in the historical record. The evidence could have been weakened by rejection of sites showing large scatter, one of several examples of bias in the available databases.

Some reversal transition paths appear to show preferred VGP longitudes, which if real can only be explained by lateral variations in the mantle or paleomagnetic artefacts. Sediments show these paths clearly, and the records appear to smooth the true field. It is irrelevant whether the transition field has a direction following a persistent VGP path or is an average of persistent directions; both require the influence of lateral variations in the mantle. Lavas show far fewer preferred paths, but this may again be a result of bias in the database resulting from rejection of flows with similar directions.

The strength of evidence for lateral variations in the paleomagnetic record lies with the similarities of several lines of evidence: the longitude bias is the same in the modern field, secular variation, paleomagnetic time average, and transition field. It requires flux to be concentrated on the same pair of longitudes, near 90° E and W. These are also zones where seismology gives high seismic velocity in the lower mantle beneath the Pacific "ring of fire", indicating a temperature or compositional anomaly and therefore potential heat flow or electrical conductivity anomaly, or topography.

The theory of lateral variations is eminently testable: all that is needed is more and better paleomagnetic data coupled to theoretical studies.

Acknowledgments. This work is partially supported by Natural Environment Research Council grant GR3/9741. Jeff Love and referees Michel Prévot and Cathy Constable provided many helpful suggestions for improvement to the original manuscript.

REFERENCES

Athanassopoulos, J., A Matuyama-Brunhes polarity reversal record: comparison between thermal and alternating field demagnetisation on ocean sediments from the north Pacific Transect, Univ. California, Santa Barbara, PhD Thesis, 1993.

Aurnou, J. M., J. L. Buttles, G. A. Neumann, and P. L. Olson, Electromagnetic core-mantle coupling and paleomagnetic reversal paths, *Geophys. Res. Lett.*, *23*, 2705–2708, 1996.

Backus, G. E., Bayesian inference in geomagnetism., *Geophys. J.*, *92*, 125–142, 1988.

Bassom, A. P., and A. M. Soward, Localised rotating convection inducted by topography, *Physica D*, *97*, 29–44, 1996.

Bell, P. I., and A. M. Soward, The influence of surface topography on rotating convection, *J. Fluid Mech.*, *313*, 147–180, 1996.

Bloxham, J., and D. Gubbins, Thermal core-mantle interactions, *Nature*, *325*, 511–513, 1987.

Bloxham, J., D. Gubbins, and A. Jackson, Geomagnetic secular variation, *Phil. Trans. Roy. Soc. Lon.*, *329*, 415–502, 1989.

Busse, F. H., and J. Wicht, A simple dynamo caused by conductivity variations, *Geophys. Astrophys. Fluid Dyn.*, *64*, 135–144, 1992.

Butler, R. F., *Paleomagnetism*, Blackwell Scientific, Boston, MA, 1992.

Camps, P., and M. Prévot, A statistical model of the fluctuations in the geomagnetic field from paleosecular variation to reversal, *Science*, *273*, 776–779, 1996.

Camps, P., M. Prévot, and R. S. Coe, Revisiting the initial sites of geomagnetic field impulses during the Steens mountain polarity reversal, *Geophys. J. Int.*, *123*, 484–506, 1995.

Clement, B. M., Geographical distribution of transitional VGP's: evidence for non-zonal equatorial symmetry during the Matuyama-Brunhes geomagnetic reversal, *Earth Planet. Sci. Lett.*, *104*, 48–58, 1991.

Constable, C. G., and R. L. Parker, Statistical of the geomagnetic secular variation for the past 5 Myr, *J. Geophys. Res.*, *93*, 11,569–11,581, 1988.

Dziewonski, A. M., and J. Woodhouse, Global images of the Earth's interior, *Science*, *236*, 37–48, 1987.

Garnier, F., C. Laj, E. Herrero-Bervera, C. Kissel, and D. M. Thomas, Preliminary determinations of Geomagnetic field intensity for the last 400 kyr from the Hawaii Scientific Drilling Project core, Big Island, Hawaii, *J. Geophys. Res.*, *101*, 11,665–11,673, 1996.

Glatzmaier, G. A., and P. H. Roberts, A three-dimensional self-consistent computer simulation of a geomagnetic field reversal, *Nature*, *377*, 203–208, 1995.

Gubbins, D., and R. Coe, Longitudinally confined geomagnetic reversal paths from non-dipolar transition fields, *Nature*, *362*, 51–53, 1993.

Gubbins, D., and P. Kelly, Persistent patterns in the geomagnetic field during the last 2.5 Myr, *Nature*, *365*, 829–832, 1993.

Gubbins, D., and P. Kelly, On the analysis of paleomagnetic secular variation, *J. Geophys. Res.*, *100*, 14955–14964, 1995.

Gubbins, D., and P. H. Roberts, Magnetohydrodynamics of

the Earth's core. Chapter 1, in *Geomagnetism Volume II*, edited by J. A. Jacobs, pp. 1–183, Academic Press, 1987.

Gubbins, D., and G. Sarson, Geomagnetic reversal transition paths from a kinematic dynamo model, *Nature*, *368*, 51–55, 1994.

Gubbins, D., and K. Zhang, Symmetry properties of the dynamo equations for paleomagnetism and geomagnetism, *Phys. Earth Planet. Int.*, *75*, 225–241, 1993.

Guyodo, Y., and J-P. Valet, Relative variations in geomagnetic intensity from sedimentary records: the past 200,000 years, *Earth Planet. Sci. Lett.*, *143*, 23–36, 1996.

Hide, R., Motions of the earth's core and mantle, and variations of the main geomagnetic field, *Science*, *157*, 55–56, 1967.

Hide, R., Self-exciting dynamos and geomagnetic polarity changes, *Nature*, *293*, 728–729, 1981.

Hide, R., and S. R. C. Malin, Novel correlations between global features of the Earth's gravitational and magnetic fields, *Nature*, *255*, 605–609, 1970.

Hoffman, K. A., Long-lived transitional states of the geomagnetic field and the two dynamo families, *Nature*, *359*, 789–794, 1992.

Hollerbach, R., and C. A. Jones, Influence of the Earth's inner core on geomagnetic fluctuations and reversals, *Nature*, *365*, 541–543, 1993.

Holt, J. W., J. L. Kirschvink, and F. Garnier, Geomagnetic field inclinations for the past 400 kyr from the 1-km core of the Hawaii Scientific Drilling Project, *J. Geophys. Res.*, *101*, 11,655–11,663, 1996.

Hulot, G., and Y. Gallet, On the interpretation of virtual geomagnetic pole (VGP) scatter curves, *Phys. Earth Planet. Int.*, *95*, 37–53, 1996.

Hulot, G., and J-L. LeMouël, A statistical approach to the Earth's main magnetic field, *Phys. Earth Planet. Int.*, *82*, 167–183, 1994.

Jacobs, J. A., *Reversals of the Earth's magnetic field*, Cambridge Univ. Press, Cambridge, 1994.

Johnson, C., and C. Constable, The time-averaged geomagnetic field as recorded by lava flows over the past 5Myr, *Geophys. J. Int.*, *122*, 489–519, 1995.

Johnson, C., and C. Constable, Paleosecular variation recorded by lava flows over the past five million years, *Phil. Trans. Roy. Soc. Lon.*, *354*, 89–141, 1996.

Kelly, P., and D. Gubbins, The geomagnetic field over the past 5 million years, *Geophys. J. Int.*, *128*, 315–330, 1997.

Kok, Y. S., and L. Tauxe, Saw-toothed pattern of relative paleointensity records and cumulative viscous remanence, 1996.

Kono, M., and O. Hiroi, Paleosecular variation of field intensities and dipole moments, *Earth Planet. Sci. Lett.*, *139*, 251–262, 1996.

Kono, M., and H. Tanaka, Mapping the Gauss coefficients to the pole and the models of paleosecular variation, *J. Geomagn. Geoelectr.*, *47*, 115–130, 1995.

Kumar, S., and P. H. Roberts, A three-dimensional kinematic dynamo, *Proc. R. Soc.*, *344*, 235–238, 1975.

Laj, C. A., A. Mazaud, M. Weeks, M. Fuller, and E. Herrero-Bervera, Geomagnetic reversal paths, *Nature*, *351*, 447, 1991.

Langereis, C. G., M. J. Dekkers, G. J. de Lange, M. Paterne, and P. J. M. van Santvoort, Magnetostratigraphy and astronomical calibration of the last 1.1 Myr from an easter Mediterranean piston core and dating of short events in the Brunhes, *Geophys. J. Int.*, *129*, 75–94, 1997.

Langereis, C. G., A. A. M. van Hoof, and P. Rochette, Longitudinal confinement of geomagnetic reversal paths as a possible sedimentary artefact, *Nature*, *358*, 226–230, 1992.

Larson, R. L., and P. Olson, Mantle plumes control magnetic reversal frequency, *Earth Planet. Sci. Lett.*, *107*, 437–447, 1991.

Lee, S., A study of the time-averaged paleomagnetic field for the last 195 million years, Australian National Univ., PhD. Thesis, Canberra, 1983.

Lee, S., and F. E. M. Lilley, On paleomagnetic data and dynamo theory, *J. Geomagn. Geoelectr.*, *38*, 797–806, 1986.

Love, J., and A. Mazaud, A database for the Matuyama-Brunhes magnetic reversal, *Phys. Earth Planet. Int.*, *103*, 207–245, 1997.

Lowrie, W., A revised magnetic polarity timescale for the Cretaceious and Cainozoic, *Philos. Trans. R. Soc. London Ser. A*, *306*, 129–136, 1982.

Mazaud, A., "Sawtooth" variation in magnetic intensity profiles and delayed acquisition of magnetization in deep sea cores, *Earth Planet. Sci. Lett.*, *139*, 379–386, 1996.

McElhinny, M. W., P. L. McFadden, and R. T. Merrill, The time averaged paleomagnetic field 0–5 Ma, *J. Geophys. Res.*, *101*, 25,007–25,027, 1996.

McFadden, P. L., and M. W. McElhinny, Variations in the Geomagnetic Field Dipole 2: Statistical Analysis of VDMs for the Past 5 Million Years, *J. Geomagn. Geoelectr.*, *34*, 163–189, 1982.

McFadden, P. L., R. T. Merrill, and M. W. McElhinny, Dipole/quadrupole family modeling of paleosecular variation, *J. Geophys. Res.*, *93* , 11,583–11,588, 1988.

McFadden, P. L., R. T. Merrill, M. W. McElhinny, and S. Lee, Reversals of the Earth's magnetic field and temporal variations of the dynamo families, *J. Geophys. Res.*, *96*, 3923–3933, 1991.

Merrill, R. T., and M. W. McElhinny, *The Earth's Magnetic Field (Its History, Origin and Planetary Perspective)*, Academic, San Diego, Calif., 1983.

Meynadier, L., and J-P. Valet, Post-depositional realignment of magnetic grains and asymmetrical saw-tooth patterns of magnetic intensity, *Earth Planet. Sci. Lett.*, *140*, 123–132, 1996.

Meynadier, L., J-P. Valet, F. C. Bassinot, N. J. Shackelton, and Y. Guyodo, Asymmetrical saw-tooth pattern of the geomagnetic field intensity from equatorial seidments in the Pacific and Indian Oceans, *Earth Planet. Sci. Lett.*, *126*, 109–127, 1994.

Olson, P., and G. Glatzmaier, Magnetoconvection and thermal coupling of the Earth's core and mantle, *Philos. Trans. R. Soc. London Ser. A*, *354*, 1–12, 1996.

Parker, E. N., Hydromagnetic Dynamo Models, *Astrophys. J.*, *122*, 293–314, 1955.

Poirier, J-P., and J-L. LeMouël, Does infiltration of core material into the lower mantle affect the observed geomagnetic field?, *Phys. Earth Planet. Int.*, *73*, 29–37, 1992.

Prévot, M., and P. Camps, Absence of preferred longitude sectors for poles from volcanic records of geomagnetic reversals, *Nature*, *366*, 53–57, 1993.

Prévot, M., M. E-M. Derder, M. McWilliams, and J. Thompson, Intensity of the earth's magnetic field: evidence for a Mesozoic dipole low, *Earth Planet. Sci. Lett.*, *95*, 129–139, 1990.

Prévot, M., E. A. Mankinen, C. S. Gromme, and R. S. Coe, How the geomagnetic field vector reverses polarity, *Nature*, *316*, 230–234, 1985.

Quidelleur, X., J-P. Valet, V. Courtillot, and G. Hulot, Long-term Geometry of the geomagnetic-field for the last 5 million years – an updated secular variation database, *Geophys. Res. Lett.*, *21*, 1639–1642, 1994.

Rochette, P., Rationale of geomagnetic reversals versus remanence recording processes in rocks: A critical review, *Earth Planet. Sci. Lett.*, *98*, 33–39, 1990.

Runcorn, S. K., Polar paths in geomagnetic reversals, *Nature*, *356*, 654–656, 1992.

Schneider, D. A., and D. V. Kent, The time-averaged palæomagnetic field, *Rev. Geophys.*, *28*, 71–96, 1990.

Shankland, T. J., J. Peyronneau, and J-P. Poirier, Electrical conductivity of the Earth's lower mantle, *Nature*, *366*, 453–455, 1993.

Tanaka, H., M. Kono, and H. Uchimura, Some global features of paleointensity in geological time, *Geophys. J. Int.*, *120*, 97–102, 1995.

Tauxe, L., and P. Hartl, 11 million years of Oligocene geomagnetic field behaviour, *Geophys. J. Int.*, *128*, 217–229, 1997.

Valet, J. -P., P. Tucholka, V. Courtillot, and L. Meynadier, Paleomagnetic constraints on the geometry of the geomagnetic field during reversals, *Nature*, *356*, 400–407, 1992.

Wilson, R. L., Permanent aspects of the earth's non-dipole magnetic field over Upper Tertiary times, *Geophys. J. R. Astron. Soc.*, *19*, 417–437, 1970.

Zhang, K., and D. Gubbins, On convection in the earth's core forced by lateral temperature variations in the lower mantle, *Geophys. J. Int.*, *108*, 247–255, 1993a.

Zhang, K., and D. Gubbins, Convection in a rotating spherical fluid shell with an inhomogeneous temperature boundary condition at infinite Prandtl number, *J. Fluid Mech.*, *250*, 209–232, 1993b.

D. Gubbins, School of Earth Sciences, University of Leeds, Leeds LS2 9JT, UK.

A One-Dimensional Map of B_s from Torsional Oscillations of the Earth's Core

Stephen Zatman[1] and Jeremy Bloxham

Department of Earth and Planetary Sciences, Harvard University, Cambridge, Massachusetts

By making certain assumptions about the core we find that we may invert a time varying model of flow at the surface of the core for the root-mean-square average on axial cylinders of the component of magnetic field which links coaxial cylinders in the core (B_s), and a coefficient of "friction" at the core-mantle boundary which may also represent excitation. The primary assumption is that equatorially symmetric, axisymmetric, zonal oscillations of the flows at the top of the core are due to decadal "torsional oscillations". We invert for r.m.s. B_s and friction within the core outside the inner core tangent cylinder from oscillations fitted to flows from 1900-1990. We find that the models of r.m.s. B_s typically have a magnitude $\sim 10^{-4}$T, with a minimum at midlatitudes, rising towards the tangent cylinder and towards a colatitude of 60° from either side. The main characteristics of models of B_s from inversions made with different assumptions tend to be fairly constant, but this is not so for the models of friction, perhaps because excitation varies between oscillations.

1. INTRODUCTION

The Earth's magnetic field, which is generated within the Earth's core by motions of highly conducting liquid iron, has been extensively mapped in the region exterior to the core. A model of the field at the Earth's surface may be inverted from magnetic observations [*Langel*, 1987]. By assuming that the Earth's mantle is an insulator one may downwardly continue the field to the core-mantle boundary (CMB), although to do so it is necessary to regularise the field so as to damp small scale structure [*Shure et al.*, 1982; *Gubbins*, 1983; *Gubbins and Bloxham*, 1985].

The field cannot be downwardly continued into the core by this method, as the core is highly conducting. Therefore, very little is known directly about the magnetic field inside the core, where the dynamo lies. However, it is possible to invert changes of the magnetic field at the CMB for models of the fluid velocity at the top of the core, although the inverted velocities are non-unique (i.e. different velocity models may fit a model of geomagnetic variation equally well), although this may be partially alleviated by incorporating extra restrictions on the flow [*Kahle et al.*, 1967; *Backus and Gilbert*, 1968; *Bloxham and Jackson*, 1992].

Dynamical arguments suggest that the decadal variation of the surface flow reflects the decadal flow in the interior, and it has been suggested that an observed oscillation of magnetic field with a period of $\simeq 60$ years may be explained by "torsional oscillations" that are independent in the z-direction (the direction parallel to the rotation axis) if the portion of the magnetic field which links axial cylinders in the core, B_s, has a magnitude of $\simeq 0.2$ mT [*Braginsky*, 1970, 1984]. Following

---

[1] *Now at USRA, NASA Goddard Space Flight Center, Greenbelt, Maryland.*

The Core-Mantle Boundary Region
Geodynamics 28
Copyright 1998 by the American Geophysical Union.

this dynamical idea, it was found that if the variation of the flow is constant on cylinders then there is a good agreement between changes in the core angular momentum and the geodetically derived discrepancy in the angular momentum of the mantle since 1900 [*Jault and LeMouël*, 1989; *Jackson et al.*, 1993; *Jackson*, 1997].

These torsional oscillations have velocities which are z-invariant, which would explain the correlation between the ΔLOD estimated from core flow models and that found geodetically. Theory suggests that decadal oscillations in the core are likely to be torsional oscillations. Inertial oscillations have much shorter timescales ($\sim$ 1 day) whilst oscillations that are magnetostrophic (a balance between Lorentz, coriolis, buoyancy and pressure forces [*Moffatt*, 1978]) are thought to have longer timescales. MC waves (involving magnetic and coriolis forces) should have periods of the order [*Gubbins and Roberts*, 1987]:

$$T \simeq \frac{2\mu_0 \rho L^2}{B^2 T_D} \qquad (1)$$

where ρ is a typical density, μ_0 is the magnetic permeability, L is the wavelength, B is a typical field strength and T_D is the length of the day. For $L = 1800$km, $\rho = 10^4$kg m^{-3} and $B = 10^{-2}$T, this indicates a period of 300 years.

We have recently developed a method of investigating the magnetic state of the interior of the Earth's core using models of flow at the surface of the core and the equations governing torsional oscillations [*Zatman and Bloxham*, 1997]. Here we describe the isolation of torsional oscillations from flow models and the inversion of such oscillations for r.m.s. B_s averaged on axial cylinders and a model of friction or excitation of torsional oscillations.

2. TORSIONAL OSCILLATIONS

In a rapidly rotating flow of liquid iron such as the Earth's core, we expect thin boundary layers to develop. We can divide the total magnetic field $\mathbf{B_t}$ into a part which includes the mainstream portion of the field and also the steady part of the boundary corrections ($\mathbf{B}$), and a part which represents the boundary correction associated with the perturbation field and flow ($\mathbf{B_b}$) so that $\mathbf{B_t} = \mathbf{B} + \mathbf{B_b}$. We divide the velocity of the core fluid with respect to the mantle ($\mathbf{v_t}$) into similar parts, so that $\mathbf{v_t} = \mathbf{v} + \mathbf{v_b}$, and also the total pressure so that $p_t = p + p + b$. This allows us to write the Navier-Stokes

equation for a Boussinesq, rotating, conducting fluid in terms of $\mathbf{v}$ and $\mathbf{B}$:

$$\rho \frac{\partial \mathbf{v}}{\partial t} + \rho \mathbf{v} \cdot \nabla \mathbf{v} + 2\rho \mathbf{\Omega} \wedge \mathbf{v} = \mathbf{f} - \nabla p + \mathbf{J} \wedge \mathbf{B} + \rho' \mathbf{g} + \eta_v \nabla^2 \mathbf{v} \qquad (2)$$

where ρ' is the deviation of local density from the average density in the core, $\mathbf{\Omega}$ the rotation rate of the reference frame with regard to an inertial frame, $\mathbf{J} = \mu_0^{-1} \nabla \wedge \mathbf{B}$ is the current density, $\mathbf{g}$ the gravitational acceleration and η_v the dynamic viscosity. $\mathbf{f}$ represents forces associated with boundary effects induced by the perturbation (e.g. the part of the electromagnetic force that depends on $\mathbf{B_b}$, or oscillations in pressure associated with the variation in $\mathbf{v_b}$). We would expect $\mathbf{f} \to 0$ as one moves from the boundary into the interior of the core.

It is thought that much of the core is in a "magnetostrophic" balance between the Coriolis force, the pressure force, the Lorentz force, and the buoyancy force. The timescale on which the magnetostrophic flow varies is expected to be long ($\simeq$ 300 years). Torsional oscillations on the other hand may allow variation on the much shorter timescales necessary to explain decadal changes in length of day (ΔLOD) [*Braginsky*, 1970, 1984]. Here, the flow is independent of z (the coordinate parallel to the rotation axis) in the bulk of the core, i.e. the surface of integration in figure 1. Buoyancy effects are neglected. The Coriolis and pressure forces balance, leaving a balance between the Lorentz force and the inertial force. This allows a much more rapid evolution of the dynamics. For this study, we are specifically interested in axisymmetric torsional oscillations, in which motion is constant and zonal on axial cylinders. We shall assume a spherical core and ignore the solid inner core.

In the Boussinesq limit density variations are neglected except in the gravitational term. Therefore, for the mainstream flow (where we may neglect $\mathbf{v_b}$):

$$\nabla \cdot \mathbf{v} \simeq 0 \qquad (3)$$

Integrating the ϕ (azimuthal) component of equation 2 on the surface of an axial cylinder $\mathbf{\Sigma}$, assuming rigid boundaries, using equation 3 and neglecting buoyancy and viscous forces in the interior of the core:

$$\int \left(\rho \frac{\partial v_\phi}{\partial t} + [\rho \mathbf{v} \cdot \nabla \mathbf{v}]_\phi - f_\phi \right) d\mathbf{\Sigma} \qquad (4)$$

$$= \int \frac{1}{\mu_0} \left[\frac{B_s B_\phi}{s} + B_s \frac{\partial B_\phi}{\partial s} + B_z \frac{\partial B_\phi}{\partial z} \right] d\mathbf{\Sigma}$$

The magnetic induction equation is:

$$\frac{\partial \mathbf{B}}{\partial t} = \nabla \wedge (\mathbf{v} \wedge \mathbf{B}) + \eta \nabla^2 \mathbf{B} \quad (5)$$

We can divide the magnetic field and velocity into a basic state and torsional oscillations:

$$\begin{aligned} \mathbf{B} &= \mathbf{B_0} + \mathbf{B_c} = \mathbf{B_0} + \mathbf{B_a} + \mathbf{B'} \\ \mathbf{v} &= \mathbf{v_0} + \mathbf{v'} \end{aligned} \quad (6)$$

where $\mathbf{B_a}$ is the part of the perturbation magnetic field ($\mathbf{B_c}$) that is associated with transport of $\mathbf{B}$ by $\mathbf{v'}$, and $\mathbf{B'}$ the part produced by the stretching of $\mathbf{B}$ by shear in $\mathbf{v'}$. We assume that the oscillations in $\mathbf{B'}$ are small, so that we can ignore terms in the induction equation that are nonlinear in oscillation variables. Assuming that $\mathbf{v'} = \mathbf{v_0'(x)}e^{\omega t}$ we can rewrite equation 5 by subtracting out the basic state and neglecting the effects of diffusion:

$$\omega \mathbf{B_c} = \nabla \wedge (\mathbf{v} \wedge \mathbf{B_c}) + \nabla \wedge (\mathbf{v'} \wedge \mathbf{B_0}) \quad (7)$$

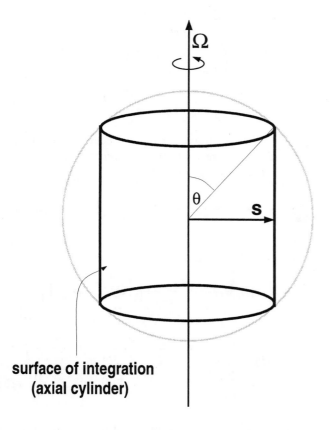

Figure 1. Surface of integration and averaging for torsional oscillations. In a Taylor state, the axial torque integrates over this surface to zero.

Scaling suggests that the first term on the right hand side may be ignored if $U/L\nu \ll 1$ where U is a typical basic state velocity, and L a typical lengthscale. If $U \simeq 3 \times 10^{-4} \mathrm{ms}^{-1}$, $2\pi/\nu \simeq 60$ years and $L \simeq r$ where r is the radius of the core, then $U/L\nu \simeq 3 \times 10^{-2}$. We shall assume that any basic state/wave interactions of the form of the first term of the right hand side may be parameterised and included in the "friction" model described below.

Neglecting the first term of the right hand side of equation 5 and assuming that $\mathbf{v'} = v_\phi'(s)e^{\omega t}\hat{\phi}$ then:

$$\begin{aligned} B_{cs} &= -\frac{\mathcal{R}}{\omega}\frac{\partial B_{0s}}{\partial \phi} \\ B_{c\phi} &= -\frac{\mathcal{R}}{\omega}\frac{\partial B_{0\phi}}{\partial \phi} + s\frac{B_{0s}}{\omega}\frac{\partial \mathcal{R}}{\partial s} \\ B_{cz} &= -\frac{\mathcal{R}}{\omega}\frac{\partial B_{0z}}{\partial \phi} \end{aligned} \quad (8)$$

where $\mathcal{R} = v_\phi'/s$. Isolating the part associated with shear of $\mathbf{v'}$:

$$\mathbf{B'} = \frac{1}{\omega}sB_{0s}\frac{\partial \mathcal{R}}{\partial s}\hat{\phi} \quad (9)$$

By removing the basic state, simplifying, integrating by parts, and using $\nabla \cdot \mathbf{B} = 0$, equation 5 becomes:

$$\begin{aligned} &\frac{\alpha\, s}{\mu_0} \int \left[B_{0z}B_\phi' \right]_{z_-}^{z_+} d\phi \\ &+ \frac{1}{\mu_0} \int \left(2\frac{B_{0s}B_\phi'}{s} + B_{0s}\frac{\partial B_\phi'}{\partial s} + B_\phi'\frac{\partial B_{0s}}{\partial s} \right) d\Sigma \\ &= \int \omega\rho s \left(\mathcal{R} + \mathcal{R}_m \right) d\Sigma - s\mathcal{R} \int F d\phi \end{aligned} \quad (10)$$

where z_+ and z_- represent the z-coordinates of the intersection of the axial cylinder with the CMB, and $\mathcal{R}_m$ is the change in spin rate of the mantle caused by the oscillation. Terms dependent on $\mathbf{B_a}$ were either removed with the basic state or canceled upon integration in ϕ. The unknown coefficient α appears because of uncertainty concerning the boundary condition at the CMB. It has been argued that if the mantle is an insulator, then as the toroidal field must drop to zero at the CMB, B_ϕ' must disappear [*Braginsky*, 1970]. However, the bottom of the mantle may have non-zero conductivity, in which case there may be toroidal B_ϕ' at the CMB. There will also, in general, be a poloidal part of B_ϕ' [*Fearn and Proctor*, 1992]. We assume that α is of $O(1)$. Our parameterisation of the boundary effect is somewhat ad-hoc, but we hope that by varying α we can understand the impact of the uncertainty in the boundary condition on the nature of the solution. We separate this boundary effect from the other boundary forces included in f_ϕ, the remainder of which is assumed to behave like a stress at the CMB, so that:

$$\int f_\phi d\Sigma \; + \; (1-\alpha)\frac{s}{\mu_0}\int \left[B_{0z}B'_\phi\right]_{z-}^{z+} d\phi$$
$$= \; s\mathcal{R}(s)\int F(s)d\phi \qquad (11)$$

We assume that the friction behaves like a stress at the CMB with the following form:

$$\int \hat{f}_\phi \, d\Sigma = s\, \mathcal{R}(s) \int F(s)\, d\phi \qquad (12)$$

where F is real, so that a negative F would imply a force per unit area that resists the shear at the core-mantle boundary. A friction of this form may represent viscous drag [*Gubbins and Roberts*, 1987], simple topographic or pressure coupling [*Hide*, 1969] or a "magnetic friction" [*Braginsky*, 1988], which can be thought of as part of the advective magnetic torque [*Stix and Roberts*, 1984] if mantle conductivity is non-zero but the mantle's magnetic diffusion timescale is short is comparison with the period of waves in the core. Recent work suggests that the diffusion timescale is very short, as the best fit lag between estimates of core angular momentum (CAM) and ΔLOD is less than one year [*Bloxham*, 1995], If there is an oscillation in $\partial v_r/\partial r$ at the CMB associated with the torsional oscillation, then the magnitude of B_r at the CMB will also vary with time, which means that equation 12 may no longer be the correct form of the friction, but as the magnitude of B_r at the CMB is not observed to vary greatly on these timescales we assume that this is a negligible effect.

$F(s)$ may also represent excitation of torsional oscillations, and other basic state/wave interactions. For the oscillations to exist there must be a mechanism to excite them. If the basic state of the core (which we suppose to be in magnetostrophic balance) evolves in such a way as to violate Taylor's constraint [*Taylor*, 1963] there will be unbalanced torques on axial cylinders, exciting torsional oscillations. In other words, as the system evolves it would excite torsional oscillations so as to leave a magnetostrophic basic state satisfying Taylor's constraint.

This presents us with two difficulties. First, we assume that the basic state is steady (or at least quasi-steady on the decadal timescale), and yet we are studying the effects of the evolution of that basic state. How-

ever, if small changes of the basic state lead to large unbalanced torques, then they may excite torsional oscillations with frequencies much higher than that of the evolution of the quasi-steady basic state.

Second, it is not at all clear that equation 12 is of the correct form for the sort of excitation which takes place in the Earth's core. For example, the excitation of a wave may involve the exchange of angular momentum between the wave and another part of the core flow, rather than between the wave and the mantle. Without a deeper knowledge of the magnetostrophic dynamics, it is hard to say what the correct form would be. However, equation 12 has the advantage of simplicity (it adds no further complication to the problem than that imposed by the core-mantle friction anyway), and also it does recover the major relevant property of excitation (i.e. it allows waves to grow).

Substituting equations 9 and 12 into equation 10, integrating, and dropping the suffix 0 for basic state magnetic field, equation 10 becomes:

$$\frac{\alpha\left\{[B_sB_z]_{z-}^{z+}\right\}}{2\left(r_c^2-s^2\right)^{\frac{1}{2}}\mu_0\omega}\frac{\partial\mathcal{R}}{\partial s}$$
$$+ \; \frac{1}{\omega\mu_0}\left(\left(\frac{3}{s}\frac{\partial\mathcal{R}}{\partial s}+\frac{\partial^2\mathcal{R}}{\partial s^2}\right)\overline{B_s^2}+\frac{\partial\mathcal{R}}{\partial s}\frac{\overline{\partial B_s^2}}{\partial s}\right)$$
$$= \; \frac{\omega\mathcal{M}\left(\mathcal{R}-\mathcal{R}_m\right)}{4\pi s\left(r_c^2-s^2\right)^{\frac{1}{2}}}-\frac{\mathcal{R}r}{r_c^2-s^2}F \qquad (13)$$

where $\mathcal{M}$ is the mass per unit width of each cylindrical annulus, $\overline{x}$ means x averaged over the cylinder, and $\{x\}$ the zonal average of x. Using

$$\overline{\frac{\partial x}{\partial s}}=\frac{\partial\overline{x}}{\partial s}-\frac{s}{r_c^2-s^2}\overline{x}+\frac{s}{2\left(r_c^2-s^2\right)}\left[\{x_{z+}\}+\{x_{z-}\}\right]$$
$$(14)$$

equation 13 becomes:

$$\frac{\left\{\alpha\,[B_sB_z]_{z-}^{z+}+\frac{s}{(r_c^2-s^2)^{\frac{1}{2}}}\left[B_{s_{z+}}^2+B_{s_{z-}}^2\right]\right\}}{2\left(r_c^2-s^2\right)^{\frac{1}{2}}\mu_0\omega}\frac{\partial\mathcal{R}}{\partial s}$$
$$+ \; \frac{1}{\omega\mu_0}\left(\left(\frac{\partial^2\mathcal{R}}{\partial s^2}+\frac{3r_c^2-4s^2}{s\left(r_c^2-s^2\right)}\frac{\partial\mathcal{R}}{\partial s}\right)+\frac{\partial\mathcal{R}}{\partial s}\frac{\partial}{\partial s}\right)\overline{B_s^2}$$
$$= \; \frac{\omega\mathcal{M}\left(\mathcal{R}+\mathcal{R}_m\right)}{4\pi s\left(r_c^2-s^2\right)^{\frac{1}{2}}}-\frac{\mathcal{R}r_c}{r_c^2-s^2}F \qquad (15)$$

Since $\mathcal{R}$ and $\mathcal{R}_m$ are complex, this equation may be divided into real and imaginary parts.

3. ASSUMPTIONS

If we have knowledge of the core flow and of B_s and B_z at the CMB, and we assume some value of α, both sides of equation 15 are left as a function of F and $\overline{B_s^2}$. Hence, we may invert for a model of F and $\overline{B_s^2}$.

In reality, we do not have a full knowledge of the flow at the core's surface. The flow is related to the secular variation of the geomagnetic field by:

$$\frac{\partial B_r}{\partial t} + \nabla_H \cdot (\mathbf{v} B_r) = 0 \qquad (16)$$

where ∇_H is the horizontal gradient operator. While one may use this to construct models of core flow at the core surface from time varying geomagnetic field models, they will be non-unique. The ambiguity may be reduced by making assumptions that place extra restrictions on the flow. Commonly made assumptions are that the flow is steady [*Gubbins*, 1982], that motions at the surface of the core are toroidal only [*Whaler*, 1980], or that the motions are geostrophic tangentially to the CMB [*Hills*, 1979; *LeMouël*, 1984]. Tangential geostrophy implies:

$$\nabla_H \cdot (\mathbf{v_H} \cos\theta) = 0 \qquad (17)$$

where θ is colatitude. For a review of the subject, see *Bloxham and Jackson* [1991]. In order to find torsional oscillations, we must use a time varying flow. The part of the flow in which we are interested - the equatorially symmetric, axisymmetric, zonal part of the flow - is both toroidal and tangentially geostrophic, although the assumption used will still have an influence through the restriction that it places on the rest of the inverted flow. We use a core flow model constructed using the tangential geostrophic constraint expanded in cubic B-splines [*de Boor*, 1978] in time.

Questions have been raised concerning the validity of steady core flow models [*Gubbins and Kelly*, 1996]. However, we expect that the time varying flows should be relatively well determined. The correlation between core angular momentum estimated from core flows and ΔLOD [*Jault and LeMouël*, 1989; *Jackson et al.*, 1993] supports our confidence in these parts of the flows.

Equation 15 depends on the quantities $\left\{[B_s B_z]_{z_-}^{z_+}\right\}$ and $\left\{B_{s_{z_+}}^2 + B_{s_{z_-}}^2\right\}$ evaluated at the CMB. For these we use estimates of these quantities for 1945 from the time dependent field model ufm1 of *Bloxham and Jackson* [1992]. The evolution of these terms evaluated from ufm1 over the timespan of the study is relatively small

compared with the overall magnitude of the terms. This is in contrast to the evolution of the velocity field: the time varying part of the velocity (which we attribute to torsional oscillations) appears to be similar in size to the steady part.

One possible source of error is that ufm1 only provides an estimate of the poloidal part of the field, as we cannot determine the toroidal part of the magnetic field at the CMB. If the bottom of the mantle has a non-zero electrical conductivity, then it is possible that there are toroidal components of B_s and B_z at the CMB from the leakage of toroidal field from the core into the mantle. In fact, it turns out that the term in $\left\{[B_s B_z]_{z_-}^{z_+}\right\}$ only affects the solution close to the tangent cylinder (the axial cylinder which grazes the inner core), and the term in $\left\{B_{s_{z_+}}^2 + B_{s_{z_-}}^2\right\}$ is only important near the equator, where the toroidal field can have little B_s for geometrical reasons (as the $\hat{s}$ direction is nearly radial). Therefore we expect the error to be small.

4. METHODS

We fit a model of one or two waves and a steady flow to the equatorially symmetric, axisymmetric zonal velocities from 1900 to 1990. We assume that the waves are of the form $v_\phi' = s\mathcal{R}(s)e^{\omega t}$ where both ω and $\mathcal{R}(s)$ are complex, allowing the waves to be growing or damped harmonic oscillations and for their phases to vary in the s direction. Since we invert for oscillations over the whole surface of the core from the pole to the equator using this parameterisation in time, we are implicitly inverting for standing waves. There is no need to enforce particular boundary conditions.

The core flow model is a spherical harmonic expansion of a poloidal-toroidal decomposition of the flow [see, for example, *Bloxham and Jackson*, 1991], i.e.:

$$\mathbf{v} = \nabla \wedge (\mathcal{T}_v \hat{\mathbf{r}}) + \nabla_h \mathcal{S}_v \qquad (18)$$

where $\hat{\mathbf{r}}$ is a unit vector in the direction of $\mathbf{r}$ and ∇_h the horizontal divergence operator. $\mathcal{T}_v$ and $\mathcal{S}_v$ are expanded in surface spherical harmonics with coefficients $s_l^m(t)$ and $t_l^m(t)$, which are then expanded in time by B-splines [*de Boor*, 1978]. Least square fits were made to estimates of the $t_1^0, t_3^0, t_5^0, t_7^0, t_9^0, t_{11}^0$ and t_{13}^0 coefficients of the velocity model at nineteen 5-year intervals (T_j) from 1900 to 1990.

We can rewrite equation 15 in the form:

$$\begin{aligned} \alpha_m + \epsilon_\alpha &= (\beta_m + \epsilon_\beta)\, F + (\gamma_m + \epsilon_\gamma)\, \overline{B_s^2} \\ &\quad + (\kappa_m + \epsilon_\kappa)\, \frac{\partial \overline{B_s^2}}{\partial s} \end{aligned} \qquad (19)$$

where $\alpha_m(s)$, $\beta_m(s)$, $\gamma_m(s)$ and $\kappa_m(s)$ are complex and are estimated from ufm1 and the fitted oscillations, and ϵ_j is the error in the estimate of j. Here F and $\overline{B_s^2}(s)$ denote the actual (unknown) friction and field rather than our estimates. As equation 19 is complex, there are two equations for each fitted oscillation. Hence we may invert for models of $F(s)$ and $\overline{B_s^2}(s)$.

F and $\overline{B_s^2}(s)$ are expanded in colatitude by 150 equally spaced cubic B-splines [de Boor, 1978]. Our estimates of κ_m/r_c are an order of magnitude smaller than those of γ_m. Therefore, the relative errors in equation 19 at different colatitudes are only important for the inversion of equation 19 when $\overline{B_s^2}(s)$ varies rapidly with s. Where $\kappa_m\ \partial\overline{B_s^2}/\partial s$ is unimportant, the inversion reduces to a simple least-squares estimate of F and $\overline{B_s^2}$ at each latitude. It turns out that different reasonable error models produce nearly identical results, so we merely assume that the variance of the error in estimates of $v_\phi(\theta) \propto s^{-1}$, that the magnitude of $\overline{B_s^2}(s)$ is around 0.2 mT and (for simplicity) that $\epsilon_\beta F$ is small compared with the other errors.

The oscillations may be inverted in several ways: both oscillations may be inverted together for a common $\overline{B_s^2}(s)$ and a common $F(s)$; both oscillations may be inverted together for a common $\overline{B_s^2}(s)$ but separate models of $F(s)$; or the oscillations may be inverted separately. Although we should expect $\overline{B_s^2}(s)$ to be identical for the two cases, this is not necessarily true for $F(s)$ as represents both core-mantle friction and excitation, which may be different for the two waves.

5. INVERSION OF FLOWS FOR OSCILLATIONS

The fit of a steady velocity and one damped harmonic wave (which we will henceforth describe as our "single wave inversion") to the axisymmetric, equatorially symmetric zonal flow is shown in figure 2. Some features of the velocity model are well fit, especially for the first half of the time-frame. but the overall fit appears to be fair at best. The variance reduction for this fit is 95.6%. The fitted frequency and damping constant is given in table 1.

The details of the fit of two damped harmonic waves and a steady flow (our "double wave inversion") are shown in table 2 and figures 3-5. The variance reduction of the fit is 98.85%. Wave A has a node at a colatitude of 48° and a rapid change of phase at 60°. Wave B has a constant decrease in phase from the equator to the pole, except for a feature near 40° which might be a poorly resolved node.

6. CONFIDENCE TESTING

We test the significance of the inverted waves from a numerical point of view (following a "Monte Carlo" philosophy): we shall make some reasonable assumptions concerning what a random flow might be like, and then simulate random flows to allow us to derive confidence levels numerically, from the proportion of simulated flow models that are better fit using our inversion technique than the observational flow model. We test our confidence that our inversion scheme provides a better fit to the observational flow model than we would expect for a randomly simulated flow. This is a different question than whether the flow actually consists of a steady flow plus one or two waves, but these questions are clearly related. Since our confidence levels will depend on the assumptions that we make for our simulated flows (simulated flows which have random t_l^0 at each time interval will tend to be worse fit than simulated flows which assume some correlation at different time intervals), we should set a high standard by simulating flows which, although random, we would expect to give rise easily to wave-like behaviour.

6.1. Confidence of Single Wave Inversion

To determine the confidence of the single wave inversion, we create a simulated velocity model by adding 19 random Gaussian velocity packets for each spherical harmonic degree, with power for each degree normalised by that of the observational field model.

Simulated velocity models of this form should provide us with a fairly strict test of confidence for our single wave inversion, first because there is no noise added (which is undoubtedly present in the observational model, even if it is predominantly non-random), second because all of the packets have the same width (which localises the power in the power spectrum close to one frequency), and third because Gaussian velocity packets can very easily add together to "simulate" an oscillation. We would expect simulated velocity models constructed from, for example, random walks to be much less likely to lead to wave-like behaviour.

Our null hypothesis is that a "random" velocity model, constructed as described above, will be at least as well fit by our single wave inversion as the observational flow model.

We assess the null hypothesis by inverting 1000 simulated velocity models. The null hypothesis fails at the 95% confidence level (it would pass at the 95.5% confidence level). The failure of the null hypothesis means that we are confident that we are resolving a real wave.

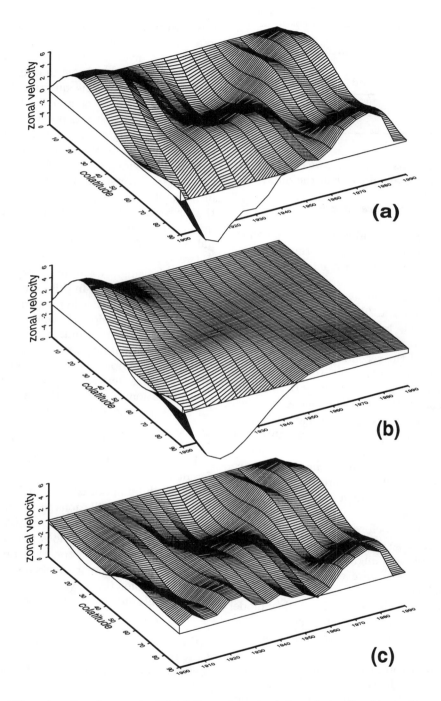

Figure 2. The axisymmetric, equatorially symmetric part of the velocity for the single wave inversion (with the steady part removed). Velocities are given in km/year. (a) is from a flow model estimated from the secular variation of the magnetic field. (b) shows a fit of one damped harmonic waves to the model shown in (a). (c) shows the residuals.

Table 1. Frequency (years^{-1}), period (years) and e^{-1} decay time (years) for the one wave fit to the flow model. A negative real part of the frequency corresponds to decay of the wave.

Frequency			
imag	real	period	decay time
0.0104	-0.0066	96.0	24.0

Given the relative strictness of our test, we can be quite confident that there is a wave present in the data.

6.2. Confidence of Double Wave Inversion

Testing confidence in the double wave inversion is slightly more complicated. It is inappropriate to compare the misfits from a double wave inversion of the observational flow model with the inversion of a random velocity model, as we have evidence that there is at least one wave in the data. Therefore, we construct simulated velocity models by adding random velocity packets to a wave. In this way, we can find the confidence of the inversion of the second wave.

We add together random Gaussian velocity packets with a width of half the period of wave B from the double wave inversion in the same manner as we used for creating the simulated velocity models for the single wave confidence test above. These are normalised for each spherical harmonic degree to have the same power as wave B. Then we add wave A from the double wave inversion. Hence, we are effectively testing our confidence in wave B, compared with a randomly created signal superimposed on wave A.

For the same reasons as before this is a fairly strict test of confidence, because again the random part of the simulated velocity model might be expected to give rise to wave-like behaviour quite easily, and no high frequency noise has been added.

Our null hypothesis in this case is that a flow model constructed as described would be at least as well fit by a double wave inversion as our observational flow model.

In this case our null hypothesis does not quite fail at the 95% confidence level: however, it would fail at the 94.3% confidence level. By changing the assumptions behind our simulated flow model we would be able to find a test which our double wave inversion would pass, but it is more instructive to take this as evidence that although there appears to be a second wave in the observational flow model, we should be suspicious of it, and we should be aware that it may not be well resolved.

6.2.1. Resolution of wave A.
It is worthwhile asking how well a double wave inversion will resolve a single wave. In other words, if there is only one wave present in a core signal but it is partially obscured by a random signal, does inverting for two waves provide a better or worse picture of that wave than inverting for one?

We can examine this question by examining simulated flows which we created for the confidence testing of our double wave inversion. We can invert these flows using both a single wave inversion and a double wave inversion. It turns out that the double wave inversion generally provided a better picture of wave A than the single wave inversion. An example of this can be seen in figure 6. The single wave inversion produces a wave that is broadly similar to wave A, but the double wave inversion produces a wave that is similar to the original wave A (wave A') and a wave that is unlike wave A (wave B'). Fortunately, the similarity between wave A' and the single wave inversion allows us to identify wave A' with wave A. In this case we have assumed that there is no second wave in the core, and wave B from the double wave inversion of the observational flow model arises merely from a random part of the signal. This is a worst-case scenario: if there actually is a second wave, then this will improve the resolution of the first wave in a double wave inversion. A description of the effects of contamination of the inverted waves on the inversions for r.m.s. B_s and F is reported in *Zatman* [1997].

7. INVERSION FOR R.M.S. B_s AND F

Figure 7 shows the models of r.m.s. B_s for various different inversions. The solution for the wave A only inversion exhibits poor convergence at colatitudes around 58° - there are oscillations in the fitted r.m.s. B_s from 56° to 60°. These can be reduced by increasing the spline density, and this does not change the solutions outside the region of poor convergence. A plot of the errors for the different inversions show that the fit is much better for the wave A only inversion than for the wave B only inversion: we believe that this is because

Table 2. Frequencies (years^{-1}), periods (years) and e^{-1} decay times (years) for the two wave fit to the flow model. A negative imaginary part of the frequency corresponds to decay of the wave.

Wave	Frequency		period	decay time
	imag	real		
A	0.0131	-0.0021	76.2	74.1
B	0.0189	-0.0017	52.7	95.7

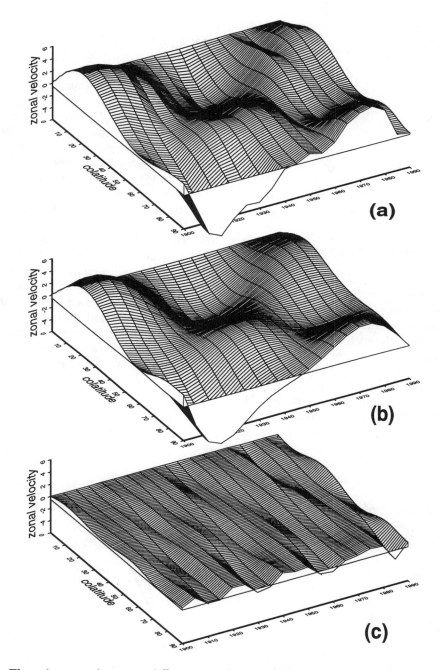

Figure 3. The axisymmetric, equatorially symmetric part of the velocity for the double wave inversion (with the steady part removed). Velocities are given in km/year. (a) is from a flow model estimated from the secular variation of the magnetic field. (b) shows a fit of two damped harmonic waves to the model shown in (a). (c) shows the residuals.

wave A is better determined. We have much greater confidence in our inversion of wave A. As the magnitude of wave A is greater than that of wave B, it is more important for the fit in the joint inversion, which consequently resembles wave A more than wave B.

Figure 8 shows the effect of excluding different portions of the core from the inversion. If the starting point of the inverted region is moved outside of the tangent cylinder, the inverted model is simply truncated - it recapitulates the "full" inversion between the tangent

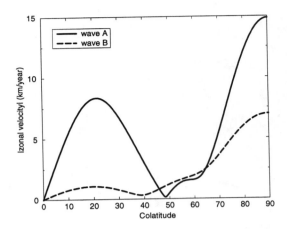

Figure 4. The amplitudes of the two fitted waves.

cylinder and equator, merely excluding the part next to the tangent cylinder. However, if the starting point is moved inside the tangent cylinder, the model diverges from that which merely excludes the tangent cylinder. Since we do not include any inner core-outer core coupling in the dynamics, we would expect our assumptions to fail within the tangent cylinder if the inner core is important in the dynamics of the tangent cylinder. This is to be expected if there is significant dynamic coupling between the inner core and the fluid within the tangent cylinder. The divergence between models that include different portions of the tangent cylinder suggests that the inner core does indeed play an important role in the dynamics there.

Figure 9 shows the effect of varying α. This corresponds with allowing some of the B'_ϕ produced by shearing of B_s within the core to leak into the bottom

of the mantle. $\alpha = 1$ would correspond to B_ϕ fully penetrating the CMB. The major effect of increasing α is to increase the fitted r.m.s. B_s near the tangent cylinder. Since, a priori, we know very little about conductivity at the base of the mantle, we should not assume a particular value of α.

Figures 10 and 11 show the coefficients of friction $F(s)$ for the different inversions. The dominant feature of the frictions from the fit to wave A is the peak close to 60°. This peak is associated with a fairly rapid change of the phase of wave A in figure 5; however, at least with the wave A only inversion, the fitted peak is narrower than the band over which the phase change occurs, and also the resolution of the input velocity and field models. This is also true for the nearby change in r.m.s. B_s that may be seen in figures 7-9. We did not include any damping in the inversions presented, so this narrowness may well be an artifact of the inversion. If the number of splines in the inversion is significantly decreased then the peak widens and flattens in such a way that the area under the peak remains approximately constant, although until the spline density is reduced by a factor of around two the peak remains largely unchanged.

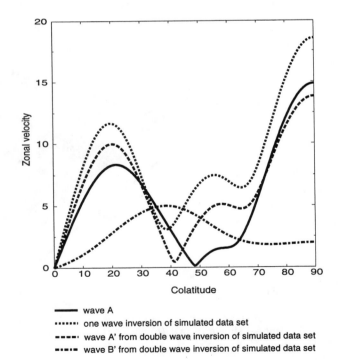

wave A
••••••• one wave inversion of simulated data set
------ wave A' from double wave inversion of simulated data set
-·-·-· wave B' from double wave inversion of simulated data set

Figure 6. Amplitudes of wave A, and the waves from the single wave and double wave inversions of a simulated velocity model where a random signal has been added to wave A.

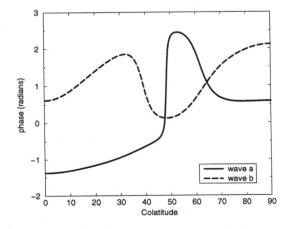

Figure 5. The phases of the two fitted waves.

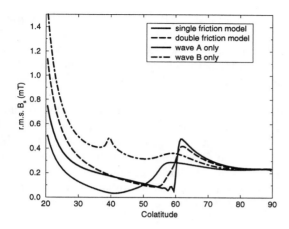

Figure 7. Root mean square B_s averaged over axial cylinders (as in figure 1). Here we only invert between the tangent cylinder (the axial cylinder tangent to the inner core) and the equator.

However, it seems reasonable to suspect that the models may not properly resolve the behavior in this area. Changing the number of splines does not significantly alter the behaviour elsewhere, which tends to be quite smooth: this is because $\kappa\, \partial \overline{B_s^2}/\partial s$ in equation 19 turns out to be small except close to 60°.

The frictions for wave A in the single and joint inversions are similar, except that the peak is tighter in the single inversion. The frictions fitted for wave B for the single and joint inversions, on the other hand, are very different, although in both cases they are negative at mid latitudes and positive closer to the tangent cylinder. This may be due to poor resolution of wave B. The single friction model is unlike any of the other models (although it does have a weak peak around 60°). This is consistent with the fact that the friction models for wave A are very different from those for wave B, so that a single friction model is unable to represent both.

It seems reasonable to ask to what extent the inverted models of r.m.s. B_s depend on the friction model. Figure 12 shows the r.m.s. B_s inverted from wave A with and without inversion of a friction model. There are clearly major differences between the models, especially approaching the tangent cylinder and in the abruptness of the increase of r.m.s. B_s at around 60°. Nonetheless there are similarities: there is a peak of r.m.s. B_s in both cases at around 60°, the drop-off is much more sharp pole-wards, and there is an increase of r.m.s. B_s close to the tangent cylinder.

If $F = 0$, then for standing waves the phase of $\mathcal{R}$ should be constant. It is clear from figure 5 that even if we correct for the node at 48° there is variation in the phase of wave A. The phase lead of the central portion is consistent with relative excitation of the central portion (or, conversely, relative damping of the poles and equator). Further discussion of the importance of constraints on the net torque for the inversion can be found in *Zatman* [1997].

8. CONCLUSIONS AND DISCUSSION

We place most confidence in our inversions for wave A alone rather than in those involving wave B, as we

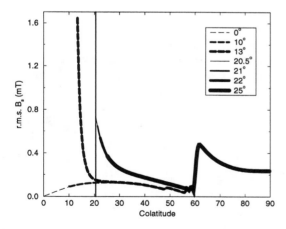

Figure 8.   Root mean square B_s averaged over axial cylinders. Here we vary the position of the inner boundary of the inverted region. Lines of different thickness mark different inversions (inner boundary set at colatitudes 0°, 10°, 13°, 20.5°, 21°, 22° and 25°). Also shown is the position of the tangent cylinder. Inversions which start at or outside the tangent cylinder lie on the same line, whereas inversions which start inside the tangent cylinder vary from this line, and may vary from each other.

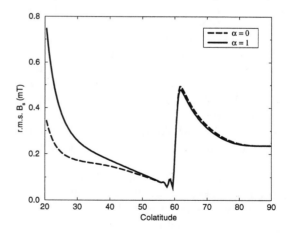

Figure 9.   Root mean square B_s averaged over axial cylinders, showing the effects of a change in α.

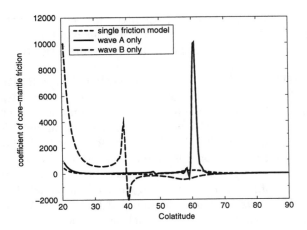

Figure 10. The coefficients of core-mantle friction fitted for the single friction inversion, and those fitted for wave A and wave B in the individual inversions.

think it is likely that wave B is relatively poorly resolved away from the equator. Instead of presenting a single model for r.m.s. B_s within the core, we prefer to note the dependence on α and the consequent variation in figure 9.

We have most confidence in the features of r.m.s. B_s that are most robust between inversions, namely a model of B_s which overall has a magnitude of $\sim 10^{-4}$ T, is at a minimum at midlatitudes, rising towards the tangent cylinder and towards a colatitude of 60° from both directions.

As there is significant variation in mantle angular momentum on decadal timescales, and in view of the relatively rapid spin down of the wave (shown in table 2), we would expect both friction and excitation to be important for the dynamics of torsional oscillations. However, the inversions of wave A with and without friction models both show the general characteristics noted above. The friction model for wave A suggests that excitation may be important, particularly in the vicinity of the peak of the model near the colatitude of 60°.

Torsional oscillations might be excited by the evolution of an underlying magnetostrophic flow. Variations in this flow would require adjustment to maintain a magnetostrophic balance. Axisymmetric torsional oscillations would be one form of adjustment, where unbalanced torques on axial cylinders would be balanced by inertia. If small changes in the underlying magnetostrophic state were to lead to large unbalanced torques, then torsional oscillations would be efficiently excited even by slight changes in the underlying state.

This is reasonable if the Lorentz torque largely cancels when integrated over axial cylinders. This would

happen in a "quasi-Taylor state", where the Lorentz torque is locally large, but cancellation occurs so that Taylor's constraint is satisfied. Small changes could then lead to large violations of Taylor's constraint, requiring adjustment via the excitement of torsional oscillations.

Another possible configuration for the geodynamo is that of a "model-Z" state [*Braginsky*, 1975, 1978, 1988], where the Lorentz torque integrated on axial cylinders is small because B_s is small rather than because cancellation of a relatively large torque occurs. In this case, friction at the core-mantle boundary balances the remaining Lorentz torque, at it continues to remain important for the field and flow even as the viscosity becomes asymptotically small, as the magnitude of B_s also becomes asymptotically small.

A model-Z dynamo could support torsional oscillations (although if core-mantle friction is high they would be heavily damped or overdamped). However, if changes in the axial torque on cylinders are primarily balanced by core-mantle friction rather than inertia, then it should be more difficult for torsional oscillations to be excited: rather, adjustments of the basic state should excite rotations of axial cylinders that would vary on the same timescale as the basic state (i.e. magnetostrophic timescales). If the magnetostrophic timescale of the core is much shorter than that expected (e.g. from equation 1), then perhaps the associated zonal motions might appear similar to the oscillations that we observe. However, given the form and period of the waves, we think that it is much more plausible that they reflect torsional oscillations.

The overall magnitude of B_s that we obtain is similar to that of the poloidal field at the CMB and to the

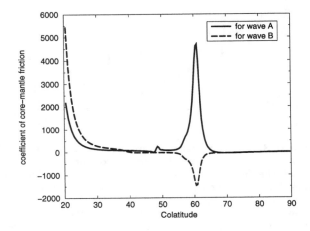

Figure 11. The coefficients of core-mantle friction fitted for wave A and wave B in the double friction inversion.

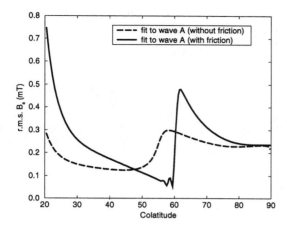

Figure 12. Fitted root mean square B_s averaged over axial cylinders, for inversions of wave A with and without a friction model.

rough estimate of Brgainsky of 0.2 mT. It seems reasonable to suppose that this is indicative of the strength of the poloidal field in the core, unless the poloidal field has a tendency to be oriented in a model-Z like configuration, in which case it might be much larger in magnitude than our r.m.s. B_s - although, as we indicate, we think that this is unlikely. *Zatman and Bloxham* [1997] suggest that the increase in magnetic field towards the tangent cylinder may be related to a patch of reversed flux on the CMB within the tangent cylinder in the northern hemisphere: such a patch requires that the field turn around within the core, which should be associated with extra B_s. Alternatively, the increase may be associated with the presence of the inner core. The increased magnitude of B_s in the band around the equator (for colatitudes between 60° and 120°) may similarly be associated with the band of high and reversed flux patches in that region.

Acknowledgments. We thank Richard Holme and Bruce Buffett for their useful comments and insight, and for improving the readability of the paper. This material is based on work supported by NASA grant NAG5-2971 and under a National Science Foundation Graduate Research Fellowship (SZ) and a Presidential Young Investigator Award (JB).

REFERENCES

Backus, G. E., and F. Gilbert, Geophys. J. R. Astron. Soc. *13*, 247-276, 1968.

Bloxham, J., Time-dependent Geomagnetic Field Models and Flow at the Core Surface, EOS trans. AGU, *76*, F164, 1995.

Bloxham, J., and A. Jackson, Fluid flow near the surface of the of Earth's outer core, Rev. Geophys., *29*, 97-120, 1991.

Bloxham, J., and A. Jackson, Time-dependent mapping of the magnetic field at the core-mantle boundary, J. Geophys. Res., *97*, 19537-19563, 1992.

Braginsky, S. I., Torsional Magnetohydrodynamic Vibrations in the Earth's Core and Variations in Day Length, Geomag. Aeron., *10*,, 3-12, 1970.

Braginsky, S. I., Nearly Axisymmetric Model of the Hydromagnetic Dynamo of the Earth I, Geomag. Aeron., *15*,, 122-128, 1975.

Braginsky, S. I., Nearly Axisymmetric Model of the Hydromagnetic Dynamo of the Earth II, Geomag. Aeron., *18*,, 225-231, 1978.

Braginsky, S. I., Short-period geomagnetic secular variation, Geophys. Astrophys. Fluid Dynam., *30*,, 1-78, 1984.

Braginsky, S. I., A Model-Z Geodynamo with Magnetic Friction, Geomag. Aeron., *28*,, 407-412, 1988.

de Boor, C., *A Practical Guide to Splines*, Springer-Verlag, New York, 1978.

Fearn, D. R., and M. R. E. Proctor, Magnetostrophic balance in non-axisymmetric, non-standard dynamo models, Geophys. Astrophys. Fluid Dynam., *67*, 117-128, 1992.

Gubbins, D., Finding core motions from magnetic observations, Phil. Trans. R. Soc. Lond. A, *73*, 641-652, 1982

Gubbins, D., and J. Bloxham, Geomagnetic Field Analysis-III. Magnetic Fields on the Core Mantle Boundary, Geophys. J. R. Astr. Soc., *80*, 695-713, 1985

Gubbins, D., Geomagnetic Field Analysis-I. Stochastic Inversion, Geophys. J. R. Astr. Soc., *73*, 641-652, 1983

Gubbins, D., and P. Kelly, A difficulty with using the frozen flux hypothesis to find steady core motions, Geophys. Res. Lett., *73*, 1825-1828, 1996

Gubbins, D., and P. H. Roberts, Magnetohydrodynamics of the Earth's Core, in *Geomagnetism, Volume 2*, edited by J. A. Jacobs, pp. 1-183, Academic Press, 1987.

Hide, R., Interaction Between the Earth's Liquid Core and Solid Mantle, Nature, *222*, 1055-1056, 1969.

Hills, R. G., Convection in the Earth's mantle due to viscous shear at the core-mantle interface and due to large-scale buoyancy, PhD. thesis, N. M. State Univ., Las Cruces, 1979.

Jackson, A., Time dependency of geostrophic core-surface motions, Phys. Earth Planet. Int., *103*, 293-311, 1997.

Jackson, A., J. Bloxham and D. Gubbins, Time-Dependent Flow at the Core Surface and Conservation of Angular Momentum in the Coupled Core-Mantle System, in Dynamics of Earth's Deep Interior and Earth Rotation, edited by J.-L. Le Mouël, D. E. Smylie and T. Herring, pp. 97-107, Geophysical Monograph 72 IUGG Volume 12, 1993.

Jault, D., and J.-L. LeMouël, The topographic torque assiciated with a tangentially geostrophic motion at the core surface and inferences on flow inside the core, Geophys. & Astrophys. Fluid Dyn., *48*, 273-296, 1989.

Kahle, A. B., R. H. Ball, and E. H. Vestine, Estimated Surface Motions of the Earth's Core, J. Geophys. Res., it 72, 4917-4925, 1967.

Langel, R. A., The Main Field, in *Geomagnetism, Volume 1*, edited by J. A. Jacobs, pp. 249-512, Academic Press, 1987.

LeMouël, J.-L., Outer core geostrophic flow and secular variation of Earth's magnetic field, Nature, *311*, 734-735, 1984

Moffatt, H. K., *Magnetic field generation in electrically conducting fluids*, C.U.P., 1978.

Shure, L., R. L. Parker, and G. E. Backus, Harmonic Splines for Geomagnetic Modelling, Phys. Earth Planet. Int., *28*, 215-299, 1982.

Stix, M., and P. H. Roberts, Time-dependent electromagnetic core-mantle coupling, Phys. Earth Planet. Int., *36*, 49-60, 1984.

Taylor, J. B., The Magnetohydrodynamics of a Rotating Fluid and the Earth's Dynamo Problem, Pro. R. Soc. Lond. A., *274*, 274-283, 1963.

Whaler, K. A., Does the whole of the Earth's core convect?, Nature, *287*, 528-530, 1980.

Zatman, S. A., Angular Momentum and the Core of the Earth, PhD. thesis, Harvard University, Cambridge, Massachusetts, 1997.

Zatman, S. A., and J. Bloxham, Torsional oscillations and the magnetic field within the Earth's core, Nature, *388*, 760-763, 1997. .

S. A. Zatman, Code 921, NASA Goddard Space Flight Center, Greenbelt, MD 20771. (e-mail: zatman@denali.gsfc.nasa.gov)

J. Bloxham, Department of Earth and Planetary Sciences, Harvard University, Cambridge, MA 02138. (e-mail: bloxham@geophysics.harvard.edu)

Numerical Dynamo Modeling: Comparison with the Earth's Magnetic Field

Weijia Kuang[1] and Jeremy Bloxham

Department of Earth and Planetary Sciences, Harvard University, Cambridge, Massachusetts

The geodynamo operates in a parameter regime that is not directly accessible to numerical models. In order to investigate the geodynamo numerically we must instead perform calculations in that part of parameter space that is numerically accessible and introduce appropriate assumptions so as to achieve a dynamical regime that is similar to that of the geodynamo. Here we show how the results of one such numerical study can be compared to the Earth's magnetic field: we describe the leading order force balance of our solution and how that balance can be used to scale our results to the regime of the geodynamo. We find encouraging similarities between our dynamo model and observations of the geomagnetic field: the field strength and spectrum at the core-mantle boundary are similar; the secular variation includes a westward drift, episodes of flux expulsion, and behaviour similar to the Pacific dipole window observed in the historical field; and the flow at the core surface is similar to that inferred from the secular variation.

1. INTRODUCTION

The geodynamo, which operates in the Earth's fluid outer core and which is responsible for the generation of the Earth's magnetic field, is particularly challenging to study. The governing equations cannot be solved analytically, and fully three-dimensional numerical models *Glatzmaier and Roberts* [1995a, 1995b, 1996a, 1996b]; *Kuang and Bloxham* [1997b, 1997a] are restricted to that part of parameter space that is numerically accessible, which is not necessarily that which is attained in the Earth's core.

Despite this, these models have enjoyed surprising success in reproducing the basic dipolar structure of the Earth's magnetic field and the westward drift of the field. In this paper, we begin by focusing on the problems that arise in attempting to relate numerical dynamo models to geomagnetic observations, and then having established the extent to which such comparisons can be made, we compare the results of our geodynamo model with observations.

The basic equations are the magnetic induction equation which describes the evolution of the magnetic field $\boldsymbol{B}$, the momentum equation which describes the evolution of the fluid flow $\boldsymbol{u}$, and the energy equation which describes the evolution of the temperature T

$$(\partial_t + \boldsymbol{u} \cdot \nabla)\boldsymbol{u} + 2\boldsymbol{\Omega} \times \boldsymbol{u} =$$
$$-\nabla p + \frac{1}{\rho_0} \boldsymbol{J} \times \boldsymbol{B} + \frac{\rho}{\rho_0} \boldsymbol{g} + \nu \nabla^2 \boldsymbol{u} \quad (1)$$

$$\partial_t \boldsymbol{B} = \nabla \times (\boldsymbol{u} \times \boldsymbol{B}) + \eta \nabla^2 \boldsymbol{B} \quad (2)$$

$$(\partial_t + \boldsymbol{u} \cdot \nabla)T = \kappa \nabla^2 T + Q + Q_{\text{diss}} \quad (3)$$

where $\boldsymbol{\Omega}$ is the Earth's rotation, p is the pressure (including the centrifugal contribution), ρ_0 is the mean density, $\boldsymbol{g}$ is the gravitational acceleration, ν is the kinematic viscosity, η and κ are the magnetic and thermal diffusivities respectively, Q is the internal heating, and Q_{diss} is the heating from dissipative processes, in this case Ohmic heating.

To non-dimensionalize the system, we chose the core radius r_0 as our lengthscale, the magnetic diffusion time

The Core-Mantle Boundary Region
Geodynamics 28
Copyright 1998 by the American Geophysical Union.

r_0^2/η as our timescale, $\mathcal{B} = \sqrt{2\Omega\mu\rho_0\eta}$ as our magnetic field scale, and $h_T r_0$ as the temperature scale (where h_T is the imposed temperature gradient at the inner core boundary), giving

$$R_o\left(\partial_t + \boldsymbol{u}\cdot\nabla\right)\boldsymbol{u} + \hat{\boldsymbol{z}}\times\boldsymbol{u} =$$
$$-\nabla p + \boldsymbol{J}\times\boldsymbol{B} + E\nabla^2\boldsymbol{u} + Ra\,\Theta\hat{\boldsymbol{r}}$$
$$\partial_t\boldsymbol{B} = \nabla\times(\boldsymbol{u}\times\boldsymbol{B}) + \nabla^2\boldsymbol{B} \qquad (4)$$
$$\left(\partial_t + \boldsymbol{u}\cdot\nabla\right)\Theta =$$
$$-\boldsymbol{u}\cdot\nabla T_0(r) + q_\kappa\nabla^2\Theta + \frac{\alpha_J}{Ra}|\boldsymbol{J}|^2 \qquad (5)$$

where Θ is the temperature relative to a well-mixed reference state $T_0(r)$. We have used the following non-dimensional numbers: the Ekman number E, the magnetic Rossby number R_o, the Prandtl number q_κ, and the heating number α_J, given by

$$E \equiv \frac{\nu}{2\Omega r_o^2} \qquad R_0 \equiv \frac{\eta}{2\Omega r_o^2}$$
$$q_\kappa \equiv \frac{\kappa}{\eta} \qquad \alpha_J \equiv \frac{r_0\alpha_T g_0}{c_p} \qquad (6)$$

where g_o is the gravitational acceleration at r_o and c_p is the specific heat capacity, and the Rayleigh number, Ra, which scales the buoyancy force that drives the flow

$$Ra \equiv \frac{\alpha_T\,g_o\,h_T\,r_o^2}{2\Omega\eta} \qquad (7)$$

There are a number of obvious advantages to nondimensionalizing the governing equations. In particular, given that the parameter regime of the Earth's core is not accessible numerically, nondimensionalizing the system helps elucidate the compromises that are inevitable in numerical geodynamo calculations.

Two nondimensional numbers are of a particular interest: the Ekman number which scales the viscous term and the magnetic Rossby number which scales the inertial terms in the momentum balance. In the Earth's core, these are both small, suggesting that viscous and inertial effects are small in the core. In the limit in which these terms vanish, the core is in a magnetostrophic balance, in which the Lorentz force, the Coriolis force and the buoyancy force balance. In an important paper, *Taylor* [1963] examined the degeneracy that must then result, and showed that for solutions to exist, the magnetic field must be configured so that the axial Lorentz torque integrated over the surface of any cylinder in the core coaxial with the rotation axis vanishes. Small departures from this configuration, which is called a Taylor state, result in torques which must be balanced at next order by re-introducing inertia, resulting in torsional oscillations about a Taylor

state. Most probably this is the dynamical regime of the geodynamo: the issue that we must face is how best to model this regime numerically.

The principal problem is the smallness of the Ekman number. Viscous dissipation controls the smallest lengthscales in the flow, and so we are severely limited by computational constraints in how small we can make the Ekman number. The critical issue is whether the Ekman numbers that are computationally feasible are sufficiently small to render viscous effects small both at leading order, to obtain a leading order magnetostrophic balance, and at higher order, so that the small departures from a Taylor state result are balanced by inertia, rather than by viscous drag, thus resulting in oscillations about a Taylor state.

Viscous effects are important in boundary layers, in particular in Ekman boundary layers at the inner core and core-mantle boundaries. Computational constraints limit the spatial resolution, even if a variable grid-spacing or spectral representation is used near the boundaries, to say 10^4m; this constrains the smallest effective Ekman number to be about 10^{-5}. With this constraint, it is not clear that we will be able to achieve our goal described above.

Smaller Ekman numbers may be achieved if a hyperviscosity is employed [*Glatzmaier and Roberts*, 1995a]. A hyperviscosity is an artificial enhancement of the viscosity with wavenumber in order to promote numerical convergence, of the form

$$\nu = \nu_0 \qquad\qquad l \le l_0$$
$$\nu = \nu_0(1 + \epsilon(l - l_0)^n) \qquad l > l_0$$

where l is spherical harmonic degree in the spectral expansion of the velocity. Similar enhancements of the other diffusivities can be employed. Here we use $l_0 = 0$, $\epsilon = 0.05$ and $n = 2$, weaker than the values $l_0 = 0$, $\epsilon = 0.075$ and $n = 3$ used by *Glatzmaier and Roberts* [1995a]. The Ekman number calculated on the basis of ν_0 will then be significantly smaller than the effective Ekman number, which is probably best calculated on the basis of the viscosity corresponding to the mean wavenumber of the flow. Additionally, *Glatzmaier and Roberts* [1995a] increased the viscosity in their calculations by a further factor of 10 near the core-mantle boundary adding further disparity between the Ekman number based on ν_0 away from the boundaries and the effective Ekman number.

In order to reduce the role that viscous effects play in our numerical model, we apply viscous stress-free boundary conditions at the rigid boundaries instead of no-slip conditions. This eliminates the Ekman boundary layers. Then, departures of the core from a Taylor state will be balanced primarily by inertia since the viscous drag at the boundaries will be zero. A small viscous drag still arises due to internal rather than boundary viscous effects, but it is much smaller, by a factor $\mathcal{O}(E^{\frac{1}{2}})$, than that which would arise at the boundaries.

However, this does place a restriction on our choice of magnetic Rossby number. Unlike the solutions of Glatzmaier and Roberts in which no-slip boundary conditions are imposed resulting in thick Ekman layers at the boundaries and in which, as a result, any oscillations about a Taylor-state are rapidly damped, we have to be concerned with resolving the period of such oscillatory motions in our solutions, because they are much more weakly damped.

This restricts us to values of the magnetic Rossby number much larger than that in the core, but values which are still small. Provided that inertia effects are unimportant at leading order in our solutions, then this restriction on our choice of magnetic Rossby number will not be serious, and we will be able to scale our results to the Earth using the leading order momentum balance. We discuss this in more detail later.

Our choice of magnetic Rossby number is consistent, however, with another requirement, namely that since we are appealing to turbulent processes to enhance the diffusivities, then the ratio of the Ekman number to the magnetic Rossby number should be unity, so that the magnetic Prandtl number (the ratio of the kinematic viscosity to the magnetic diffusivity) is unity. In fact, given that dynamo calculations are limited, as we have discussed above, to an effective Ekman number of $\mathcal{O}(10^{-5})$, the magnetic Rossby number is similarly limited if the requirement that the magnetic Prandtl number is unity is to be honoured.

Given the choice between having viscous effects play too strong a role but being able to choose a reasonable magnetic Rossby number on the one hand and having inertial effects playing too strong a role but being able to reduce viscous effects to an inconsequential level on the other, we believe that the latter is the better choice.

We use the following values of the non-dimensional numbers: Rayleigh number $Ra = 1.5 \times 10^4$, Ekman number $E = 2 \times 10^{-5}$, magnetic Rossby number $R_0 = 2 \times 10^{-5}$, Prandtl number $q/\kappa = 1.0$, and heating number $\alpha_J = 0.5$.

2. COMPARISON WITH THE EARTH'S FIELD: THEORY AND SCALING

The solution that we describe here has been followed for over eight magnetic decay periods. The solution was slow to evolve to its current configuration, taking roughly three magnetic decay periods for the transient

behaviour to relax fully. During this transient period, the field reversed once and underwent several excursions. However, for the last five decay times the field has settled into a predominantly dipolar field configuration with no indications of possible reversal or excursion. The structure of the flow within the core can be inferred, in part, from Plate 1 in which we show the density variations that drive the convective flow.

In order to compare this solution with the Earth's magnetic field we need to re-dimensionalize the results. This is not entirely straightforward since the solution, as we have discussed, does not correspond to the Earth's nondimensional numbers. Thus, we cannot simply ascribe values for the dimensional parameters (such as the viscosity, the rotation rate, etc.) and match both the values that are appropriate for the Earth and the set of nondimensional numbers that we have used. This is inescapable in any numerical dynamo calculation with current limits on computational resources.

However, we can nonetheless make a sensible scaling to the Earth's field by appealing to the dynamical balance that holds in this solution and in the Earth's core. As we have discussed, the leading order balance in the Earth's core is a magnetostrophic balance. We find the same balance holds in our solution (Kuang and Bloxham, ms. in preparation). Thus at leading order, in both the Earth's core and in our solution we have

$$\hat{z} \times u = -\nabla p + J \times B + Ra\,\Theta\hat{r} \qquad (8)$$

To re-dimensionalize the magnetic field we need simply to multiply by the magnetic field scale $\mathcal{B} = \sqrt{2\Omega\mu\rho_0\eta}$. $\mathcal{B}$ depends on the rotation rate Ω, the magnetic permeability μ, the mean density ρ_0 and the magnetic diffusivity η, each of which we can set equal to their respective estimates in the Earth's core. In so doing, we have also set the magnetic decay time, enabling us to re-dimensionalize the timescale in our solution.

The Rayleigh number, which we specify and which is defined by (7), depends upon the coefficient of thermal expansion α_T, the gravitational acceleration g_0, the radius of the core r_0, the rotation rate Ω, the magnetic diffusivity η, and the temperature gradient at inner core-boundary h_T. Thus we can calculate the heat flux for our solution in dimensional units by using our specified value of the Rayleigh number and setting the remaining parameters equal to their value in the Earth's core. This corresponds to a super-adiabatic heat flux of roughly 10^{12} W, which is geophysically reasonable.

The values of the parameters that we have used above in re-scaling our results are not consistent with our choice of magnetic Rossby number and, if the magnetic

Prandtl number is one, with our choice of Ekman number. This is not a problem if we consider the leading order behavior of our solution, since as we have shown the leading order force balance does not depend on either inertial or viscous effects, and these are the only terms that are scaled by these nondimensional numbers. However, if we wish to examine higher order effects then we have to be more careful: for example, with this scaling we cannot, unless we appeal to an asymptotic theory enabling us to extrapolate our results, estimate either the period of oscillations about a Taylor state or the damping time of such oscillations.

It is perhaps surprising that the magnetic Rossby number, which is the ratio of the rotation period to the magnetic decay time, is unimportant at leading order. Even though, in dimensional terms, we are studying the evolution of the system on the magnetic decay timescale and even though the Coriolis force depends on the rotation rate, the ratio of these is unimportant at leading order. This is easily understood, however, when one realizes that this ratio is only important in mapping the timescale of inertial oscillations that depend on the rotation period to the magnetic decay time, which is the timescale that we have chosen to use to nondimensionalize the system. In effect, we can identify two separate roles that the rotation rate plays in our solution: at leading order it plays a role in determining the magnetic field strength; and at higher order it plays a role in determining the timescale of inertial oscillations. The first role we model correctly, the second we do not.

3. COMPARISON WITH THE EARTH'S FIELD: RESULTS

3.1. The Main Field

In Plate 2 we show the radial magnetic field at the core-mantle boundary from our dynamo model, and for comparison the average field inferred from observations over the last 150 years, from *Bloxham and Jackson* [1992]. The field from our dynamo model is similar to that inferred from observations, with a predominantly dipolar structure, several patches of reverse flux, and concentration of the flux at higher latitudes into two lobes that are positioned nearly symmetrically about the magnetic equator. The rms radial field strength is 3.05 mT, compared to 3.04 mT for the magnetic field in 1980 inferred from *Magsat* [*Gubbins and Bloxham*, 1985] and 2.85 mT for the average field strength over the last 150 years. The energy in the radial field at the core-mantle boundary is partitioned slightly more than 80% into odd modes (i.e. modes that are antisymmetric about the equator) compared to that inferred from observations where the odd modes account for roughly

75% of the energy (with both models truncated at degree 14).

We show, in Plate 3, the power spectrum of the field at the core-mantle boundary over the last 1500 years of this run and of the field inferred from observations over the last 300 years. The spectra are similar, though the average power in degrees 3–6 from the dynamo model is less than that inferred from recent observations, although of these only for degrees 4 and 5 does the dynamo model average spectrum not fall within the observed variability of the observed spectrum. This lack of power in these degrees is also apparent in Plate 2 where the field from the dynamo model is a little less complicated than the observed field.

A number of features in the recent observed field are absent from the dynamo model: we do not see small patches of very low flux near the geographic poles in the dynamo model, though such features are persistent in the observed field over the last few hundred years; patches of reversed flux are confined in the dynamo model to a belt within 30° of the equator, whilst in the observed field they also occur at higher latitudes; and the maxima of radial flux are slightly weaker in the dynamo model than in the observed field. The dynamo model, of course, represents a gross simplification of the Earth's core and it is possible that adding additional complexity to the model may result in some of these features appearing, but it is also possible that they will result simply from varying the Rayleigh number, or even from just allowing the solution to evolve for a longer period of time.

3.2. The Secular Variation

We plot the instantaneous secular variation of the dynamo model field in Plate 4 (for the same epoch as Plate 2). The rms secular variation, again truncated to degree 14, is roughly 900 nT/yr, or about half that inferred from observations by *Bloxham and Jackson* [1992]. However, the strength of the secular variation at the core-mantle boundary is poorly constrained by observations, with different models varying by a factor of 3 while providing good fits to the data. On the other hand, the vigour of the secular variation might be reduced in our dynamo model because the magnetic Rossby number is too large so that torsional oscillations have too long a period.

Interestingly, the secular variation shows a pronounced asymmetry between the western half of the plot where the secular variation is vigorous and the eastern half where it is considerably weaker. Note that this asymmetry is also apparent in Plate 2 where the field is substantially more complicated in the western than in the eastern hemisphere. This asymmetry is obviously not

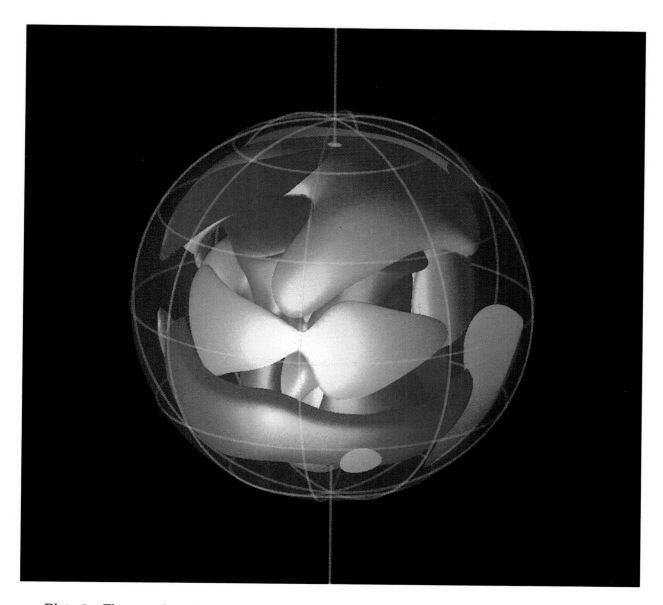

Plate 1. The non-spherically symmetric part of the density variations that drives the flow. The orange colored isosurface encloses less dense (hotter) fluid, the blue colored isosurface encloses denser (cooler) fluid.

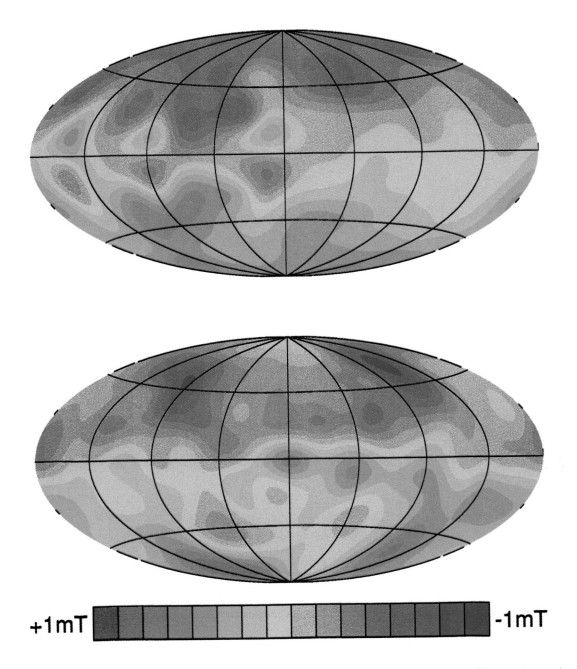

+1mT   -1mT

Plate 2. The radial component of the magnetic field at the core-mantle boundary. The top image is from the dynamo model; the lower image the average field for the period 1840–1990 from the model of *Bloxham and Jackson* [1992].

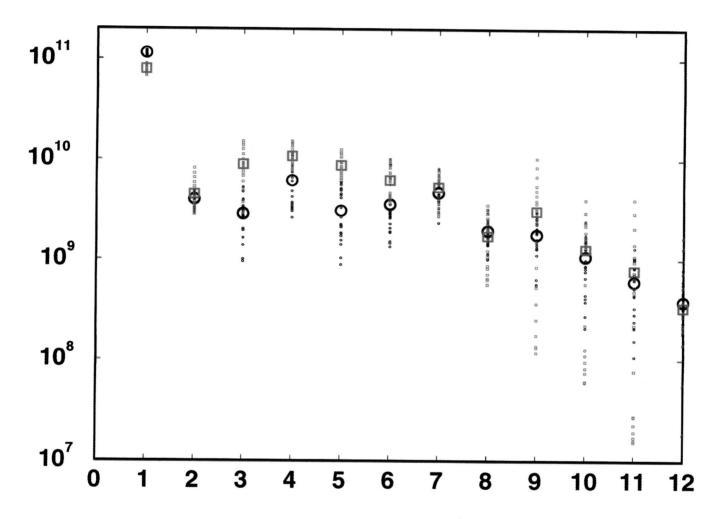

Plate 3.   The power spectrum at the core-mantle boundary as a function of spherical harmonic degree. The scale is in units of $(nT)^2$. The small black circles are from the dynamo model over a 1500 year period, sampled every 75 years, with the larger circles showing the mean over the interval. The red squares are from the field models of *Bloxham and Jackson* [1992] sampled every ten years, with the larger squares showing the mean.

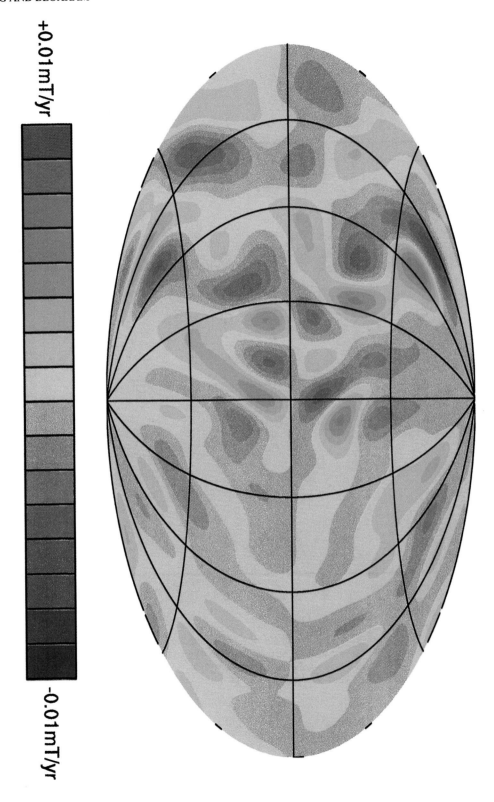

+0.01mT/yr

-0.01mT/yr

Plate 4. The radial component of the secular variation at the core-mantle boundary.

an axisymmetric effect, and so it is unlikely to be an artifact due to torsional oscillations of too long a period. It is similar to recent behavior seen in the observed field: currently the field is simpler and the secular variation weaker beneath the Pacific hemisphere, as first noted by *Fisk* [1931]. There has long been evidence that this Pacific dipole window or Pacific secular variation low has persisted over a much longer timescale (order 10^6 years) than that covered by direct observations [*Doell and Cox*, 1961]; however, not all later studies have supported that result, as recently reviewed by *McElhinny et al.* [1996]. *Runcorn* [1992] hypothesized that this anomalous behaviour of the recent observed field beneath the Pacific and its possible existence over much longer time scales results from an electrically highly conducting layer at the base of the mantle which screens the secular variation.

Our dynamo model has homogeneous boundary conditions, and a uniformly conducting layer at the base of the mantle. An interesting question then is examine for how long this pattern of secular variation persists in the dynamo model. This, and other questions, are most usefully addressed by examining the secular variation over a longer time interval. In Plate 5 we show the evolution of our dynamo model over a 675 year interval; Plates 2 and 4 correspond to 150 years after the start of this sequence.

A number of features are apparent. First, we see a clear ingredient of westward drift in the secular variation, especially in equatorial regions beneath the western hemisphere. The rate of westward drift is about $0.2°$/year, similar to that observed for the present geomagnetic field. However, as is the case for the Earth, the secular variation is more complicated than a simple westward drift, and includes features, such as changes in null flux curve topology, that cannot be explained by simple westward advection of the field. In particular, at the eastern edge of the western half of the plots (in other words near the center of the plots) we observe two cycles of flux expulsion, in which two adjacent regions of oppositely signed flux emerge, grow in intensity, drift westward, and eventually decay. The first cycle begins between years 75 and 150; the second around year 375. At the end of the sequence there does not appear to be a further repetition of this cycle.

This process of flux expulsion and strong secular variation in one hemisphere appears to be related to the simple field morphology and low secular variation in the other hemisphere. At the start of the sequence of plots in Plate 5, and again at the end, the east-west asymmetry in the field and secular variation is much weaker. The Pacific dipole window in the dynamo

model is a short-lived feature that lasts for a little more than 500 years.

The last 300 years of evolution of the Earth's field is similar in many respects. As mentioned, beneath the Pacific secular variation is low, but beneath the Indian and Atlantic Oceans the secular variation is much stronger, with at least one episode of flux expulsion occurring beneath the Indian Ocean in the first half of this century [*Bloxham and Gubbins*, 1985]. It is possible that the patch of reversed flux currently beneath South America represents a previous episode of flux expulsion.

An important question is whether the Pacific dipole window is persistent over long timescales. The paleomagnetic evidence until recently increasingly seemed to suggest that it is not [*McElhinny et al.*, 1996], in which case it is a temporary feature similar to that observed in our dynamo model, but the most recent evidence indicates that anomalous field behaviour is a long standing feature of the field in this region [*Johnson and Constable*, 1997]. If it is longer-lived in the Earth, then it will be interesting to see whether longer such episodes are found in the evolution of the dynamo model, or, if not, whether it then provides evidence of lateral heterogeneity at the base of the mantle. Presently, given that the paleomagnetic evidence is equivocal, and that our dynamo model yields a dipole window of duration at least that demanded by historical observations, one cannot conclude on geomagnetic grounds that there is a special conductivity structure at the base of the mantle beneath the Pacific hemisphere.

3.3. Flux Conservation

Apart from the northern and southern hemisphere null-flux patches (i.e. the flux patches delineated by the main magnetic equator), the topology of null-flux patches varies through the 750 year sequence shown in Plate 4. In fact, none of the other patches survives for more than 300 years before either decaying or being assimilated into one of the hemispheric patches. As a result it is not useful to plot the flux through these patches as a function of time, since diffusion is clearly operating. We will address the implications of this for determining the flow at the core surface from the geomagnetic secular variation elsewhere.

3.4. Core Surface Flow

In Figure 1 we show the flow at the core surface (again at the same epoch as Plates 2 and 4), and for comparison a model of the steady part of the flow over the last 150 years from *Bloxham* [1992]. The rms flow speed

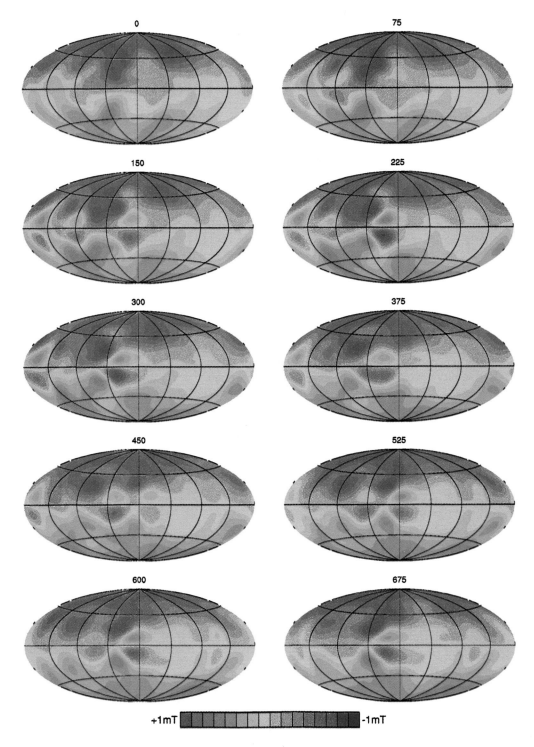

Plate 5.   The radial component of the magnetic field at the core-mantle boundary over a 675 year period.

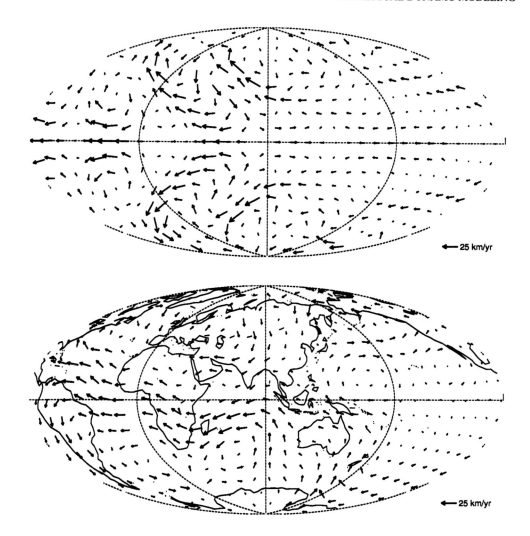

Figure 1. The fluid flow immediately beneath the core-mantle boundary. The top image is from the dynamo model; the lower image is from *Bloxham* [1992].

from the dynamo model is 11.2 km/yr with 95% of the kinetic energy in modes that are symmetric about the equator, compared to 9.6 km/yr with 77% of the kinetic energy in symmetric modes for the flow based on observations. Given that some uncertainty attains determinations of the flow from observations, these differences are most likely not significant.

The pattern of flow is similar. The strong east-west hemispherical difference in the field discussed above is also apparent in the flow: the flow speed beneath the eastern hemisphere of each plot (i.e. beneath the Pacific) is much weaker than beneath the western hemisphere. Where the flow is strong, it is characterized by a westward equatorial jet in each case, and large gyres centered at mid-latitude.

Although it would be interesting to address the steadiness of the flow in the dynamo model, we do not believe that a meaningful comparison can be made with geomagnetic observations for the reasons outlined earlier, specifically that unsteady flow, at least on the timescale of torsional oscillations, may not be accurately represented in the dynamo model on account of the too large value that we have had to adopt for the magnetic Rossby number.

4. CONCLUSIONS

We have shown in this paper how the results of our dynamo model, which necessarily employs a choice of nondimensional parameters different from those that are believed to pertain in the Earth, can nonetheless be scaled to provide geophysically meaningful results. This scaling is possible because our dynamo model operates in a quasi-Taylor state dynamical regime as is believed to be the case for the geodynamo.

Our dynamo model agrees well with geomagnetic observations: the field strength at the core-mantle boundary is very similar, the morphology of the field and its secular variation has several important characteristics in common with the earth's field, and the flow at the core surface is similar.

During the evolution of the dynamo model, one particular episode is very similar to the recent evolution of the geomagnetic field. We have seen the opening, and then some 500 years later, the closing of a Pacific dipole window. Simultaneously, we observe intense sec-[1] ular variation in the other hemisphere, characterized in particular by flux expulsion.

Much remains to be done, including an examination of how the behaviour of our model depends upon our choice of Rayleigh number. So far, all the calculations have been for a Rayleigh number $Ra = 15\,000$, which corresponds to a super-adiabatic heat flux of roughly 10^{12} W. A heat flux as much as a factor of 10 larger cannot be ruled out geophysically.

We have not undertaken a comparison with paleomagnetic observations for the simple reason that our dynamo model only spans order 10^5 years, which is too short for a useful comparison. To exploit fully the wealth of paleomagnetic observations we anticipate that it will be necessary to develop a parameterized version of our dynamo model which could track the long-term evolution of the field much more efficiently.

Acknowledgments. This work was supported by NSF Grants EAR95-26914 and EAR93-17156. We are very grateful to Gary Glatzmaier and Paul Roberts for many useful discussions about their work.

REFERENCES

Bloxham, J., The steady part of the secular variation of the Earth's magnetic field, *J. Geophys. Res., 97*, 19565–19579, 1992.

Bloxham, J., and D. Gubbins, The secular variation of the Earth's magnetic field, *Nature, 317*, 777–781, 1985.

Bloxham, J., and A. Jackson, Time-dependent mapping of the magnetic field at the core-mantle boundary, *J. Geophys. Res., 97*, 19537–19563, 1992.

Doell, R., and A. Cox, Palaeomagnetism of Hawaiian lava flows, *Nature, 192*, 645–646, 1961.

Fisk, H., Isopors and isoporic motion, *Inter. Geodet. Geophys. Un., Terr. Mag. Electr. Sec. Bull.*, pp. 280–292, 1931.

Glatzmaier, G., and P. Roberts, A three-dimensional convective dynamo solution with rotating and finitely conducting inner core and mantle, *Phys. Earth Planet. Inter., 91*, 63–75, 1995a.

Glatzmaier, G., and P. Roberts, A three-dimensional self-consistent computer simulation of a geomagnetic field reversal, *Nature, 377*, 203–209, 1995b.

Glatzmaier, G., and P. Roberts, An anelastic evolutionary geodynamo simulation driven by compositional and thermal convection, *Physica D, 97*, 81–94, 1996a.

Glatzmaier, G., and P. Roberts, Rotation and magnetism of Earth's inner core, *Science, 274*, 1887–1890, 1996b.

Gubbins, D., and J. Bloxham, Geomagnetic field analysis - III. Magnetic fields on the core-mantle boundary, *Geophys. J. Roy. Astr. Soc., 80*, 695–713, 1985.

Johnson, C., and C. Constable, The time-averaged geomagnetic field: global and regional biases for 0–5 ma, *Geophys. J. Int., 131*, 643–666, 1997.

Kuang, W., and J. Bloxham, An Earth-like numerical dynamo model, *Nature, 389*, 371–374, 1997a.

Kuang, W., and J. Bloxham, On the dynamics of the topographical core-mantle coupling, *Phys. Earth Planet. Inter., 99*, 289–294, 1997b.

McElhinny, M., P. McFadden, and R. Merrill, The myth of the Pacific dipole window, *Earth Planet. Sci. Lett., 143*, 13–22, 1996.

Runcorn, S., Polar paths in geomagnetic reversals, *Nature, 356*, 654–656, 1992.

Taylor, J., The magnetohydrodynamics of a rotating fluid and the Earth's dynamo problem, *Proc. Roy. Soc. Lond., A274*, 274–283, 1963.

W. Kuang, AVS Inc., 300 Fifth Avenue, Waltham, MA 02154. (e-mail: rkuang@avs.com)

J. Bloxham, Department of Earth and Planetary Sciences, Harvard University, Cambridge, MA 02138. (e-mail: bloxham@geophysics.harvard.edu)

[1] *Now at AVS Inc., Waltham, Massachuestts.*

Geodynamically Consistent Seismic Velocity Predictions at the Base of the Mantle

I. Sidorin and Michael Gurnis

Seismological Laboratory, California Institute of Technology

A model of thermoelastic properties for a chemically homogeneous adiabatic lower mantle is calculated. Constraints provided by this model are used in convection models to study dynamics of a chemically distinct layer at the bottom of the mantle. We find that the layer must be at least 2% denser than the overlying mantle to survive for a geologically significant period of time. Realistic decrease with depth of the thermal expansivity increases layer stability but is unable to prevent it from entrainment. Seismic velocities are computed for an assumed composition by applying the thermal and compositional perturbations obtained in convection simulations to the adiabatic values. The predicted velocity jump at the top of the chemical layer is closer to the CMB in the cold regions than in the hot. The elevation of the discontinuity above CMB in the cold regions decreases with increasing thermal expansivity and increases with increasing density contrast, while in the hot regions we find that the opposite is true. If the density contrast is small, the layer may vanish under downwellings. However, whenever the layer is present in the downwelling regions, it also exists under the upwellings. For a 4% density contrast and realistic values of expansivity, we find that the layer must be more than 400 km thick on average to be consistent with the seismically observed depth of the discontinuity. A simple chemical layer cannot be used to interpret the D" discontinuity: the required change in composition is large and must be complex, since enrichment in any single mineral probably cannot provide the required impedance contrast. A simple chemical layer cannot explain the spatial intermittance of the discontinuity.

1. INTRODUCTION

The D" region at the base of the mantle has been the focus of considerable seismological, geochemical and geodynamical study (see the review by *Loper and Lay,* [1995]). Although our knowledge of the region has dramatically improved, we still lack a comprehensive understanding of the processes responsible for the structures at the bottom of the mantle. Moreover, we lack

a unified theory which is needed to reconcile diverse observations.

Some of the most detailed and relevant observations concerning the D" region come from seismology. Most significant, perhaps, are the results of high resolution tomography inversions [*Grand,* 1994; *van der Hilst et al.,* 1997] which show that distinct seismically fast anomalies can be connected to subducted slabs. Though differing in details, the recent tomographic models agree at large scales at the bottom of the mantle. The most striking features of both models are well defined fast velocity anomalies under the Americas and Southern Asia which can be linked to the subducted Farallon and Tethys plates [*Grand et al.,* 1997].

The Core-Mantle Boundary Region
Geodynamics 28
Copyright 1998 by the American Geophysical Union.

Another source of information on D" is the analysis of the travel times of various core phases (including traversing, reflected and diffracted phases). These studies provide more localized information on the bottom of the mantle. Significant findings include a triplication caused by an apparent 2-3% velocity jump a few hundred kilometers above the CMB [*Lay and Helmberger*, 1983] and an ultra-low velocity zone (ULVZ) at the bottom 40 kilometers or less of the mantle [*Garnero and Helmberger*, 1995; *Garnero and Helmberger*, 1996]. The most likely cause of the ULVZ is partial melting [*Williams and Garnero*, 1996], but other interpretations, such as chemical anomalies cannot be ruled out. While the model of *Lay and Helmberger* [1983] is a viable interpretation of the observed triplication at the top of D", an alternative model has been suggested by *Ding* [1997] who demonstrated that the observed triplication can be explained by a smaller velocity jump (1%) than previously suggested (2-3%), provided that the vertical gradient of velocity is sufficiently high as may exist atop an old cold slab with a diffuse boundary layer.

The complexity of the region, its remoteness from the surface, and its ultimate importance in the global dynamics of the Earth call for an interdisciplinary approach to its exploration. Several such studies have been undertaken. *Christensen and Hofmann* [1994] studied mixing and transport in a thermo-chemical model of convection and predicted isotopic systematics of mantle derived basalts. *Forte and Peltier* [1989] compared seismically inferred CMB topography with that produced by a global flow model using buoyancy constrained by seismic tomography. Several studies attempted to predict temperature or density variations in the mantle from tomographic models [e.g., *Hager et al.*, 1985; *Yuen et al.*, 1993] or seismic velocities from dynamic models [e.g., *Davies and Gurnis*, 1986]. However, all previous studies generally use simple scalings between temperature and seismic velocity while a comprehensive interrelation between the dynamic modeling and mantle composition and thermoelastic properties is required.

It is obvious that the increasing resolution of tomography, more sophisticated dynamical models, and rapidly developing techniques in mineral physics experiments which provide elastic properties at higher pressures and temperatures, require a systematic approach to the problem. Such a systematic approach should allow us to close the gap between seismological inversions and dynamic models.

In this work we address the problem of the dynamics of the D" region and develop tools for testing numerical convection models against seismological observations by predicting seismic velocities from convection models. We first study the dynamics of a chemically distinct layer at the bottom of the mantle and explore the conditions under which such layer would survive for extended periods of time. Then we determine the properties of this layer which are consistent with fundamental seismological observations of the CMB region.

To make the dynamic simulation and inferred seismic velocities consistent, a coherent adiabatic reference model of thermoelastic parameters as a function of depth is required for the whole mantle. Since the focus of this study is the bottom of the mantle, we start by computing an adiabatic model for a chemically homogeneous lower mantle, using currently existing mineral physics data and geochemical constraints. Where necessary, the upper mantle parameters can be evaluated by interpolating between the lower mantle and surface values.

A convection model is then formulated with a bottom layer of specified physical properties. The convection in the mantle is then simulated numerically using parameters consistent with the adiabatic 1D model and the corresponding properties (such as density) of the bottom layer. The temperature field and the spatial distribution of the distinct material, obtained from the simulation, are then used to adjust the adiabatic elastic parameters to account for the effects of non-adiabaticity. This approach provides a dynamically consistent two-dimensional field of seismic velocities. Thus, the calculated seismic velocities are closely linked to the geodynamic model through the adiabatic 1D parameters and properties of the distinct material.

2. ADIABATIC MODEL FOR CHEMICALLY HOMOGENEOUS LOWER MANTLE

The approach we used to calculate the adiabatic model for a chemically homogeneous lower mantle is similar to that of *Zhao and Anderson* [1994]. A $(Mg,Fe)SiO_3$ perovskite – $(Mg,Fe)O$ magnesiowüstite lower mantle composition is assumed with given molar partitioning coefficient $\chi_{Pv} = Pv/(Pv + Mw)$ and a molar fraction of iron $\chi_{Fe} = Fe/(Fe + Mg)$. The values of the adiabatic bulk modulus, K_S, rigidity, G, and density, ρ, at a given foot temperature, T_f, of the adiabat (i.e. temperature at depth $z = 0$), are calculated separately for (Mg,Fe)SiO$_3$ and (Mg,Fe)O using equations [*Burdick and Anderson, 1975*]:

$$K_S(P_0, T_f) = K_{S_0} \exp\left\{-\int_{T_s}^{T_f} \alpha(T)\bar{\delta}_S dT\right\} \quad (1)$$

$$G(P_0, T_f) = G_0 \exp\left\{-\int_{T_s}^{T_f} \alpha(T)\bar{\Gamma} dT\right\} \quad (2)$$

$$\rho(P_0, T_f) = \rho_0 \exp\left\{-\int_{T_s}^{T_f} \alpha(T) dT\right\}, \quad (3)$$

where $T_s = 300K$ is the surface temperature, $P_0 = 0$ is the surface pressure, $K_{S_0} = K_S(P_0, T_s)$, $G_0 = G(P_0, T_s)$ and $\rho_0 = \rho(P_0, T_s)$ are the ambient adiabatic bulk modulus, rigidity and density, respectively, $\alpha(T) \equiv 1/V(\partial V/\partial T)_P$ is the coefficient of thermal expansion at $P = 0$, $\bar{\delta}_S \equiv (\partial \ln K_S/\partial \ln \rho)_P$ and $\bar{\Gamma} \equiv (\partial \ln G/\partial \ln \rho)_P$ are average dimensionless anharmonic parameters at $P = 0$ (taken as a mean value of the corresponding ambient and high temperature parameters).

The major difference between our calculations and the study of *Zhao and Anderson* [1994] is in the approach to the pressure and temperature dependence of the coefficient of thermal expansion, α. The coefficient of thermal expansion has a significant influence on both the dynamics of convection [*Yuen et al., 1991*] and the seismic velocity calculation and is the major parameter linking the convection simulation and seismic velocity prediction in our study. This is why an accurate analysis of this important property as a function of pressure, temperature and composition is required.

The values of α for magnesiowüstite reported by various authors [e.g., *Chopelas, 1996; Chopelas and Boehler, 1992; Duffy and Ahrens, 1993; Isaak et al., 1989; Isaak et al., 1990; Suzuki, 1975*] seem to agree rather well with each other. Though the current state of the experimental techniques is not able to provide accurate measurements at simultaneous high pressures and temperatures appropriate for the bottom of the mantle, the values at intermediate pressures and temperatures can be extrapolated to the required conditions with a certain degree of uncertainty. Fortunately, there seems to be general agreement on the first-order temperature and pressure dependence of α for (Mg,Fe)O [*Anderson et al., 1992; Chopelas and Boehler, 1992*]. For temperature dependence of the coefficient of thermal expansion of magnesiowüstite at $P = 0$ in equations (1)-(3) we use the data from *Isaak et al.* [1989].

However, the database of experimental measurements of α for silicate perovskite, (Mg,Fe)SiO$_3$, is sparse and inconsistent even at ambient conditions. An analysis of various reports shows that the range of reported values seem to have a bimodal nature, clustering around low values of about 1.7×10^{-5} K^{-1} [*Wang et al., 1994; Chopelas, 1996; Stacey, 1996*] and high values about 4.0×10^{-5} K^{-1} [*Knittle et al., 1986; Mao et al., 1991; Patel et al., 1996*].

In this study, the low values of α for (Mg,Fe)SiO$_3$ are favored for the following reasons. First, such values were supported by different experimental techniques. The values of the coefficient of thermal expansion obtained in large volume press experiments [*Funamori and Yagi, 1993; Wang et al., 1994; Utsumi et al., 1995*] agree reasonably well with each other and the results of Raman spectroscopy measurements [*Chopelas, 1996*]. Secondly, such choice is supported by a statistical analysis of existing P-V-T data [*Jackson and Rigden, 1996*] and theoretical thermodynamic considerations [*Anderson et al., 1995; Stacey, 1996*]. Besides, it has been suggested [*Wang et al., 1994*] that the early measurements of expansivity of (Mg,Fe)SiO$_3$ may have been made outside the stability field and so may not be correct. Taking the huge uncertainty in measured values, we follow the approach of *Anderson et al.* [1995] and compute α from thermoelastic parameters [*Anderson and Masuda, 1994*] rather than use it as an input parameter. So the choice between the low and high values of α is, in fact, the choice between the values of the corresponding thermoelastic parameters.

The computation begins with the *Suzuki et al.* [1979] equation:

$$\frac{\triangle V}{V_0} = \frac{\left[1 + 2\tilde{k} - \left(1 - \frac{4\tilde{k}E_{TH}(\Theta/T)}{Q}\right)^{1/2}\right]}{2\tilde{k}a_v} - 1, \quad (4)$$

where V_0 is the molar volume; $\tilde{k} = (1/2)(K_T' - 1)$, where $K_T' = \partial K_T/\partial P$ is the pressure derivative of the bulk modulus; $a_v = V(T_s)/V(0)$; $Q = K_{T_0}V_0/\gamma_0$, where K_{T_0} is the ambient isothermal bulk modulus, γ_0 is the ambient pressure Grüneisen ratio; $E_{TH}(\Theta/T)$ is the Debye energy and Θ is the Debye temperature. The temperature dependence of the ambient pressure coefficient of thermal expansion, $\alpha(P_0, T)$ is then obtained as a temperature derivative of (4).

After the values of K_S, G, and ρ at temperature T_f are known, their high pressure values can be calculated using the third-order Birch-Murnaghan equation of state [*Birch, 1952*]:

$$P = 3K_S(P_0, T_f)f(1 + 2f)^{5/2}\left(1 - \frac{3}{2}(4 - K_S')f\right), \quad (5)$$

where $f = \frac{1}{2}[\rho(P,T)/\rho(P_0,T_f)]^{2/3} - 1$ is the finite strain. Then the elastic parameters of the adiabatically compressed material can be calculated as follows [Sammis et al., 1970; Davies and Dziewonski, 1975]:

$$K_S(P,T) = K_S(P_0,T_f)(1+2f)^{5/2}[1 + (3K_S'-5)f] \tag{6}$$

$$G(P,T) = G(P_0,T_f)(1+2f)^{5/2}\{1 + [3G'K_S(P_0,T_f)/G(P_0,T_f)-5]f\} \tag{7}$$

$$\rho(P,T) = \rho(P_0,T_f)(1+2f)^{3/2}. \tag{8}$$

The elastic properties of the composite material are calculated by taking a Reuss-Voigt-Hill (RVH) average of the corresponding values for perovskite and magnesiowüstite:

$$M = \left\{ \frac{M_{Pv}M_{Mw}\left[\chi_{Pv}^{vol.}M_{Pv} + (1-\chi_{Pv}^{vol.})M_{Mw}\right]}{\chi_{Pv}^{vol.}M_{Mw} + (1-\chi_{Pv}^{vol.})M_{Pv}} \right\}^{1/2},$$

where M is either K_S, G or ρ and $\chi_{Pv}^{vol.}$ is the volume fraction of perovskite that can be calculated from the molar fraction χ_{Pv}.

The adiabatic temperature in the mantle is calculated using equation

$$\frac{dT}{dz} = \frac{T\gamma\rho g}{K_S}, \tag{9}$$

where γ is the Grüneisen parameter. It is generally accepted that the values of γ are not very sensitive to temperature above Θ [D. L. Anderson, 1988; Anderson et al., 1996]. The effect of compression on γ is given by [Anderson et al., 1993]

$$\gamma = \gamma_0\eta^q, \tag{10}$$

where $\eta = \rho(P_0,T_f)/\rho(P,T)$ and γ_0 is the value at $\eta = 1$. The dimensionless parameter q has values close to 1 for most minerals but is itself a function of pressure and temperature.

We separately calculate the Grüneisen parameters for perovskite and magnesiowüstite and then take the RVH average to represent the Grüneisen parameter of the compound material which is used in the equation for temperature (9).

To calculate the coefficient of thermal expansion along the mantle adiabat, the calculated isobars $\alpha(P_0,T)$ must first be converted to isochores $\alpha(\eta=1,T)$. This is done by integrating the identity

$$\left(\frac{\partial\alpha}{\partial T}\right)_V = \left(\frac{\partial\alpha}{\partial T}\right)_P - \bar{\delta}_T\alpha^2, \tag{11}$$

obtained by Anderson and Masuda [1994]. The Anderson - Grüneisen parameter, δ_T, varies with temperature below the Debye temperature [e.g., Isaak et al., 1990], so we use the average value, $\bar{\delta}_T$, of the ambient and high temperature values in the equation above. After that the coefficient of thermal expansion at a given temperature, T, and compression, η, can be calculated using the following equation, obtained by Anderson et al. [1992]:

$$\alpha(T,\eta) = \alpha(T,\eta=1)\exp\left[-\frac{\delta_{T_0}}{k}(1-\eta^k)\right], \tag{12}$$

where δ_{T_0} is the high-temperature value of the Anderson - Grüneisen parameter at zero pressure, and k is a dimensionless thermoelastic parameter controlling the dependence of δ_T on compression [Anderson and Issak, 1993]:

$$\delta_T = \delta_{T_0}\eta^k. \tag{13}$$

We use equations (12) and (13) to calculate α and δ_T, respectively, along the adiabat. The calculations are performed separately for perovskite and magnesiowüstite and a RVH average is taken to represent the lower mantle values.

The pressure dependence of the adiabatic Anderson - Grüneisen parameter, $\delta_S \equiv (\partial\ln K_S/\partial\ln\rho)_P$, has received significant attention recently. This parameter controls the sensitivity of the adiabatic bulk modulus, K_S, to temperature and is essential in evaluation of the temperature dependence of the compressional seismic velocity. The discussion on the matter was stimulated by an observation of apparent inconsistency of seismic tomography results with the mineral physics data that was first noted by D. L. Anderson [1987] from the analysis of seismic velocities variation in tomographic models. The ratio of the relative shear velocity lateral variations to the compressional velocity variations, $(\partial\ln V_S/\partial\ln V_P)_P$, calculated from seismic tomography models, is about a factor of 2 larger than the estimates from zero-pressure mineral physics data, and the separate analysis of P and S velocity variations gives Γ values at the bottom of the mantle similar to the zero-pressure values (≈ 5.8), while requires significantly lower values of δ_S (≈ 1.8) than the zero-pressure measurements (≈ 4.0) [D. L. Anderson, 1987].

The problem of δ_S pressure dependence was later addressed by Agnon and Bukowinski [1990] and Isaak et al. [1992]. Both studies provided further support for the relative insensitivity of Γ to pressure and a substantial decrease in δ_S.

Anderson [1967] showed an empirical relationship

$$\delta_S \approx \delta_T - \gamma, \tag{14}$$

valid for most minerals above the Debye temperature. From the analysis of the high temperature data on NaCl and KCl, *Yamamoto et al.* [1987] suggested that the following form of the relationship is more appropriate:

$$\delta_S \approx \delta_T - 1.4\gamma. \tag{15}$$

We use the relationship (15) to calculate δ_S along the adiabat. Parameter Γ is assumed to be constant along the adiabat and equal to 5.25, consistent with high-temperature values for MgO [*Isaak et al.*, 1989; *Isaak et al.*, 1992].

Following the described procedure, we calculate adiabatic profiles of K_S, G, ρ, T, α, γ, δ_T, and δ_S for a homogeneous lower mantle. Significant tradeoffs exist between the composition, foot temperature of the adiabat, T_f, and zero-pressure values of the coefficient of thermal expansion, α [*Bukowinski and Wolf*, 1990; *Hemley et al.*, 1992; *Stixrude et al.*, 1992; *Zhao and Anderson*, 1994; *Stacey*, 1996]. The coefficient of thermal expansion, in turn, depends on several thermodynamic parameters, of which the most important are the Grüneisen ratio, γ, and the Anderson - Grüneisen parameter, δ_T [*Anderson et al.*, 1995; *Stacey*, 1996].

Parameters χ_{Pv}, χ_{Fe}, T_f were systematically varied in the limits $0.5 - 1.0$, $0.08 - 0.12$ and $1400 - 2100K$, respectively, in search for the best fit to PREM in the $800 - 2600km$ depth range. The top and bottom regions of the lower mantle were excluded because the effects of non-adiabaticity may be significant [e.g., *Butler and Anderson*, 1978]. The resulting preferred values are $\chi_{Pv} = 0.55$, $\chi_{Fe} = 0.11$ and $T_f = 1750K$, which are in agreement with the results of *Stacey* [1996]. However, this solution is obviously non-unique due to the existing tradeoffs and significant uncertainties in the values of input mineral physical properties and wide bounds of geochemical constraints. The corresponding profiles of density, ρ, and elastic moduli, K_S and G, are given in Figure 1 and the calculated seismic velocities

$$V_P = \sqrt{\frac{K_S + \frac{4}{3}G}{\rho}} \tag{16}$$

and

$$V_S = \sqrt{\frac{G}{\rho}} \tag{17}$$

are given in Figure 2. The modeling results (ρ, K_S and G) have systematically steeper slopes than PREM. Though this may be a drawback of the low-order equation of state, it does not necessarily imply that the calculated along an adiabat thermoelastic parameters are systematically wrong, since weakly super-adiabatic radial gradients of temperature cannot be ruled out for the lower mantle [*Bukowinski and Wolf*, 1990].

Several assumptions were used in the calculations. First, all parameters involved, except ρ and G, are probably not very sensitive to iron content [*Mao et al.*, 1991; *Wang et al.*, 1994]. So in most cases we used the properties of the Mg end members for both (Mg,Fe)O and (Mg,Fe)SiO$_3$ (Table 1). Second, we assumed non-equal partitioning of iron between magnesiowüstite and perovskite, consistent with the study of *Kesson and Fitz Gerald* [1992]. The iron partitioning coefficients for (Mg,Fe)SiO$_3$ and (Mg,Fe)O, respectively, were taken as

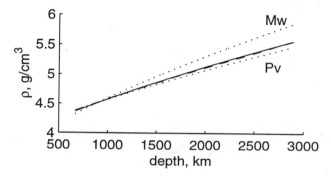

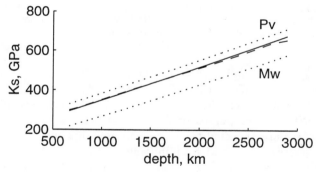

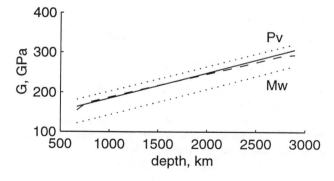

Figure 1. Elastic moduli and density calculated along a mantle adiabat. Dotted lines give values for magnesiowüstite and silicate perovskite; solid lines represent the RVH average values; dashed lines indicate PREM values for reference.

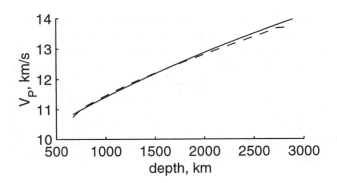

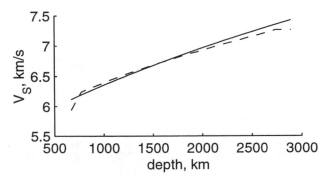

Figure 2. Seismic velocities calculated along a mantle adiabat. Solid lines represent the RVH average values; dashed lines indicate PREM values for reference.

$\chi_{Fe}^{Pv} = 0.07$ and $\chi_{Fe}^{Mw} = 1.727\chi_{Fe} - 0.016$ according to Figure 2 of *Kesson and Fitz Gerald* [1992].

We neglected the possible effect of the presence of high-pressure phases of Al and Ca-bearing minerals. These are expected to be in small amounts and in perovskite structure with a minor effect on the properties of the lower mantle [e.g., *Jeanloz and Knittle*, 1986; *Zhao and Anderson*, 1994].

The isothermal bulk modulus, K_{T_0}, in equation (4) was calculated from the adiabatic bulk modulus:

$$K_T = K_S/(1 + \alpha\gamma T) \qquad (18)$$

and it was assumed that $K_T' \approx K_S'$ [*D. L. Anderson*, 1987; *Isaak*, 1993].

When calculating the pressure dependence of (10) and (13), we took into account the pressure and temperature effects on the values of parameter q [*Anderson et al.*, 1993] and temperature effect on the values of parameter k [*Anderson et al.*, 1992] for magnesiowüstite, while constant values of q and k were used for silicate perovskite due to the absence of data.

By taking a RVH average, we assumed that various properties of a composite material are bounded by

the corresponding properties of the constituents. While this is true for density and elastic moduli [*Watt et al.*, 1976], this issue has not been thoroughly studied for the coefficient of thermal expansion, α, and the thermoelastic parameters δ_T and γ. *Anderson et al.* [1996] argued that γ of the magnesiowüstite-perovskite mix must be bounded by the corresponding values of the constituents. We expect this to be also true for the α and δ_T. The calculated values of δ_T, γ and δ_S along the lower mantle adiabat are given in Figure 3.

For a successful dynamic model, values of the coefficient of thermal expansion are required throughout the mantle. We used a function in the form

$$\alpha(z) = \frac{\alpha_0}{1 + az^b} \qquad (19)$$

to fit the calculated values of the coefficient of thermal expansion along the mantle adiabat (Figure 4). In equation (19) α_0 is the value of expansivity at zero depth, z is depth normalized by the radius of the Earth, and a and b are the fitting parameters to be determined. For the surface value of expansivity, α_0 we take the zero-pressure value of olivine at $T = 1750K$, which is about 4.5×10^{-5} K^{-1} [*Duffy and Anderson*, 1989]. The best fit is found for $a = 10.5$, $b = 0.85$.

3. GEODYNAMIC MODEL

Thermo-chemical convection with two materials, each with distinct properties, is computed in a 2D cylindrical coordinate system (r, ϕ) for a variety of cases. An incompressible flow model (Boussinesq approximation) is used in the simulation. The system is governed by the following non-dimensional equations [*Zhong and Gurnis*, 1993], representing conservation of momentum

$$\nabla \cdot (\mu\nabla u) = -\nabla p + \frac{1}{\zeta^3}\alpha(r)Ra\left(T - \frac{1}{\alpha(r)}B\,C\right)\hat{r}, \qquad (20)$$

energy

$$\frac{\partial T}{\partial t} = -(u \cdot \nabla)T + \nabla^2 T + H, \qquad (21)$$

and mass

$$\nabla \cdot u = 0. \qquad (22)$$

In the above equations, $\hat{r}$ is a unit vector in the radial direction, r is dimensionless radius, u is dimensionless velocity, p is dimensionless pressure, T is dimensionless temperature, t is dimensionless time, and ζ is dimensionless depth of the core-mantle boundary. These parameters are related to the corresponding dimensional

Table 1. Thermoelastic parameters.

	(Mg,Fe)O (mw)	(Mg,Fe)SiO₃ (pv)	SiO₂ (st)	Reference
V_0, cm³	$11.25 + 1.0\chi_{\mathrm{Fe}}^{\mathrm{Mw}}$	$24.447 + 1.143\chi_{\mathrm{Fe}}^{\mathrm{Pv}}$	14.014	1, 2
K_{S_0}, GPa	$163 - 8\chi_{\mathrm{Fe}}^{\mathrm{Mw}}$	264	305	3, 4, 2
G_0, GPa	$131 - 77\chi_{\mathrm{Fe}}^{\mathrm{Mw}}$	177.3	217	3, 4, 2
K_S'	3.8[a]	4.0	5.3	5, 6, 2
G'	1.7[a]	1.6	1.8	5, 6, 2
Θ, K	945	1020	1192	7, 8, 3
γ_0	1.52	1.4	-	9, 8
δ_{T_0}	5.0	4.5	-	7, 8
$\bar{\delta}_T$	4.9	5.0	-	7, 8
$\bar{\delta}_S$	3.15	2.7	2.7[b]	7, 10
Γ	5.25	5.25[c]	5.25[c]	11
$\bar{\Gamma}$	5.1	5.1[c]	5.1[c]	7
k	$k(T)$[d]	1.5	-	12, 13
q	$q(\eta, T)$[e]	1	-	14, 8

References: 1, *Saxena* [1996]; 2, *Li et al.* [1996]; 3, *Duffy and Anderson* [1989]; 4, *Yeganeh-Haeri* [1994]; 5, *Isaak* [1993]; 6, *Zhao and Anderson* [1994]; 7, *Isaak et al.* [1989]; 8, *Anderson et al.* [1996]; 9, *Chopelas* [1996]; 10, *Bukowinski and Wolf* [1990]; 11, *Isaak et al.* [1992]; 12, *Anderson et al.* [1992]; 13, *Anderson and Masuda* [1994]; 14, *Anderson et al.* [1993].
[a]Calculated mid-mantle value is used.
[b]Assumed to be the same as for (Mg,Fe)SiO₃.
[c]Assumed to be the same as for MgO.
[d]A specific value is taken at any given temperature.
[e]A specific value is taken at any given compression and temperature.

values by means of the following scalings (asterisk denotes dimensional values):

$$r^* = R_0 r, \tag{23}$$

$$u^* = \frac{\kappa}{R_0} u, \tag{24}$$

$$p^* = \frac{\kappa \bar{\mu}}{R_0^2} p, \tag{25}$$

$$T^* = T_S + \triangle T\, T, \tag{26}$$

$$t^* = \frac{R_0^2}{\kappa} t, \tag{27}$$

where R_0 is the radius of the Earth, κ is the thermal diffusivity, $\bar{\mu}$ is the volume averaged mantle viscosity, T_S is the temperature at the surface, and $\triangle T$ is the temperature increase across the mantle (Table 2).

Dynamic viscosity, μ, and coefficient of thermal expansion, α, in equation (20) are non-dimensionalized using the volume averaged values as characteristic scales:

$$\mu^* = \bar{\mu}\mu, \tag{28}$$
$$\alpha^* = \bar{\alpha}\alpha, \tag{29}$$

where $\bar{\mu}$ and $\bar{\alpha}$ are volume averaged viscosity and expansivity of the mantle, respectively. We take $\bar{\mu} =$

$10^{21.5}$ Pa·s, consistent with the inversion results of *Mitrovica and Forte* [1997]. The value of the volume averaged coefficient of thermal expansion, $\bar{\alpha}$, is calculated from the 1D adiabatic model presented above. In terms of the dimensionless radius, r, the depth dependence (19) of α can be written as:

$$\alpha(r) = \frac{4.5 \times 10^{-5}}{1 + 10.5(1 - r)^{0.85}}. \tag{30}$$

Taking the volume average, we have:

$$\bar{\alpha} = \frac{1}{V} \iint \log \alpha\, dV = 1.535 \times 10^{-5} K^{-1}. \tag{31}$$

For the purpose of convenience we will drop the '*' for dimensional values in all consequent equations.

The effects of internal heating are not addressed in this study and only bottom-heated models are considered, so that $H = 0$ in (21). Dimensionless parameter Ra in (20) is the thermal Rayleigh number which characterizes the vigor of convection in such a system. It is given by

$$Ra = \frac{g\bar{\alpha}\rho_0 \triangle T D^3}{\kappa \bar{\mu}}, \tag{32}$$

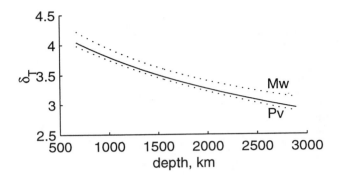

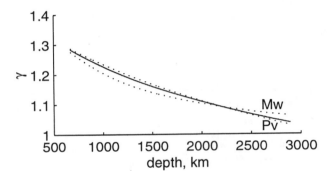

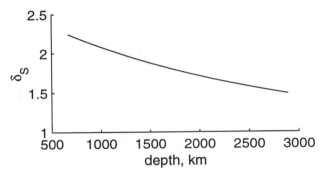

Figure 3. Thermoelastic parameters δ_T, γ and δ_S, calculated along a mantle adiabat. Dotted lines give values for magnesiowüstite and silicate perovskite; solid lines represent the RVH average values. The values of δ_S were calculated using equation (15).

where g is the gravitational acceleration, ρ_0 is the density of the ambient material, and D is the depth of the core-mantle boundary. The anomalous material has density $\rho_0 + \triangle\rho_0$ and its distribution is described by the composition function, $C(r, \phi)$, which takes values from 0 (ambient material) to 1 (anomalous material). The effect of the introduced anomalous material on the flow is characterized by the compositional Rayleigh number, Ra_c, given by:

$$Ra_c = \frac{g\triangle\rho_0 D^3}{\kappa\bar{\mu}}. \qquad (33)$$

Parameter B in (20) is the ratio of the two Rayleigh numbers, $B = Ra_c/Ra$. It characterizes the relative importance of the compositional and thermal buoyancy forces and is given by

$$B = \frac{\triangle\rho_0}{\rho_0\bar{\alpha}\triangle T}. \qquad (34)$$

The model domain is represented by a half-annulus (Figure 5) with the inner radius corresponding to the core-mantle boundary and the outer radius corresponding to the surface of the Earth. Plates are modeled by imposing velocity boundary conditions on the top, as shown in Figure 5a. Free slip conditions are used at the bottom and side walls of the domain. The value of imposed surface velocity was determined from a free-slip calculation with no temperature dependent viscosity to ensure that the plate neither speeds up nor slows down the flow from what would be expected in convection with a free slip top [*Gurnis and Davies*, 1986]. A velocity overshoot is added in the back-arc basin to initiate subduction and prevent the slab from being sucked up under the overriding plate [*Christensen*, 1996; *Davies*, 1997]. The trench migration velocity, U_{trench}, equal to

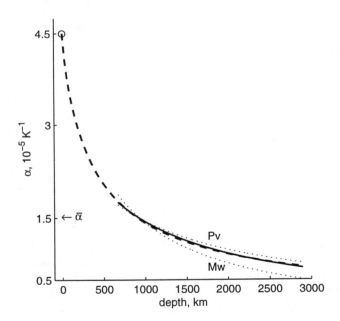

Figure 4. Coefficient of thermal expansion along a mantle adiabat. Dotted lines give values for magnesiowüstite and silicate perovskite; solid lines represent the RVH average values. The open circle represents the high-temperature value for olivine [*Duffy and Anderson*, 1989]. The dashed line gives the best fit using (19). The intercept on the vertical axis shows the volume averaged value used in dynamic models.

Table 2. Values of parameters used in convection simulation

Parameter	Symbol	Value
Radius of the Earth	R_0	6371 km
Depth of the CMB	D	2891 km
Gravity acceleration	g	10 m/s^2
Average mantle density	ρ_0	4.0 g/cm^3
Average mantle viscosity	$\bar{\mu}$	10$^{21.5}$ Pa·s
Thermal diffusivity	κ	10^{-6} m^2/s
Average thermal expansivity	$\bar{\alpha}$	1.535×10^{-5} K^{-1}
Temperature increase across the mantle	$\triangle T$	2900 K
Maximum velocity of subducting plate	U_{plate}	5 cm/yr
Maximum trench migration velocity	U_{trench}	0.5 cm/yr

the velocity of the overriding plate, is set to 10% of the surface velocity, U_{plate}, of the subducting plate.

The temperature is initially uniform throughout the interior of the mantle with superimposed top and bottom thermal boundary layers. The bottom boundary layer is 100 million years old. The thermal boundary layer for oceanic lithosphere is calculated using an infinite half-space cooling model with a velocity of $U_{plate} = 5$ cm/yr. The age of the lithosphere at the initial trench location is about 160 million years. To facilitate the detachment of the sinking slab from the overriding plate, the temperature on top of the overriding plate is set to the mantle interior temperature [*Christensen*, 1996; *Davies*, 1997]. The temperature variation across the bottom thermal boundary layer is taken equal to the variation across the lithosphere. The temperature in the mantle interior should correspond to the foot temperature of the mantle adiabat, $T_f = 1750$K, calculated above. This constraint and the choice of temperature variation across the bottom thermal boundary gives the value of the non-adiabatic temperature increase across the mantle, $\triangle T = 2900$K.

The imposed velocity of the subducting plate linearly increases from $0.05U_{plate}$ to U_{plate} over a time of 190 million years, roughly the time it takes the slab to reach the bottom of the mantle with such a velocity. Accordingly, the velocity of the overriding plate, which is equal to the velocity of the trench, increases from $0.05U_{trench}$ to U_{trench} over the same time interval. Such a gradual increase in imposed velocity helps to initiate subduction and is well justified physically, since the driving buoyancy force gradually increases with the increasing volume of subducted material. This is also consistent with a fully dynamic model of oceanic plates where plate margins are modeled with faults and surface velocities are model outcomes [*Zhong and Gurnis*, 1995].

A layer of distinct material with initial thickness d_0^{layer} is introduced at the bottom of the mantle. The material interface is initially represented by a chain of

$N = 2000$ particles, extending from the left side-wall to the right. Each particle is a passive tracer and is characterized by its coordinates (r_i^{int}, ϕ_i^{int}) in the system. Though this chain is not closed, the no-flux conditions at the bottom and the side-walls produce an effectively all-encompassing boundary. At any time a point lying above the chain is assumed to have an ambient composition, while a point lying below the chain is assumed to have a distinct composition.

The temperature and depth dependence of the dimensionless viscosity, μ, is given by

$$\mu = A\mu_0(r) \exp\left\{ \frac{c_1}{c_2 + \theta} - \frac{c_1}{c_2 + 0.5} \right\}. \quad (35)$$

The depth dependent part, $\mu_0(r)$, includes a factor of 10 increase across the 670 km depth:

$$\mu_0(r) = \begin{cases} 1 & \text{if } r > 0.89 \\ 10 & \text{if } r < 0.89. \end{cases} \quad (36)$$

Parameters c_1 and c_2 in (35) control the range of viscosity variation with temperature and the stiffness of the cold slab. We use $c_1 = 17.22$, which corresponds to an activation energy of 415 kJ·mol^{-1}, and $c_2 = 0.64$. A cutoff value of 10^3 is used. These parameters lead to 3 orders of magnitude in viscosity variation due to the temperature variations across the top boundary layer and another 2 orders of magnitude due to temperature variations in the bottom boundary layer.

The normalization parameter A in (35) is chosen so that the following equality holds when averaged over the entire duration of the simulation:

$$\frac{1}{V} \int\int \log\mu dV = 0, \quad (37)$$

where V is the volume of the model domain. We find $A \approx 0.16$. Figure 5b shows viscosity of the flow as the slab penetrates into the lower mantle.

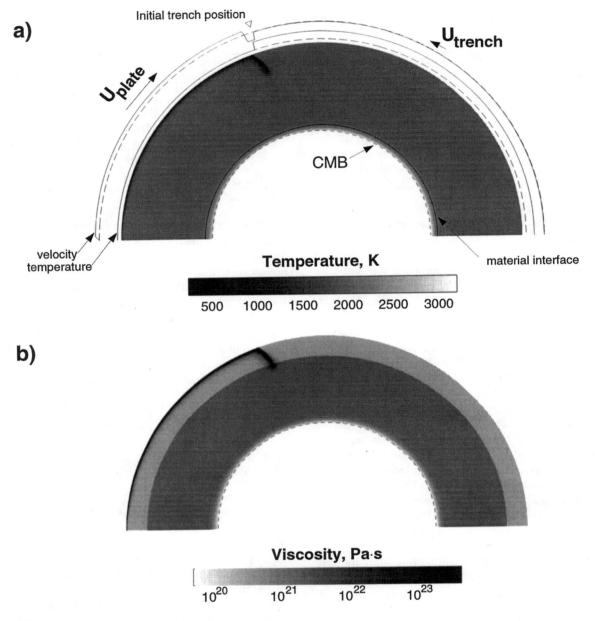

Figure 5. Dynamic model setup. a) Temperature field after 100 million years — just as the slab penetrates into the lower mantle. The value of the velocity overshoot in the back-arc region was scaled by a factor of 0.5 to give a clearer view of imposed plate velocities. b) Viscosity after 100 million years.

Equations (20)-(22) are solved using a finite-element code ConMan [*King et al.*, 1990], as modified by *Zhong and Gurnis* [1993] for a cylindrical geometry. The finite elements mesh has 300 elements in the azimuthal direction and 100 elements in the vertical. The mesh is refined both vertically and horizontally to increase the resolution in the thermal boundary layers and in the area of subduction.

The particles in the chain, representing the material interface, are advected using a third-order accurate predictor-corrector method. Whenever the distance between two adjacent particles exceeds a specified limit, δ_{max}, a new particle is introduced into the chain and placed between the two particles, so that the resolution of the boundary never falls below δ_{max}. In our calculations we use $\delta_{max} = 13$ km. An integration routine is used to calculate the ratio of each material in the finite element mesh. The associated compositional buoyancy is then fed back into the momentum equation (20). The accuracy of the material tracking algorithm was tested using steady-state divergence free analytic stream functions. The relative change in the volume, occupied by the distinct material did not exceed 1% after two overturn times, which is quite satisfactory for our purposes.

With all obvious limitations, introduction of compositional buoyancy into a convective system provides a nearly perfect resolution of the material interface, unlike solving an advective-diffusion equation or tracer particle methods. This jump in properties is essential for modeling seismic discontinuities. Moreover, the method does not seem to be vulnerable to the spurious settling problem encountered in various studies for the tracer particle techniques [*Gurnis*, 1986; *Christensen and Hofmann*, 1994; *van Keken et al.*, 1997].

We compute twelve different models of thermo - chemical convection, varying in the form of depth dependence of the coefficient of thermal expansion, $\alpha(r)$, the properties of the material in the bottom layer and the layer thickness, d_0^{layer} (Table 3). All studied cases have the same average thermal Rayleigh number. This is ensured by the normalization of the non-dimensional coefficient of thermal expansion, requiring its volume average to be 1. In models with depth dependent thermal expansivity α varies with depth according to the adiabatic 1D model (30).

4. COMPUTING 2D VELOCITY PERTURBATIONS

To estimate the effect of the super-adiabatic temperatures and the inclusion of the chemically distinct material on the adiabatic 1D seismic velocities we use a Taylor series expansion of equations (16) and (17) restricted to linear terms:

$$
\begin{aligned}
\triangle V_P &= \frac{\partial V_P}{\partial K_s} dK_s + \frac{\partial V_P}{\partial G} dG + \frac{\partial V_P}{\partial \rho} d\rho \\
&= \frac{\partial V_P}{\partial K_S} \left(\frac{\partial K_S}{\partial T} dT + \frac{\partial K_S}{\partial C} dC \right) \\
&\quad + \frac{\partial V_P}{\partial G} \left(\frac{\partial G}{\partial T} dT + \frac{\partial G}{\partial C} dC \right) \\
&\quad + \frac{\partial V_P}{\partial \rho} \left(\frac{\partial \rho}{\partial T} dT + \frac{\partial \rho}{\partial C} dC \right) \\
&= \frac{1}{2} \frac{K_{S_0}}{\rho_0 V_{P_0}} (\delta_S \alpha_0 \triangle T_{na} + (\epsilon_{K_S} - 1)C) \\
&\quad + \frac{2}{3} \frac{G_0}{\rho_o V_{P_0}} (\Gamma \alpha_0 \triangle T_{na} + (\epsilon_G - 1)C)
\end{aligned}
$$

$$
-\frac{1}{2} V_{P_0} (\alpha_0 \triangle T_{na} + (\epsilon_\rho - 1)C) , \quad (38)
$$

$$
\begin{aligned}
\triangle V_S &= \frac{\partial V_S}{\partial G} dG + \frac{\partial V_S}{\partial \rho} d\rho \\
&= \frac{1}{2} \frac{G_0}{\rho_o V_{S_0}} (\Gamma \alpha_0 \triangle T_{na} + (\epsilon_G - 1)C) \\
&\quad - \frac{1}{2} V_{S_0} (\alpha_0 \triangle T_{na} + (\epsilon_\rho - 1)C) , \quad (39)
\end{aligned}
$$

where $\epsilon_\rho = \rho_d/\rho_a$, $\epsilon_{K_S} = K_{S_d}/K_{S_a}$, $\epsilon_G = G_d/G_a$ are the elastic parameters ratios of the chemically distinct and the ambient materials, $\triangle T_{na}$ is the non-adiabatic temperature, C is the concentration of the distinct material and zero subscripts refer to the adiabatic 1D values. Values of δ_S at any given depth are taken consistent with the thermodynamic estimates along the adiabat (Figure 3). The value of Γ is assumed to be constant with depth and equal to 5.25.

The non-adiabatic temperature, $\triangle T_{na}$, and the concentration of the distinct material, C, are obtained from the convection simulation. The non-adiabatic temperature is calculated as

$$
\triangle T_{na} = T - T_f, \quad (40)
$$

where T_f is the foot temperature of the adiabat and T is the temperature from convection computations.

5. RESULTS

The convection simulations have been integrated for 400 million years in each case. We now discuss the details of the results for a nominal case M9 with $\triangle \rho / \rho = 4\%$ and α varying with depth according with (30). The slab starts its descent into the lower mantle at a very steep dip angle and stays almost vertical until its tip reaches the bottom. After that the lower part of the slab levels off and continues along the top of the bottom layer. The dip of the slab is reduced and some folding occurs. The slab has a substantial influence on the morphology of the bottom layer, depressing the material below and pushing it aside, towards the upwelling regions. This results in significant topography of the material interface, which is raised in the upwelling regions on both sides of the slab and depressed below the slab.

Results of the convection simulations for models M1, M2, M9 and M10 after 400 million years are given in Figure 6. In order to estimate the influence of model parameters (α, B, and d_0^{layer}) on the morphology of the bottom layer we computed the RMS topography of the

Table 3. Values of parameters used in convection simulation

Convection Model	α	d_0^{layer}, km	$\triangle\rho_0/\rho_0$	$\triangle d_{RMS}(400\text{Ma})$, km
M1	constant	200	0%	2750
M2	variable	200	0%	2017
M3	constant	200	1%	1622
M4	variable	200	1%	521
M5	constant	200	2%	346
M6	variable	200	2%	147
M7	variable	200	3%	118
M8	constant	200	4%	154
M9	variable	200	4%	103
M10	variable	400	4%	114
M11	variable	200	5%	91
M12	variable	200	6%	78

For models with variable α the depth dependence of the coefficient of thermal expansion is consistent with the adiabatic 1D calculations (30). For constant α cases the volume averaged value of α, given by (31) is used.

layer for all model cases, using

$$\triangle d_{RMS} = \left[\frac{1}{N} \sum_{i=1}^{N} \left(r_i^{\text{int}} - d_0^{\text{layer}} - R_0 + D \right)^2 \right]^{\frac{1}{2}} . \quad (41)$$

The parameter $\triangle d_{RMS}$ characterizes the radial disturbance of the layer and is equal to zero for a uniform layer and increases with topography. The values of $\triangle d_{RMS}$ for all cases after 400 million years are given in Table 3. Figure 7 gives the dependence of $\triangle d_{RMS}$ on the thermal expansion and the density contrast of the bottom material.

The topography of the layer is significantly reduced when the value of α at the bottom of the mantle is decreased. However, the effect of α on the slab dynamics is minor.

As Figure 7 shows, the stability of the layer dramatically increases when its density anomaly increases to about 2%. In all cases with $\triangle\rho_0/\rho_0 < 2\%$ most of the bottom layer was already destroyed after 400 million years of integration through entrainment of the material by plumes. Though smaller values of α increase the survival time of the layer, the decrease in α alone, without a density anomaly in the bottom material, is not able to keep the layer intact for long periods.

Figure 8 gives the values of $\triangle d_{RMS}$ as a function of time for six different cases. The topography of the material interface is controlled by two opposing processes. In subduction areas, the weight of the cold slab and the stress transmitted by the slab from the surface tend to depress the material boundary. Away from the slab, in the upwelling regions, the growing bottom thermal boundary layer increases the thermal buoyancy

of the material, tending to elevate the boundary. The topography rapidly increases for about 100-200 million years as the slab descends to the bottom of the mantle and depresses the chemical layer. Then topography stays approximately constant for some time until the layer heats up and thermal buoyancy becomes substantial. A 1% chemical density contrast is obviously not enough to balance this buoyancy and the basal material is rapidly entrained by the upwellings after about 300 million years of model integration. For a 2% density contrast, at least after the period of integration used in our study, the layer survives at the bottom. However, all the material is swept towards the upwelling regions away from the core-mantle boundary below the slab.

An interesting change in dynamics occurs when the thickness of the layer is increased and a sufficient density contrast is applied. With an initial thickness of 400 km and a density contrast of 4% the slab fails to thin the layer significantly. The increased temperature drop across the bottom thermal boundary layer has a destabilizing influence [*Lenardic and Kaula*, 1994] and leads to the initiation of an upwelling below the slab (Figure 6d). This initiated upwelling increases the buoyancy of the layer below the slab, thus partially reducing the depression of the interface. However, this buoyancy increase is unable to reverse the effect of the slab and produce elevated topography below the slab.

The bottom panels in Figures 6 show seismic velocities through the radial profiles indicated in the top panels. The values of the adiabatic 1D velocities were corrected for the effects of temperature and chemistry using the procedure presented in Section 4. For cases M1 and M2 ($\triangle\rho_0/\rho_0 = 0$) the composition of the basal layer

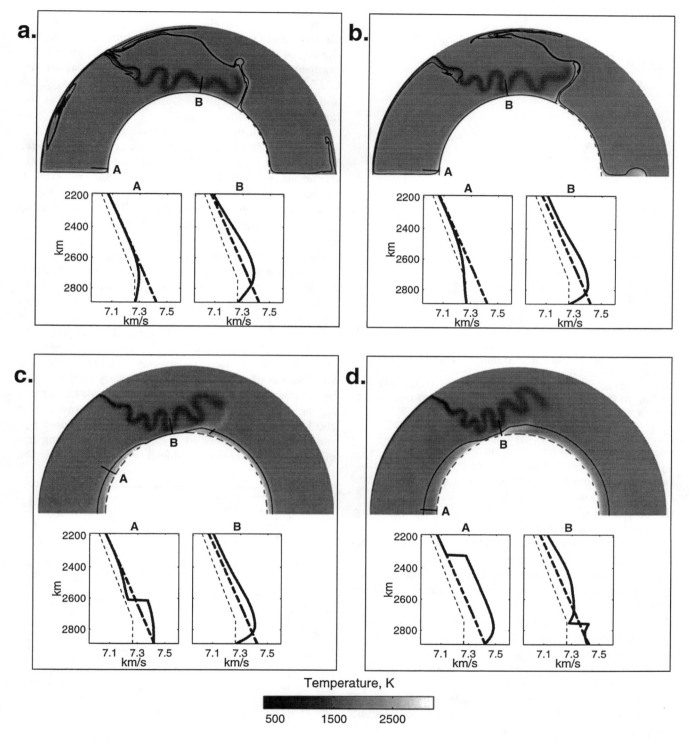

Figure 6. Four cases of convection simulation after 400 million years: a) model M1; b) model M2; c) model M9; d) model M10. Top panels show the temperature field and material interface (thin solid line). The core-mantle boundary is outlined by a dashed line. Bottom panels present seismic velocities (solid lines) computed for the cross-sections indicated in the top panels. For models M1 and M2 the basal material is assumed to be identical to the overlying mantle. For models M9 and M10 a stishovite and iron enrichment is assumed (composition indicated by a triangle in Figure 9). Cross-sections labeled 'A' correspond to upwelling regions, 'B' - to downwelling regions. The adiabatic 1D model (thick dashed lines) and PREM values (thin dashed lines) values are given for reference.

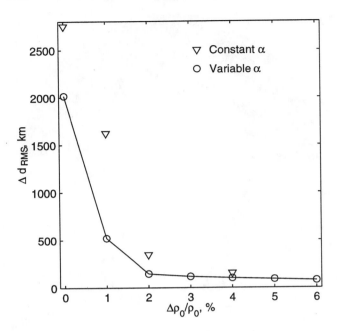

Figure 7. Radial RMS disturbance of the material interface after 400 million years. Circles represent models with the coefficient of thermal expansion varying with depth according to the adiabatic 1D model. Triangles correspond to models with constant α.

was identical to the overlying mantle ($\epsilon_\rho = 1$, $\epsilon_{K_S} = 1$, $\epsilon_G = 1$). For cases M9 and M10 ($\triangle\rho_0/\rho_0 = 4\%$) a distinct composition was assumed for the basal layer, with $\epsilon_\rho = 1.04$, $\epsilon_{K_S} = 1.047$ and $\epsilon_G = 1.079$. This choice of the properties of the basal layer and the corresponding composition will be discussed below.

We will first discuss cross-sections in the downwelling regions (cross-sections labeled 'B'). For all models, except M10, where the layer was initially 400 km thick, V_S remains nearly adiabatic down to a depth of about 2200 km, where it starts to increase due to the slab thermal anomaly. In model M10, the slab rests higher in the mantle, since the thickness of the bottom chemical layer is higher. The shear velocity for this case starts to deviate from its adiabatic values at about 2000 km depth. The fast velocity anomaly then continues to increase with depth, reaching its maximum at depth of 2200-2500 km, corresponding to the middle of the slab. At greater depths the trend reverses and V_S returns to its adiabatic values. Both the increase and decrease take place over a depth range of about 200 km, as the slab is diffuse. Following the attainment of adiabatic values, velocity starts to decrease with a large gradient due to the super-adiabatic temperatures of the bottom thermal boundary layer. In model M10 (Figure 6d) a

discontinuous velocity increase by about 2% occurs at the top of the layer in the region of a negative velocity gradient. The elevation of this discontinuity above the CMB is about 150 km. For models with $\triangle\rho_0/\rho_0 < 4\%$ and $d_0^{layer} = 200$ km no discontinuity is observed as the layer vanishes below the slab.

In the upwelling regions (cross-sections labeled 'A'), V_S remains adiabatic down to depth of about 2500 km. It then starts to decrease because of the super-adiabatic temperatures of the bottom thermal boundary layer. A discontinuous increase by about 2% occurs at the top of the chemical layer in models M9-M10. The elevation of the discontinuity above the CMB is about 290 km for model M9 and about 590 km for model M10 — much higher than for the downwelling regions.

Lower basal values of α increase the height of the discontinuity in the downwelling regions, while decreasing it in the upwelling regions. The influence of the density contrast increase is similar to the influence of α reduction. In terms of providing a given topography in the downwelling regions, a tradeoff exists between the layer density contrast, its average thickness, and the basal values of α. Although a 2% density contrast may be sufficient to prevent the layer from entrainment, a higher density contrast and a thickness of more than 200 km are required to ensure that the layer does not vanish in the downwelling regions. An interesting implication of the dynamics of such systems is that if the

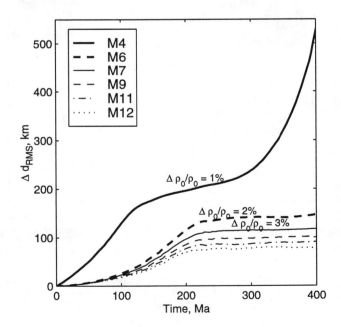

Figure 8. Disturbance of the basal layer as a function of time.

layer survives for significant periods of time it becomes hot. This leads to high temperature gradients across the material interface in the downwelling regions, so that the discontinuous seismic velocity change occurs in the area of negative velocity gradient.

6. DISCUSSION

A chemical layer at the bottom of the mantle has been proposed in several studies [e.g., *Davies and Gurnis*, 1986; *Christensen and Hofmann*, 1994; *Wysession*, 1996] as causing the seismic discontinuity at the top of D". Using both previous and new results, is such an interpretation of the seismology consistent with the dynamics of the mantle and mineralogical and geochemical constraints?

Seismic body waves have been interpreted in terms of a 2 − 3% discontinuous velocity increase 150 to 350 km above the core-mantle boundary with a median height of 250 km in many regions of the world [*Loper and Lay*, 1995]. There are, however, reports of the height of the discontinuity as low as 130 km [*Vidale and Benz*, 1993] and as high as 450 km [*Kendall and Shearer*, 1994]. One of the controversial issues about the discontinuity is the problem of its intermittent nature (see discussion by *Loper and Lay* [1995]). It is still not clear if the discontinuity is a global feature [*Nataf and Houard*, 1993] or if it only exists in some places in the world. The discontinuity has been observed in most seismically fast (cold) regions at the base of the mantle but it has also been reported in some slow (hot) regions [*Kendall and Shearer*, 1994]. While it is still difficult to make firm conclusions about the correlation of the discontinuity with the location of upwelling and downwelling regions, an interpretation of this velocity jump should be able to provide some explanation to the absence of the discontinuity in seismically slow regions and why its detection in those regions would be complicated.

Dynamics of a dense layer at the base of a convection system has previously been studied [*Davies and Gurnis*, 1986; *Gurnis*, 1986; *Hansen and Yuen*, 1989; *Olson and Kincaid*, 1991]. All of the previous studies and our new work demonstrate the same qualitative influence of convection on the morphology of the layer. The layer is depressed below the cold downwellings and is swept towards hot regions where it may or may not be entrained by the upwellings. Independent of the layer formation, all dynamic models predict that the layer would be thinnest in the downwelling regions and thickest below the upwellings. This means that if the seismic discontinuity in D" occurs on top of a chemical layer,

this layer must be at least 200-250 km thick since this is the height of the observed velocity jump in most seismically fast regions and regions which are most likely regions of downwellings.

If a layer of sufficient thickness is formed at the bottom of the mantle, under what conditions would the layer survive for a geologically long time without substantial recharge? This problem was studied by *Davies and Gurnis* [1986] in convection models with constant viscosity. It was found that an initially 300 km thick layer which is 2-3% heavier than the surrounding mantle would survive for at least 250 million years. Laboratory experiments [*Olson and Kincaid*, 1991] suggest that at least 2% density anomaly is required to prevent the layer from immediate overturn. A layer with a higher density contrast would be stable against mixing on a geophysically significant time scale. *Sleep* [1988] formulated a single model of the entrainment of a dense layer and concluded that an approximately 6% density anomaly is required to prevent the layer from entrainment. *Kellogg and King* [1993] gave a similar estimate (3-6%) and noted that a low viscosity bottom layer would retard entrainment, reducing the density anomaly, required for layer stability. *Hansen and Yuen* [1989] emphasized the importance of depth dependent thermal expansion on the dynamics of the dense layer and suggested that the decrease in α with depth increases layer stability. Our results suggest a minimum required density contrast of 2% and agree with most previous studies.

The effect of internal heating on the stability of the layer was not addressed in this study. *Christensen and Hofmann* [1994] showed that varying the amount of internal heating in their models did not significantly affect the rate of entrainment of the layer. *Tackley* [1998] demonstrated that when the basal heating is completely excluded, the rate of entrainment is considerably reduced. Such a model is unlikely true for the Earth as high temperature gradients probably exist at the base of the mantle [*Williams*, 1998]. However this issue requires a more detailed study.

Another important limitation of this study is a two-dimensional approximation. Results of *Tackley* [1998] show that the rate of entrainment is similar in equivalent 2D and 3D models. So while the dynamics in three dimensions can be qualitatively different from two-dimensional convection, our conclusions about the layer stability requirements will probably hold for three dimensions.

If the material is efficiently entrained by the upwellings, considerable recharge of the layer is required so that the layer can exist for long periods of geological time. The sources of such recharge may include products of reactions between the core and mantle [*Knittle and Jeanloz*, 1991; *Kellogg and King*, 1993] or rock subducted from the surface [*Gurnis*, 1986; *Christensen and Hofmann*, 1994].

Modeling results of *Kellogg and King* [1993] indicate that if the layer is recharged from the core a density contrast of 3-6% is required for the material in the layer. Otherwise, entrainment processes dominate and the layer is destroyed. Such a mechanism would create a layer less than 100 km thick throughout Earth history - about half of what would be required to explain the D" discontinuity. If the material of the layer is intermixed by convection with the material above, creating a thicker layer with an intermediate density [*Kellogg and King*, 1993], the resulting layer would probably have a diffuse boundary and would be unable to produce seismic reflections.

Recharge by subduction was studied by *Gurnis* [1986] and *Christensen and Hofmann* [1994]. The problem with this scenario is that at the CMB only the original crustal layer of the subducting slab is expected to be denser than the surrounding material. This posteclogite phase of basalt comprising the oceanic crust was estimated by *Christensen and Hofmann* [1994] to be about 2% denser than the surrounding material under the conditions relevant at the CMB. Convection models with constant viscosity demonstrated that separation of this dense phase from the rest of the slab is required in order to produce some chemical pool at the bottom of the mantle [*Gurnis*, 1986]. Incorporating the layered structure into numerical models, *Richards and Davies* [1989] suggested that no separation between the original crust and depleted lithosphere would occur in the upper mantle and the transition zone. This result was corroborated by *Gaherty and Hager* [1994], who showed that slab dynamics in the upper part of the mantle is controlled by the thermal buoyancy and not the density differences associated with chemical lamination of the slab. However, *Christensen and Hofmann* [1994] suggested that the separation of the dense phase from the slab may occur at the bottom of the mantle and hypothesized that such separation would indeed produce a layer of post-eclogitic rock on top of the core-mantle boundary. It was found, however, that only a small fraction of the subducted crust would accumulate at the bottom of the mantle, while the rest would be remixed into the mantle. This fraction strongly depends on the Rayleigh number and is less than 2% for $Ra \sim 10^7$ [*Christensen and Hofmann*, 1994]. Using the present day rate of crust subduction of 20 km^3/yr we can esti-

mate that such a process would require 80-100 billion years to form a layer 200-250 km thick. Although the mantle dynamics was likely to be different in the early Earth history, we doubt that such mechanism could produce sufficient amount of material to form the D" layer. Moreover, because of intensive mixing of the subducted crust a rather diffuse boundary is expected.

If a chemical layer with a thickness which matches the height of the D" discontinuity exists, what could the layer be comprised of? The material has to satisfy the following two properties: 1) in order to survive for a geologically significant period of time, it must be at least 2% denser than the surrounding material and 2) in order to match seismological constraints, it must have seismic velocities 2-3% higher than a lower mantle of "ambient" composition. Iron enrichment is the most effective mechanism for a density increase. However, Fe enrichment of the magnesiowüstite and silicate perovskite lower mantle results in lower velocities of the resulting assemblage. A possible exception is the low-spin Fe^{2+} [*Gaffney and Anderson*, 1973]. A spin-pairing transition in FeO was suggested by several studies [e.g., *Strens*, 1969; *Gaffney and Anderson*, 1973; *Sherman*, 1988]. However, not only are the elastic properties of the low-spin phase unknown, the transition pressure itself is a subject of an ongoing discussion [e.g., *Sherman and Jansen*, 1995]. In a recent study *Cohen et al.* [1997] predicted from first-principles calculations that a magnetic collapse can occur in iron under the lower mantle conditions. However, no experimental evidence for such transitions has yet been found.

The subducted post-eclogite ocean crust, as estimated by *Christensen and Hofmann* [1994], is about 2% denser than the magnesiowüstite and silicate perovskite lower mantle and, as estimated by *Wysession* [1996], has a 2-3% faster seismic velocity. However, as was discussed above, a sufficiently thick layer of subducted ocean crust at the base of the mantle is unlikely.

Vidale and Benz [1993] estimated that a 25% enrichment in stishovite of the basal layer could explain the seismic data. Using a third-order Birch-Murnaghan equation of state we calculated the properties of SiO$_2$ stishovite at pressures 250 km above the core mantle boundary. The estimated density of stishovite at that depth is about 3% lower than the density of a magnesiowüstite and silicate perovskite assemblage with $\chi_{Pv} = 0.55$ and $\chi_{Fe} = 0.11$, which is the composition that provides a good fit to PREM, as estimated in our study. Several post-rutile structures have been proposed for SiO$_2$ and two were experimentally identified. These are a CaCl$_2$-type structure [*Tsuchida and Yagi*, 1989] and a α-PbO$_2$-type structure [*Dubrovinsky et al,* 1997; *Karki et al.,* 1997a; *Karki et al.,* 1997b]. However, the volume change associated with each of the associated transitions is less than 1% [*Tsuchida and Yagi,* 1989; *Karki et al.,* 1997a]. This means that even if all silica at the base of the mantle is transformed into CaCl₂ structure, as suggested by *Kingma et al.* [1995], or even into α-PbO₂ structure, it will remain less dense than the magnesiowüstite and silicate perovskite assemblage. It is obvious that silica enrichment alone cannot provide a dynamically stable layer and some iron enrichment is also required.

Any extra iron in D" is likely to enter FeO, though some amount of FeSi alloy may also be present [*Knittle and Jeanloz,* 1991]. So iron enrichment is equivalent to the increase in FeO content. This means that if any high pressure phases of FeO exist, their properties may significantly affect the amount of FeO required to dynamically stabilize the layer. Two phase transitions in FeO at high pressures have been reported. One, taking place around 16 GPa, has been associated with a distortion of the rock-salt-type cell into a rhombohedral cell [*Zou et al.,* 1980; *Yagi et al.,* 1985]. Another transition observed at pressures about 70 GPa [*Jeanloz and Ahrens,* 1980; *Knittle and Jeanloz,* 1986; *Yagi et al.,* 1988] is interpreted as a transition to a NiAs phase [*Fei and Mao,* 1994]. The phase transition at 70 GPa is accompanied by a density increase of at least 10-16% [*Jackson and Ringwood,* 1981]. However, the elastic moduli are similar for the low pressure and high pressure polymorphs, so that velocity is expected to decrease in the transition [*Jeanloz and Ahrens,* 1980]. A solid solution of MgO and FeO may have a significantly different phase diagram. Shock compression of Mg₀.₆Fe₀.₄O to 200 GPa [*Vassiliou and Ahrens,* 1982] did not find a convincing evidence for any phase change similar to the phase transition of FeO at 70 GPa. This questions the existence of high pressure modifications of magnesiowüstite in the lower mantle and suggests that the density and elastic moduli of any such phases must be similar to the low-pressure phase.

We calculate the influence of stishovite and iron enrichment on the density and shear velocity of the adiabatic 1D model (Figure 9). The filled circle represents the composition of the computed adiabatic 1D model (a silicate perovskite and magnesiowüstite assemblage with $\chi_{Pv} = 0.55$ and $\chi_{Fe} = 0.11$). The solid contours indicate the influence of the change in iron content or stishovite enrichment on the density of the assemblage, while the dashed contours indicate the effect on the shear velocity. The light shaded area shows the range of compositions consistent with the requirements for a dynamic stability of the chemical layer

($\triangle\rho_0/\rho_0 \geq 2\%$). The medium shaded region represents the range of compositions consistent with a seismologically observed shear velocity increase by 1-3%. The intersection of the two regions, the dark shaded region, represents the range of compositions that are both dynamically and seismologically consistent. According to this analysis, 25-60% by volume of stishovite, accompanied by a significant increase of χ_{Fe} could produce a layer that would provide the required velocity jump and be dynamically stable. Due to the very small volume change and predicted average shear modulus softening of 10-30% estimated by *Jeanloz* [1989] for rutile→CaCl₂ transition, transformation of silica into one of the mentioned above post-rutile phases probably would not affect the estimates significantly or considerably reduce the required amount of silica.

The source of the extra SiO₂ and FeO is not clear. Several possibilities may exist. It was first suggested by *Birch* [1952] that silicates would break down into simple oxides under high pressures. A later study by

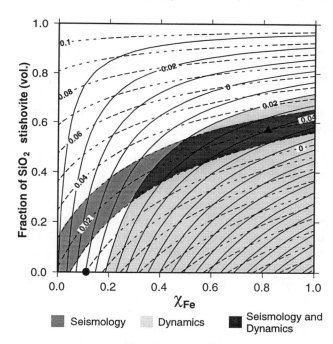

Figure 9. Calculated relative change in density (solid contours) and shear velocity (dashed contours) of the adiabatic 1D model, resulting from a stishovite enrichment and change in Fe content. The light shaded area indicate the range of dynamically consistent compositions; the medium shade indicates seismologically consistent composition; the range of compositions that are consistent with both seismology and dynamics is marked by the dark shade. The filled circle indicates the composition of the adiabatic 1D model. The filled triangle marks the composition used to calculate seismic velocities for dynamic models with $\triangle\rho_0/\rho_0 = 0.04$.

Stixrude and Bukowinski [1990] supported the dissociation of silicate perovskite to SiO_2 and $(Mg,Fe)O$ under the D" conditions. Such dissociation would produce a seismic velocity increase by about 3%, as estimated by *Wysession* [1996]. *Stixrude and Cohen* [1993] argue against such decomposition under lower mantle conditions, but more the recent experimental work of *Meade et al.* [1995] shows evidence for a dissociation of silicate perovskite into an assemblage of perovskite and mixed oxides. However, estimates show that even if SiO_2 is transformed into one of the known post-rutile structures, perovskite remains denser than the isochemical mixture of oxides [*Tsuchida and Yagi*, 1989]. This means that silicate perovskite is expected to be stable under the lower mantle conditions, unless some denser phases of silica and magnesiowüstite exist.

Chemical reactions between the core and the mantle material may be another source of SiO_2 and FeO in D" [*Knittle and Jeanloz*, 1991], but, as was discussed above, this mechanism is unlikely to be sufficiently efficient to create enough reaction products to generate a layer of required thickness.

The layer may also have been created in the process of the Earth differentiation. A model of inhomogeneous accretion predicts a layer of refractory material at the top of the CMB, since this material is expected to be about 2% denser than normal mantle [*Ruff and Anderson*, 1980]. However, as estimated by *Ruff and Anderson*, the refractories would have about 5% lower seismic velocity, so such composition is an unlikely candidate for a D" chemical layer consistent with the seismic discontinuity at its top.

The stishovite and iron enriched composition used for computing the shear velocity cross-sections (Figure 6) is marked in Figure 9 with a filled triangle. It corresponds to $\chi_{Fe} = 0.82$ and 58% (vol.) of stishovite. This composition provides a 2% velocity jump at the top of the chemical layer, and its density contrast (4%) is consistent with the value of $\Delta\rho_0/\rho_0$ used in the dynamic modeling. However, the only case that predicts the height of the discontinuity consistent with seismological observations in the seismically fast (cold) regions is model M10, where the layer was originally 400 km thick. All other cases predict a discontinuity too close (less than 100 km) to the CMB or the discontinuity is missing. This implies that, unless even higher density contrast is assumed for the basal material, the chemical layer must be about 400 km thick on average to be consistent with seismological observations.

What possible reasons may complicate seismological detection of the top of such a dense chemical layer in seismically slow regions? Source-receiver pair geography may play some role [*Kendall and Shearer*, 1994]. It is also very possible that the seismically slow regions in D" where no evidence for the discontinuity exists are chemically different from the rest of the basal material. One possibility is iron-rich phases coming from the core. This material, when mixed with the dense and seismically fast basal layer, would reduce its velocity and this reduction may be sufficient to neutralize the jump at the top of the basal layer. Hot upwellings may facilitate this process. Another possible mechanism that may potentially complicate detection is defocusing of seismic waves by a convex top boundary of the layer, expected in the upwelling regions. However, our models show a very smooth material interface with very low curvature. Such an effect is unlikely to play any significant role.

A viable alternative to a chemical layer is a phase transition. This interpretation of the seismic discontinuity is favored by *Nataf and Houard* [1993] and was considered by *Wysession* [1996]. No relevant phase change has been experimentally confirmed. But the hypothetical possibilities include the breakdown of silicate perovskite into the constituent oxides [*Wysession*, 1996] or some, not yet known, transition in perovskite or magnesiowüstite.

7. CONCLUSIONS

Our current state of knowledge on D" does not allow us to accept or rule out a chemical origin of the seismic discontinuity at the top of the D" region. Tighter constraints are required on the high pressure properties of minerals relevant to the lower mantle, the phase diagrams and the change of volume and elastic moduli associated with various transitions. We still poorly understand the extent of material exchange between the core and the mantle and the scope and rate of chemical reactions at the CMB, as well as the evolution of the Earth as a whole. However, dynamic modeling, seismic observations and general considerations argue against a simple chemical layer. Some of the arguments are the following:

1. Dynamic models suggest that at least a 2% density contrast is required in order for a chemical layer to survive for geologically significant periods of time. This, combined with a seismic observations of a 2-3% velocity increase, requires a very large impedance change. A complex change in composition is probably required, since enrichment or depletion in a single mineral is unlikely to

be able to produce the required impedance contrast.

2. An even higher density contrast (4% or more) and initial thickness of about 400 km are required to ensure the layer exists under the downwellings. A mechanism able to form a layer of such thickness throughout the history of the Earth is not known. Estimates show that neither subducted material, nor the products of reactions with the core can supply enough material. Refractories emplaced from the core, although they may have the required density contrast, are expected to have lower velocities than normal mantle and thus are unlikely candidates.

3. Dynamics of a dense layer implies that the material interface in the downwelling regions is accompanied by a high vertical temperature gradient. This means that the discontinuous velocity increase will occur in a region with negative vertical gradient of seismic velocity. Such a model would require a higher velocity jump (> 3%) than the *Lay and Helmberger* [1983] model in order to produce sufficiently strong reflections, consistent with seismic observations. This requires an even higher impedance contrast.

4. It is difficult to explain why a seismic discontinuity has not been confirmed on top of D" in any seismically slow (hot) region of the world. A chemical layer of significant thickness in the downwelling regions, completely thinned out in the upwelling regions, is not dynamically plausible. Dynamic models predict a very smooth material interface in the upwelling regions, so any defocusing effects cannot be important.

5. A simple chemical layer cannot explain other seismological observations relevant to D", such as the high heterogeneity or anisotropy.

If D" is a chemical layer, it is most likely chemically heterogeneous and its dynamics is much more complex than shown by our models or any other models to date. Its morphology is probably influenced by other processes so far ignored and, perhaps, still unknown processes that are yet to be discovered.

Acknowledgments. We are very grateful to Don Anderson and Don Helmberger for stimulating discussions. We also thank Don Anderson and an anonymous reviewer for reading the manuscript and providing a number of valuable comments. This research was funded by NSF grant EAR-9629279. This represents contribution number 6217 of the Division of Geological and Planetary Sciences, California Institute of Technology.

REFERENCES

Agnon, A., and M. S. T. Bukowinski, δ_S at high pressure and $d \ln V_S / d \ln V_P$ in the lower mantle, *Geophys. Res. Lett.*, *17*, 1,149-1,152, 1990.

Anderson, D. L., A seismic equation of state II. Shear properties and thermodynamics of the lower mantle, *Phys. Earth Planet. Inter.*, *45*, 307-323, 1987.

Anderson, D. L., Temperature and pressure derivatives of elastic constants with application to the mantle, *J. Geophys. Res.*, *93*, 4,688-4,700, 1988.

Anderson, O. L., Equation for thermal expansivity in planetary interiors, *J. Geophys. Res.*, *72*, 3,661-3,668, 1967.

Anderson, O. L., and D. G. Isaak, The dependence of the Anderson-Grüneisen parameter δ_T upon compression at extreme conditions, *J. Phys. Chem. Solids*, *54*, 221-227, 1993.

Anderson, O. L. and K. Masuda, A thermodynamic method for computing thermal expansivity, α, versus T along isobars for silicate perovskite, *Phys. Earth Planet. Inter.*, *85*, 227-236, 1994.

Anderson, O. L., K. Masuda, and D. Guo, Pure silicate perovskite and the PREM lower mantle model: a thermodynamic analysis, *Phys. Earth Planet. Inter.*, *89*, 35-49, 1995.

Anderson, O. L., K. Masuda, and D. G. Isaak, Limits on the value of δ_T and γ for MgSiO$_3$ perovskite, *Phys. Earth Planet. Inter.*, *98*, 31-46, 1996.

Anderson, O. L., H. Oda, and D. Isaak, A model for the computation of thermal expansivity at high compression and high temperatures: MgO as an example, *Geophys. Res. Lett.*, *19*, 1,987-1,990, 1992.

Anderson, O. L., H. Oda, A. Chopelas, and D. G. Isaak, A thermodynamic theory of the Grüneisen ratio at extreme conditions: MgO as an example, *Phys. Chem. Miner.*, *19*, 369-380, 1993.

Birch, F., Elasticity and constitution of the Earth's interior, *J. Geophys. Res.*, *57*, 227-286, 1952.

Bukowinski, M. S., and G. H. Wolf, Thermodynamically consistent decompression: Implications for lower mantle composition, *J. Geophys. Res.*, *95*, 12,583-12,593, 1990.

Burdick, L., and D. L. Anderson, Interpretation of velocity profiles of the mantle, *J. Geophys. Res.*, *80*, 1,070-1,074, 1975.

Butler, R., and D. L. Anderson, Equation of state fits to the lower mantle and outer core, *Phys. Earth Planet. Inter.*, *17*, 147-162, 1978.

Chopelas, A., Thermal expansivity of lower mantle phases MgO and MgSiO$_3$ perovskite at high pressure derived from vibrational spectroscopy, *Phys. Earth Planet. Inter.*, *98*, 3-15, 1996.

Chopelas, A., and R. Boehler, Thermal expansivity in the lower mantle, *Geophys. Res. Lett.*, *19*, 1,983-1,986, 1992.

Christensen, U. R., The influence of trench migration on slab penetration into the lower mantle, *Earth Planet. Sci. Lett.*, *140*, 27-39, 1996.

Christensen, U. R., and A. W. Hofmann, Segregation of subducted oceanic crust in the convecting mantle, *J. Geophys. Res.*, *99*, 19,867-19,884, 1994.

Cohen, R. E., I. I. Mazin, and D. G. Isaak, Magnetic collapse in transition metal oxides at high pressure: Implications for the Earth, *Science*, *275*, 654-657, 1997.

Davies, G. F., Incorporating plate into mantle convection models (abstract). In: AGU Chapman Conference. The History and Dynamics of Global Plate Motions, p. 42, 1997.

Davies, G. F., and A. M. Dziewonski, Homogeneity and constitution of the Earth's lower mantle and outer core, *Phys. Earth Planet. Inter.*, *10*, 336-343, 1975.

Davies, G. F., and M. Gurnis, Interaction of mantle dregs with convection: Lateral heterogeneity at the core-mantle boundary, *Geophys. Res. Lett.*, *13*, 1,517-1,520, 1986.

Ding, X., High resolution studies of deep Earth structure, Ph.D. thesis, 108 pp., Calif. Inst. of Technology, June 1997.

Dubrovinsky, L. S., S. K. Saxena, P. Lazor, R. Ahuja, O. Eriksson, J. M. Wills, and B. Johansson, Experimental and theoretical identification of a new high-pressure phase of silica, *Nature*, *388*, 362-365, 1997.

Duffy, T. S. and T. J. Ahrens, Thermal expansion of mantle and core materials at very high pressures, *Geophys. Res. Lett.*, *20*, 1,103-1,106, 1993.

Duffy, T. S., and D. L. Anderson, Seismic velocities in mantle minerals and the mineralogy of the upper mantle, *J. Geophys. Res.*, *94*, 1,895-1,912, 1989.

Fei, Y., and H. K. Mao, In situ determination of the NiAs phase of FeO at high pressure and temperature, *Science*, *266*, 1678-1680, 1994.

Forte, A. M. and W. R. Peltier, Core-mantle boundary topography and whole-mantle convection, *Geophys. Res. Lett.*, *16*, 621-624, 1989.

Funamori, N., and T. Yagi, High pressure and high temperature in situ X-ray observation of $MgSiO_3$ perovskite under lower mantle conditions, *Geophys. Res. Lett.*, *20*, 387-390, 1993.

Gaffney, E. S., and D. L. Anderson, Effect of low-spin Fe^{2+} on the composition of the lower mantle, *J. Geophys. Res.*, *78*, 7,005-7,014, 1973.

Gaherty, J. B., and B. H. Hager, Compositional vs. thermal buoyancy and the evolution of subducted lithosphere, *Geophys. Res. Lett.*, *21*, 141-144, 1994.

Garnero, E. J., and D. V. Helmberger, A very slow basal layer underlying large-scale low-velocity anomalies in the lower mantle beneath the Pacific: evidence from core phases, *Phys. Earth Planet. Inter.*, *91*, 161-176, 1995.

Garnero, E. J., and D. V. Helmberger, Seismic detection of a thin laterally varying boundary layer at the base of the mantle beneath the central-Pacific, *Geophys. Res. Lett.*, *23*, 977-980, 1996.

Grand, S. P., Mantle shear structure beneath the Americas and surrounding oceans, *J. Geophys. Res.*, *99*, 11,591-11,622, 1994.

Grand, S. P., R. D. van der Hilst, and S. Widiyantoro, Global seismic tomography: a snapshot of convection in the Earth, *GSA Today*, *7*, 1-7, 1997.

Gurnis, M., The effects of chemical density differences on convective mixing in the Earth's mantle, *J. Geophys. Res.*, *91*, 11,407-11,419, 1986.

Gurnis, M. and G. F. Davies, Mixing in numerical models of mantle convection incorporating plate kinematics, *J. Geophys. Res.*, *91*, 6,375-6,395, 1986.

Hager, B. H., R. W. Clayton, M. A. Richards, R. P. Comer, and A. M. Dziewonski, Lower mantle heterogeneity, dynamic topography and the geoid, *Nature*, *313*, 541-545, 1985.

Hansen, U., and D. A. Yuen, Dynamical influences from thermal-chemical instabilities at the core-mantle boundary, *Geophys. Res. Lett.*, *16*, 629-632, 1989.

Hemley, R. J., L. Stixrude, Y. Fei, and H. K. Mao, 1992. Constraints on lower mantle composition from $P - V - T$ measurements of $(Fe,Mg)SiO_3$-perovskite and $(Fe,Mg)O$. In: Y. Syono and M H. Manghnani (Editors), High-pressure research: Application to Earth and Planetary Sciences, American Geophysical Union, Washington, DC, pp. 183-189.

Isaak, D. G., The mixed P, T derivatives of elastic moduli and implications on extrapolating throughout Earth's mantle, *Phys. Earth Planet. Inter.*, *80*, 37-48, 1993.

Isaak, D. G., O. L. Anderson, and T. Goto, Measured elastic moduli of single-crystal MgO up to 1800 K, *Phys. Chem. Minerals*, *16*, 704-713, 1989.

Isaak, D. G., R. E. Cohen, and M. J. Mehl, Calculated elastic and thermal properties of MgO at high pressures and temperatures, *J. Geophys. Res.*, *95*, 7,055-7,067, 1990.

Isaak, D. G, O. L. Anderson, and R. E. Cohen, The relationship between shear and compressional velocities at high pressures: Reconciliation of seismic tomography and mineral physics, *Geophys. Res. Lett.*, *19*, 741-744, 1992.

Jackson, I., and S. M. Rigden, Analysis of $P - V - T$ data: Constraints on the thermoelastic properties of high-pressure minerals, *Phys. Earth Planet. Inter.*, *96*, 85-112, 1996.

Jackson, I., and A. E. Ringwood, High-pressure polymorphism of the iron oxides, *Geophys. J. R. Astron. Soc.*, *64*, 767-783, 1981.

Jeanloz, R., Phase transitions in the mantle, *Nature*, *340*, 184, 1989.

Jeanloz, R., and T. J. Ahrens, Equations of state of FeO and CaO, *Geophys. J. R. Astron. Soc.*, *62*, 505-528, 1980.

Jeanloz, R., and E. Knittle, 1986. Reduction of mantle and core properties to a standard state by adiabatic decompression. In: S. K. Saxena (Editor), Chemistry and physics of terrestrial planets. Springer, New York, pp. 275-309.

Karki, B. B., M. C. Warren, L. Stixrude, G. J. Ackland, and J. Crain, *Ab initio* studies of high-pressure structural transformations in silica, *Phys. Rev. B*, *55*, 3,465-3,471, 1997a.

Karki, B. B., M. C. Warren, L. Stixrude, G. J. Ackland, and J. Crain, *Ab initio* studies of high-pressure structural transformations in silica, *Phys. Rev. B*, *56*, 2,884-2,884, 1997b.

Kellogg, L. H., and S. D. King, Effect of mantle plumes on the growth of D" by reaction between the core and mantle, *Geophys. Res. Lett.*, *20*, 379-382, 1993.

Kendall, J.-M., and P. M. Shearer, Lateral variations in D" thickness from long-period shear wave data, *J. Geophys. Res.*, *99*, 11,575-11,590, 1994.

Kesson, S. E. and J. D. Fitz Gerald, Partitioning of MgO, FeO, NiO, MnO and Cr_2O_3 between magnesian silicate perovskite and magnesiowüstite: implications for the origin of inclusions in diamond and the composition of the lower mantle, *Earth Planet. Sci. Lett.*, *111*, 229-240, 1992.

King, S. D., A. Raefsky, and B. H. Hager, ConMan: Vectorizing a finite element code for incompressible two-dimensional convection in the Earth's mantle, *Phys. Earth Planet. Inter., 59*, 195-207, 1990.

Kingma, K. J., R. E. Cohen, R. J. Hemley, and H.K. Mao, Transformation of stishovite to a denser phase at lower-mantle pressures, *Nature, 374*, 243-245, 1995.

Knittle, E. and R. Jeanloz, High-pressure metallization of FeO and implications for the Earth's core, *Geophys. Res. Lett., 13*, 1,541-1,544, 1986.

Knittle, E. and R. Jeanloz, Earth's core-mantle boundary: Results of experiments at high pressures and temperatures, *Science, 251*, 1,438-1,443, 1991.

Knittle, E., R. Jeanloz, and G. L. Smith, Thermal expansion of silicate perovskite and stratification of the Earth's mantle, *Nature, 319*, 214-216, 1986.

Lay, T., and D. V. Helmberger, A lower mantle *S* wave triplication and the shear velocity structure of D", *Geophys. J. R. Astron. Soc., 75*, 799-838, 1983.

Lenardic, A., and W. M. Kaula, Tectonic plates, D" structure, and the nature of mantle plumes, *J. Geophys. Res., 99*, 15,697-15,708, 1994.

Li, B., S. M. Rigden, and R. C. Liebermann, Elasticity of stishovite at high pressure, *Phys. Earth Planet. Inter., 96*, 113-127, 1996.

Loper, D. E., and T. Lay, The core-mantle boundary region, *J. Geophys. Res., 100*, 6,397-6,420, 1995.

Mao, H. K., R. J. Hemley, Y. Fei, J. Shu, L. Chen, A. P. Jephcoat, W. Wu, and W. A. Basset, Effect of pressure, temperature, and composition on lattice parameters and density of (Fe,Mg)SiO3-perovskites to 30 GPa, *J. Geophys. Res., 96*, 8,069-8,079, 1991.

Meade, C., H. K. Mao, and J. Hu, High temperature phase transition and dissociation of (Mg,Fe)SiO3 perovskite at lower mantle pressures, *Science, 268*, 1,743-1,745, 1995.

Mitrovica, J.X., and A.M. Forte, Radial profile of mantle viscosity: Results from the joint inversion of convection and postglacial rebound observables, *J. Geophys. Res., 102*, 2,751-2,769, 1997.

Nataf, H-C., and S. Houard, Seismic discontinuity at the top of D": A world-wide feature?, *Geophys. Res. Lett., 20*, 2,371-2,374, 1993.

Olson, P., and C. Kincaid, Experiments on the interaction of thermal convection and compositional layering at the base of the mantle, *J. Geophys. Res., 96*, 4,347-4,354, 1991.

Patel, A., G. D. Price, M. Matsui, J. Brodholt, and R. Howarth, A computer simulation approach to the high pressure thermoelasticity of MgSiO3 perovskite, *Phys. Earth Planet. Inter., 98*, 55-63, 1996.

Richards, M. A., and G. F. Davies, On the separation of relatively buoyant components from subducted lithosphere, *Geophys. Res. Lett., 16*, 831-834, 1989.

Ruff, L., and D. L. Anderson, Core formation, evolution, and convection: A geophysical model, *Phys. Earth Planet. Inter., 21*, 181-201, 1980.

Sammis, C., D. L. Anderson, and T. Jordan, Application of isotropic finite strain theory to ultrasonic and seismological data, *J. Geophys. Res., 75*, 4,478-4,480, 1970.

Saxena, S. K., Earth mineralogical model: Gibbs free energy minimization computation in the system MgO-FeO-SiO2, *Geochim. Cosmochim. Acta, 60*, 2,379-2,395, 1996.

Sherman, D. M., High-spin to low-spin transition of iron(II) oxides at high pressures: Possible effects on the physics and chemistry of the lower mantle. In: S. Ghose et al. (Editors), Structural and Magnetic Phase Transitions in Minerals, *Adv. Phys. Geochem., 7*, 113-128, 1988.

Sherman, D. M., and H. J. F. Jansen, First-principles prediction of the high-pressure phase transition and electronic structure of FeO: Implications for the chemistry of the lower mantle and core, *Geophys. Res. Lett., 22*, 1,001-1,004, 1995.

Sleep, N. H., Gradual entrainment of a chemical layer at the base of the mantle by overlying convection, *Geophys. J. Int., 95*, 437-447, 1988.

Stacey, F. D., Thermoelasticity of (Mg,Fe)SiO3 perovskite and a comparison with the lower mantle, *Phys. Earth Planet. Inter., 98*, 65-77, 1996.

Stixrude, L., and M. S. T. Bukowinski, Fundamental thermodynamic relations and silicate melting with implications for the constitution of D", *J. Geophys. Res., 95*, 19,311-19,325, 1990.

Stixrude, L., and R. E. Cohen, Stability of orthorhombic MgSiO3 perovskite in the Earth's lower mantle, *Nature, 364*, 613-615, 1993.

Stixrude, L., R. J. Hemley, Y. Fei, and H. K. Mao, Thermoelasticity of silicate perovskite and magnesiowüstite and stratification of the Earth's mantle, *Science, 257*, 1,099-1,101, 1992.

Strens, R. G. J., The nature and geophysical importance of spin-pairing in minerals of iron (II). In: S. K. Runcorn (Editor), The Physics and Chemistry of Minerals and Rocks, John Wiley, New York, pp. 213-220, 1969.

Suzuki, I., Thermal expansion of periclase and olivine, and their anharmonic properties, *J. Phys. Earth, 23*, 145-159, 1975.

Suzuki, I., S. Okajima, and K. Seya, Thermal expansion of single-crystal manganosite, *J. Phys. Earth, 27*, 63-69, 1979.

Tackley, P., Three-dimensional simulations of mantle convection with a thermochemical CMB boundary layer: D"?, *this volume*, 1998.

Tsuchida, Y., and T. Yagi, A new, post-stishovite high-pressure polymorph of silica, *Nature, 340*, 217-220, 1989.

Utsumi, W., N. Funamori, T. Yagi, E. Ito, T. Kikegawa, and O. Shimomura, Thermal expansivity of MgSiO3 perovskite under high pressures up to 20 GPa, *Geophys. Res. Lett., 22*, 1,005-1,008, 1995.

van der Hilst, R. D., S. Widiyantoro, and E. R. Engdahl, Evidence for deep mantle circulation from global tomography, *Nature, 386*, 578-584, 1997.

van Keken, P.E., S.D. King, H. Schmeling, U.R. Christensen, D. Neumeister, and M.-P. Doin, A comparison of methods for the modeling of thermochemical convection, *J. Geophys. Res., 102*, 22,477-22,495, 1997.

Vassiliou, M. S., and T. J. Ahrens, The equation of state of Mg0.6Fe0.4O to 200 GPa, *Geophys. Res. Lett., 9*, 127-130, 1982.

Vidale, J. E., and H. M. Benz, Seismological mapping of fine structure near the base of the Earth's mantle, *Nature, 361*, 529-532, 1993.

Wang, Y., D. J. Weidner, R. C. Liebermann, and Y. Zhao, $P - V - T$ equation of state of (Mg,Fe)SiO3 perovskite: constraints on composition of the lower mantle, *Phys. Earth Planet. Inter., 83*, 13-40, 1994.

Watt, J. P., G. F. Davies, and R. J. O'Connell, The elastic

properties of composite materials, *Rev. Geophys. Space Phys., 14,* 541-563, 1976.

Williams, Q., The temperature contrast across D", *this volume,* 1998.

Williams, Q., and E. J. Garnero, Seismic evidence for partial melt at the base of Earth's mantle, *Science, 273,* 1,528-1,530, 1996.

Wysession, M. E., 1996. Imaging cold rock at the base of the mantle: The sometimes fate of slabs? In: Subduction: Top to Bottom, Geophysical Monograph 96, AGU, Washington, DC, pp. 369-384.

Yagi, T., T. Suzuki, and S. Akimoto, Static compression of wüstite ($Fe_{0.98}O$) to 120 GPa, *J. Geophys. Res., 90,* 8,784-8,788, 1985.

Yagi, T., K. Fukuoka, H. Takei, and Y. Syono, Shock compression of wüstite, *Geophys. Res. Lett., 15,* 816-819, 1988.

Yamamoto, S., I. Ohno, and O. L. Anderson, High temperature elasticity of sodium chloride, *J. Phys. Chem. Solids, 48,* 143-151, 1987.

Yeganeh-Haeri, A., Synthesis and re-investigation of the elastic properties of single-crystal magnesium silicate perovskite, *Phys. Earth Planet. Inter., 87,* 111-121, 1994.

Yuen, D. A., O. Cadek, A. Chopelas, and C. Matyska, Geophysical inferences of thermal-chemical structures in the lower mantle, *Geophys. Res. Lett., 20,* 899-902, 1993.

Yuen, D. A., A. M. Leitch, and U. Hansen, 1991. Dynamical influences of pressure-dependent thermal expansivity on mantle convection. In: R. Sabadini and K. Lambeck (Editors), Glacial Isostasy, Sea-level and Mantle Rheology, Kluwer, Dordrecht, pp. 663-701.

Zhao, Y., and D. L. Anderson, Mineral physics constraints on the chemical composition of the Earth's lower mantle, *Phys. Earth Planet. Inter., 85,* 273-202, 1994.

Zhong, S., and M. Gurnis, Dynamic feedback between a continentlike raft and thermal convection, *J. Geophys. Res.,* 98, 12,219-12,232, 1993.

Zhong, S., and M. Gurnis, Mantle convection with plates and mobile, faulted plate margins, *Science, 267,* 838-843, 1995.

Zou, G. T., H. K. Mao, P. M. Bell, and D. Virgo, High-pressure experiments on the iron oxide wüstite ($Fe_{1-x}O$), *Year Book Carnegie Inst. Washington, 79,* 374-376, 1980.

Igor Sidorin and Michael Gurnis, Seismological Laboratory

252-21, California Institute of Technology, Pasadena, CA 91125. (e-mail: igor@gps.caltech.edu; gurnis@caltech.edu)

Three-Dimensional Simulations of Mantle Convection with a Thermo-Chemical Basal Boundary Layer: D"?

Paul J. Tackley

Department of Earth and Space Sciences, University of California, Los Angeles, CA

Numerical simulations in both two- and three-dimensions are performed to investigate the hypothesis that D" is a thermo-chemical boundary layer. A layer of dense, compositionally-distinct material above the CMB reduces the characteristic horizontal lengthscales and interior mantle temperature. In three-dimensions, the layer forms a spoke pattern, with entrainment in up-welling cylinders, even for systems which are heated entirely from within. Long-term stability of the layer requires rather high density contrasts if the Boussinesq approximation is assumed, but geophysically reasonable density contrasts using a compressible formulation with depth-dependent properties. Temperature-dependent viscosity and internal heating also promote greater layer stability. Even if the core heat flux is zero, the layer can become several 100 K higher than the mantle above it, due to trapped radiogenically-produced heat. The combination of high deep-mantle viscosity and other depth-dependent parameters allows a stable layer with very high topography, leading to huge, thermo-chemical 'megaplumes' extending at least half way through the lower mantle. These stationary megaplumes are stable for long geological times and may correspond to the 'megaplumes' imaged tomo-graphically under Africa and the Pacific, as well as acting as a geochemical reservoir and providing an explanation for various other geophysical observations. The dense layer increases the lateral heterogeneity in density and probably seismic velocity in the deep mantle, although the signature of the layer may be partly masked by cancellation of thermal and chemical effects. The interaction of thermal convection with a preexisting dense, homo-geneous layer does not cause short-wavelength heterogeneity *per se* (although sharp edges may be generated), and thus some additional mechanism, such as the introduction of fresh heterogeneity by slabs, must be active to explain seismic observations of short-scale D" variation. Unfortunately, these mechanisms tend to produce a layer boundary too gradual to produce a strong seismic reflection, perhaps leading to a phase change as the prefered explanation of the "Lay discontinuity".

INTRODUCTION

The anomalous D" layer at the base of the mantle has been probed by a range of seismic and other geophysical observations (for recent reviews see [*Loper and Lay*, 1995; *Wysession*, 1996]). The layer is characterized by a shear (and possibly P-) wave discontinuity 150 to 450 km above the core-mantle boundary (CMB), a reduced or negative ve-locity gradient below the discontinuity, and lateral variabil-ity on a range of range of lengthscales from ~250 km (e.g., [*Vidale and Benz*, 1993]) to several 1000s km (e.g., [*Su et al.*, 1994; *Masters et al.*, 1996]). While a negative velocity gradient may be expected for a thermal boundary layer, the shear wave discontinuity cannot be generated in this

The Core-Mantle Boundary Region
Geodynamics 28
Copyright 1998 by the American Geophysical Union.

manner, leading to the suggestion of a chemically-distinct dense layer, or a phase transition [*Stixrude and Bukowinski*, 1990; *Nataf and Houard*, 1993]. Recent observations also suggest the presence of a very thin, ultra low-velocity zone immediately above the CMB [*Garnero and Helmberger*, 1996].

In this paper, we investigate the possibility that D" is both chemical and thermal in origin. How could such a dense chemical layer arise? Three possible origins have been proposed: (i) it could be primordial, i.e., formed during the initial differentiation of the Earth, (ii) it could be formed by chemical reactions with the core, producing an iron-rich layer [*Knittle and Jeanloz*, 1989], and (iii) it could be formed by subducted oceanic crust which may segregate from the residuum component when slabs reach the CMB.

The dynamical consequences of a dense chemical layer have previously been investigated using two-dimensional simulations. The formation of plumes from a thermo-chemical boundary layer with temperature-dependent viscosity was modeled by [*Christensen*, 1984], who found that the dense material is readily incorporated in the plume if the ratio of chemical to thermal buoyancy B is less than one, but remained as a stable layer if this ratio is greater than one. This finding was later corroborated by laboratory experiments of two-layered convection [*Olson and Kincaid*, 1991], in which rapid overturn and mixing occurred for values of B less than one, but stable layering and long-term survival occurred for B greater than one.

If the material remains as a stable layer, a number of things occur. Firstly, the material tends to pile up under upwellings and become thinned under downwellings, with the amount of topography dependent on the Buoyancy ratio. [*Davies and Gurnis*, 1986] found that a dense layer with a density contrast of 2% could result in a discontinuous distribution of material (i.e., zero thickness in places), and cause a bottom topography of several km amplitude, with the CMB depressed under upwellings, the opposite to what is expected from purely thermal convection. These results were corroborated by [*Hansen and Yuen*, 1988] and [*Hansen and Yuen*, 1989] who modeled a more time-dependent system, and also found that the layer could be maintained by a dynamical balance between entrainment and replenishment from the core. Secondly, the dense layer does not survive indefinitely, but is slowly entrained in upwellings, a process which was studied analytically by [*Sleep*, 1988]. Indeed, the analytical models predict that the layer should be completely entrained over geological time, unless its density contrast is at least 6%.

The above models all assume a preexisting layer. Other two-dimensional calculations and one laboratory study have addressed the dynamical feasibility of the two proposed modes of ongoing formation. Can the dense, eclogitic crust of a slab separate out from the thicker, residuum layer when the slab reaches the base of the mantle? This has commonly been modeling by injecting into downwellings two layers of different-density tracers representing oceanic crust and lithosphere, and examining their relative trajectories when they reach the CMB. When the viscosity is constant, the two slab components are unable to separate, and the chemical variations thus give only a small perturbation to the thermal flow [*Gurnis*, 1986]. However, when viscosity is temperature-dependent, segregation of the denser crustal component can indeed occur in the low-viscosity lower boundary layer [*Christensen*, 1989; *Olson and Kincaid*, 1991], and remain there for sufficiently long to explain observed lead and Nd isotope ratios [*Christensen and Hofmann*, 1994]. However, simple mass balance calculations [*Sidorin and Gurnis*, 1998] indicate that the total volume of oceanic crust subducted since the formation of Earth may be too little to account for the observed thickness of the D" layer, particularly since the numerical calculations [*Christensen and Hofmann*, 1994] suggest that only ~20% of the subducted crustal material remains in the layer.

The growth of D" by chemical reactions with the core was modeled by [*Kellogg and King*, 1993], who found that for a buoyancy ratio larger than about 2, a stable layer could form. The required buoyancy ratio is larger than that found in previous studies of a preexisting dense layer probably because the diffusive nature of the formation process causes the composition to change gradually from one extreme to the other, rather than the abrupt change modeled in the studies discussed earlier. A problem with this mode of D" formation is that the reactions are likely to occur only in a very thin layer right at the CMB, with chemical diffusion governing the upwards migration of the material, a very slow process indeed [*Stevenson*, 1993].

In this paper we consider the influence of three-dimensionality on the dynamics of mantle convection with a thermo-chemical D". For these first 3-D results we do not attempt to model the formation of the layer, but simply assume a preexisting layer, and focus on the basic thermal and chemical structures generated, the influence of realistic material properties (e.g., depth-dependent viscosity and other parameters, temperature-dependent viscosity), and the lateral spectrum of heterogeneity at different depths, a quantity which can be directly compared to the heterogeneity spectra from seismic observations.

MODEL

As is common in studies of mantle convection, we make the infinite-Prandtl number approximation and the anelastic-liquid approximation [*Jarvis and McKenzie*, 1980]. The equations, nondimensionalized to the mantle depth (D), thermal diffusion timescale (D^2/κ, where κ=thermal diffusivity) and superadiabatic temperature drop, ΔT_{sa}, are those of continuity:

$$\nabla \cdot (\overline{\rho} \underline{v}) = 0 \qquad (1)$$

conservation of momentum:

$$\nabla \bullet \underline{\underline{\tau}} - \nabla p = Ra.(\overline{\alpha\rho}T - BC)\ \hat{\underline{z}}$$
$$\tau_{ij} = \eta(v_{i,j} + v_{j,i} - \tfrac{2}{3}v_{k,k}\delta_{ij}) \tag{2}$$

conservation of energy:

$$\overline{\rho}\overline{C}_p \frac{DT}{Dt} = -Di_s\overline{\alpha\rho}Tv_z + \nabla \bullet (\overline{k}\nabla T) + \overline{\rho}\overline{H} + \frac{Di_s}{Ra}\tau_{ij}v_{i,j} \tag{3}$$

and conservation of composition:

$$\frac{DC}{Dt} = 0 \tag{4}$$

where the surface dissipation number Di_s is given by:

$$Di_s = \frac{\alpha_s g D}{C_{p,s}} \tag{5}$$

v, p, T, C, τ, and η are velocity, dynamic pressure, absolute temperature, composition, deviatoric stress, and dynamic viscosity, respectively, $\underline{z}$ is a unit vector in the vertical direction, and the barred ρ, α, C_p, κ, and k ($=\rho C_p\kappa$) are depth-dependent reference state parameters density, thermal expansivity, heat capacity, thermal diffusivity and thermal conductivity, respectively. These are discussed later. The subscript 's' indicates the surface value, and the vertical coordinate z runs from 0 at the base to 1 at the surface. The Boussinesq approximation is recovered by setting $Di=0$ and all depth-dependent properties to 1. The Rayleigh number quoted is based on parameters at the surface, with viscosity (when it is temperature-dependent) defined at the reference adiabat:

$$Ra_s = \frac{\rho_s g \alpha_s \Delta T_{sa} D^3}{\eta(T_{as},0)\kappa_s} \tag{6}$$

where g is the gravitational acceleration, and T_{as} is the surface temperature of the reference adiabat. The Rayleigh number based on properties at the CMB is also quoted.

The chemical composition C varies from 0 to 1. The buoyancy parameter B expresses the ratio of compositional to thermal buoyancy in the system:

$$B = \frac{\Delta\rho_c}{\rho_s\alpha_s\Delta T_{sa}} \tag{7}$$

where $\Delta\rho_c$ is the density difference due to compositional variation, assumed for simplicity to be constant throughout the mantle. The above definition is based on surface parameters; the value based on CMB parameters is also

quoted. Note that in equations (1) to (3), chemical variations are neglected except in the buoyancy term of the momentum equation. This is consistent with the Boussinesq or anelastic approximations in which density variations from the reference value are assumed small except in the buoyancy term of the momentum equation, and is justified since B is of order 1. The chemical Rayleigh number Rc used by some authors is simply given by:

$$Rc = Ra.B \tag{8}$$

Viscosity is either constant ($\eta=1$), depth-dependent, or temperature- and depth-dependent. For depth-dependent-viscosity cases, the viscosity increases exponentially by a factor of 10 across the mantle depth and discontinuously by a factor of 30 across the 660 km discontinuity (at $z=0.7716$):

$$\eta(z) = \exp[(1-z)\ln 10].[1 + 29H(0.7716 - z)] \tag{9}$$

where H(x) is the Heavyside step function. The fully temperature- and depth-dependent viscosity law is chosen to give a similar depth-variation along the reference adiabat, plus a strong temperature dependence.

$$\eta(T,z) = A_0 \exp\left[\frac{27.631 + 11(1-z)}{T + T_{off}}\right] \tag{10}$$

where A_0 is calculated such that $\eta(T_{as},1) = 1.0$. T_{off} is the temperature offset added to the nondimensional temperature to reduce the otherwise extreme viscosity variation across the upper boundary, and is taken to be 0.88 for compressible cases (this means that the nondimensional surface temperature used in the viscosity law is 1.0, since the surface temperature boundary condition is 0.12). The nondimensional activation energy of 27.631 is the activation energy of dry olivine [Karato et al., 1986], nondimensionalized to the specified temperature scale.   To prevent a completely rigid lid from forming [Solomatov, 1995; Ratcliff et al., 1997], the maximum viscosity is clipped at 100. Although such clipping may in principle also limit the deep mantle viscosity, in practice the system heats up appreciably when temperature-dependent viscosity is used, so that high viscosity values are not reached in the deep mantle, and this clipping only occurs in the lithosphere. This variation would be reduced if non-Newtonian rheology, which may be important in some regions of the mantle [Karato and Wu, 1993], were included, but this is a topic for the future.

REFERENCE STATE

A simple thermodynamic model is used to calculate $\rho(z)$ and $\alpha(z)$, with C_p assumed constant, and thermal conductivity $k(z)$ assumed to vary with the fourth power of density

[*Anderson*, 1987; *Osako and Ito*, 1991]. The equations are [*Anderson et al.*, 1992; *Duffy and Ahrens*, 1993; *Tackley*, 1996]:

$$\frac{\partial \overline{T}}{\partial z} = -Di.\overline{T} = -Di_s \frac{\overline{\alpha}}{\overline{c}_p}\overline{T}\,;\, \frac{\partial \overline{\rho}}{\partial z} = -K\overline{\rho} = K_0 \frac{\overline{\alpha}}{\overline{c}_p}\left(\frac{\gamma_0}{\gamma}\right)\overline{\rho} \quad (11)$$

$$Di = \frac{\alpha g D}{c_p}\,;\quad K = \frac{\alpha g D}{\gamma c_p} = \frac{Di}{\gamma} \quad (12)$$

$$\gamma\rho = \text{constant} \quad (13)$$

$$\alpha = \alpha_0 \exp\left[-\frac{\delta_{T0}}{n}\left(1 - \left(\frac{\rho_0}{\rho}\right)^n\right)\right] \quad (14)$$

where γ is the Gruneisen parameter, Di is the dissipation number and K is the compressibility number, $\delta_{T0} = 6.0$, and $n=1.4$. For the Boussinesq cases, $Di=K=0$, giving constant coefficients. Thus, in order to construct the non-dimensional reference state, the following surface quantities must be specified: Di_s, T_{as}, K_s and γ_s. This leads to depth variations of ρ, α, and κ, which are all assumed to be 1.0 at the surface. Phase changes at 410 and 660 km depth require specifying a different reference state for each mineral assemblage (from 0-410, 410-660, and 660-2890 km depth). For simplicity, the parameters are taken to be the same, except for the surface density ρ_0 and surface adiabat T_{as} which are chosen to give the PREM density jumps and appropriate adiabatic temperature jumps (due to latent heat) across the two phase transitions.

Table 1 lists assumed parameters for Boussinesq cases (note that since the Boussinesq adiabat is an isotherm, $\Delta T = \Delta T_{sa}$), and Table 2 lists parameters for compressible cases, which were chosen to give good fits to PREM density [*Dziewonski and Anderson*, 1981] and thermal expansivity from recent thermodynamic analyses of the lower mantle [*Anderson et al.*, 1992; *Chopelas and Boehler*, 1992; *Duffy and Ahrens*, 1993]. The resulting depth variation of density, temperature, thermal expansivity, thermal conductivity and thermal diffusivity is illustrated in Figure 1.

Dimensional values are also given to allow quantitative comparison of the results with observations. Typical conversions between nondimensional and dimensional heat fluxes and velocities are also given. The Earth's mantle heat flux [*Stacey*, 1992; *Pollack et al.*, 1993] corresponds to 24-28 nondimensional units.

The radiogenic heating rate (assumed constant with time) is chosen so that internally-heated cases have the same surface heat flow as basally-heated cases. For comparison, the heating rate in carbonaceous chondrites, which are usually thought to be representative of mantle material [*Stacey*, 1992] is 5.2×10^{-12} W kg^{-1}, corresponding to 23.2 nondi-

Table 1. Thermodynamic parameters: Boussinesq cases

Parameter	Symbol	Value	Units	Non-D value
Input and Derived Parameters				
Depth of mantle	D	2890	km	1.0
Temperature drop	ΔT	2500	K	1.0
Expansivity	α	2.5×10^{-5}	K^{-1}	1.0
Conductivity	k	3.0	$Wm^{-1} K^{-1}$	1.0
Heat capacity	Cp	1200	$J\,kg^{-1}K^{-1}$	1.0
Density	ρ	4000	$kg\,m^{-3}$	1.0
Rayleigh number	Ra	-	-	10^6
Reference visc.	η_{ref}	10^{23}	Pa s	1.0
Nominal Dimensional Conversions				
Velocity	v	1	$cm\,yr^{-1}$	1468
Time	t	424	Gyr	1.0
Heat flux	F	2.6	$mW\,m^{-2}$	1.0
Cond. heat flux	F_{cond}	2.6	$mW\,m^{-2}$	1.0

mensional units, which should be reduced to 13.5 nondimensional units in Cartesian geometry to take account of the different surface area:volume ratio [*Tackley*, 1996], but could also be increased to simulate the secular cooling of the mantle, which resembles internal heating both mathematically (in the energy equation) and physically (in its effect on convection) [*Weinstein and Olson*, 1990]. Note that, in the compressible case, the nondimensional heat flux is 2.42 times higher than the Nusselt number, with the latter defined as the ratio of actual heat flux to the conductive heat flux.

NUMERICAL METHOD

The exact details of the numerical method and its benchmarking are described elsewhere [*Tackley*, 1994] and are only briefly summarized here. The instantaneous velocity and pressure fields given by (1)-(3) are calculated by a finite difference (control volume [*Patankar*, 1980]) multigrid technique [*Brandt*, 1982; *Press et al.*, 1992], with vertical grid refinement in the upper and lower boundary layers. Timestepping (equation (4)) is performed explicitly, using the Multidimensional Positive-Definite Advection Transport Algorithm of [*Smolarkiewicz*, 1984] for advection, and second order finite-differences for diffusion, viscous dissipation and adiabatic heating or cooling. Steps of typically 0.5-0.9 the Courant condition are used. The method is well suited to parallel computers, and the presented results were obtained on Intel Paragon and Cray T3E at San Diego Supercomputer Center and the Cray T3E at NASA Goddard Space Flight Center. Benchmarks and accuracy are described in detail in Appendix B of [*Tackley*, 1994].

Phase Transitions. The anomalous buoyancy associated with phase transition deflection is represented as a sheet mass anomaly at the appropriate depth. This technique, previously used by [*Tackley et al.*, 1993; *Tackley et al.*, 1994; *Bunge et al.*, 1997], is similar to the "effective α"

Table 2. Thermodynamic parameters: compressible cases

Parameter	Symbol	Value	Units	Non-D value
Input and Derived Parameters				
Depth of mantle	D	2890	km	1.0
Surface Dissipation number	Di_S	-	-	1.2
Mean dissipation #	$<Di>$	-	-	0.441
Surface Gruneisen parameter	γ_S	-	-	1.091
Superadiabatic temperature drop	ΔT_{sa}	2500	K	1.0
Surface temperature	T_s	300	K	0.12
CMB Temperature	T_c	3700	K	1.48
Expansivity: surface	α_s	5.0×10^{-5}	K^{-1}	1.0
Conductivity: surface	k_s	3.0	$Wm^{-1}K^{-1}$	1.0
Heat capacity	C_p	1200	$J\,kg^{-1}K^{-1}$	1.0
Density: surface	ρ_s	3300	$kg\,m^{-3}$	1.0
Density: CMB	ρ_b	5467	$kg\,m^{-3}$	1.657
400 density jump	$\Delta\rho_{400}$	280	$kg\,m^{-3}$	0.6784
660 density jump	$\Delta\rho_{660}$	400	$kg\,m^{-3}$	0.9696
Clapeyron slope: 400	γ_{400}	3.0	$MPa\,K^{-1}$	0.0668
Clapeyron slope: 660	γ_{660}	-3.0	$MPa\,K^{-1}$	0.0668
Surface Ra	Ra_s	-	-	10^7
Basal Ra	Ra_b	-	-	3.894×10^3
Reference viscosity	η_{ref}	1.3×10^{22}	Pa s	1.0
Nominal Dimensional Conversions				
Velocity	v	1	$cm\,yr^{-1}$	1211
Time	t	350	Gyr	1.0
Heat flux	F	3.53	$mW\,m^{-2}$	1.0
Conductive heat flux	F_{cond}	8.575	$mW\,m^{-2}$	2.43

approach [*Christensen and Yuen*, 1985] shrunk into a discontinuity. Latent heat release is included in the reference adiabat and is treated numerically by advecting the superadiabatic temperature, rather than absolute temperature. The phase transition implementation has been validated by comparing to previous results, both in a 2-D Cartesian box [*Christensen and Yuen*, 1985], and 3-D spherical shell [*Tackley et al.*, 1994; *Bunge et al.*, 1997].

Chemical field. The denser material is represented by a swarm of tracer particles. For statistical reasons, several 10s of tracers are required per cell [*Christensen and Hofmann*, 1994]. A tracer particle method has the advantages over a continuum field method of having zero chemical diffusion and an ability to represent filaments of material smaller than the grid scale; and the advantages over a marker chain method of being easy to implement (particularly in 3-D) and not growing exponentially in computational cost when the boundary gets complicated [*van Keken et al.*, 1997]. Tracers are advected using a fourth-order Runge-Kutta technique. Chemical buoyancy is computed at each timestep by converting the tracers to the continuum C field using a cell-counting method.

Accuracy becomes an important concern in thermo-chemical convection calculations. [*van Keken et al.*, 1997] present a benchmark for the entrainment of a dense layer at the base of the mantle, and find that for accurate results, the

Reference State

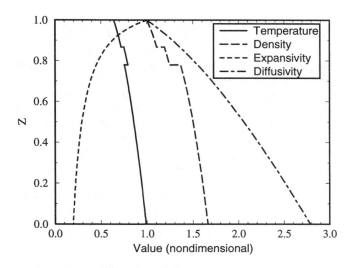

Figure 1. Depth variation of nondimensional reference state parameters temperature, density, thermal expansivity, and thermal diffusivity. The base of the mantle is at $z=0$, with the top at $z=1$. Discontinuous jumps are due to the phase transitions at 410 and 660 km depth.

Thermo–chemical Benchmark #2

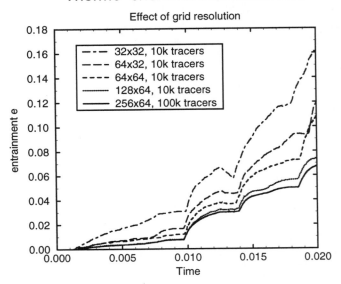

Figure 2. Entrainment vs. time for the thermo-chemical benchmark #2 of [*van Keken et al.*, 1997] (compare with their Figure 12), calculated with various grid resolutions and number of tracer particles. This benchmark models the entrainment of a dense chemical layer. Entrainment 'e' is defined as the fraction of the compositionally dense material that is entrained above $z=0.2$. The higher resolution results (128x64 or 256x64) are compatible with the high resolution results reported in [*van Keken et al.*, 1997]. Lower resolution results in a higher rates of entrainment, although the convective pattern looks qualitatively similar.

grid resolution must be appreciably higher than that required for thermal convection, and that even then, results computed using different codes diverge after a fairly short time interval. A test of the present code against this benchmark for the period during which all codes obtained the same solution (up to time 0.02) is shown in Figure 2 (directly comparable to Figure 12 in [*van Keken et al.*, 1997]). This figure shows the amount of entrainment vs. time calculated using different resolutions, where the cells are equally-sized in the x-direction but refined in the boundary layers by up to a factor of ~3 in the z-direction. The highest resolution cases (128x64 and 256x64) give similar entrainment vs. time curves to those obtained in the high resolution cases reported in [*van Keken et al.*, 1997], whereas lower resolution leads to unrealistically high entrainment by up to a factor of ~2.5, although the thermal and chemical fields look qualitatively similar. The results did not appear to be particularly sensitive to the number of tracers used.

A resolution test has also been performed for a 2-D version of the Boussinesq, basally-heated case presented later in this paper. Plate 1 compares results computed using a

128x32 grid and 25,000 tracers with results computed using a 512x64 grid and 200,000 tracers. As can be seen, the convective patterns look qualitatively the same. However, the rate of entrainment is significantly higher in the lower resolution case, which is why it is shown at an earlier time. One reason for this innacuracy may be the low tracer density (i.e., #tracers per cell) in the filaments of entrained material, the effect of which must be quantitatively determined in the future.

Another aspect of accuracy is the cumulative error in advecting tracer particles, which can be measured by advecting a single particle in a prescribed, steady-state velocity field, for a length of time which should return it to exactly the same position [*van Keken et al.*, 1997]. For the tests described above, and most of the presented results, first-order velocity interpolation was used, giving a cumulative position error of $O(10^{-3})$. When the velocity interpolation was upgraded to second-order, the error dropped to $O(10^{-5})$.

From these tests, it is concluded that: (i) high grid resolution is important if a *quantitatively* accurate entrainment rate is central to the problem being studied, but (ii) the general *qualitative* behavior of the system, e.g., large-scale structure and dynamics, general appearance, is robust with respect to grid resolution. This is also evident in the results presented in figures 8 to 12 in [*van Keken et al.*, 1997].

Since the resources necessary to calculate a suite of 3-D convection calculations with the resolution necessary to accurately capture entrainment rate are not yet available, we must for the moment be satisfied with results which are robust wrt. the general structure and processes, but may overestimate entrainment.

Experimental Procedure

An 8x1 (2-D) or 8x8x1 (3-D) box with periodic side boundaries and free-slip horizontal boundaries was used, with the top boundary isothermal and the bottom boundary either isothermal (basally-heated cases) or insulating (completely internally-heated cases). For each combination of thermodynamic and heating parameters, an isochemical case (i.e., with no dense chemical layer) was run until it reached statistically steady-state; this provided the temperature field for initializing chemically-layered cases. The chemical field was initialized with a dense chemical layer of thickness 0.1 (290 km) at the base of the mantle. Simulations were run until the system had adjusted to the presence of the dense layer, or in many cases, until the layer was completely entrained. The simulation time was typically in the range 0.03-0.07 nondimensional time units (scaling to real time is discussed later), requiring 10-20,000 numerical time steps.

Three sets of cases were performed: (i) Boussinesq, constant properties, no phase transitions. (ii) Compressible, depth-dependent viscosity and other properties, with phase transitions. (iii) As (ii), with a viscosity that is dependent on temperature as well as depth. Experiments were heated

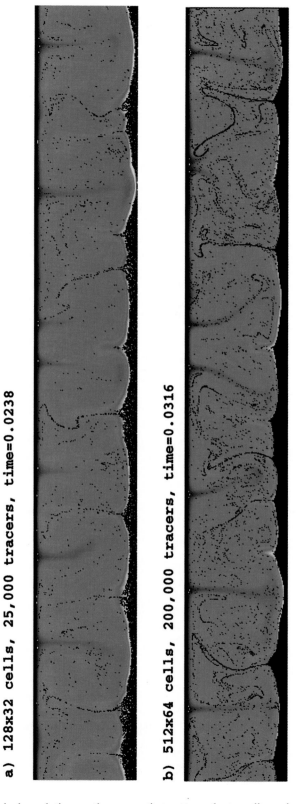

a) 128x32 cells, 25,000 tracers, time=0.0238

b) 512x64 cells, 200,000 tracers, time=0.0316

Plate 1. The effect of numerical resolution on the convection pattern: the two-dimensional, Boussinesq case with B=1.0, calculated with two different resolutions and number of tracer particles. (a) 128 by 32 cells, 25000 tracers, time=0.0238, and (b) 512 by 64 cells, 200,000 tracers, time=0.0316. The convection patterns look qualitatively identical, even though there are large quantitative differences in the rate of entrainment.

Table 3. Simulations performed.

Formulation	F	$v_{surf, rms}$	Phase Buoyancy Parameter P						
			0.03125	0.0625	0.125	0.25	0.5	1.0	2.0
Bouss, B	19.4	910					•	•	◊
Bouss, I	19.4	280				◊	•	◊	◊
Compr, B	17.5	1150				◊	•	◊	
Compr, I	18.0	870			◊	•	◊		
C+η(T), B	15.8	620		◊	◊	•	◊		
C+η(T), I	18.0	750	◊	◊	•	◊			

'◊' indicates a 2-D simulation only, '•' indicates both 3-D and 2-D simulations, 'B' means basally-heated, 'I' means internally-heated, F is the heat flux, and $v_{surf,rms}$ is the rms. surface velocity.

either 100% from below, or 100% from within, in order to elucidate the basic physics associated with the different heating modes. The radiogenic heating rate for the internally-heated cases was chosen to give the same surface heat flow as the equivalent isochemical basally-heated case. This allows the same temperature scale to be used for all cases, which is important when temperature-dependent viscosity and phase changes are included. For each set, a suite of 2-D calculations with varying chemical buoyancy ratio B were performed. A subset of the cases were then run in 3-D.

RESULTS

The simulations performed are summarized in Table 3. The 3-D cases were chosen to be the ones with the smallest B for which the dense layer survived for a significant length of time, i.e., a reduction of B by a factor of 2 would result in rapid overturn and mixing of the dense layer, as seen in [*Olson and Kincaid*, 1991]. For reasons of length this paper will focus only on six 3-D cases, one for each parameter combination. R.m.s. surface velocities (Table 3) all scale to less than 1 cm/year using the conversions listed in Tables 1 and 2, suggesting that the Rayleigh number should be increased by an order of magnitude to reach the Earth-like parameter range, although a smaller increase is indicated by the heat fluxes. The low surface velocities also affect the scaling of times to dimensional values: a conversion based simply on thermal diffusion timescales (as in Tables 1 and 2) would lead to run times in the range 10-30 Gyr, numbers which should perhaps be reduced by a factor of 5 to reflect the lower than Earth-like velocities.

Boussinesq Cases

Plate 2 shows the time evolution of the basally-heated case with B=1.0, showing the initial state, a time when the system has adjusted to the presence of the layer, and a later time when the layer has been mostly entrained. Plate 3 shows the initial state and a well-adjusted state for the internally-heated case. The presence of the dense layer clearly has substantial effects on the pattern of convection and on the temperature distribution, in addition to displaying its own interesting dynamics.

Thermal planform. The chemically-stable layer at the base of the mantle changes the planform and horizontal lengthscales of the convection, dramatically so in the basally-heated case. In that case, the planform changes from one characterized by upwelling and downwelling sheets (for isochemical convection), to one dominated by upwelling and downwelling plumes. In addition, the horizontal lengthscale has become much smaller, i.e., the upwellings and downwellings are now much more closely spaced. This can be understood by considering that due to the dense stable layer, the 'effective' lower boundary condition for the upper ~0.9 of the mantle has changed from being free-slip to rigid: the convective pattern observed is thus characteristic of rigid-lid convection [*Tackley*, 1993; *Ratcliff et al.*, 1997]. In internally-heated convection, the dynamics are dominated by the upper boundary layer; the lower boundary condition does not have much effect, and so the effect on planform is much lower, although still noticeable.

Chemical structure and entrainment. In both cases the dense chemical material is thinned under downwellings and thickened under upwellings, such that the thick regions form an interconnected series of ridges. This pattern is commonly known as the spoke pattern in thermal convection experiments, and has apparently also been observed in laboratory experiments with a dense chemical layer [*U.R. Christensen, personal communication*]. Entrainment of the layer occurs in cylindrical plumes which form at the intersection of these ridges. Plate 4 shows a zoomed-in view of the chemical and thermal fields for the two cases, highlighting this cylindrical entrainment.

It is perhaps surprising that internally-heated system also displays cylindrical entrainment. In a 100% internally-heated system, there are no active upwellings: the downwellings are usually cylindrical and the upwelling is in the form of a general distributed return flow. Despite this lack of active upwellings, the chemical field displays very similar dynamics, forming a spoke pattern and cylindrical entrainment plumes. One effect which may be partly responsible for this is that the lower layer heats up, due to radiogenic heat sources and the inability of this volumetric heating to escape by means other than rather slow thermal conduction, and may thus become a heat source for active upwellings. How rapid is this heating up? The chondritic abundance of

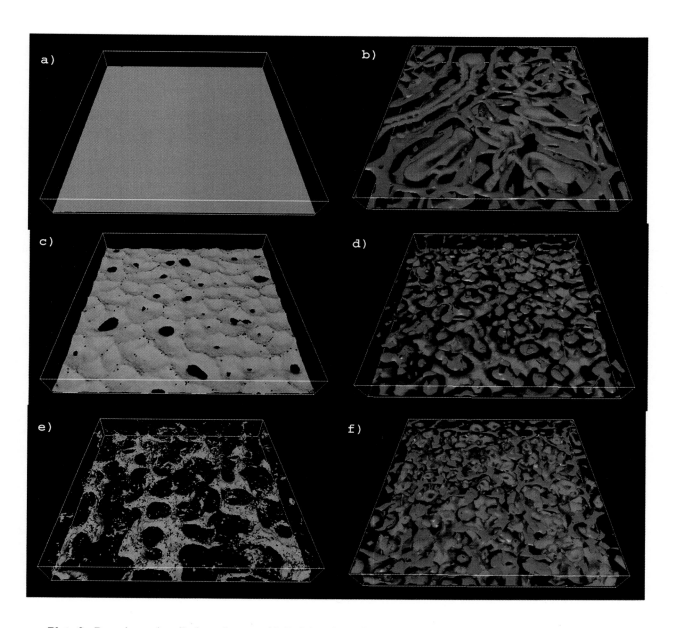

Plate 2. Boussinesq, basally-heated case with B=1.0 at three different times. The left parts (a., c., and e.) show an isocontour of the compositional field, corresponding to C=0.3, and the right parts (b., d., and f.) show isosurfaces of the residual temperature field, showing where the temperature is 0.1 hotter (red) or 0.1 colder (blue) than the horizontal average. The times are t=0 (a. and b.), t=0.0345 (c. and d.) and t=0.0693 (e. and f.).

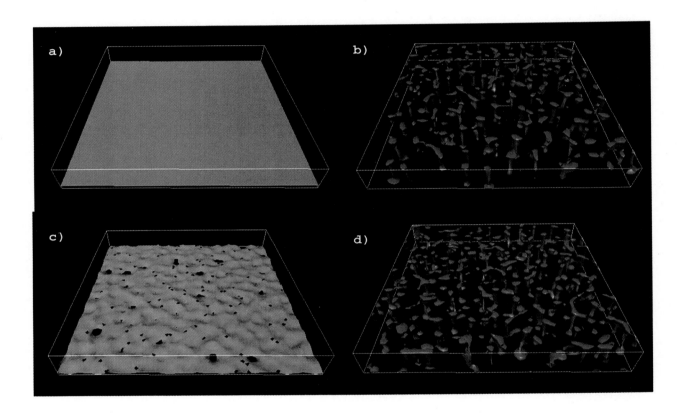

Plate 3. Boussinesq, internally-heated case with B=0.5 at two different times: the initial state (a. and b.), and t=0.0463 (c. and d.). Compositional isocontour C=0.3 (a. and c.), and residual temperature isocontour t=-0.1 (b. and d.).

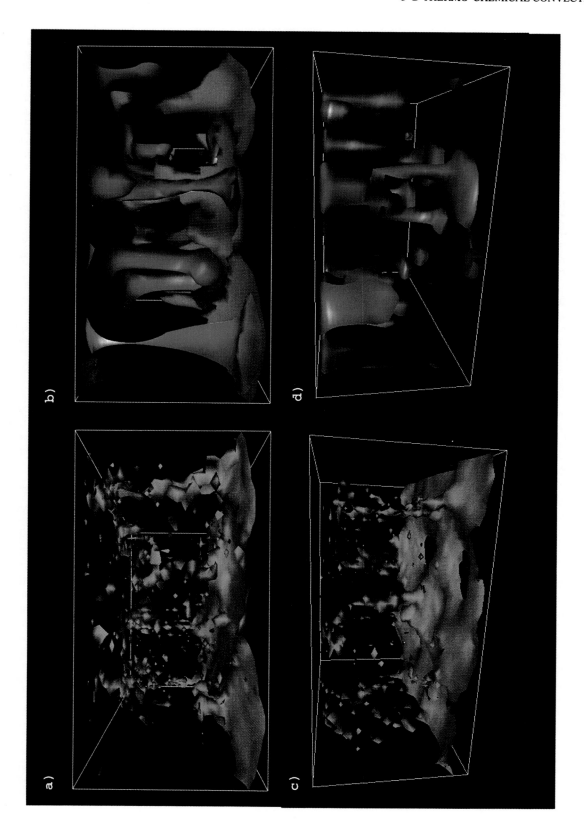

Plate 4. Zoom in views of the Boussinesq basally-heated case (top) and internally-heated case (bottom) showing cylindrical entrainment features. Composition isocontour $C=0.08$ (a.) and $C=0.1$ (c.). Residual temperature isocontours are $T=0.1$ (red) and $T=-0.1$ (blue).

heat-producing elements would cause a temperature rise of 136 K per billion years, based on current radioactive decay rates. Although this rate is small, it may be important if the layer can survive for the age of the Earth, particularly since the rate of heat production was higher in the past.

The rate of entrainment is much lower in the 100% internally-heated cases than in the 100% basally-heated cases, for the same buoyancy parameter B. For example, with B=0.5, the dense layer experiences rapid overturn and mixing for basal heating, but is stable and survives for a long time with 100% internal heating. This is not surprising since there are no active upwellings to entrain the dense material. In addition, the temperature drop across the mantle, ΔT, is lower in the internally-heated case even though surface heat flow is the same: if this lower ΔT were used in the calculation of B, a larger B would result. If ΔT over the chemical layer were used instead of total ΔT, an even larger difference would be found. At first glance this finding would appear to be at odds with [Christensen and Hofmann, 1994], who found that the fraction of internal heating did not affect the rate of entrainment of a dense layer. The main differences are probably that all their cases included mixed heating from below and within so that active upwellings always existed, and that the internal heating was varied over a smaller parameter range. When basal heating is switched off altogether, entrainment of the layer is considerably slower.

Temperature profiles. The dense layer also has a strong effect on the horizontally-averaged temperature profile (geotherm), as illustrated in Figure 3, which compares temperature profiles for isochemical and layered cases. With basal heating (Figure 3(a)), the dense layer acts as a thermal blanket over the core, reducing the interior temperature by up to 0.17 (nominally 425 K), and increasing the temperature drop over the lower thermal boundary layer by a corresponding amount. This larger ΔT results in stronger upwellings which are able to entrain the dense layer more effectively. Thus, the rate of entrainment increases as time goes on. At a later time (dashed line) the layer has been substantially entrained, the thermal blanketing is lower, and the internal temperature is rising again.

With pure internal heating (Figure 3(b)), the effect on internal temperature is minimal, but a new region of high temperatures is observed in the dense layer. This is due to trapped radiogenically-produced heat, as discussed above, and results in horizontally-averaged temperature anomalies of up to 0.16 (nominally 400K). The internal temperature profile is subadiabatic, as expected for substantially internally-heated convection [Schubert, 1992; Parmentier et al., 1994; Tackley, 1996].

Compressible, Depth-Dependent Properties

Structures. The thermal and compositional fields for basally- and internally-heated cases at times when the system has adjusted to the presence of the dense layer are shown in Plate 5. Initial temperature distributions are not shown because they look very similar to the illustrated temperature distributions, except for features being somewhat narrower in the deep mantle. The convective planforms are very stable and the long wavelength structure does not change much with time, as is characteristic for convection with strongly depth-dependent properties [Balachandar et al., 1992; Hansen et al., 1993; Zhang and Yuen, 1995; Bunge et al., 1996]. The phase transitions do not appear to be having much effect on the convection, probably due to the viscosity increase at 660 km depth [Monnereau and Rabinowicz, 1996; Bunge et al., 1996].

In contrast to the Boussinesq, constant-property cases, (i) the dense material does not have a large effect on the thermal structure, and (ii) the dense material is swept into very large piles (or ridges in the internally-heated case), which extend vertically most of the way through the lower mantle. These two effects are no doubt related: since the dense material is swept into discrete piles it no longer acts as a thermal blanket over the core, and it no longer acts as an effectively rigid boundary condition on the flow above.

Large topography. Why does the dense material form large, thick piles instead of a continuous, varying-thickness layer as in the constant-property cases? An obvious explanation might be that the buoyancy parameters B_{surf} are lower in these cases, at 0.5 and 0.25 compared to 1.0 and 0.5, which would allow thermal buoyancy to lift the material further. However, it is probably more appropriate to calculate B using the density and thermal expansivity near the CMB, since this is where the dense material resides. This results in B_{cmb} of 1.5 and 0.75, actually larger than the constant properties cases.

The main reason for the increased topography is the low local Rayleigh number in the deep mantle (Ra_{cmb}=4x10^3) due to depth-dependent viscosity, thermal diffusivity, and thermal expansivity. The effect of Ra on dense layer topography is illustrated in Plate 6, for a two-dimensional, constant-properties scenario with B=1.0. At Ra=10^6, a continuous, varying-thickness layer is formed, but at Ra=10^4, discrete piles are formed, somewhat similar to the piles in the 3-D compressible cases. The reason for this increased topography is the increased stresses associated with the thicker upwellings and downwellings in the lower Ra case. (Note that with the conventional thermal nondimensionalization, nondimensional stresses *increase* with increasing Ra, even though *dimensional* stresses decrease with increasing Ra). If the stress scaled as the thickness of the upwellings and downwellings, i.e., approximately $Ra^{-1/3}$, this would predict ~4.5 times as much topography in the lower Ra case, which appears to be an upper bound on what is observed.

Temperature profile. As with the three-dimensional structure, the dense material does not have a large effect on the horizontally-averaged temperature (Figure 3(c) and (d)). The lower boundary layer in the basally-heated case extends over the lower ~0.25 (700 km) of the mantle, rather larger

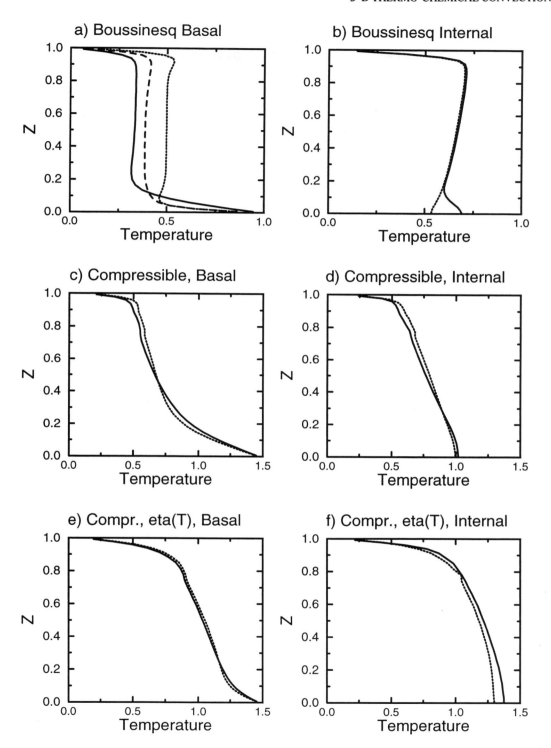

Figure 3. Temperature profiles (geotherms) for the six 3-D cases reported. Dotted lines indicate initial temperature profiles (i.e., for isochemical cases), whereas solid lines indicates the temperature profiles for the times illustrated in the Figures 4, 5, 8 and 10. For the Boussinesq, basal case, an additional dashed line shows the profile for the third time illustrated in Plate 2.

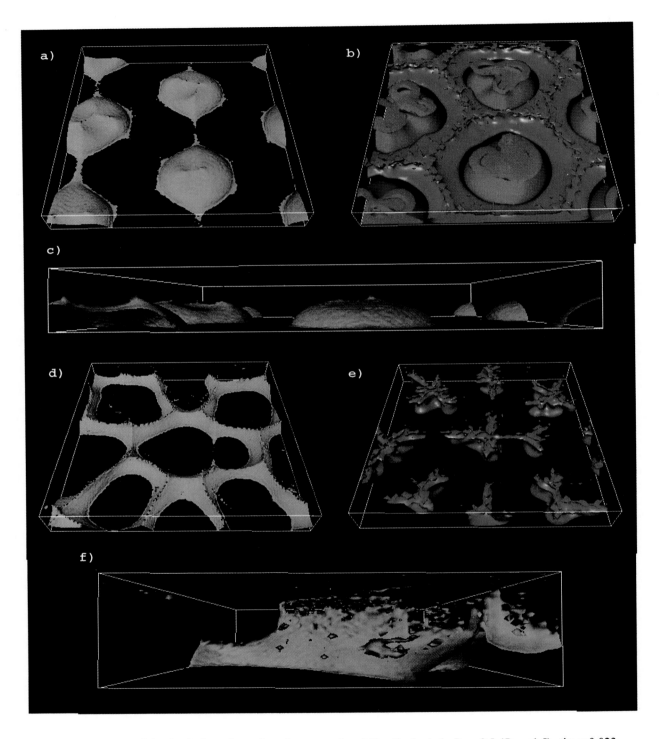

Plate 5. Compressible, depth-dependent viscosity cases. (a.-c.) Basally-heated, B_{surf}=0.5 (B_{cmb}=1.5), time=0.023, and (d.-f.) internally-heated, B_{surf}=0.25 (B_{cmb}=0.75), time=0.026. Compositional isocontour C=0.3 (green, a., c., d., and f.) and residual temperature isocontours 0.1 (red) and -0.1 (blue).

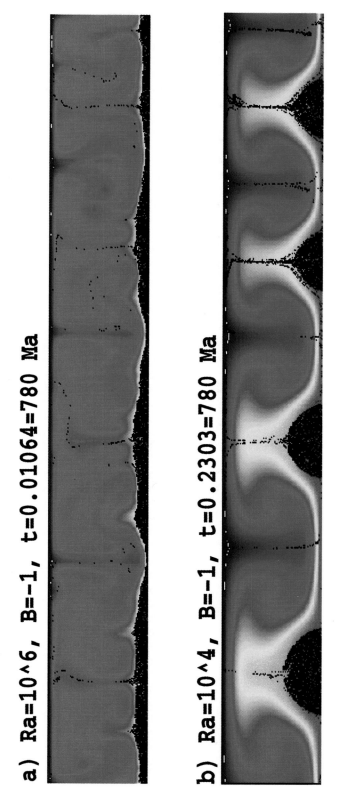

Plate 6. Effect of Rayleigh number on the topography of the dense layer, for two-dimensional, basally-heated Boussinesq cases with B=0.1. (a) Ra=10^6, t=0.0106, (b) Ra=10^4, t=0.2303. Note that when scaled by the velocities, these two times are almost identical.

than commonly assumed, but consistent with previous calculations of convection with reasonable depth-dependent properties [*Balachandar et al.*, 1992; *Tackley*, 1996].

Temperature-Dependent Viscosity

Structures. Thermal and compositional fields for basally- and internally-heated cases at times when the system has adjusted to the presence of the dense layer are shown in Plate 7. As with the depth-dependent viscosity cases, large piles of dense material are formed, and the dense material does not have a huge effect on overall large-scale thermal structure.

The buoyancy parameters needed to get long-term stability are reduced by a factor of two compared to the depth-dependent viscosity cases, at B_{surf}=0.25 (B_{cmb}=0.75) and B_{surf}=0.125 (B_{cmb}=0.375). The two main reasons for this are probably (i) due to the high viscosity upper boundary layer the interior temperature is higher, reducing the temperature drop over the CMB boundary layer, hence the thermal buoyancy available for upwellings to entrain the dense material, and (ii) due to the temperature-dependence of viscosity, the dense material is less viscous than the mantle above, and thus more difficult to entrain [*Kellogg and King*, 1993]. It has also been suggested that dense D" material may be more viscous than the mantle above, also making it more difficult to entrain [*Manga*, 1996]. However, any intrinsic compositional effects are likely to be overwhelmed by the very strong dependence of viscosity on temperature [*Sleep*, 1988].

The internally-heated planform perhaps now looks the more 'Earth-like' of the two, with long-wavelength structure and chemical structure that is dominated by a single pile with ridge-like extension. The chemical pile has heated up due to the radiogenic heat sources, and results in a hot temperature structure which closely mimics the chemical structure. The density effects of the thermal and composition fields largely cancel out.

The temperature profile (Figure 3 (e) and (f)) is again little changed by the chemical field, the main effect being an increase in temperatures in the internally-heated case, presumably due to the inhibiting effect of the dense material on thermal convection, particularly within the dense pile.

Spectral Heterogeneity Maps

A useful tool for making statistical comparisons between convection models and seismic tomographic models is the Spectral Heterogeneity Map (SHM), first introduced by [*Tackley et al.*, 1994]. These show the lateral spectrum of a field (e.g., temperature or seismic velocity) as a function of depth, in the form of a contour plot. SHMs for four of the 3-D cases presented are shown in Figures 4 and 5. The 'spatial frequency', plotted on the horizontal axis, is simply the inverse wavelength

$$sf = 1 / \lambda \qquad (15)$$

such that the fundamental mode of the 8x8x1 box has a spatial frequency of 0.125, and a spatial frequency of 1 corresponds to a wavelength of ~3000 km, corresponding roughly to spherical harmonic *L*=13.

These figures show the thermal and compositional fields, and also the total density anomaly field, which is a combination of the two, given by:

$$\rho = \overline{\alpha \rho}.T - B_s.C \qquad (16)$$

It would also be interesting to plot the seismic velocity field, for direct comparison with seismic tomography. However, the scaling factor between temperature and seismic velocity is somewhat uncertain in the deep mantle, and the scaling factor between composition and seismic velocity is arbitrary since we have not assigned any particular composition to this field.

The Boussinesq, basally-heated, isochemical case (Figure 4 (a)) has a strong (and equal) concentration of heterogeneity in the upper and lower thermal boundary layers, with little in-between, as is normal for this configuration [*Jarvis and Peltier*, 1986]. Adding the dense chemical layer (part (c)) results in much more heterogeneity in the lower TBL than in the upper one, and shifts the peak to shorter wavelengths (higher spatial frequency). The chemical field (part (e)) displays strong heterogeneity at the base of the mantle, as expected. The density field (part (g)) displays heterogeneity in both the upper and lower boundary layers, with the lower boundary layer being somewhat stronger.

Two interesting points arise from the above plots. (i) The vertical extent of D" density heterogeneity is less than that of either chemical or thermal heterogeneity, i.e., estimating the thickness of D" from the density field would result in a smaller estimate than by using the thermal or compositional field individually. This is because the system arranges itself in such a way that thermal and compositional densities tend to cancel each other out, particularly near the top of the compositional layer. (ii) Whereas temperature heterogeneity is zero right at the CMB (because of the isothermal boundary condition in basally-heated cases), compositional heterogeneity reaches a peak at this depth. This could, in principle, be used as a way of distinguishing thermal from compositional heterogeneity near the base of D".

The internally-heated Boussinesq case displays similar trends to the basally-heated case, except that the dominant heterogeneity is shorter-wavelength and is less affected by the dense layer, and the thermal heterogeneity does not go to zero at the CMB because the boundary condition is zero flux rather than isothermal. An isothermal boundary condition is more realistic for the Earth, however, so point (ii) above still stands.

Due to length limitations, only the compressible, temperature-dependent viscosity cases are shown (Figure 5). In both cases, long-wavelength heterogeneity dominates. In the basally-heated result (left column), the heterogeneity associated with the lower TBL now extends through a large

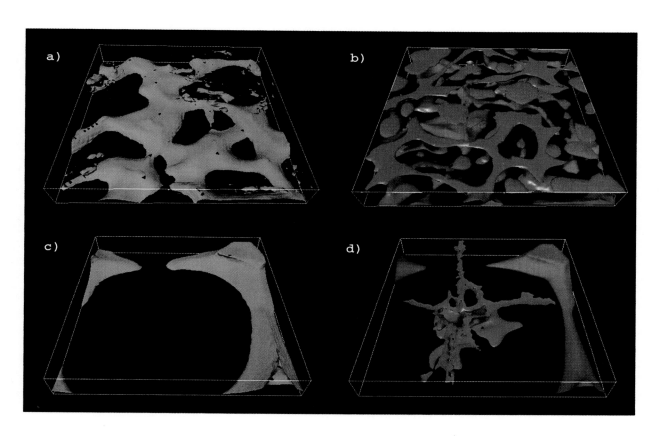

Plate 7. Compressible, temperature- and depth-dependent viscosity cases. (a.-b.) Basally-heated, $B_{surf}=0.25$ ($B_{cmb}=0.75$), time=0.0321, and (c.-d.) internally-heated, $B_{surf}=0.125$ ($B_{cmb}=0.375$), time=0.0367. Compositional isocontour $C=0.3$ (green, a. & c.) and residual temperature isocontours 0.1 (red) and -0.1 (blue).

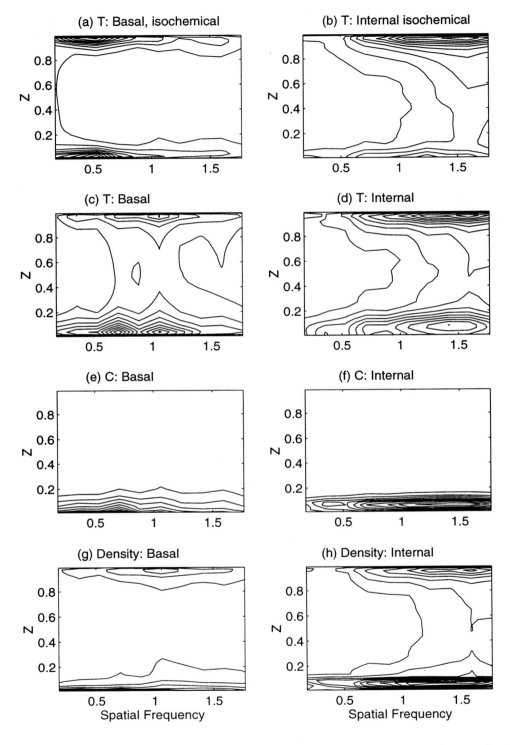

Figure 4. Spectral heterogeneity maps (i.e., contour plots of the rms. amplitude of heterogeneity as a function of depth and horizontal spatial frequency) for the Boussinesq cases. (a., c., e., and g) basally-heated case, and (b., d., f., and h.) internally-heated case. (a. and b.) show the initial temperature field (for isochemical convection), the other parts show fields for the times illustrated previously: (c. and d.) temperature field, (e. and f.) chemical field, (g. and h.) total density. A spatial frequency of 1 is roughly equivalent to spherical harmonic degree 13.

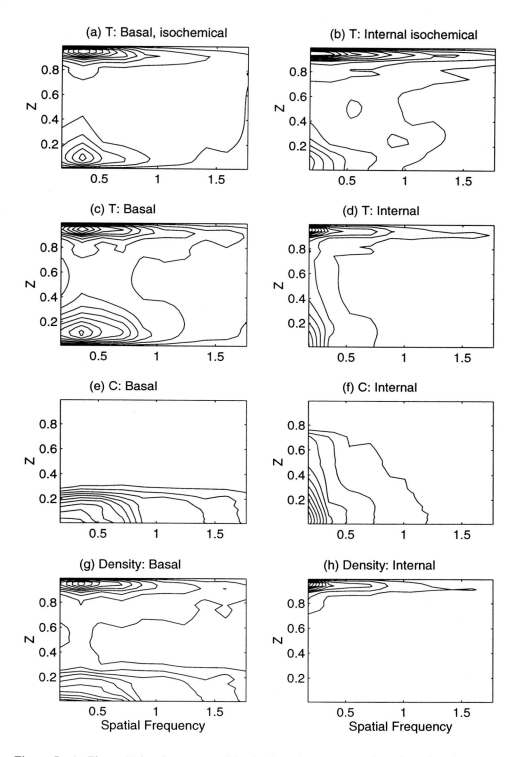

(a) T: Basal, isochemical

(b) T: Internal isochemical

(c) T: Basal

(d) T: Internal

(e) C: Basal

(f) C: Internal

(g) Density: Basal

(h) Density: Internal

Spatial Frequency

Spatial Frequency

Figure 5. As Figure 11 but for compressible, depth- and temperature-dependent viscosity cases.

fraction of the lower mantle, consistent with the geotherm plot (Figure 3(e)), and is slightly increased by the addition of the dense material. The chemical heterogeneity also extends to mid way through the lower mantle, but density heterogeneity is more confined due to the cancellation of thermal and chemical buoyancy in the mid-mantle. The internally-heated case displays the same trends, with cancelation of thermal and chemical buoyancy effects so effective that the signature of the dense layer is lost in the density SHM. The compressible cases with depth-dependent viscosity (not illustrated here) have similar SHM except that the upper boundary layer has a lower amplitude.

The compressible cases all display an abrupt change in the amplitude of heterogeneity and of the lateral spectrum at mid-lower-mantle depth. This is compatible with seismic tomographic models, for example [Su et al., 1994], which also display an increase in amplitude and reddening of the spectrum at this depth [Tackley et al., 1994; Schubert and Tackley, 1995].

DISCUSSION

Thermo-Chemical Stationary Megaplumes

The compressible simulations display large piles of material which extend much of the way through the lower mantle and can survive over geological time. For example, the internally-heated, temperature-dependent-viscosity case is illustrated in Plate 7 at a time of 0.0367 (which corresponds to somewhere between 1 and 13 Gyr depending on whether the r.m.s. surface velocity or thermal diffusion is used for the dimensional scaling), and shows very little entrainment.

"Megaplumes" which may be compatible with these structures are observed under Africa and the Pacific in global seismic tomographic models [Su et al., 1994; Masters et al., 1996; Grand et al., 1997; van der Hilst et al., 1997]. These megaplumes extend vertically at least half way through the lower mantle but then appear to stop, with no obvious low-velocity vertical continuation through the rest of the mantle, a surprising behavior if they are interpreted purely as hot thermal structures, which should rise without inhibition throughout the mantle, perhaps thinning as they rise due to decreasing viscosity. The present results suggest that megaplumes may be both chemical and thermal in origin, with their topography maintained by a balance between positive thermal buoyancy and negative chemical buoyancy, thus explaining why they do not move upwards. Internal circulation occurs within the megaplumes, and they may heat up due to radiogenic heat trapped over geological time, as well as heat conducted through the CMB. The regions above these megaplumes appear to be preferential locations for hotspots. In the proposed scenario, plumes would arise off the top of these megaplumes and might entrain some of the chemically anomalous material.

In addition to (i) the apparent stationarity of the mega-plumes discussed above, this thermo-chemical hypothesis might also explain: (ii) The location of a geochemical reservoir, perhaps the high-^{3}He/^{4}He reservoir, and perhaps containing the heat-producing elements which appear to be missing in the MORB source. The material is slowly entrained from its top by 'hotspot' plumes and/or the background mantle flow. (iii) The large variation of seismic velocities in the deep mantle, apparently too large to be explained purely by thermal effects [Wysession, 1996], an observation which has already led to the inference of chemical variations [Yuen et al., 1993; Cadek et al., 1994; Yuen et al., 1996]. (iv) The breakdown in scaling between shear and compressional velocities in the deep mantle [Bolton and Masters, 1996; Robertson and Woodhouse, 1996]. The main breakdown appears localized to a region under the Pacific [Bolton and Masters, 1996], the location of a 'megaplume'. (v) The change in character of seismic heterogeneity at mid-lower-mantle depths, as seen by a change in lateral spectrum [Tackley et al., 1994] and a broadening of the radial correlation function [Jordan et al., 1993]. (vi) The best fit to the geoid is obtained when the only buoyancy forces are provided by downwelling slabs [Ricard et al., 1993], suggesting that slow regions may be neutrally buoyant.

Density Contrast Required for Stable Layering

While the present study does not attempt to closely define the transition from rapid overturn to stable layering (taking Buoyancy ratios that are separated by a factor of two), it is useful to consider what constrains it does place on the required density contrast. The density contrasts associated with various Buoyancy ratios for Boussinesq and compressible scenarios are listed in Table 4.

For the Boussinesq cases, the answer depends proportionally on what the average mantle thermal expansivity is taken to be, and on what temperature drop is assumed. For the values listed in Table 1, the results indicate that for basal heating a density contrast of 6.25% results in stable layering, whereas a density contrast of 3.125% results in rapid overturn.

Internal heating reduces the required density contrast, as does compressibility (mainly because of the decrease of thermal expansivity with depth), and temperature-dependent viscosity. For the case with all three of these components, a density contrast of only 0.94% of CMB density, or 1.56% of surface density, is sufficient for long-term stability of the chemical layer.

How 'long-term' is the layer's stability? [Sleep, 1988] calculates analytically that a 6% density contrast is required for the layer to remain over the age of the Earth, assuming $\alpha = 2 \times 10^{-5}$. This would reduce to 3% if a more realistic deep-mantle α of 1×10^{-5} were used. The numerical results in this present study, which may substantially overestimate the entrainment rate (see earlier discussion) suggest that a dense layer can survive over geological timescales with a smaller density contrast than this, although this needs to be

Table 4. Equivalent density contrasts for buoyancy ratios (B)

B_{surf}	Boussinesq		Compressible			
	$\Delta\rho$ (kg/m^3)	$\Delta\rho$ (%)	$\Delta\rho$ (kg/m^3)	$\Delta\rho/\rho_{surf}$ (%)	$\Delta\rho/\rho_{cmb}$ (%)	B_{cmb}
1.0	250	6.25	412.5	12.5	7.5	3.0
0.5	125	3.125	206.25	6.25	3.75	1.5
0.25	62.5	1.562	103.125	3.125	1.875	0.75
0.125	31.25	0.781	51.5625	1.5625	0.9375	0.375

more specifically and accurately addressed in the future. The density contrast also affects the topography on the top of the layer: lower density contrasts result in larger topography.

CONCLUSIONS

Simple 3-D Boussinesq, constant-properties simulations of mantle convection with a preexisting dense chemical layer at the base indicate that, if the dense layer has a sufficient density contrast to be stable for long periods (i.e., ~5% or more): (i) the layer forms a spoke pattern, with an interconnected network of ridges, (ii) entrainment occurs in cylindrical plumes rising from the intersections of these ridges, even in 100% internally-heated cases, (iii) the layer heats up even with zero basal heat flux because of trapped radiogenically-produced heat, (iv) thermal blanketing of the core reduces the interior mantle temperature, (v) the layer acts as an effective rigid lower boundary condition, reducing the horizontal lengthscales of the thermal convection, and (vi) internal heating reduces the ability of the thermal convection to entrain the layer.

With depth-dependent viscosity and other properties included within the framework of a compressible formulation, (i) the density contrast required for long-term stable layering is reduced, and (ii) material may be swept into large, thermo-chemical megaplumes extending ~half way through the lower mantle. Temperature-dependent viscosity further reduces the density contrast required for stable layering.

Examination of the depth variation of thermal, compositional, and density heterogeneity, and the lateral spectrum of heterogeneity at different depths using spectral heterogeneity maps (SHMs) indicates that (i) the dense layer increases the lateral heterogeneity of thermal and total density fields in the deep mantle, (ii) thermal and compositional buoyancy (density) contributions cancel out in the upper part of the dense layer, (iii) while thermal heterogeneity is zero at the CMB due to the isothermal boundary condition, chemical heterogeneity peaks, providing a potential means of distinguishing the two, and (iv) for the compressible cases a dramatic change in the lateral heterogeneity takes place at depths corresponding to the top of the thermo-chemical megaplumes. This change in heterogeneity is also apparent in seismic topographic models.

The thermo-chemical megaplumes may provide an explanation for various geophysical observations as dis-

cussed earlier, but do not explain D", since (i) the discontinuity topography (where it is observed) varies by much less, from ~150 to ~450 km; and (ii) the discontinuity is still observed under inferred mantle downwellings [*Kendall and Shearer*, 1994], incompatible with numerical results where it is swept completely aside. Smaller topography could be recovered by increasing the density contrast. However, mineralogical constraints [*Sidorin and Gurnis*, 1998] indicate that large density contrasts are physically implausible. Another possibility that would reduce the topography is that the viscosity in the deep mantle is lower than assumed here (hence the local Rayleigh number is higher), due to a smaller increase with depth or a lower near-surface "reference" value. This is quite possible since the deep mantle viscosity is not well constrained by postglacial rebound or geoid modeling approaches.

Thus, there are various difficulties with explaining D" in terms of a thermo-chemical boundary layer. Firstly, as discussed above, with a small, mineralogically-feasible density contrast and high deep-mantle viscosity, the topography generated on the top of the layer will be much larger than seismically observed. Even if reasonable density and topography are possible, there some difficulties which center around obtaining a sharp seismic reflection while simultaneously generating heterogeneity down to small lengthscales. If the layer were primordial, it might be able to preserve a fairly sharp boundary over geological time (if the layer survives at all [*Sleep*, 1988]), but short-wavelength lateral variations would be difficult to generate since the lateral variations tend to have the same scale as the driving thermal field, as illustrated by the presented calculations. If, on the other hand, the layer were dynamically generated through the segregation of oceanic crust or reactions with the core, short-scale lateral variations are expected but the transition from the dense material to normal mantle is probably too gradual [*Kellogg and King*, 1993; *Christensen and Hofmann*, 1994] to generate a seismic reflection.

The true situation is probably more complicated than can be modeled with a single compositional field. For example, D" could be partly primordial and partly dynamically-generated, resulting in a complex interplay of compositional and thermal effects, although the density problem still exists. One appealing possibility is that the seismic discontinuity is unrelated to chemistry but is generated by a phase transition [*Stixrude and Bukowinski*, 1990; *Nataf and Houard*, 1993]. This is particularly appealing since it allows the existence of thermo-chemical megaplumes to explain various geophysical observations, in addition to the "Lay discontinuity".

Future work must address these problems and possibilities in detail, and determine what seismic signatures are produced from various dynamically-produced thermo-chemical structures. The detailed dynamics of potential slab segregation at the CMB [*Olson and Kincaid*, 1991] and the fate of the two differentiated components needs further evaluation, as does the question of long-term survivability of a dense layer for geophysically reasonable density contrasts and other parameters. A greater understanding of the relevant mineral physics, including the possible existence of a phase transition, will clearly also play an essential role.

Acknowledgments. Simulations were performed on the Intel Paragon and Cray T3E at San Diego Supercomputer Center and the Cray T3E at NASA Goddard Space Flight Center. Supported by NSF, NASA and IGPP/LANL.

REFERENCES

Anderson, D.L., A seismic equation of state .2. Shear properties and thermodynamics of the lower mantle, *Phys. Earth Planet. Inter.*, *45*, 307-323, 1987.

Anderson, O.L., H. Oda, and D. Isaak, A model for the computation of thermal expansivity at high compression and high temperatures - MgO as an example, *Geophys. Res. Lett.*, *19*, 1987-1990, 1992.

Balachandar, S., D.A. Yuen, and D. Reuteler, Time-dependent 3-dimensional compressible convection with depth-dependent properties, *Geophys. Res. Lett.*, *19*, 2247-2250, 1992.

Bolton, H., and G. Masters, A region of anomalous dlnVs/dlnVp in the deep mantle, *Eos Trans. AGU, Fall Meeting Suppl.*, *77*, 697, 1996.

Brandt, A., Guide to multigrid development, *Lecture Notes In Mathematics*, *960*, 220-312, 1982.

Bunge, H.P., M.A. Richards, and J.R. Baumgardner, Effect of depth-dependent viscosity on the planform of mantle convection, *Nature*, *379*, 436-438, 1996.

Bunge, H.P., M.A. Richards, and J.R. Baumgardner, A sensitivity study of 3-dimensional spherical mantle convection at 10^8 Rayleigh number - Effects of depth-dependent viscosity, heating mode, and an endothermic phase change, *J. Geophys. Res.*, *102*, 11991-12007, 1997.

Cadek, O., D.A. Yuen, V. Steinbach, A. Chopelas, and C. Matyska, Lower mantle thermal structure deduced from seismic tomography, mineral physics and numerical modeling, *Earth Planet. Sci. Lett.*, *121*, 385-402, 1994.

Chopelas, A., and R. Boehler, Thermal expansivity in the lower mantle, *Geophys. Res. Lett.*, *19*, 1983-1986, 1992.

Christensen, U., Instability of a hot boundary layer and initiation of thermo-chemical plumes, *Annales Geophysicae*, *2*, 311-319, 1984.

Christensen, U.R., Models of mantle convection - one or several layers, *Phil. Trans. R. Soc. London A*, *328*, 417-424, 1989.

Christensen, U.R., and A.W. Hofmann, Segregation of subducted oceanic crust In the convecting mantle, *J. Geophys. Res.*, *99*, 19867-19884, 1994.

Christensen, U.R., and D.A. Yuen, Layered convection induced by phase transitions, *J. Geophys. Res.*, *90*, 291-300, 1985.

Davies, G.F., and M. Gurnis, Interaction of mantle dregs with convection - lateral heterogeneity at the core-mantle boundary, *Geophys. Res. Lett.*, *13*, 1517-1520, 1986.

Duffy, T.S., and T.J. Ahrens, Thermal expansion of mantle and core materials at very high pressures, *Geophys. Res. Lett.*, *20*, 1103-1106, 1993.

Dziewonski, A.M., and D.L. Anderson, Preliminary reference Earth model, *Phys. Earth Planet. Inter.*, *25*, 297-356, 1981.

Garnero, E.J., and D.V. Helmberger, Seismic detection of a thin laterally varying boundary-layer at the base of the mantle beneath the central Pacific, *Geophys. Res. Lett.*, *23*, 977-980, 1996.

Grand, S.P., R.D. van der Hilst, and S. Widiyantoro, Global seismic tomography: A snapshot of convection in the Earth, *GSA Today*, *7*, 1-7, 1997.

Gurnis, M., The effects of chemical density differences on convective mixing in the Earth's mantle, *J. Geophys. Res.*, *91*, 1407-1419, 1986.

Hansen, U., and D.A. Yuen, Numerical simulations of thermal-chemical instabilities at the core mantle boundary, *Nature*, *334*, 237-240, 1988.

Hansen, U., and D.A. Yuen, Dynamical influences from thermal-chemical instabilities at the core-mantle boundary, *Geophys. Res. Lett.*, *16*, 629-632, 1989.

Hansen, U., D.A. Yuen, S.E. Kroening, and T.B. Larsen, Dynamic consequences of depth-dependent thermal expansivity and viscosity on mantle circulations and thermal structure, *Phys. Earth Planet. Inter.*, *77*, 205-223, 1993.

Jarvis, G.T., and D.P. McKenzie, Convection in a compressible fluid with infinite Prandtl number, *J. Fluid Mech.*, *96*, 515-583, 1980.

Jarvis, G.T., and W.R. Peltier, Lateral heterogeneity in the convecting mantle, *J. Geophys. Res.*, *91*, 435-451, 1986.

Jordan, T.H., P. Puster, G.A. Glatzmaier, and P.J. Tackley, Comparisons between seismic Earth structures and mantle flow models based on radial correlation functions, *Science*, *261*, 1427-1431, 1993.

Karato, S., and P. Wu, Rheology of the upper mantle - a synthesis, *Science*, *260*, 771-778, 1993.

Karato, S.I., M.S. Paterson, and J.D. Fitzgerald, Rheology of synthetic olivine aggregates - Influence of grain-size and water, *J. Geophys. Res.*, *91*, 8151-8176, 1986.

Kellogg, L.H., and S.D. King, Effect of mantle plumes on the growth of D" by reaction between the core and mantle, *Geophys. Res. Lett.*, *20*, 379-382, 1993.

Kendall, J.M., and P.M. Shearer, Lateral variations in D" thickness from long-period shear wave data, *J. Geophys. Res.*, *99*, 11575-11590, 1994.

Knittle, E., and R. Jeanloz, Simulating the core-mantle boundary - an experimental study of high-pressure reactions between silicates and liquid iron, *Geophys. Res. Lett.*, *16*, 609-612, 1989.

Loper, D.E., and T. Lay, The core-mantle boundary region, *J. Geophys. Res.*, *100*, 6397-6420, 1995.

Manga, M., Mixing of heterogeneities in the mantle - Effect of viscosity differences, *Geophys. Res. Lett.*, *23*, 403-406, 1996.

Masters, G., S. Johnson, G. Laske, and H. Bolton, A shear-velocity model of the mantle, *Phil. Trans. R. Soc. London A*, *354*, 1385-1410, 1996.

Monnereau, M., and M. Rabinowicz, Is the 670 km phase transition able to layer the Earth's convection in a mantle with depth-dependent viscosity?, *Geophys. Res. Lett.*, *23*, 1001-1004, 1996.

Nataf, H.C., and S. Houard, Seismic discontinuity at the top of D" - a worldwide feature?, *Geophys. Res. Lett.*, *20*, 2371-2374, 1993.

Olson, P., and C. Kincaid, Experiments on the interaction of thermal convection and compositional layering at the base of the mantle, *J. Geophys. Res.*, *96*, 4347-4354, 1991.

Osako, M., and E. Ito, Thermal diffusivity of MgSiO$_3$ perovskite, *Geophys. Res. Lett.*, *18*, 239-242, 1991.

Parmentier, E.M., C. Sotin, and B.J. Travis, Turbulent 3-D thermal convection in an infinite Prandtl number, volumetrically heated fluid - Implications for mantle dynamics, *Geophys. J. Int.*, *116*, 241-251, 1994.

Patankar, S.V., *Numerical Heat Transfer and Fluid Flow*, Hemisphere Publishing Corporation, New York, 1980.

Pollack, H.N., S.J. Hurter, and J.R. Johnson, Heat flow from the Earth's interior - Analysis of the global data set, *Rev. Geophys.*, *31*, 267-280, 1993.

Press, W.H., S.A. Teulolsky, W.T. Vetterling, and B.P. Flannery, *Numerical Recipes*, Cambridge University Press, Cambridge, UK, 1992.

Ratcliff, J.T., P.J. Tackley, G. Schubert, and A. Zebib, Transitions in thermal convection with strongly variable viscosity, *Phys. Earth Planet. Inter.*, *102*, 201-212, 1997.

Ricard, Y., M. Richards, C. Lithgow-Bertelloni, and Y. LeStunff, A geodynamic model of mantle density heterogeneity, *J. Geophys. Res.*, *98*, 21895-21909, 1993.

Robertson, G.S., and J.H. Woodhouse, dlnVs/dlnVp in the lower mantle and its mineral physics implications, *Eos Trans. AGU, Fall Meeting Suppl.*, *77*, 697, 1996.

Schubert, G., Numerical models of mantle convection, *Ann. Rev. Fluid Mech.*, *24*, 359-394, 1992.

Schubert, G., and P.J. Tackley, Mantle dynamics: The strong control of the spinel-perovskite transition at a depth of 660 km, *J. Geodynamics*, *20*, 417-428, 1995.

Sidorin, I., and M. Gurnis, Geodynamically consistent seismic velocity predictions at the base of the mantle, in *Observational and theoretical constraints on the core-mantle boundary region (this volume)*, edited by M. Gurnis, American Geophysical Union, 1998.

Sleep, N.H., Gradual entrainment of a chemical layer at the base of the mantle by overlying convection, *Geophys. J. Oxford*, *95*, 437-447, 1988.

Smolarkiewicz, P.K., A fully multidimensional positive definite advection transport algorithm with small implicit diffusion, *J. Comp. Phys.*, *54*, 325-362, 1984.

Solomatov, V.S., Scaling of temperature-dependent and stress-dependent viscosity convection, *Phys. Fluids*, *7*, 266-274, 1995.

Stacey, F.D., *Physics of the Earth*, Brookfield Press, Kenmore, Queensland, Australia, 1992.

Stevenson, D.J., Why D" is unlikely to be caused by core-mantle interactions (abstract), *Eos Trans. AGU, Fall Meeting Suppl.*, *74*, 51, 1993.

Stixrude, L., and M.S.T. Bukowinski, Fundamental thermodynamic relations and silicate melting with implications for the constitution of D", *J. Geophys. Res.*, *95*, 19311-19325, 1990.

Su, W.J., R.L. Woodward, and A.M. Dziewonski, Degree-12 model of shear velocity heterogeneity in the mantle, *J. Geophys. Res.*, *99*, 6945-6980, 1994.

Tackley, P.J., Effects of strongly temperature-dependent viscosity on time-dependent, 3-dimensional models of mantle convection, *Geophys. Res. Lett.*, *20*, 2187-2190, 1993.

Tackley, P.J., Three-dimensional models of mantle convection: Influence of phase transitions and temperature-dependent viscosity, Ph.D. Thesis, California Institute of Technology, Pasadena, 1994.

Tackley, P.J., Effects of strongly variable viscosity on three-dimensional compressible convection in planetary mantles, *J. Geophys. Res.*, *101*, 3311-3332, 1996.

Tackley, P.J., D.J. Stevenson, G.A. Glatzmaier, and G. Schubert, Effects of an endothermic phase transition at 670 km depth in a spherical model of convection in the Earth's mantle, *Nature*, *361*, 699-704, 1993.

Tackley, P.J., D.J. Stevenson, G.A. Glatzmaier, and G. Schubert, Effects of multiple phase transitions in a 3-dimensional spherical model of convection in Earth's mantle, *J. Geophys. Res.*, *99*, 15877-15901, 1994.

van der Hilst, R.D., S. Widiyantoro, and E.R. Engdahl, Evidence for deep mantle circulation from global tomography, *Nature*, *386*, 578-584, 1997.

van Keken, P.E., S.D. King, H. Schmeling, U.R. Christensen, D. Neumeister, and M.-P. Doin, A comparison of methods for the modeling of thermochemical convection, *J. Geophys. Res.*, *102*, 22477-22496, 1997.

Vidale, J.E., and H.M. Benz, Seismological mapping of fine structure near the base of the Earth's mantle, *Nature*, *361*, 529-532, 1993.

Weinstein, S.A., and P. Olson, Planforms in thermal convection with internal heat sources at large Rayleigh and Prandtl numbers, *Geophys. Res. Lett.*, *17*, 239-242, 1990.

Wysession, M.E., Imaging cold rock at the base of the mantle: The sometimes fate of slabs?, *AGU Geophysical Monograph*, *96*, 369-384, 1996.

Yuen, D.A., O. Cadek, A. Chopelas, and C. Matyska, Geophysical inferences of thermal-chemical structures in the lower mantle, *Geophys. Res. Lett.*, *20*, 899-902, 1993.

Yuen, D.A., O. Cadek, P. van Keken, D.M. Reuteler, H. Kyvlova, and B.A. Schroeder, Combined results from mineral physics, tomography and mantle convection and their implications on global geodynamics, in *Seismic modeling of Earth structure*, edited by E. Boschi, G. Ekström, and A. Morelli, pp. 463-505, Instituto Nationale di Geofisica, Italy, 1996.

Zhang, S.X., and D.A. Yuen, The influences of lower mantle viscosity stratification on 3-D spherical-shell mantle convection, *Earth Planet. Sci. Lett.*, *132*, 157-166, 1995.

P.J. Tackley, Department of Earth and Space Sciences, University of California, Los Angeles, 405 Hilgard Avenue, Los Angeles, CA 90095-1567

The EDGES of the Mantle

Don L. Anderson

Seismological Laboratory, Division of Geological and Planetary Sciences, California Institute of Technology, Pasadena, California

The core-mantle boundary region is often considered to be the source of narrow upwellings which drive or influence plate motions and continental breakup, fuel large igneous provinces and generate volcanic chains. The plume hypothesis has influenced most fields of geochemistry, petrology, geodynamics and mantle evolution. The key axioms underlying the plume paradigm are identified: Axioms are self-evident truths and are seldom stated explicitly. When they are, they are sometimes not so self-evident. The role of the surface boundary layer is discussed in connection with large igneous provinces and volcanic chains. Partial melting is the expected natural state of the upper mantle and only abnormally high seismic velocities imply absence of melting (slabs, cratons). Plume theoreticians have underestimated the average temperature of the mantle and have overestimated melting temperatures. Extensive melting in the upper mantle does not require abnormal temperatures or plumes. The dynamics and chemistry of midplate volcanism are explainable by near-surface processes. "Midplate" volcanism starts at plate boundaries or discontinuities, at lithospheric "edges", in regions of extension generated by plate processes. The chemistry of "hotspot" basalts implies contamination by processes and materials that occur near the surface of the Earth. A buoyant metasomatised layer at the top of the mantle, near the melting point, removes the need to import heat and chemical inhomogeneity from D" to explain midplate volcanism. D" is an interesting and important region, but any connection to surface processes or chemistry is speculative.

INTRODUCTION

"It is only after you have come to know
the surface of things
that you can venture to seek what is underneath . . .
The surface is already so vast and rich and various
that it more than suffices
to saturate the mind with information and meanings."
Mr. Palomar
Italo Calvino (1983)

"But then, to what end" said Candide,
"was the world formed?
"To make us mad," said Martin.
Candide

The Core-Mantle Boundary Region
Geodynamics 28
Copyright 1998 by the American Geophysical Union.

The core-mantle boundary (CMB) region is one of the most important boundaries in the Earth. The changes in density and viscosity across the CMB are larger than at the surface. The dross of mantle and core processes collect there. Molten metal contacts crystalline oxide there. Temperature variations of hundreds of degrees are damped to near zero. But, because of pressure, buoyancy is hard to come by, and heat is transmitted readily by conduction. D" and E' are important and fascinating parts of the CMB neighborhood. But does D" drive the plates and fill up the asthenosphere with giant plume heads and control the chemistry of ocean islands? Or does outer Earth control its own fate? Do narrow volcanic roots extend to CMB or only to the base of a cracked plate? Are volcanic chains part of the surface mosaic, or are they imposed from deep in the Earth? Does Earth dynamics start and end at D"? To many Earth scientists, the concepts of hotspots, plumes, primitive mantle and D" are inextricably intertwined in a thread that weaves throughout the tapestry of mantle dynamics, evolution and chemistry. Herein I explore the connections.

Some of the assumptions underlying global Earth dynamics are treated as axiomatic; these are collected in Appendix I. The role of plumes vs. plates is discussed in Appendix 2. The focus on the deep mantle by geochemists is treated as an exercise in deductive logic in Appendix 3. The issues of partial melting, passive ridges, reference frames, radioactivity of the lower mantle and linear volcanic chains are discussed. These have all been considered by some authors to be relevant to the base of mantle. Each one of these "side issues" is particularly compelling to some in formulating views of the Earth. For example, the fixed hotspot reference frame has convinced some that hotspots are due to deep mantle plumes rather than propagating cracks or leaky transform faults. The passivity of ridges and mass balance calculations have convinced some that a deep source is required for ocean island basalts. Many linear volcanic chains, however, are due to surface tectonics and lithospheric architecture. Only when these are understood can we "venture to seek what is underneath . . .".

THE BASE OF THE MANTLE

It is unlikely that the lower mantle is entirely homogeneous. A compositionally distinct D" is expected on several grounds:

1. During accretion, material denser than the mantle, but less dense than the core, will settle into D".

2. As the core cools and the inner core grows, light material which can no longer be held in solution, or suspension, will rise to the top of the core and may become part of the lower mantle.

3. The Earth accreted, in part, in the inhomogeneous accretion mode. The deep mantle may have more Ca-Al-Ti-rich refractory material than the top of the mantle [Ruff and Anderson, 1980].

4. The early formed nucleus of a planet may be able to retain more volatiles than the outer parts. Later impacts result in complete devolatization. On the other hand, the final assembly of a planet may involve a volatile-rich late veneer, brought in from the far reaches of the solar system. A big planet can retain low molecular weight species.

5. Subduction on a small planet, or the sinking of dense cumulates, may have brought the products of near-surface fractionation to the CMB, even if this is not possible today.

6. FeO-rich phases may be able to separate from MgO-rich phases and, perhaps, form a layer at the base of the mantle.

Many of these processes make the CMB denser than normal lower mantle, and stable stratification is the likely result. In this situation, the chemically distinct layer will not be uniform in properties or thickness. The top is expected to have high relief and the layer may be missing in places as it is swept toward mantle upwellings. Because of the large density contrast between the mantle and core, and the thermal and viscosity properties of the core, the bottom

of D" is expected to be isothermal and low relief. A high thermal gradient and a silicate-iron slurry may make a thin, low shear velocity, layer at the boundary. In Dante's "Inferno", DIS is the boundary between Upper and Nether Hell, and this is an appropriate name for the base of the D" layer.

The intensely slow region (DIS), recently found at the base of the mantle [Garnero and Helmberger, 1995], may be another analogy between the top and bottom of the mantle; both are thermal boundary layers, both collect material of extreme density and both may be close to or above the melting point. The upper part of the Earth contains thick accumulations of basalt or residual peridotite. These are called continents, plateaus, swells and cratons. D" can be expected to be equally variable.

Whether or not D" is important for the understanding of mantle dynamics and geochemistry depends on which of the following statements is more nearly true:

"Plumes are secondary flow superposed on mantle wide plate scale flow [Davies, 1988]."

or

"Hot upwellings are part of the main mantle flow. Surface expressions of volcanism are controlled by plate tectonics, subduction and rifting."

This issue is discussed in Appendix 2. Plates and cracks certainly modulate expressions of magmatism at the Earth's surface; the question is, can D" do so as well? Are volcanoes caused by concentrated jets of hot fluid, or are they the result of lithospheric failure? Is DIS relevant to surface tectonics and volcanism?

LOW VELOCITIES AND MELTING

The most plausible explanation for the occurrence of intensely slow regions of the mantle is partial melting or the presence of a fluid phase [Anderson and Sammis, 1970; Anderson and Bass, 1984]. Partial melting, dehydration, dislocation relaxation and change in crystal orientation are the main mechanisms by which extremely low seismic velocities can be generated [Anderson, 1989; Minster and Anderson, 1981].

Dislocation relaxation is a viable velocity reduction mechanism and, formally, has the same effect as thin grain boundary melts [Anderson, 1989]. If the fluid does not wet grain boundaries, and occurs in spherical or cylindrical inclusions, then small amounts of melt do not cause large reductions in velocity. If at lower mantle pressures (in DIS) the viscosity of silicate melts is very high, or if they do not wet grain boundaries, then extensive melting is required to substantially lower the seismic velocity. According to Zerr et al. [1997] melting of silicates in D" is unlikely, but this is still an open question.

An alternative to partial melting in DIS is the incorporation of molten Fe. If metallic fluids wet silicate grain boundaries, then the melt should drain out. Fe-melts, however, may be trapped in silicates. Although Fe may not diffuse or percolate far up into the base of the mantle, a silicate slurry at the top of the core, formed by stopping and entrainment, may be able to trap metal as it joins the base of the mantle. DIS may represent a transition region between mantle and core and therefore be intermediate in density and seismic velocity. The lowermost layer of the mantle is usually thought of as a mantle silicate but it may be a metal-oxide mix with very high density.

A large part of the upper mantle above 300 km has such low seismic velocities than partial melting is implied [Anderson and Bass, 1984]. It is only cratons and slabs that appear to be subsolidus. Oceanic mantle and mantle under tectonically active areas have seismic velocities that are lower than any combination of stable minerals at reasonable temperature. Lithospheric architecture and volcanic regions show excellent correlation with upper mantle seismic velocities [Scrivner and Anderson, 1992; Wen and Anderson, 1995a,b, 1997]. There is very poor correlation with the lower mantle [Ray and Anderson, 1994].

Cratonic roots remain dry and buoyant (and strong) because they are deeper (~200 km) than the dehydration depth in slabs and are protected from refluxing [Polet and Anderson, 1995]. The MORB reservoir is probably deeper than 200 km, except at ridges, and is immune from contamination, remaining depleted, dry and homogeneous [Anderson, 1996]. Slabs sinking below 200 km are efficiently demetasomatized and do not contaminate the MORB reservoir or craton roots. On the other hand, MORB mantle rising to the surface at newly extending rifts will react with the shallow mantle and recover some of the recycled material, thereby generating enriched magmas, common at the inception of continental rifting, seafloor spreading and island arc formation. Conditions near the top of the mantle need to be understood before the base of the mantle is implicated in magmagenesis. If variations in temperature of 100°C are associated with plate tectonics (slab cooling, continental insulation, craton roots), and variations of tens of degrees in melting temperature are plausible (water content, fertility, composition) and if extents and volumes of melting depend on focusing and lithospheric thickness as well as temperature, then there is little need to look deeper than the upper mantle to explain the various styles of magmatism.

The existence of a widespread partially molten layer in the upper mantle [Anderson and Sammis, 1970] eliminates the need to import deep hot material into this region to explain magmatism. It is conditions in the lithosphere that mainly control the locations of volcanoes. Magmatism attributed to hotspots and convective processes in the mantle may be opportunistic invasions of magma from an already partially molten mantle [Bailey, 1992]. It is hard to construct a geotherm that does not intersect the damp solidus [White and Wyllie, 1992]. The seismic LID acts as a strong impenetrable barrier to melt migration. In brief, melting is the expected and observed normal state of the upper mantle, volcanoes are probably the result of lithospheric cracking and melt focusing [Vogt, 1974a,b],

and shallow mantle geochemistry is appropriate for midplate volcanoes [Anderson, 1996]. Midplate volcanism is a normal result of plate tectonics. Plates are not rigid or homogeneous or uncracked.

It is commonly assumed that deep strong plumes are required to explain large igneous provinces (LIP). However, these are usually found at lithospheric discontinuities, such as craton boundaries, or at new ridges, triple junctions, and migrating ridges. These regions focus the flow and are weak points of the lithosphere. Since the upper mantle is partially molten almost everywhere, a conclusion obtained both from seismology and experimental petrology, and magmas are less dense than the overlying mantle and crust, only small extension is required to allow ascent of magma to the surface. Using the heights of volcanoes as a guide to the pressure differential (~1 kb) and using estimates of magma viscosity (~103 c.g.s.) a crack of the order of 1 cm in width is all that is required to generate the inferred flow rates of large igneous provinces (~10-7 km3/sec). The heights of volcanoes imply a magma source near the base of the lithosphere [Vogt, 1974]. A crack 1 cm wide implies a strain of only 10-8 if it is the result of extension over 1000 km. Gravitational stresses associated with topography, slab pull or slab roll-back may be the source of extension. The absence of uplift prior to LIP eruption rules out the thermal plume explanation [e.g., Kamo et al., 1996; Fedorenko et al., 1996]. Other problems with current plume models for generating LIPs are summarized in Cordery et al. [1997].

TEMPERATURES IN THE MANTLE

Seismic velocities and properties of upper mantle phase changes are the main geophysical constraints on mantle temperature. A potential temperature of 1400°C is an appropriate average temperature [Anderson, 1989; Anderson and Bass, 1984]. A range of about 200°C is expected from normal plate tectonic processes [Anderson, 1998]. McKenzie and Bickle [1988] adopt a "normal" mantle potential temperature of 1280°C. Evidence for higher temperatures is taken as evidence for plumes. The plume hypothesis, then, to some extent, is based on a straw man. Mantle temperatures are higher than in the "standard model" and "hotspot" magmas probably have lower melting temperatures than assumed because of the presence of volatiles. The highest temperature magmas (komatiites, picrites) may only require 1400°C in order to melt [Stone et al., 1997; Parman et al., 1997], well within the range of "normal" mantle temperatures.

MASS BALANCE

The large-ion lithophile (LIL) elements are partitioned strongly into the crust and upper mantle. Although the crust is only 0.5% of the mass of the mantle, it contains more than 50% of some of the most incompatible elements. Almost all of the most incompatible elements are in the crust and upper mantle. These include U, Th, Pb, Ba, Cs, K, I and Cl [Anderson, 1989; Rudnick and Fountain, 1995; Dérvelle, et al., 1992].

Highly incompatible elements such as I, Cl and 40Ar are strongly concentrated into the atmosphere, oceans and pelagic sediments. U, Th, K, Ce and Pb are highly concentrated into the crust and require extraction from 60 to 100% of the mantle. There is no mass balance argument that requires only the upper mantle be depleted in LIL to form the crust or that only the fertile MORB reservoir is depleted. Most of the mantle, in fact, appears depleted and there is evidence of strong upward concentration of the LIL elements. Lead isotopes and ratios such as U/Nb and Pb/Ce show that no sampled reservoir is primordial or unfractionated [Hofmann et al, 1986]. The lower mantle, the refractory residue of upper mantle and crustal extraction, apparently is not sampled by current processes. In contrast to the MORB reservoir, it is predicted to be both depleted and infertile.

The volatile elements are depleted in the Earth by two orders of magnitude relative to chondritic abundances. The abundance of elements such as H, Cl and I suggest that volatiles could have been brought in by a late veneer representing 1% or less of the mass of the Earth [Anders and Owen, 1977; Dérvelle, et al., 1992]. The high content of siderophiles in the upper mantle plus the chondritic relative ratios of the trace siderophiles suggest that they too were brought in as a late veneer [Dreibus and Wänke, 1989]. Much of the atmosphere and oceans could have been from a late veneer. The same may hold true for the noble gas budget of the Earth. Deep sea sediments today are high in noble gases because of the high concentration of interplanetary dust particles (IDPs). The sediments are retentive of trapped 3He but not 4He [Farley, 1995]. Therefore sediments are subducted with 3He/4He ratios comparable to those found in Loihi basalts. Today's IDPs flux cannot account for today's volcanic 3He flux but neither quantity is constant in time [Anderson, 1993]. Most of the late veneer probably accreted prior to 2 Ga. IDPs are remarkably retentive of noble gases and it is the fate of the fluid in sediments after subduction that matters. Noble gases are transported by fluids and are trapped in secondary minerals and fluid-filled inclusions. Loosely bound 10Be and OH make their way into the mantle. The high 3He/4He ratios and high iodine contents of Loihi basalts suggests that a pelagic source may be involved. There is no evidence, however, that Loihi basalts are particularly high in 3He. The high 3He/4He ratios of Loihi basalts apparently reflect low 4He abundances, and therefore, a U-depleted refractory reservoir [Anderson, D. L., The Helium Paradoxes, submitted to Earth and Planetary Science Letters for publication, 1997].

3He plays an important role in discussions of mantle evolution. The presence of 3He in the mantle must be explained since it is difficult to accrete and retain the light volatile elements. Both midocean ridge basalts and hotspot magmas contain 3He. Either this is brought into the Earth during the primary high-temperature accretion, or as a late veneer. The easiest place and time to degas the Earth is at the surface during initial accretion. The easiest time to incorporate volatiles is during a drawn out late accretion stage.

The other aspect of helium geochemistry is the high 3He/4He ratio of some OIB such as at Hawaii, Loihi, Iceland and Samoa. OIB typically have two orders of magnitude less 3He than MORB, and midocean ridge volcanism supplies more than 300 times the amount of 3He provided to the atmosphere by hotspots [Sano and Williams, 1996]. A late veneer and cold subduction can explain the presence of 3He in the mantle but high 3He/4He ratios imply, in addition, a low U and Th storage tank for OIB gases. The lithosphere is such a storage tank [Anderson, D. L., The Helium Paradoxes, submitted to Earth and Planetary Science Letters for publication, 1997]. Thus, another argument for a deep plume source is brought into question.

RADIOACTIVITY OF LOWER MANTLE

Mass balance, heat flow and Pb-isotope data suggest that most of the lower mantle is depleted in the heat producing elements [Birch, 1965; Patterson and Tatsumoto, 1964; Spohn and Schubert, 1991; Anderson, 1989]. This was probably accomplished by upward concentration of volatile and incompatible elements during accretion [Anderson, 1989]. The crust contains more than 50% of the terrestrial inventory of U, Th and K and the atmosphere contains about 70% of the 40Ar produced over the age of the Earth from 40K [Rudnick and Fountain, 1995; Wedepohl, 1995; Anderson, 1989]. These observations are consistent with a strongly differentiated and degassed Earth and with a lower mantle that is infertile (low in basaltic elements such as Al, Ca and Ti), depleted (low in U, Th, K, Ba, etc.), and degassed. The various paradoxes that occur in geochemical mass balance calculations (e.g., the Pb-paradox) imply that there is an unsampled reservoir [e.g., Hofmann et al., 1986]. I suggest that this is the lower mantle.

REFERENCE FRAMES

Oceanic plates move rapidly in the paleomagnetic and paleogeographic reference frames. The plates are part of the surface boundary layer and are expected to be mobile. Markings on the surface are not expected to provide a good reference system. For the same reason, D" is not a good reference system; not only is it a boundary layer but it has very low viscosity and motions may be fast because of the small area. Plumes arising from D" are not expected to define a fixed reference system, particularly after having risen through various shear layers and phase boundaries in the overlying mantle. The relative fixity of hotspots is not an argument for a deep-mantle plume origin or for a source at the base of a high-viscosity layer.

Ridges and trenches migrate. It is of interest to compare motions of candidate fixed points, such as hotspots, to other reference frames which are known to be flexible.

Continents move episodically; stationary periods occur when they are over downwellings. Relative motions of slowly moving continents provide a standard to which hotspots reference frames can be compared. These are comparable to inferred motions of hotspots.

The hotspot hypothesis as a means of determining an absolute reference frame for lithospheric plate motions seems to be somewhat of a tail-wagging-the-dog, since the volumes of material produced by volcanoes is small in comparison to the amount produced at other tensional features in the plate tectonic system [Kaula, 1975]. An absolute reference frame defined by minimizing the transitional motion of plate boundaries differs by only 0.6 cm/yr from a frame defined by hotspot traces and by 0.4 cm/yr from a frame defined by drag forces on plates [Kaula, 1975]. Although ridges are expected to be mobile, there is a remarkable spatial correlation of ridges with hotspots, suggesting a relation between the two. Many hotspots which are not currently on ridges were on-ridge within the past 60 Ma. A spreading ridge creates space and updrafts, and permits extensive melting of rising mantle. Even at linear ridges the induced convection is three-dimensional, with some regions, both along- and off-axis delivering more melts than others. Some of these productive regions have been called "hotspots" with the implication that they differ from normal ridge induced convection and that they have deep roots. In spite of their expected mobility, a reference frame based on ridges is just as stationary as one based on hotspots [Kaula, 1975]. The validity of the fixed hotspot reference frame is in dispute [Stock and Molnar, 1987; Cande et al., 1995; Norton, 1995]. Very few island chains are either parallel or have consistent age progressions with other tracks on the same plate.

The Hawaiian is the best studied volcanic chain. It is parallel to some features on the Nazca and Antarctic plates and some of the boundaries of the Pacific plate, but there is a notable absence of parallel island chains. The Louisville chain, an oft quoted analog of the Hawaiian chain, in fact is not a fixed magmatic center [Norton, 1995]. It appears to terminate on the Eltanin Fracture system although the actual location of the Louisville "hotspot" is in dispute. The Hawaiian melting anomaly is not fixed relative to other hotspots [Cande et al., 1995; Norton, 1995].

NON-PLUME MODELS

Recent discussion about hotspots suppose that they are the result of convective phenomena; i.e., intense narrow upwellings. The surface locations of volcanoes, however, may be almost entirely due to lithospheric conditions and have little to do with deep small-scale convective processes (Appendix 2). Propagating cracks were discussed some time ago [Turcotte and Oxburgh, 1978] but dismissed because of the perceived fixity of hotspots. The propagating crack idea was based on homogeneous plates and uniform mantle. Detailed gravity and magnetic maps, however, show an extensive fabric to the seafloor [Smith and Sandwell, 1997].

Island and seamount chains are often along, or parallel to, fracture zones or are near subduction zones (Samoa, Juan Fernandez, San Feliz, Galapagos) where plate tearing stresses are highest. Some hotspots occur along abandoned ridges or where oceanic plates are fragmenting (E and NE Pacific). Many island chains originated at a spreading ridge and all continental flood basalts occur at cratonic boundaries. These regions tend to focus melts resulting in larger volumes of magma than in an extending situation involving uniform plates. Other large igneous provinces occur at new ridges or triple junctions, and at times of plate reorganization. These all indicate lithospheric control of volcano locations, and a role for transient drainage effects.

The opening of leaky transform faults, or the propagation of cracks toward the interior of a plate from a ridge or a subduction zone, can also cause linear volcanic chains. Such cracks may explain hotlines and rejuvenated volcanism as well as non-parallel tracks. Rates and durations of magmatism depend on stress conditions in the plate. These in turn depend on boundary conditions such as slab pull, ridge push and thermal contraction, and local conditions such as volcanic loads. If plates act as stress guides, areas of maximum tensile stress will tend to be fixed relative to the boundary configuration; stagnation points in mantle flow will be as well. Volcanic chains tend to propagate counter to the direction of plate motion. Lithospheric tears due to lateral changes in subduction dip (Nazca plate, Samoa) or trend (Galapagos) also propagate counter to the plate motion, and therefore appear to be relatively fixed. The Easter "hotline" appears to be propagating counter to plate motions [Kruse et al., 1997].

Time progressive magmatism is a characteristic of ridge "propagators", leaky transform faults, crack propagation and progressive opening of oceans, such as the south to north unzipping of the North Atlantic. Nevertheless, the existence of age progressive volcanism is often taken as evidence for a mantle plume. Most long volcanic chains do not exhibit age progression or are not parallel to plate motions [e.g., Okal and Batiza, 1987; McHone, 1996b; Anderson, 1996; Vogt, 1995]. The ubiquitous presence of melt in the low-velocity zone, ponding of melt at the base of the lithosphere and the rapid extrusion of melt through narrow conduits suggests that only moderate extension is required to form large and transient igneous provinces. The hydrostatic head inferred from volcano heights and the thickness of the lithosphere, combined with estimates of magma viscosity, is consistent with inferred rapid extrusion rates. The composition of near-surface and subducted materials, and the shallow mantle are appropriate for midplate volcanics [Anderson, 1996]. The concept of a non-rigid plate plus a partially molten upper mantle (at "normal" mantle temperatures), is an alternative to plumes.

THE PASSIVE RIDGE DIVERSION

It was once believed that ridges and trenches were the up- and down-wellings of mantle convection. It then became clear that ridges were passive; plate tectonic forces caused the plates to separate and mantle passively upwelled to fill the cracks. Ridges migrate about the Earth's surface and become annihilated as they approach trenches. The idea took hold that ridges were entirely passive; that active, convective upwellings were elsewhere; and that ridges could form essentially anywhere.

However, ridges and trenches are the largest source and sink of material on Earth. Slabs cause entrainment and cooling of the mantle and control the locations of the most significant downwellings in the mantle. Plates are strong and have high viscosity and contain an element of chemical buoyancy. They are thicker and colder than a normal thermal boundary layer and have a larger influence on mantle thermal properties than a simple unstable TBL. The locations of subduction zones are fundamental in controlling the planform of mantle convection. Trenches provide a good reference frame, comparable to hotspots [Kaula, 1975; Chase, 1978].

Upwelling under thin lithosphere causes adiabatic melting and creates buoyancy. Flow is focused toward ridges. The mantle in the vicinity of a ridge becomes a sink for mantle flow and "active" upwellings are attracted to the ridge. Although the ridge, from a plate tectonic point of view, may be passive, and entirely driven by "ridge-push" and "slab-pull", it will certainly influence the locations of the major mantle convective upwellings. It is no accident that most hotspots occur on young lithosphere (most of the others being at lithospheric discontinuities such as transform faults, old ridges and craton boundaries). This is where the plate is thin and under tensile stress, and where upwelling mantle and melts are focused. It is also no accident that most of the world's ridges (and hotspots) overlie mantle that has not experienced subductive cooling since at least the breakup of Pangea [Wen and Anderson, 1995a]. Ridges (and hotspots) occur over hotter than average mantle, and mantle not cooled from below [Anderson, 1996]. Spreading ridges not only induce localized upwelling by plate divergence, adiabatic melting and focusing (upslope migration) but they also provide sites where the lithosphere is most vulnerable to penetration.

Therefore, while ridges are technically passive and they do migrate and "jump", they seldom migrate far or fast. In fact, the "fixed-ridge" assumption provides as good a reference system as the "fixed hotspot" assumption [Kaula, 1975]. Ridges and trenches can annihilate, but overridden ridges show up as regions of low seismic velocity and extensive volcanism; for example, Western North America. Subducted ridges may be responsible for some volcanic chains. Slab windows are implicated in many regions of continental magmatism.

Long-lived ridges are no longer processing chemically diverse mantle that may have initially been at the top of the system (the perisphere). New ridges, slowly spreading ridges, migrating ridges and continental rifts all sample mantle that is enriched relative to that providing N-MORB

at rapidly spreading, and long-lived, ridges. New ridges and triple junctions form when there is a major plate reorganization, such as the breakup of Pangea, and it is at these times and places that there occur transient episodes of enriched non-MORB magmatism. These are the times and places where the shallow mantle is being sampled and it is at later times that the deeper MORB reservoir becomes available, and the shallow enriched layer becomes depleted or attenuated [Anderson, 1996]. The ultimate source of the enriched layer is probably refractory residue permeated by fluids from dehydrating subducting slabs and trapped interstitial residual melts. The MORB reservoir is generally thought to be well mixed because of its homogeneity. It, in fact, is probably depleted and homogeneous because it is deep, and protected from the contaminating effects of subduction. Low-melting and shallow constituents have been removed by prior magmatism. MORB are depleted and homogeneous because they do not encounter pre-existing crust, lithosphere or sediments as they rise. The anomalous mantle that appears at the onset of rifting is often attributed to a plume or plume head. It could as well be shallow mantle, and probably is.

A similar idea has been put forth by Jason Phipps Morgan [1997]. He proposes a "compositional lithosphere", the refractory residue of melting at midocean ridges. Such a shallow infertile (refractory) explains the age dependence of oceanic bathymetry and the puzzle that plume flux does not depend on age of the lithosphere. This model is similar to the perisphere model except that the perisphere is formed by billions of years of melt extraction and a chemical buoyant layer may also be a relic of crystallization of a magma ocean. Additionally, the perisphere is the ultimate repository of slab fluids which carry the OIB recycled signatures and trapped small volume melts such as lamproites and kimberlites. Carbonatites may derive from this region. A shallow enriched layer, sampled at new rifts and at the onset of seafloor spreading, explains temporal and spatial relations between enriched and depleted magmas and predicts a MORB reservoir which does not extend to the base of the plate except at spreading ridges. "Plume heads", and "fossil plume heads" have been used in a similar way but the underlying assumption is that, otherwise, only MORB would come from the upper mantle (Appendix 1). A refractory layer also extends the conductive geotherm, making it possible to reduce the size of the melting column and increase "normal" mantle temperatures over those assumed by McKenzie and Bickle [1988].

THE PLUME HYPOTHESIS

In order for narrow thermal plumes to rise almost 3000 km from D" or DIS through a convecting, internally heated mantle without deflection or cooling, they must be strong. In the laboratory, the plume fluid must be removed from the system, heated externally and then reinjected; narrow plumes with low viscosity do not arise spontaneously in a heated fluid [Nataf, 1991].

Morgan [1972a] required 20 plumes, 150 km in diameter, rising at 2 m/year in order to influence the motion of surface plates and dominate over ordinary convection. He calculated a plume volume flux of 500 km3/yr and this served to drive the asthenosphere at 5 cm/year. The heat flow associated with plumes was calculated to be half the total heat flow of the Earth. If plumes were much weaker, ridges would close up and the return flow associated with plate tectonics would dominate plume flow.

The mass fluxes and heat flow associated with plumes have now been estimated [Davies, 1988] and are only a fraction of those required by Morgan [1972a]. The minimum plume flux inferred by Morgan is about two orders of magnitude higher than the observed flux and about 25 times the theoretical flux from D" [Stacey and Loper, 1983]. The strong plume theory has therefore been falsified, and no weak plume theory has been put forward that has the attributes needed if plumes are to be a significant form of convection.

Narrow plumes, or jets, are an unusual form of convection [Tackley, 1995, 1996], usually simulated in the laboratory by injecting hot fluids at the base of a tank of non-convecting fluid, or by puncturing a membrane which has isolated a layer of hot fluid. In systems that are internally heated, or convecting, small instabilities tend to get swept toward the main upwelling and isolated independent narrow upwellings do not exist. Normal thermal convection tends to buffer the temperature rise in the TBL so that narrow plume-like instabilities do not form [Nataf, 1991].

It is generally assumed that thermal instabilities in a deep thermal boundary layer are the only possible source of upwelling but this is not true (Appendix 2). Most of the globe's magmatism is due to passive, or opportunistic upwelling, or due to flux induced lowering of the melting point. Lithospheric architecture (craton boundaries, fracture zones, etc.) generate focusing and vertical motions (i.e., upwellings). Adiabatic melting and bubble exsolution contribute to buoyancy once an upwelling gets close to the surface. Focusing of magma, updip or toward thin lithosphere, and 3D effects can increase the volume of magmatism without increasing temperature or extent of melting. If melting of the upper mantle is widespread, as implied by the low seismic velocities, then the existence and magnitude of extrusion may be entirely a function of lithospheric conditions, and time (transient vs. steady-state). Finally, convection can be driven by cooling from above, in the absence of heating from below [King and Anderson, 1995].

Modern plume models attribute massive volcanism to localized, and deep, processes that are independent of plate motions and mantle convection. They are not so easily falsified as Morgan's plume theory since they are less specific and testable. The idealized plume has two components: plume heads which are supposedly responsible for very short-lived massive igneous events, and narrow plume tails which generate long-lived hotspot tracks. The plume head heats the lithosphere and causes or localizes continental

breakup. Plume head magmatism is predicted to be very brief [Campbell and Griffiths, 1990; Richards et al., 1989].

This idealized behavior is not observed. Large igneous provinces (LIP) are seldom accompanied by hotspot tracks. Most hotspot tracks do not terminate in LIP. LIP are always deposited at major lithospheric boundaries at times of plate reorganization or major plate boundary changes. The lithosphere is presumably under extension at the time. LIP are often unaccompanied by precursory uplift. This observation alone falsifies the plume hypothesis since uplift is an unavoidable consequence of insertion of hot material into the upper mantle. Rift induced circulation or lateral temperature gradients, however, can drive a large amount of mantle material through the melting field [King and Anderson, 1995].

EDGES

Excess volcanism at margins does not correlate with the presence of hotspots [Mutter, 1993; Holbrook and Kelemen, 1993] but with proximity to thick cratonic lithosphere (e.g., Brazil, Congo, Greenland, north Australian margin), ancient sutures (Narmada-Son, Cochambanba-Paraná, etc.) or rapid pull-apart [Hopper et al., 1992]. Basalt thickness in the Red Sea and Afar contradict a plume influence [Altherr et al., 1988]. The initiation and focusing of igneous activity at lithospheric discontinuities, particularly cratonic boundaries, continental margins, ridges and triple junctions, is an argument against deep mantle plumes, and an argument for plate control. Some LIP are the result of passive plate pull-apart. Rapid lithospheric extension induces short-lived small-scale convection in the mantle, resulting in a pulse of magmatism at the inception of seafloor spreading. Crustal stretching, a slow process, is not adequate and is not required. In plate theory, magmatism should not be uniform all along newly formed ridges [Mutter et al., 1988; Sawyer and Harry, 1991].

The correlation of CFB to cratons and lithospheric discontinuities is much stronger than to hotspot tracks. Every CFB province is adjacent to an Archean Craton or on a shear zone [Anderson, 1996]. Many hotspot tracks are along, or influenced by, fracture zones, sutures and ridges. The lithosphere plays a major role in magma ascent, mantle convection and LIP [Mutter et al., 1988; Mutter, 1993; Zehnder et al., 1990; Bailey, 1992]. Edge-driven gyres and eddies (EDGE) are a form of convection driven by lateral temperature gradients [King and Anderson, 1995].

"Hotspots" which are related to pre-existing fractures, sutures or lithospheric boundaries [A. S. Davis, 1989] include Line, MacDonald, Louisville, Cobb, Bowie, Yellowstone, Galapagos, Bermuda, Canary, St. Helena (Cameroon), Cape Verde, E. Australia, Erebus, Easter and the New England Seamounts. Pre-existing fracture zones or sutures are implicated in Deccan, North Atlantic, Ascension, Cape Verde, Columbia River and Paraná volcanism. The Line Islands, Socorro Island and the Manihiki Plateau may be related to ridge jumps. The Hess and Shatsky Rises appear to be related to migrating triple junctions. These rises bracket the Emperor seamount chain and all three features terminate at the Mendocino fracture zone. "Midplate" volcanoes are not in random places. They are concentrated at "edges" such as margins, sutures, mobile belts and fracture zones.

LINEAR VOLCANIC CHAINS

Linear volcanic chains in the Pacific were the motivation for the hotspot and plume theories [Wilson, 1963; Morgan, 1972a,b] and the idea that the deep mantle was involved in their genesis. Viewed as the results of thermal instabilities, the logical place for them to originate was a thermal boundary layer in D". Since D" is almost 3000 km from the surface and plumes must penetrate through a convecting mantle and multiple changes in phase and viscosity, they should certainly emerge at random places on the surface. In order to test the deep mantle or D" origin of volcanic chains, it is important to place them in a tectonic context. Many midplate volcanoes occur on pre-existing boundaries and can be attributed to changes in plate stress or separation that are ongoing plate tectonic controlled processes.

The active or recently active volcanoes in the eastern Pacific are mostly within 600-800 km of coastlines of the Americas, usually a trench. Some are related to recent ridge jumps (Socorro). Juan Fernandez and San Felix bound the flat slab section of the Nazca plate and fall at the ends of the volcano gap in Chile. They seem clearly to be related to tearing of the plate due to dip angle changes. The same is true for the Samoan chain. This chain is at the northern end of the Tonga trench where the dip of the Pacific plate changes from steep to horizontal. This region is where the convergence rate changes from the fastest rate in the world to zero. The Galapagos appear to be due to fragmentation of the young portions of the Nazca plate, due to complex trench geometry.

Well defined hotspot tracks with consistent age progressions are uncommon. Many seamount chains are active long after passage over a conjectured hotspot, are simultaneously active over their entire length or have trends inconsistent with plate motions. Examples include Line, Marquesas, Samoa, Cruiser, Marshall, Marcus-Wake, Bowie, and Cameroon. The median duration for volcanic chains in the Pacific is only 15 million years [Pringle, 1993]. Hotspot tracks often start and end at times of plate reorganization. Many isolated seamounts with no obvious connection to chains share geochemical characteristics with long-lived hotspots such as Hawaii. There are only a few old, or long-lived, age-progressive volcanic chains. Most of these were built on ridges. The Musician hotspot track contains only 6 seamounts and has a total duration of 13 million years; it terminated at a FZ at a time of Pacific plate reorganization [Pringle, 1993]. No LIP is involved. These

are typical characteristics of "hotspot tracks" and are not readily explained by plume theories. They are consistent with stress in the plate-controlling onset (and shutdown) of volcanism.

Most long hotspot tracks (aseismic ridges) were built on young oceanic lithosphere, implying control by ridges. Some (90° E ridge) are clearly related to long linear fracture zones, or age offsets on the seafloor. The possibility that the Emperor chain was built on a pre-existing fracture zone or ridge is hard to test because of the lack of magnetic anomalies during the long Cretaceous normal polarity interval. Part of it is bounded by deep troughs, not a swell. It terminates at a trench and a fracture zone (Mendocino).

It appears that the orientation of island chains is controlled by pre-existing features, such as fracture zones, transform faults and ridges, and by divergent pulls of subduction zones and that volcanoes are a better guide to stress conditions than to plate motions. If far field stresses control the motions of plates, and stresses in the interiors of plates, then regions favorable for volcanism (extensile strains) will be relatively fixed with respect to the boundaries. Volcanism will propagate counter to the plate motion, modulated by stresses of volcanic loads and orientation of fracture zones.

THE RELATION TO CONTINENTAL TECTONICS

Midocean ridges are not uniform sheets of rising magma. They exhibit a basic three-dimensionality [Morgan and Forsyth, 1988]. Ridges are segmented by offsets and fracture zones and by variations in magma flux. When magma supply rates are particularly high a hotspot is often invoked. The spacings of major fracture zones and gravity anomalies in the oceans are comparable to upper mantle depths. The spacings of volcanoes and small-scale fracture patterns are comparable to lithospheric thicknesses [Vogt, 1974a]. Large fracture zones can often be traced inland onto adjacent continents. Continental tectonics appear to impress itself onto the structure of a newly opening ocean [Sykes, 1978]. Rarely do volcanic chains originate in the middle of an ocean. Ridges, fracture zones and shear zones or mobile belts are implicated at the starting end. Large volcanoes on pre-existing lithosphere contribute their own stresses and may cause self-propagating fractures. Large volcanoes probably stop growing because of hydrostatic head considerations not because they drift off of a hotspot [Vogt, 1974b]. Some island chains stop when they reach a major fracture zone (Emperors, Louisville, Socorro).

Seamount chains in the Atlantic are related to leaky transform faults or extensional fractures, rather than to hotspots [McHone and Butler, 1984; Smith, A. D., 1993; Meyers and Rosendahl, 1991]. The Canaries are on the continental passive margin, near the border of the Eurasian and African plates. Geochronological and geophysical data do not support a hotspot origin for the Canary Islands. The periods of magmatic activity correspond to the three main tensional phases in the nearby Atlas mountains [Anquita and Hernan, 1975]. Geoid anomalies and bathymetry provide no support for any swell, lithospheric heating or mantle plume [Filmer and McNutt, 1989]. The entire Canary chain has been active in recent times. A propagating fracture model is consistent with the data. The Azores owe their origin to a plate reorganization [Searle, 1980]. The Bouvet and Conrad "hotspots" are related to fracture zones of the same names, near the South America, Africa and Antarctic plate stable triple junction [Sclater et al., 1976]. Walvis ridge and Rio Grande rise may be the result of early transform faults with components of extension [Sykes, 1978; Martin, 1987]. Deccan magmatism is related to reactivation of the Narmada-Son lineament, a 1000 km long active feature that splits the Indian Peninsula [Verma and Banerjee, 1992]. Other hotspot tracks appear to be overridden, or "ghost" ridges [Sutherland, 1983]. None of the midplate igneous activity and topographic swell patterns in the western North Atlantic and eastern North America can be easily reconciled with simple plume concepts [Vogt, 1991]. Global or regional plate reorganization seems to be implicated. The numerous plateaus in the Atlantic and Indian Oceans are coast-parallel, about 1000 km offshore and were active when the ocean was only 2000-3000 km broad. They appear to be EDGE effects (Vogt, 1991).

Many aseismic ridges and seamount chains in the Atlantic are extensions of shear zones in the adjacent continents. The extension of the Walvis Ridge corresponds to the southern boundary of the Congo craton, the Damara tectonic belt, and a line of alkalic basalts. Canaries, Cape Verde [McNutt, 1988] and the Cameroons are on trends of cross-African rift zones. The Rio Grande rise is a continuation of the Cochabamba-Paraná shear zone [Eyles and Eyles, 1993]. Intracratonic rift zones accommodated deformation in the continents during the early opening of the Atlantic and apparently also affected oceanic lithosphere.

The Red Sea-Afar-Gulf of Aden region appears to be the passive response to plate tectonic forces, in particular the rotation of Arabia from the African plate [Coleman, 1993]. The Red Sea and Gulf of Aden are rifts propagating toward the Afar depression. The ocean crust is thin near the Afar and is abnormally thick at points along the Red Sea far removed from the conjectured plume influence. These observations are all in conflict with the plume hypothesis. Mantle upwelling, concomitant uplift and timing of magmatism all follow the passive rifting scenario. Premagmatic uplift at "hotspots" is rare, a clear violation of the thermal plume hypothesis.

Giret and Lameyre [1985] point out that the timing of the New England, Nigerian, Afar and Kerguelan magmatic provinces are not consistent with plume theory. They propose that melting is triggered by systems of lithosphere shear openings. The Monteregean Hills-White Mountains province is not easily reconciled with plume theory [Foland

et al., 1988; McHone, 1996a,b], but is consistent with rejuvenation of lithospheric fractures.

Other proposed hotspot tracks associated with lithospheric cracks do not exhibit the shallow reheating expected for a mantle plume [Filmer and McNutt, 1989]. The NW African plate hotspots (Canaries, Madeira, Cape Verde) show episodic, non-time progressive volcanism and seem to be tectonically related to convergence of Africa and Europe and to lithospheric features such as fracture zones and continent-ocean margins [Schmincke, 1982; McNutt, 1988; Vogt, 1974b].

Bowie seamount is not a hotspot; it is part of a crude alignment of seamounts (Bowie-Kodiak) with an origin at the Tuzo Wilson Volcanic Field (TWVF), a complex diffuse, leaky transform-ridge triple junction [Allan et al., 1993]. Geological and geochemical data contradict the Bowie plume hypothesis. There is no evidence of an enriched plume component in the seamount magmas; they are indistinguishable from MORB. The purported deep mantle plume imaged tomographically [Nataf and van de Car, 1993] is to the north of the Bowie Seamount and far to the northeast of the origin of the seamount chain. The deep mantle plume should be to the SE of the seamounts.

A seismic image of the Iceland mantle was recently presented [Wolfe, Solomon and van de Car, 1997]. A cone-shaped anomaly is apparent which has the appearance of upflow being focused toward the ridge by passive spreading. No plume head is evident. Thus, this may be a passive rather than active upwelling. The seismic experiment, as designed, cannot distinguish between the competing hypotheses for the North Atlantic; passive upwelling, edge-driven convection or plume. The shape and inferred depth extent (400 km) need to be checked by less vertical rays since along-ray smearing can give a similar image. The North Atlantic is a new and narrow ocean and is bounded by thick Archean cratons. It has not yet evolved to a mature spreading center. It was recently covered and insulated by thick lithosphere. Iceland may be a double-EDGE feature, similar to Bermuda, Rio Grande, Kerguelen, etc. rises when these were active (i.e., in a new narrow ocean, close to cratons).

The Chagos-Laccadive ridge, the conjectured volcanic trace of the Réunion hotspot, was built on very thin lithosphere, near a spreading ridge-transform junction [Martin, 1987]. The 90°E ridge is on or near fracture zones and is only 5° from the 85°E ridge. Many hotspot tracks and fracture zones are only about 600 km apart (Bouvet, Shona; 85°E, 90°E ridges) about the depth of the upper mantle. This is very close together for plumes from CMB. The proposed hotspots at Ascension and Circe are close to the mid-Atlantic ridge and the Ascension FZ and are a likely result of small-scale convection associated with the temperature gradient across the FZ [Freedman and Parsons, 1990].

Along some ridge segments, quite substantial magmatism has occurred apparently without the influence of crustal stretching, elevated mantle temperatures or proximity to a

264   ANDERSON

hotspot [Ellam, 1991a,b]. Lithospheric conditions apparently can affect the style and amount of eruption, as well as the location.

These are just a few examples of plate tectonic, rather than convective, control of what have been called "hotspot tracks." A fixed anomaly in the mantle does not appear necessary, any more so than is required to explain the relatively slow motions of ridges and trenches. This lengthy discussion of surface effects seems necessary since linear volcanic chains and "fixity" of hotspots are two of the main arguments that have been advanced to support a D" origin.

If cracks, or tears, propagate away from the consuming trench, as seems to be the case for San Felix, Juan Fernandez and Samoa, then the volcanic chain will propagate counter to the plate motion and may appear fixed in the trench reference frame, which Kaula [1975] has already shown to be slowly moving. If interior stresses in a plate are controlled by the geometry of the surrounding boundaries then the maximum tensile stress region may also appear relatively fixed.

It should be pointed out that there are few unambiguous examples of time progression in volcanic chains and long-lived progressions are extremely rare. The propagating crack, or leaky transform, models of volcanic chains should be studied in more detail. They seem to explain several paradoxes of the plume hypothesis (hot lines, many hotspots on same track, rejuvenated volcanism, absence of swells and geoid highs, non-age progression, absence of parallelism, track offsets, etc.). On a more local scale, volcanoes in a group or chain are often controlled by a fracture grid [Vogt, 1974a].

CHEMISTRY OF PLUMES

At one time, the chemistry of ocean island and continental flood basalts was thought to reflect primitive undifferentiated mantle. The location of this primordial reservoir was placed below the "convecting layer", it being thought that only in the lower ("non-convecting") mantle could primordial material survive for the age of the Earth, and only the lower mantle could serve as a fixed reference system. It was subsequently found that isotopically "primitive" basalts were just part of a continuum that extended from enriched to depleted, relative to expected primitive compositions [e.g., Anderson, 1989]. Much of the isotopic and trace element chemistry of ocean island basalts (OIB) can be attributed to pelagic and continental sediments, altered oceanic crust and seawater contamination. The oxidization state and oxygen isotopic signatures point toward a surface origin for at least some components in so-called hotspot magmas. OIB chemistry is often intermediate between MORB and island arc volcanics (IAV). It is now generally agreed that, for most trace element and isotopic systems, the geochemistry of OIB reflects recycling, primarily of oceanic crust and sediments. This material, of course, is provided to the

system from the top and the question arises, how deep is it circulated before contributing to OIB magmatism? Continental flood basalts (CFB) show evidence for interaction with the underlying lithosphere, crust and sediments. Even the noble gas composition of "hotspot" magmas and xenoliths exhibit strong contamination by the atmosphere or seawater or seawater infiltrated crust and upper mantle [Farley and Poreda, 1993].

Nevertheless, in spite of this strong evidence for near-surface sources of the non-MORB characteristics of OIB and CFB, the old idea that "hotspot" magmas require a deep mantle source has survived. Material is believed to be recycled deep into the mantle, heated up and then brought to the surface in narrow plumes. The widespread evidence for OIB-like material at ridges, rifts and island arcs is attributed to pollution from a point source (plumes), channeling by and to ridges, long distance channeling by "grooves" in the lithosphere, flattened plume heads, incubating plumes and fossil plumes. The evidence is even more consistent with a universal enriched layer underlying the lithosphere, and widely available at new rifts, spreading centers and off-axis volcanoes [Anderson, 1996]. This enriched layer has been called the "perisphere" (all-around, close) to distinguish it from the lithosphere and asthenosphere, which are rheological terms. Models invoking a deep source for OIB and CFB tacitly assume that the whole upper mantle is depleted (yet fertile) and suitable only for providing MORB (Appendix 3).

The *coup de grace* for the chemical part of the plume hypothesis is the widespread occurrence of basalts identical to OIB in extending terranes in all tectonic environments, including local extension at consuming margins [e.g., numerous references in Smellie et al., 1994 and Anderson, 1996, 1998]. OIB-like basalts are associated with cessation of subduction, slab window formation, slab roll-back, thin spots in the lithosphere and extensional tectonism in general. Quite often no precursory uplift is observed, ruling out a thermal anomaly. In this context, the alternatives to plume, or thermal, magmatism include passive upwelling, convective partial melting, displacive upwelling, flux induced melting, and flow driven by lateral temperature gradients [Mutter et al., 1988; King and Anderson, 1995].

To my knowledge, there is no evidence that suggests that any basalt source has been in contact with the core or with the lower mantle. Although partition coefficients (and the siderophile nature of elements) depend on temperature and pressure, it is likely that mantle which has been in contact with the molten core or a perovskite lower mantle would have chemical characteristics quite unlike mantle that has not. There is abundant evidence, however, that much of the geochemistry of OIB was acquired at or near the surface. The main remaining argument for a primordial source is the high 3He/4He ratio of some basalts, assuming that it is the 3He that is in excess rather than 4He (or U and Th) that is in deficit.

THE PB-PARADOX AND THE HE-PARADOX

Lead isotopes, and other source specific ratios such as U/Nb, Pb/Ce and Lu/Hf show that a primordial reservoir has not been sampled and may not exist [Bichert-Toft and Albarede, 1997; Hofmann, et al., 1986]. The presence of such a reservoir is fundamental in mass balance box models that attribute the continental crust to extraction of all of the incompatible elements out of the top 30% of the mantle. Modern mass balance calculations imply that most or all of the mantle has been processed to form the crust and upper mantle [Anderson, 1989; Hofmann, et al., 1986; Zhang and Zindler, 1989; Bichert-Toft and Albarede, 1997].

Plumes play a central role in noble gas models of the mantle. Some hotspots have a subset of their eruption products with 3He/4He ratios in excess of MORB. These have been assumed to contain excess 3He. If so, calculations suggest that the source mantle must be enriched in helium by two-orders of magnitude compared to other less volatile species [Kellogg and Wasserburg, 1990; Porcelli and Wasserburg, 1995]. A gas-rich primordial lower mantle has been proposed.

However, hotspots have low 3He abundances and low ratios of He to other noble gases and CO2. The low He/Ne, He/CO2 ... ratios of high 3He/4He ratio magmas indicate that these do not represent degassed magmas but may be related to the vesicles in degassing magmas. High 3He/4He apparently represents a deficit in 4He and therefore in U and Th. The lithosphere is a low U, Th environment and CO2-He rich fluid-filled inclusions are common in lithospheric xenoliths. There seems to be no support for the standard assumptions that high 3He/3He ratios imply excess 3He and that hotspot magmas are low in 3He because they have been extensively degassed compared to MORB. There is therefore no support for the conjecture that high 3He/4He ratio basalts are from a deep primordial undegassed reservoir. The number of atoms of 3He associated with "hotspot" volcanoes and basalts is trivial compared with those at ridges and island arcs [Anderson, D. L., The Helium Paradoxes, submitted to EPSL for publication, 1997].

DISCUSSION

Hopefully, it is clear by now why so much attention has been paid to surface tectonics. All the tectonic and geochemical features discussed have been attributed by one author or another to deep mantle processes including plumes and D" instabilities. Sometimes correlations are claimed between surface processes or chemistry and deep mantle tomography, or magnetic field variations. The correlations of hotspot locations (but not volume) with lithospheric age, fracture zones and cratonic boundaries are expected if plate architecture, rather than deep mantle convection, controls the locations of magmatism (Appendix 2). The deep Earth correlations, however, have received much more attention. Ridges, slabs and cratons provide a template which controls the planform of mantle convection. Past subduction controls the large scale temperature, melting point and chemical structure of the mantle. Past ridges and rifts control the fertility and volatile content of the upper mantle. Thickening plates, slabs and lateral temperature gradients due to lithospheric architecture probably provide adequate stresses to drive the plates, and to provide the extending stresses conducive to extrusion. These near surface and plate tectonic effects must be understood and removed [Wen and Anderson, 1995a,b] before the role of unrelated deep mantle processes on surface processes can be understood (Appendix 1.G).

The assumptions that ridges are entirely passive, that the ridge reservoir fills up the whole mantle, that melting requires importation of deep heat, that magma volumes depend only on temperature, that the hotspot reference frame is fundamental, that recycling of surface contaminants is deep and that the lower mantle is more primitive and fertile than the upper mantle underlie much of the current deep Earth literature (Appendix 1). The prevailing idea that the deep mantle readily communicates with the surface through narrow plumes is based on the above assumptions. The fact that these assumptions ("axioms") are not generally discussed is the motivation for the emphasis in the present paper.

Normal mantle geothermal and melting curves are such that partial melting is the normal state of the upper mantle. The exceptions are in regions of refractory dry peridotite (craton roots) and cold advected geotherms (slabs). Regions which have experienced melt extraction or refractory cumulate formation, such as the perisphere or compositional lithosphere, may dominate the shallow mantle and these regions may also be subsolidus. Melts migrate to thin regions of the lithosphere and to lithospheric discontinuities. Lithospheric strength and compression keep melts in the mantle until there is extension and development of centimeter to meter-size cracks at which point rapid drainage, transient, events occur. When stress conditions are appropriate, the magma itself may participate in opening these cracks by magma fracturing. Some extending regions evolve to rifts and spreading ridges. The basic three dimensionality of passive upwellings gives regions with excess volcanism, often at plate boundaries, or edges. Lithospheric architecture imposes itself on the opportunistic upwellings and on the mantle flow. Some volcanic chains are on old lithospheric flaws, fracture zones, transform faults, abandoned ridges and sutures. Others are on new or reactivated cracks near plate boundaries. As extension (magmatism) evolves to rift and to drift the magma chemistry changes from shallow contaminated (hotspot) to deep "pure" (MORB). Newly opening areas (North Atlantic) may be expressing a temporal phase rather than a spatial feature (deep active upwellings). This is a self-consistent topside alternative to D" driven volcanism.

The questions now are: which end of the mantle is the tail and which is the dog, or are they on separate animals? Are ocean islands the result of fluid dynamics, intense hot jets and thermal instabilities deep in the system, or are they the result of cracks in the LID, opened by far field and nearby stresses?

Are volcanoes permitted by the lithosphere (passive or opportunistic upwellings driven by the hydrostatic head from a partially molten asthenosphere, regenerated by subduction) or are they the tops of narrow convective features from the bottom of the system? "To what end" do we look? Is the mantle a Top Down system, or Bottoms Up? Do we so understand the surface of things that we can venture to seek what is underneath?

> *"All I want to do," I said*
> *"is open a few specific eyes.*
> *Warn a couple of people*
> *about the path they're treading."*
>   Dick Francis
>   The Edge (1989)

Acknowledgements. I appreciate the input of Lianxing Wen and Javier Favela. Dave Sandwell was generous with unpublished data and maps. This research was supported by NSF Grant EAR 92-18390, Contribution No. 6207, Division of Geological and Planetary Sciences, California Institute of Technology.

APPENDIX 1

The Plume Paradigm

A paradigm is the infrastructure of a culture of ideas and thought processes that is shared by the practitioners in a given discipline. It includes assumptions, techniques, language and defense mechanisms. Every paradigm has paradoxes; as paradoxes multiply a paradigm can become vulnerable. It can only be overthrown from the outside and, ironically, only when the tenets become unfalsifiable (Just-So Stories are not falsifiable). It is useful to periodically collect together the key assumptions or axioms of a current paradigm, to see if they can shed light on the current paradoxes. The key axioms of the plume paradigm are:

Physical (P)

1. Midplate volcanoes require a convective instability that arises from a thermal boundary layer at the base of the system (rather than opportunistic utilization of lithospheric fractures, sutures, or tensile stress regimes).

2. Mantle melting (away from ridges and island arcs) requires the importation of heat from a deep TBL (rather than being a natural state of the upper mantle, utilizing normal temperature and chemical variations).

3. An instantaneous steady state exists between heat productivity and heat flow (some geochemical models use this assumption and conclude that there must be a primordial undifferentiated reservoir; i.e., lower mantle).

4. Volume of melting depends only on mantle temperature.

5. The Earth accreted cold and there is no primordial heat (a corollary of P.3).

Kinematic (K)

1. The hotspot reference frame is fixed, implying a deep source.

2. Volcanic chains on a plate are age progressive concentric circles.

3. Ridges are passive and can move about at will.

4. The source of ocean ridge volcanoes is shallow.

5. The upper mantle provides midocean ridge basalts and can only provide midocean ridge basalts; ocean island basalts require a source external to the convecting upper mantle.

6. Hotspots occur at random places and times, unrelated to plate tectonics and large scale convective flow.

Geometric (G)

1. The dimensions of midplate volcanoes, and widths of volcanic chains are related to dimensions of convective features below the plate (rather than to cracks or other lithospheric features).

2. The dimensions of Large Igneous Provinces and associated volcanics and of subsequently uplifted regions are related to dimensions that plume heads acquire as they rise 3000 km (rather than being the convective scales of the upper mantle).

Chemical (C)

1. The crust is a product of the upper mantle and only the upper mantle.

2. The crust and the upper mantle are complementary.

3. Non-MORB characteristics of ocean island and other "hotspot" volcanics are either primitive (undifferentiated mantle) or recycled material that has been taken deep into the mantle.

4. High $3He/4He$ ratios imply excess $3He$ and require a deep undegassed or primordial reservoir (rather than implying a deficiency in $4He$ and therefore, U and Th).

5. The characteristics acquired by plates at the surface (sediments, seawater alterations, oxidization state) are carried deep into the mantle (rather than stripped out of the slab at shallow depth).

6. The lower mantle differs chemically from the upper mantle.

7. Volatiles are incorporated into the bulk of the Earth in chondritic abundances, rather than as a late veneer.

Few readers will believe most of these axioms, as stated, but unstated they underlie much of modern deep Earth science. The widespread belief in deep mantle plumes rests on the validity of these statements. Sometimes beliefs outlive the axiomatic framework they were based on. It is useful, therefore, to isolate the key assumptions and reevaluate them as data and theory advance.

APPENDIX 2

Plume vs. Plates

Large igneous provinces, lateral variations in mantle temperature, geoid highs, hotspot tracks, rapid sea-floor spreading and rises in sea level have all been attributed to plumes rising from the core-mantle boundary. Plumes, as used in the current literature, are secondary forms of convection superposed on, and independent of, the plate scale flow. In a convecting Earth, heated from within, there are broad scale upwellings, but plumes are a hypothetical smaller scale phenomenon, thought to arise from a thermal boundary layer at the base of the mantle, driven by core heat. The presence of a broader scale of convection can destroy this layer, or advect instabilities toward the broader upwelling, so very specific circumstances are required in order for a plume to rise rapidly through a convecting mantle. Broadscale flow, and narrow cold downwellings, can also remove heat from the core, a role generally attributed exclusively to narrow, thermal upwelling plumes. The presence of compressibility, pressure-dependent properties and phase changes also affect the ability of a plume to form or to rise rapidly through the mantle. Plates and slabs also impose a large wavelength structure on mantle convection. Plate-related phenomena and normal mantle convection can cause all of the effects that have been attributed to narrow deep plumes. Subduction and instabilities in the thermal boundary layers of plates can introduce thermal and chemical inhomogeneities into the mantle. These thermal anomalies are of the order of 200°C or greater. Mantle unaffected by subduction for a long period of time will be hotter than mantle elsewhere.

Plate-generated forces (slab pull, ridge push, trench suction, slab roll-back) create extension and plate separation which can induce upwellings and adiabatic melting. Boundary effects can penetrate deeply into plate interiors, causing basins or uplift, or extension. Changes in plate motions and boundary conditions can cause extension across fracture zones and transform faults or pre-existing sutures. Leaky transform faults can open progressively, giving the appearance of time-progressive hotspot tracks. New ridges, or supercontinental breakup, can cause sea level to rise.

Large Igneous Provinces such as continental flood basalt provinces (CFB) and oceanic plateaus, typically occur at triple junctions or ridges, and occur at times of plate reorganizations, ridge jumps or continental breakup. Lithospheric architecture can cause convection and can focus melts. Edge-driven convection is a dynamic alternative to the passive thermal plume hypothesis.

The question is, how can one distinguish these passive and active plate-related processes from the conjectured deep mantle hydrodynamic instabilities? Plume theories are local. A plume must be focused on one part of the lithosphere for a long period of time; long sustained heating, stretching and thinning are implied. Eruption associated with the plume head should be rapid and follow a long period of uplift [Campbell and Griffiths, 1990]. In many large igneous provinces, the evidence for uplift, heating or stretching is missing. In fact, many CFB occur on top of deep sedimentary sequences and in depressions.

There is evidence for high temperatures and high degrees of melting far from the influence of any "plume" [Johnson and Dick, 1992], suggesting large-scale thermal conditions, or hot regions [Anderson, 1993; 1996].

Flood basalt provinces may be generated in response to regional tectonic forces and do not require external influences such as plumes [Ziegler, 1992; Dickinson, 1997; Smith, A.D., 1993]. Some CFB provinces are in back-arc basin locations (CRB) and sites of ridge-trench collisions or slab-windows. Some are associated with margins having experienced a long history of subduction (Gondwana Mesozoic CFB provinces). The shallow mantle in these regions has been fluxed with volatiles and sediments. All CFB provinces are associated with pre-existing lithospheric fractures, sutures, continent-ocean boundaries or craton edges.

Convective and plate motions can be driven or influenced by variations in lithospheric thickness. No critical Rayleigh number is associated with convection driven by lateral temperature gradients at the top of the system. Mantle upwellings will focus at lithospheric discontinuities [King and Anderson, 1995].

Plates therefore play several roles in mantle dynamics and magmatic activity. Without plates, one tends to focus on convective control of surface magmatism. It is likely that lithospheric architecture and stress-state of the surface layers are more important in controlling the locations of volcanoes that a deep thermal boundary layer, such as D" [e.g., Vogt, 1974b].

There are two endmember views of the driving forces of plate tectonics, and the planform of mantle convection. In one view the plates drive themselves by plate tectonic forces (ridge push, slab pull) and the planform of mantle convection is controlled by the plates. Stresses in the plates form a slowly evolving pattern as long as the ridge and trench systems are slowly evolving. Volcanism is expected to be focused at regions of extension (ridges, leaky transforms, lithosphere boundaries) and mountain building and island arcs are at converging boundaries. If hotspots and hotspot tracks are newly extending regions, and ridges are long-lived extending regions, then they both should be slowly moving until boundary forces change by, for example, ridge-trench interactions. In this view, it is no surprise that there is a close spatial relationship between hotspots and ridges and that most hotspots start at ridges, or end up as ridges.

In the other endmember view, convection drives plate tectonics and, in particular controls the locations of "midplate" volcanoes (i.e., hotspots). D" controls the locations of volcanoes and the near-surface simply modulates the "plume flux". Motions of hotspots are due to motions of plumes.

APPENDIX 3

Deductive Logic

The main geochemical argument for a deep source for non-MORB is as follows:
     MORB are from the upper mantle
     OIB are not-MORB
     Therefore OIB do not come from the upper mantle.
Symbolically:
     If M then UM
     OIB are not-M
     So if OIB then not-UM
That is, only MORB can come from the upper mantle. This is the well known Modus Moron fallacy [Margaris, 1967]. It is this fallacy that has led to proposals that non-MORB come from the continental lithosphere (technically outside the "convecting upper mantle"), the lower mantle, D" or the transition region.

It is this fallacy plus the view that the MORB reservoir must be immediately below the plate, if ridges are passive, that has given support to the geochemical aspects of the lower mantle reservoir and plume hypotheses. The geochemical signatures of OIB include MORB, sediments, oceanic crust and, for the noble gases, seawater and atmospheric components. All of these, of course, are near the surface.

The limitations of geochemical approaches in determining source provenances are reasonable well-known but not widely appreciated. The reason for dwelling on this obvious point is the following; the only reason for invoking a deep mantle, or plume source, in much of the current geological and petrological literature is the presence of a non-MORB basalt or component in a basalt. That is to say, in many papers, a non-MORB component (or an OIB-component) is considered a sufficient condition to invoke a deep-mantle plume [e.g., Hart et al., 1997; Wendt et al., 1997; Arndt et al., 1997]. Necessary conditions, such as precursory uplift, or high temperatures, are almost always overlooked. Secondary characteristics such as rifting or time progression are sometimes discussed.

To some extent, D" plumes entered the geochemical literature because of the need for a non-MORB source and the belief that the upper mantle could only provide MORB. However, the plume head aspect of the plume hypothesis makes this assumption self-contradictory. If plume heads are of order 1000 km in radius, the whole central Atlantic upper mantle is contaminated by plumes and ridges will not have access to the MORB reservoir. Likewise, the large plume heads thought to be responsible for the Atlantic bordering CFB would displace a large part of the "MORB asthenosphere". In the perisphere model, a thin enriched layer is available everywhere at the onset of rifting and seafloor spreading until it is replaced by the process of extensive and long-sustained magmatism. Enriched magmas represent transient magmatism; ocean ridge magmatism is the steady-state. One does not need to import the plume components into the upper mantle from the deep mantle; they are already there. On the contrary, one needs to eliminate the shallow enriched layer in order for uncontaminated MORB to be the dominant magma. The lack of sediments, old crust and thick lithosphere at ridges make MORB less susceptible to contamination than "midplate" volcanoes.

REFERENCES

Acton, G. D., and R. G. Gordon, Paleomagnetic tests of Pacific plate reconstructions and implications for motion between hotspots, *Science, 263,* 1246-1254, 1994.
Allan, J. F., R. L. Chase, B. Cousens, P. J. Michael, M. P. Gorton, and S. D. Scott, The Tuzo-Wilson-Volcanic-Field, Ne Pacific - Alkaline volcanism at a complex, diffuse, transform-trench-ridge triple junction, *J. Geophys. Res. Solid Earth, 98,* 22367-22387, 1993.
Altherr, R., F. Henjis-Kunst, H. Puchelt, and A. Baumann, Volcanic activity in the Red Sea axial trough - evidence for a large mantle diapir?, Tectonophysics, 150, 121-133, 1988.
Anders, E., and T. Owen, Mars and Earth, *Science, 198,* 453-465, 1977.
Anderson, D. L., Thermally induced phase changes, lateral heterogeneity of the mantle, continental roots and deep slab anomalies, *J. Geophys. Res., 92,* 13,968-13,980, 1987.
Anderson, D. L., *Theory of the Earth,* Blackwell Scientific Publications, Boston, 1-366, 1989.
Anderson, D. L., 3He from the mantle; primordial signal or cosmic dust?, *Science, 261,* 170-176, 1993.
Anderson, D. L., Superplumes or Supercontinents?, *Geology, 22,* 39-42, 1994a.
Anderson, D. L., Superplumes or supercontinents?: Reply to Dr. Sheridan, *Geology, 22,* 763-765, 1994b.
Anderson, D. L., Superplumes or Supercontinents? Reply to Dr. Kent, *Geology, 22,* 1055-1056, 1994c.
Anderson, D. L., Enriched asthenosphere and depleted plumes, *Inter. Geol. Rev., 38,* 1-21, 1996.
Anderson, D. L., The scales of mantle convection, in press, Tectonophysics, 1998.
Anderson, D. L., and J. D. Bass, Mineralogy and composition of the upper mantle, *Geophys. Res. Lett., 11,* 637-640, 1984.
Anderson, D. L., and C. G. Sammis, Partial melting in the upper mantle, Phys. *Earth Planet. Int., 3,* 41-50, 1970.
Anquita, F., and F. Hernan, A propagating fracture model vs. a hotspot origin for the Canary Islands, *Earth Planet. Sci. Lett., 27,* 11-19, 1975.
Arndt, N. T., A. C. Kerr, J. Tarney, Dynamic melting in plume heads, *Earth Planet. Sci. Lett., 146,* 289-301, 1997.
Bailey, D. K., *Episodic alkaline igneous aivity across Africa in*

Magmatism and the Causes of Continental Breakup, Geological Society, London, 1992.

Bichert-Toft, J., and F. Albarede, The Lu-Hf isotope geocemistry of chondrites, in press, Earth Planet. Sci. Lett., 1997.

Birch, F., Speculations on the Earth's thermal history, *Geol. Soc. Am. Bull., 76,* 133-154, 1965.

Campbell, I. H., and R. W. Griffiths, Implications of mantle plume structure for the evolution of flood basalts, *Earth Planet. Sci. Lett., 99,* 79-93, 1990.

Cande, S. C., C. A. Raymond, J. Stock, and W. F. Haxby, Geophysics of the Pitman Fracture-Zone and Pacific-Antarctic plate motions during the Cenozoic, *Science, 270,* 947-953, 1995.

Chase, C., Extension behind island arcs and motions relative to hotspots, *J. Geophys. Res., 83,* 5385-5387, 1978.

Chase, C. G., Asthenospheric counterflow: a kinematic model, *Geophys. J. R. Astron. Soc., 56,* 1-18, 1979.

Coffin, M. F., and O. Eldholm, Large igneous provinces - crustal structure, dimensions, and external consequences, *Rev. Geophys., 32,* 1-36, 1994.

Coleman, R. G., *Geological Evolution of the Red Sea,* Oxford University Press, New York, 1-186, 1993.

Cordery, M. J., G. F. Davies, I. H. Campbell, Genesis of flood basalts from eclogite-bearing mantle plumes, *J. Geophys. Res. Solid Earth, 102,* 20-179-20,197, 1997.

Davies, G. F., Ocean bathymetry and mantle convection 2. Small-scale flow, *J. Geophys. Res., 93,* 10481-10488, 1988.

Davis, A. S., et al., Ratak Chain- Marshall Islands, *J. Geophys. Res., 94,* 5757, 1989.

Dérvelle, B., G. Dreibus, and A. Jambon, Iodine abundances in oceanic basalts, *Earth Planet. Sci. Lett., 108,* 217-227, 1992.

Dickinson W. R., Tectonic implications of Cenozoic volcanism in coastal California, *GSA Bull., 109,* 936-954, 1997.

Dreibus, G., and H. Wänke, Supply and loss of volatile constituents during accretion of the terrestrial planets, in *Origin and Evolution of Planetary and Satellite Atmospheres,* edited by S. K. Atreya et al., pp. 881, University of Arizona Press, Tucson, 1989.

Ellam, R., Iceland melting dynamics, *Nature, 351,* 191-192, 1991a.

Ellam, R., The rift narrows, *Nature, 351,* 525-562, 1991b.

Eyles, N., and C. H. Eyles, Glacial geologic confirmation of an intraplate boundary in the parana basin of Brazil, *Geology, 21,* 459-462, 1993.

Farley, K. A., Cenozoic variations in the flux of interplanetary dust recorded by 3He in a deep-sea sediment, *Nature, 376,* 153-156, 1995.

Farley, K. A., and R. J. Poreda, Mantle neon and atmospheric contamination, *Earth Planet. Sci. Lett., 114,* 325-339, 1993.

Fedorenko, V., et al., Petrogenesis of the Siberian floor basalt sequences, *Int. Geol. Rev., 38,* 99-135, 1996.

Filmer, P. E., and M. R. McNutt, Geoid anomalies over the Canary Island group, *Marine Geophys. Res., 11,* 77-87, 1989.

Foland, K. A., et al., Nd and Sr isotopic signatures of Mesozoic plutons in northeastern No. America, *Geology, 16,* 684-687, 1988.

Freedman, A. P., and B. Parsons, Geoid anomalies over two South Atlantic fracture zones, *Earth Planet. Sci. Lett., 100,* 18-41, 1990.

Garnero, E. J., and D. V. Helmberger, Seismic detection of a thin laterally varying boundary layer at the base of the mantle beneath the central-Pacific, *Geophys. Res. Lett., 23,* 977-980, 1996.

Giret, A., and J. Lameyre, Inverted alkaline-tholeiitic sequences related to lithospheric thickness in the evolution of continental rifts and oceanic islands, *J. Afr. Earth Sci., 3,* 261-268, 1985.

Handschumacher, D., Formation of the Emperor Seamount Chain, *Nature, 244,* 150-152, 1973.

Hardeback, J., and D. L. Anderson, Eustasy as a test of a Cretaceous superplume hypothesis, *Earth Planet. Sci. Lett., 137,* 101-108, 1996.

Hart, S. R., J. Plusztajn, W. LeMasurier, D. Rex, Hobbs Coast Cenozoic volcanism: Implications for the West Antarctic rift system, *Chem. Geol., 138,* 223-248, 1997.

Heller, P. W., D. L. Anderson, and C. L. Angevine, Is the middle cretaceous pulse of rapid seafloor spreading real or necessary?, *Geology, 24,* 491-494, 1996.

Hofmann, A. W., K. P. Jochum, M. Seufert, and W. M. White, Nb and Pb in oceanic basalts: new constraints on mantle evolution, *Earth Planet. Sci. Lett., 79,* 33-45, 1986.

Holbrook, W. S., and P. B. Kelemen, Large igneous province on the US Atlantic margin and implications for magmatism during continental breakup, *Nature, 364,* 433-436, 1993.

Hopper, J. R., J. C. Mutter, R. L. Larson, C. Z. Mutter, and N. A. S. Group, Magmatism and rift margin evolution: Evidence from NW Australia, *Geology, 20,* 853-857, 1992.

Jacobsen, S., and G. J. Wasserburg, The mean age of mantle and crustal reservoirs, *J. Geophys. Res., 84,* 218-234, 1979.

Johnson, K. M., and H. J. B. Dick, Open system melting and temporal and spatial variation of peridotite and basalt at the Atlantis II fracture zone, *J. Geophys. Res., 97,* 9219-9241, 1992.

Kamo, S. L., G. K. Czamanske, T. E. Krogh, A minimum U/Pb Age for Siberian flood basalt volcanism, *Geochim. Cosmochim. Acta, 60,* 3505-3511, 1996.

Kaula, W., Absolute plate motions by boundary velocity minimizations, *J. Geophys. Res., 80,* 244-248, 1975.

Kellogg, L. H., and G. J. Wasserburg, The role of plumes in mantle helium fluxes, *Earth Planet. Sci. Lett., 99,* 276-289, 1990.

King, S. D., and D. L. Anderson, An alternative mechanism of flood basalt formation, *Earth Planet. Sci. Lett., 136,* 269-279, 1995.

Knittle, E., and R. Jeanloz, Earth's core-mantle boundary - Results of experiments at high-pressures and temperatures, *Science, 251,* 1438-1443, 1991.

Kruse, S., A. Liu, D. Naar and R. Duncan, Effective elastic thickness of the lithosphere along the Easter seamount chain, *J. Geophys. Res., 102,* 27,305-27,317, 1997.

Larson, R. L., Latest pulse of Earth: Evidence for a mid-Cretaceous superplume, *Geology, 19,* 547-550, 1991.

Margaris, A., *First order mathematical logic,* Dover Publications, Inc., New York, pp. 211, 1967.

Martin, A. K., Plate reorganizations around Southern Africa, hot-spots and extinctions, *Tectonophysics, 142,* 309-316, 1987.

McHone, J. G., Broad-terrane Jurassic flood basalts across northeastern North America, *Geology, 24,* 289-384, 1996a.

McHone, J. G., Constraints on the mantle plume model for Mesozoic alkaline intrusions in Northeastern North-America, *Can. Mineral., 34,* 325-334, 1996b.

McHone, J. G., and J. R. Butler, Mesozoic igneous provinces of New England the opening of the North Atlantic Ocean, *Geol. Soc. Am. Bull., 95,* 757-765, 1984.

McKenzie, D., and M. Bickle, The volume and composition of melt generated by extension of the lithosphere, *J. Petrol., 29,* 625-679, 1988.

McNutt, M., Thermal and mechanical properties of the Cape Verde Rise, *J. Geophys. Res., 93,* 2784-2794, 1988.

Meyers, J. B., and B. Rosendahl, Seismic reflection character of the Cameroon volcanic line: evidence for uplifted oceanic crust., *Geology, 19,* 1072-1076, 1991.

Minster, J. B., and D. L. Anderson, A model of dislocation-controlled rheology for the mantle, *Philos. Trans. Roy. Soc. of London, 299,* 319-356, 1981.

Morgan, J. P., The Generation of a compositional lithosphere by midocean ridge melting and its effect on subsequent off-axis hotspot upwelling and melting, *Earth Planet. Sci,.Lett., 146,* 213-232, 1997.

Morgan, J. P., and D. W. Forsyth, Three-dimensional flow and temperature perturbations due to a transform affect, *J. Geophys. Res., 93,* 2955-2966, 1988.

Morgan, W. J., Deep mantle convection plumes and plate motions, *Am. Assoc. Pet. Geol. Bull., 56,* 203-213, 1972a.

Morgan, W. J., Plate motions and deep mantle convection, in *Studies in Earth and Space Sciences (Hess Volume),* vol. 132, edited by R. Shagam et al., pp. 7-122, Mem. Geol. Soc. Am., 1972b.

Mutter, J. C., Margins declassified, *Nature, 364,* 393-394, 1993.

Mutter, J. C., S. R. Buck, and C. M. Zehnder, Convective partial melting, 1. A model for the formation of thick basaltic sequences during the initiation of spreading, *J. Geophys. Res., 93,* 1031-1048, 1988.

Nataf, H.-C., and J. van de Car, Seismological detection of a mantle plume, *Nature, 364,* 115-120, 1993.

Nataf, H. C., Mantle convection, plates and hotspots, *Tectonophysics, 187,* 361-371, 1991.

Norton, I. O., Plate motions in the North Pacific - the 43 Ma nonevent, *Tectonics, 14,* 1080-1094, 1995.

Okal, E., and R. Batiza, Hotspots: The first 25 years, in *Seamounts, Islands and Atolls,* vol. Geophysical Monograph 43, edited by B. H. Keating, P. Fryer, R. Batiza and G. W. Boehlert, pp. 405, American Geophysical Union, Washington, D. C., 1987.

Parman, S. W., J. C. Dann, T. L. Grove, M. J. deWit, Emplacement conditions of komatiite magmas from the 3.49 Ga Komati formation, Barberton Greenstone Belt, South Africa, *Earth Planet. Sci. Lett., 150,* 303-323, 1997.

Patterson, C., and M. Tatsumoto, The significance of lead isotopes in detrital feldspar, *Geochim. Cosmochim. Acta, 28,* 1-22, 1964.

Pick, T., and L. Tauxe, Geomagnetic palaeointensities during the Cretaceous normal superchron measured using submarine glass, *Nature, 366,* 238-242, 1993.

Poirier, J. P., and J. L. le Mouël, Does infiltration of core material into the lower mantle affect the observed geomagnetic field?, *Phys. Earth Planet. Int., 73,* 29-37, 1992.

Polet, J. and D. L. Anderson, Depth extent of cratons as inferred from tomographic studies, *Geology, 23,* 205-208, 1995.

Porcelli, D., and G. J. Wasserburg, Mass transfer of helium, neon, argon, and xenon through a steady-state upper mantle, *Geochim. Cosmochim. Acta, 59,* 4921-4937, 1995.

Pringle, M., Age Progressive Volcanism in the Musicians Seamounts, in *The Mesozoic Pacific: Geology, Tectonics and Volcanism,* pp. 187, American Geophysical Union Geophys. Mono. 77, 1993.

Ray, T. W., and D. L. Anderson, Spherical disharmonics in the Earth sciences and the spatial solution: Ridges, hotspots, slabs, geochemistdry and tomography correlations, *J. Geophys. Res., 99,* 9605-9614, 1994.

Richards, M., R. Duncan, and V. Courtillot, Flood basalts and hotspot tracts: Plume heads and tails, *Science, 246,* 103-108, 1989.

Ringwood, A. E., Slab-mantle interactions, *Chem. Geol., 82,* 187-207, 1990.

Ringwood, A. E., Role of the transition zone and 660 km discontinuity in mantle dynamics, *Phys. Earth Planet. Int., 86,* 5-24, 1994.

Rudnick, R. L., and D. M. Fountain, Nature and composition of the continental-crust - a Lower Crustal Perspective, *Rev. Geophys., 33,* 267-309, 1995.

Ruff, L. J., and D. L. Anderson, Core formation, evolution, and convection: A geophysical model, *Phys. Earth Planet. Int., 21,* 181-201, 1980.

Sano, Y. and S. N. Williams, Fluxes of mantle and subducted carbon along convergent plate boundaries, *Geophys. Res. Lett., 23,* 2749-2752, 1996.

Sawyer, D. S., and D. L. Harry, Dynamic modeling of divergent margin formation: Application to the U. S. Atlantic Margin, *Marine Geol., 102,* 29-42, 1991.

Schmincke, H. U., Volcanic and chemical evolution of the Canary Islands, in *Geology of the NW African Continental Margin,* edited by U. V. Rad, R. Hinz, M. Sarnthern and E. Seibold, Springer-Verlag, Berlin, 1982.

Sclater, J. D., C. Bowin, R. Hey, H. Hoskins, J. Pierce, J. Pillips, and C. Tapscott, The Bouvet triple junction, *J. Geophys. Res., 81,* 1857-1869, 1976.

Scrivner, C., and D. L. Anderson, The effect of post Pangea subduction on global mantle tomography and convection, *Geophys. Res. Lett., 19,* 1053-1056, 1992.

Searle, R., Tectonic pattern of the Azores spreading center and triple junction, *Earth Planet. Sci. Lett., 51,* 415-434, 1980.

Smellie, J. L. (ed), *Volcanism Associated with Extension at Consuming Plate Margins,* Geol. Soc. London, Sp. Publ. No. 81, pp. 293, 1994.

Smith, A. D., The continental mantle as a source for hotspot volcanism, *Terra Nova, 5,* 452-460, 1993.

Smith, W., and D. Sandwell, Measured and estimated seafloor topography, *World Data Center A,* Publication RP-1, 1997.

Spohn, T., and G. Schubert, Thermal equilibration of the Earth following a giant impact, *Geophys. J. Int., 107,* 163-170, 1991.

Stacey, F. D., and D. Loper, The thermal boundary layer interpretation of D", *Phys. Earth Planet. Int., 33,* 45-55, 1983.

Stock, J., and P. Molnar, Revised history of early Tertiary plate motion in the southwest Pacific, *Nature, 325,* 495-499, 1987.

Stone, W. E., E. Deloule, M. S. Larson and C. M. Lesher, Evidence for hydrous high-MgO melts in the Precambrian, *Geology, 25,* 143-146, 1997.

Sutherland, F. L., Timing, trace and origin of basaltic migration in eastern Australia, *Nature, 305,* 123-126, 1983.

Sykes, L. R., Intraplate seismicity, reactivation of preexisting zones of weakness, alkaline magmatism, and other tectonism postdating continental fragmentation, *Rev. Geophys. Space Phys., 16,* 621-688, 1978.

Tackley, P. J., On the penetration of an endothermic phase transition by upwellings and downwellings, *J. Geophys. Res., 100,* 15,477-15,488, 1995.

Tackley, P. J., On the ability of phase-transitions and viscosity layering to induce long-wavelength heterogeneity in the mantle, *Geophys. Res. Lett., 15,* 1985-1988, 1996.

Tarduno, J. A., and J. Gee, Large-scale motion between Pacific and Atlantic hotspots, *Nature, 378,* 477-480, 1995.

Turcotte, D. L., and E. R. Oxburgh, Intra-plate volvanism, *Phil. Trans. R. Soc. Lond., 288,* 561-579, 1978.

Verma, R. K., and P. Banerjee, Nature of the continental crust along the Narmada-Son Lineament inferred from gravity and deep seismic sounding data, *Tectonophysics, 202,* 375-397, 1992.

Vogt, P. R., Volcano height and plate thickness, *Earth Planet. Sci. Lett., 23,* 337-348, 1974a.

Vogt, P. R., Volcano spacing, fractures, and thickness of the lithosphere, *Earth Planet. Sci. Lett., 21,* 235-252, 1974b.

Vogt, P. R., Bermuda and Appalachian-Labrador rises, *Geology, 19,* 41-44, 1991.

Wedepohl, W. E., The composition of the continental crust, *Geochim. Cosmochim. Acta, 59,* 1217-1232, 1995.

Wen, L., and D. L. Anderson, Slabs, hotspots, cratons and mantle convection revealed from residual seismic tomography in the upper mantle, *Phys. Earth Planet. Int., 99,* 131-143, 1995a.

Wen, L., and D. L. Anderson, The fate of slabs inferred from seismic tomography and 130 million years of subduction, *Earth Planet. Sci.Lett., 133,* 185-198, 1995b.

Wen, L., and D. L. Anderson, Layered mantle convection: A model for geoid and topography, *Earth Planet. Sci. Lett., 146,* 367-377, 1997.

Wendt, J., M. Regelous, K. Collerson, A. Ewart, Evidence for a contribution from two mantle plumes to island-arc lavas from northern Tonga, *Geology, 25,* 611-614, 1997.

Wilson, J. T., A possible origin of the Hawaiian Islands, *Can. J. Physics, 41,* 863-870, 1963.

White, B. S. and P. J. Wyllie, Solid reactions in synthetic lherzolite-H2O-CO2 from 20-30 kb, *J. Volcan. Geotherm. Res., 50,* 117-130, 1992.

Wolfe, C. J., I. T. Bjarnason, J. C. Vandecar, and S. C. Solomon, Seismic structure of the Iceland mantle plume, *Nature, 385,* 245-247, 1997.

Zehnder, C., J. C. Mutter, and P. Buhl, Deep seismic and geochemical constraints on the nature of rift-induced magmatism during breakup of the No. Atlantic, *Tectonophysics, 173,* 545-565, 1990.

Zerr, A., G. Serghious, and R. Boehler, Melting of CaSiO3 perovskite to 430 kb and first in-situ measurements of lower mantle eutectic temperature, *Geophys. Res. Lett., 24,* 909-912, 1997.

Zhang, Y., and A. Zindler, Noble gas constraints on the evolution of the Earth's atmosphere, *J. Geophys. Res., 94,* 13,719-13,737, 1989.

Ziegler, P. A., European Cenozoic rift˙ systems, *Tectonophysics, 208,* 91-111, 1992.

Don L. Anderson, Seismological Laboratory 252-21, Division of Geological and Planetary Sciences, California Institute of Technology, Pasadena, CA 91125.

The D″ Discontinuity and its Implications

Michael E. Wysession[1], Thorne Lay[2], Justin Revenaugh[2], Quentin Williams[2], Edward J. Garnero[3], Raymond Jeanloz[3], and Louise H. Kellogg[4]

A synthesis is presented of the results from over 40 different studies of a discontinuous increase in seismic velocity, referred to as the "D″ discontinuity", that occurs about 250 km above the core-mantle boundary. This discontinuity is seen in many regions around the world, both with P and S waves. The D″ discontinuity has a mean apparent depth of about 250 - 265 km for both P and S waves. While many more non-observations (where no discontinuity is observable above the seismic noise levels) exist for P waves than for S waves, both P and S non-observations and observations are distributed with little obvious geographical coherence. The D″ discontinuity is often modeled as a precipitous increase in seismic velocity of about 2.5-3.0%, though most data can accommodate a transition width of up to 50 or 75 km. The D″ discontinuity is usually modeled with lower than average velocities above the discontinuity, but the velocity gradient beneath the discontinuity has been modeled with negative, positive, or zero velocity gradients. There is a strong indication that the base of D″ displays a strong decrease in seismic velocities. There is no significant correlation between large-scale seismic velocities at the base of the mantle and (1) the occurrence or absence of a D″ discontinuity, or (2) a variation in the thickness of the observed depth of the discontinuity. Available seismic evidence leaves several possibilities open for D″ discontinuity origins. Ponding of the remnants of subducted slabs may provide a large-enough decrease in temperature to generate a 3% velocity increase. A mineralogical phase change in a major silicate constituent, perhaps caused by thermal or chemical anomalies, might provide a significant velocity increase. In both cases, a mechanism must exist to cause significant topography on the discontinuity. A third possibility is of significant heterogeneity, either as a stable chemical boundary layer or in the form of laminar layers (perhaps from core-mantle reactions or of ancient ocean crust) that are scattered about the base of the mantle.

[1]Department of Earth and Planetary Sciences, Washington University, St. Louis, MO
[2]Institute of Tectonics and Earth Sciences Department, University of California, Santa Cruz, CA
[3]Department of Geology and Geophysics, University of California, Berkeley, CA
[4]Department of Geology, University of California, Davis, CA

The Core-Mantle Boundary Region
Geodynamics 28
Copyright 1998 by the American Geophysical Union.

1. INTRODUCTION

Few seismic observations have revolutionized our concept of the deep Earth like the discovery of a discontinuous increase in velocity a few hundred kilometers above the core-mantle boundary (CMB). This shattered the prevailing paradigm that reduced velocity gradients predominate at the base of the mantle, and led to further investigations of lateral heterogeneity in the D″ region, the lower 200-300 km of the mantle. The D″ discontinuity, with its variations in structure and geographical distribution, remains a vital component of any synthesis of knowledge about the CMB, and poses intriguing implications for mineral physics and geodynamical modeling of CMB boundary layer processes.

The discovery of the D″ discontinuity modified the classic definition of D″. Many geoscientists currently use the discontinuous velocity *increase* to define the top of the D″ region, though this name was introduced by *Bullen* [1949] for the region of a *decrease* in the seismic velocity gradients at the base of the mantle, an observation made previously by *Gutenberg* [1913], *Witte* [1932], *Dahm* [1934], *Gutenberg and Richter* [1939], and *Jeffreys* [1939]. In keeping with this tradition, we will use the term "D″" to refer to the region of the base of the mantle that displays a velocity gradient less than that of the lower mantle above it (D′), and the "D″ discontinuity" as the anomalous increase in seismic velocities that is reported to occur at a depth that often but not always coincides with the top of D″. While evidence of inhomogeneous structure and decreasing velocity gradients at the base of the mantle accumulated for decades [*Cleary*, 1974], the possibility of a discontinuity with increasing velocity at the top of D″ was not suggested until the 1970's. *Mitchell and Helmberger* [1973] modeled ScS amplitudes and waveforms using a 4-7% S-velocity (V_S) increase about 40-70 km above the CMB, while a series of detailed P-wave apparent velocity measurements between 75° and 95° suggested the presence of a 1.5% P-velocity (V_P) increase about 180 km above the CMB [*Wright*, 1973; *Wright and Lyons*, 1975, 1981]. Both sets of observations were subtle and contentious; for example, many observations appear to require negative velocity gradients above the CMB rather than the strong increase proposed by *Mitchell and Helmberger* [1973], and seismic anisotropy may provide a better explanation for these data [*Lay et al.*, 1998]. The breakthrough came when *Lay and Helmberger* [1983] presented definitive evidence for a rapid 2.75% increase in V_S 250 to 300 km above the CMB, and *Wright et al.* [1985] showed waveform data supporting the existence of a 3% V_P increase 180 km above the CMB. This velocity increase has been studied in many subsequent investigations, listed in Table 1, and been the focus of several reviews [*Young and Lay*, 1987a; *Lay*, 1989, 1995; *Loper and Lay*, 1995; *Weber et al.*, 1996].

Interpretation of the D″ discontinuity in terms of mineral physics and geodynamics remains uncertain, although the seismological models provide guidance for possible thermochemical scenarios. The D″ discontinuity may involve a change in chemistry [*Bullen*, 1949], either of primordial origin, or growing from continued mantle differentiation [*Christensen and Hofmann*, 1994] and/or core-mantle reactions [*Knittle and Jeanloz*, 1989, 1991]. The D″ discontinuity may represent a transition in flow regime, deformational mechanisms, mineral alignment and/or structural fabric of the base of the mantle [e.g., *Lay et al.*, 1998]. It may be an isochemical phase change: the instability boundary for lower mantle constituents [*Stixrude and Bukowinski*, 1992]. There may also be a thermal component to it, related to the accumulation of subducted lithosphere [*Silver et al.*, 1988].

A better understanding of the geophysical processes taking place in the CMB region will require continued seismological observations of the D″ discontinuity and other properties of the lowermost mantle, such as seismic anisotropy [*Maupin*, 1994; *Vinnik et al.*, 1995; *Kendall and Silver*, 1996; *Lay et al.*, 1998], CMB scattering [*Bataille et al.*, 1990], large-scale lateral variations [*Dziewonski et al.*, 1996], and the extent of the ultra-low velocity zone (ULVZ) [*Garnero et al.*, 1998]. This paper aims to synthesize our current knowledge of the D″ discontinuity, its magnitude, depth and sharpness, the extent of its geographical distribution, and the structure above and below it. We also describe possible physical interpretations of the seismic data.

2. SEISMIC OBSERVATIONS

Abrupt velocity increases with depth produce triplications in seismic wavefields, involving three arrivals at certain distances that correspond to energy turning above the velocity increase, reflecting off of it, and turning below it. Observations of triplication arrivals and pre-critical reflections are the fundamental basis for D″ velocity models with rapid velocity increases. Arrival times of the triplication phases and their amplitudes and waveshapes provide constraints on the location and structure of the velocity increase. Usually these are measured relative to direct P or S phases or core-reflections. In some cases triplicated arrivals are directly observable on individual seismograms, though stacking of many seismograms is often required. In addition, there are many other seismic observations of structure near the base of the mantle, usually with large-scale resolution, that are compatible with, but not uniquely diagnostic of the presence of a fast velocity layer below the discontinuity. These include studies of ScS-S and PcP-P travel times and amplitudes, core-diffracted Sdiff and Pdiff waves, PKP-AB - PKP-DF times, and tomographic imaging.

2.1. Direct Observations: PdP and SdS

Detection of the D″ discontinuity involves observations of the SdS and PdP phases. Depending upon the distance travelled, these phases correspond to different wavefield interactions with the D″ discontinuity. At distances less than 70° ($\Delta < 70°$), PdP and SdS are simple (pre-critical) reflections off of the discontinuity. At greater distances PdP and SdS involve two separate phases. Pbc and Sbc reflect off of the D″ discontinuity, while Pcd and Scd turn below the discontinuity, being very dependent upon the velocity gradient within D″ (Pab and Sab are the phases that bottom above D″) (Figure 1). Before the cross-over distance of 82° (for a 600 km deep earthquake), the phases Scd and Sbc both arrive between S and ScS. For most distances less than the cross-over point Sbc and Scd arrive too close in time to be distinguished from each other, and appear as a single arrival. At greater distances Sbc and Scd become increasingly separated. Most studies use observations of the composite PdP or SdS phases at $\Delta = 65°-80°$ [e.g., *Lay et al.*, 1997; *Kendall and Shearer*, 1994; *Weber*, 1993]. However, studies of the pre-critical extension of the Pbc or Sbc branch are

Table 1. List of studies identifying a discontinuous velocity increase at the base of the mantle.

Reference	Location	δlnV (%)	Height above CMB (km)	dV/dz	Data Type	Dist. (deg)	Comments
Baumgardt [1989]	Arctic Sea	2.75	344		Pcd	72-87	Variable depth on discontinuity
Ding & Helmberger [1997]	Central America	3	200	N	Scd	70-96	No P discontinuity
Gaherty and Lay [1992]	E. Siberia	2.75	290	N	Scd	65-95	SGLE fits 80% of data.
Garnero et al. [1988]	Mid Pacific	2	280	N	S-SKS-SKKS	84-94	Disc. disappears laterally
Garnero et al. [1993]	Mid-Pacific	2.4	180	N	Scd	73-82	Model SGHP
	Western Pacific	2.75	280	N	Scd	73-82	Fits models SYL1
Garnero & Lay [1997]	Alaska	*	250-275	N	Scd	45-105	Thick D" correlates with
	E. Siberia		225-250	N			stronger anisotropy
	* (SHcd increase - 2-3%; SVcd increase - 0.5-2%)						
Houard & Nataf [1992]	Franz Josef Land		300		PdP	76-88	
Houard & Nataf [1993]	NW Siberia	2.8-3	250-300		Pcd	75-82	Regional topography on disc.
Kendall & Shearer [1994]	N. Central Asia	2.75	140-370	N	Scd	63-74	Mean = 270 from 19 obs
	Alaska	2.75	160-375	N	Scd	63-74	Mean = 296 from 16 obs (Aleutians: 375 from 6 obs)
	Arctic	2.75	170-330	N	Scd	63-74	Mean = 268 from 20 obs
	Australasia	2.75	100-430	N	Scd	63-74	Mean = 245 from 27 obs (E. Austral.: 300 from 5 obs)
Kendall & Nangini [1996]	Carib. (10N,60-85W)	2.75	250	N	Scd	73-99	Models SKNA1, SKNA2
	Carib. (20N,80-90W)	2.45	290	0	Scd	73-99	No disc. at 25°N, 65-80°W
Krüger et al. [1993]	Severnaya Zem.	0.5-2.	131-178		PdP DB	67-68	Large vars. over short distances
Krüger et al. [1995]	North Pole	1-2	200		PdP DB	84	
	Svalbard	1-2	260		PdP DB	69-70	
	Baffin Island	1-2	180		PdP DB	82	
	Foxe Basin	1-2	260		PdP DB	73	
Lay & Helmberger [1983]	Alaska	2.75	280	0,P	SdS	70-95	Model SLHO
	Caribbean	2.75	253	0,P	SdS		Model SLHA
	NW Siberia	2.75	320	0,P	SdS		Model SLHE
Lay & Young [1991]	Alaska	2.1	175		S+Sdiff	70-105	Both SH and SV are modeled
Lay et al. [1997a]	Eurasia, Alaska		200-300	N	Scd	65-95	ΔVs of ±4% in top 50 km of D" and/or disc. topogr. of ± 50 km
Matzel et al. [1996]	Alaska	3.0	291	N	SH (Scd)	70-106	SV flat, no discontinuity
Mitchell & Helmberger [1973]	Caribbean, Alaska	4-7	40-70		ScS/S	60-75	From ScS/S amps., ScS-S times
Nataf & Houard [1993]					binned p	70-98	Global feature
Nataf & Breger [1996]	Gulf of Alaska	3	300		Pbc stack	24	Δρ = +3%, disc thickness ≤ 8 km
Neuberg & Wahr [1991]	Tasman Sea		none		Pbc	35	No D" discontinuity
	NE Australia		320		Pcd stack	70	
Olivieri et al. [1997]	Antarctic plate	3	315	N	SdS	66-82	No disc. at a nearby spot
Revenaugh & Jordan [1991b]	NW, W, SW Pacific	2.75	270-340		ScSn stack		Disc. seen in 7 regions, but not in 6 others; ρ increase of 1.7%
Scherbaum et al. [1997]	Novaya Zem.		389		PdP	68	Correlation of thin D" with fast
	Severnaya Zem.		139		PdP	68	D" Vp
Schimmel & Paulssen [1996]	S. Fiji Basin		180		Sbc stack	13-26	D" topography of ± 30-50 km with λ = 1200-1600 km
Shibutani et al. [1993]	Western Pacific	1.1-1.6	289	P	PdP	67-83	No discontinuity observed for a spot beneath the N. Pacific
Thomas & Weber [1997]	Nansen, NE Siberia	3	286-293		Pcd	60-100	D" topography of 10-100 km with λ = 400 - 1200 km
Valenzuela&Wysession [1997]	NE Siberia	4.6	210	N	SHdiff	95-102	
	ECentral Pacific	3.4	185	N	SHdiff	105-120	

Table 1. List of studies identifying a discontinuous velocity increase at the base of the mantle.

Reference	Location	δlnV (%)	Height above CMB (km)	dV/dz	Data Type	Dist. (deg)	Comments
Vidale & Benz [1993a]	Northern Pacific	0			ScP,PcP stack	36-48	No discontinuity
Vidale & Benz [1993b]	Arctic Sea	1.5	130	0	Pcd stack	92-103	
Weber & Davis [1990]	NW Siberia	3	293	P	PdP	73-80	Models PWDK, SWDK
		2	293	P	SdS		
Weber [1993]	Nansen Basin	3	279	P	Pcd	74-79	Only 74 of 255 high S/N events
	Kara Sea, NW Siberia	3	286	P	Pcd		show Pcd
	NW Siberia	2.3-2.6	281-316	0,P	Scd		No Scd at Nansen B. + Kara Sea
Weber & Körnig [1992]	North Atlantic Ridge	2-3	250		Pcd (ISC)	70-80	Eight distinct regions found
	NCentral Siberia	2-3	230		Pcd (ISC)	70-80	with little evidence of Pcd
	Japan	2-3	180		Pcd (ISC)	70-80	
	central Mid-Atlantic	2-3	260,310		Pcd (ISC)	70-80	
	Ecuador coast	2-3	340		Pcd (ISC)	70-80	
Weber [1994]	Northern Siberia	3	205		Pcd	74-79	Thin fast layer, perhaps lamellar
Wright et al. [1985]	Indonesia, Pacific	2.8	163	N	Pcd	78-92	
Yamada & Nakanishi [1996]	Micronesia		170		PdP stack	~65-70	
Young & Lay [1987b]	India	2.75	280	N	Scd	75-100	SYL1, λ = 1200 - 1600 km
Young & Lay [1990]	Alaska	2.75	243	N	Scd	60-100	SYLO, λ of var. is < 500 km

made at shorter distances [*Schimmel and Paulssen*, 1996; *L. Breger and H.-C. Nataf*, personal communication], and of the *Pcd* branch alone at great distances [*Vidale and Benz*, 1993b]. Energy diffracted around the D″ discontinuity at Δ > 100°, predicted by *Schlittenhardt* [1986], has also been observed [i.e., *Kendall and Nangini*, 1996]. Table 1 includes the distances of data used.

Figure 2 shows an example of *SdS* data and synthetics generated using the model SYLO. Note that at the greater distances *SdS* appears as an extended shoulder of the *S* arrival, but is a distinct phase at closer distances. The clear *SdS* arrivals (mostly *Scd* energy) allow individual records to be analyzed, either through modeling with synthetic seismograms [*Lay and Helmberger*, 1983; *Young and Lay*, 1987b; *Young and Lay*, 1990; *Gaherty and Lay*, 1992; *Garnero et al.*, 1993; *Matzel et al.*, 1996; *Ding and Helmberger*, 1997], through a phase-stripping technique to remove the *S* and *ScS* phases [*Kendall and Shearer*, 1994; *Kendall and Shearer*, 1995], or through a combination of both [*Kendall and Nangini*, 1996]. While most *SdS* studies use energy at periods greater than 10 s, most *PdP* studies are done at periods of about 1 s where *PdP* is often below noise levels. Stacking techniques using array data enhance the *PdP* signal [*Weber and Davis*, 1990; *Yamada and Nakanishi*, 1996; *Houard and Nataf*, 1992; *Weber*, 1993; *Vidale and Benz*, 1993a,b]. *PdP* arrivals are identified on vespagrams, that stack a set of array records at different ray parameters. An example is shown in Figure 3 from *Weber et al.* [1996]. Another form of stacking is the double beam method [*Krüger et al.*, 1993, 1995, 1996; *Sherbaum et al.*, 1997; *Thomas and*

Weber, 1997], where records are combined from both seismometer arrays (e.g., the Gräfenberg or NORSAR arrays) and source arrays (e.g., nuclear tests in Nevada and Russia).

There are very few other types of seismic data that are able to resolve a D″ discontinuity. *Revenaugh and Jordan* [1991b] stack long-period multiple-*ScS* reverberations to enhance *SdS* and quantify the impedance contrast across the discontinuity, finding a density increase of 1.7% for an assumed V_S increase of 2.75%. *Valenzuela and Wysession* [1997] use azimuthal profiles of core-diffracted *Pdiff* and *Sdiff* waves in two ways: (1) *Pdiff* and *Sdiff* travel time slownesses vary as functions of frequency because they sample the lowermost mantle at different depths, and so are sensitive to a D″ discontinuity, and (2) the amplitude decrease of *Sdiff* as a function of frequency is indicative of radial structure, resolving a D″ discontinuity in two CMB regions.

2.2. Indirect Seismic Evidence

A velocity increase of 2-3% atop D″ is large enough that, even given the lower-than-average velocities above the D″ discontinuity found in many models, the net effect can still be a substantial increase in average lower mantle velocity relative to standard reference models. Many studies lacking the resolution to detect the discontinuity have identified heterogeneous regions of fast seismic velocities at the base of the mantle. Globally averaged models suggest that average D″ velocities are less than expected for a projection of adiabatic gradients from the rest of the lower mantle (region D′). This is seen in studies of normal modes [*Kumagai et al.*,

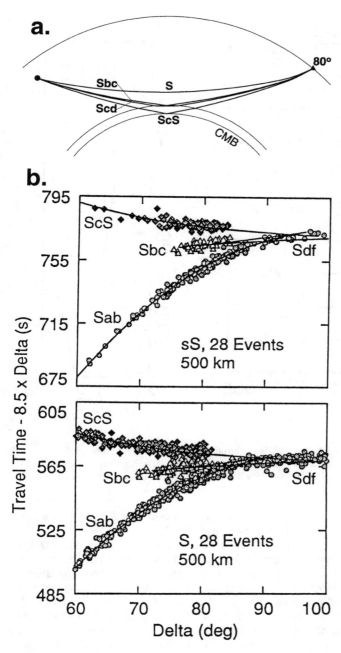

Figure 1. (a) The geometrical paths of the *ScS* precursors, *Sbc* and *Scd*, that result from a discontinuous increase in seismic velocity about 250 km above the core-mantle boundary. Shown here for an earthquake-station distance of 80°, the *Sbc* phase is a reflection off of the "D'' discontinuity", while the *Scd* phase is a refraction just below the discontinuity. An analogous situation occurs for the *PcP* precursors *Pbc* and *Pcd*. The two phases are indistinguishable at large distances and are often referred to together as *SdS* (or *PdP*). (b) An example from *Young and Lay* [1990], showing the different travel time curves for the arrivals of *S* (circles), *SdS* (triangles), and *ScS* (diamonds). The data are from Kuril, Japan and Izu-Bonin earthquakes, converted to a common source depth of 500 km.

1992], and travel times [*Kennett et al.*, 1995], as well as models that combine several seismic data sets like PREM [*Dziewonski and Anderson*, 1981]. Yet, anomalously fast regions of D''are found in studies of the travel times of *ScS* [*e.g., Lavely et al.*, 1986; *Woodward and Masters*, 1991; *Wysession et al.*, 1994], *PcP* [*Zhu and Wysession*, 1997], *PKP* [*Creager and Jordan*, 1986; *Song and Helmberger*, 1993; *McSweeney and Creager*, 1994; *Sylvander and Souriau*, 1996], *Sdiff* and *Pdiff* [*Wysession et al.*, 1992; *Souriau and Poupinet*, 1994; *Wysession*, 1996a; *Valenzuela and Wysession*, 1997; *Kuo and Wu*, 1997], and combinations of phases in tomographic inversions [*Inoue et al.*, 1990; *Tanimoto*, 1990; *Pulliam et al.*, 1993; *Su et al.*, 1994; *Vasco et al.*, 1994; *Masters et al.*, 1996; *Dziewonski et al.*, 1996; *Li and Romanowicz*, 1996; *Grand et al.*, 1997, *van der Hilst et al.*, 1997]. The variations found by these studies can exceed a 3% increase in average D'' velocity relative to reference earth models, and so are compatible with models of a D'' discontinuity. An attempt to correlate the locations of the turning points of observed *PdP* and *SdS* phases with some of these velocity models will be discussed later.

3. STRUCTURE OF VELOCITY MODELS

While models of the D'' discontinuity show a large degree of variability, they tend to have many characteristics in common. Several models for *S*-velocity structures are shown in Figure 4, and many others are described in Table 1. Most models are variations of the original SLHO model of *Lay and Helmberger* [1983], which had a 2.75% discontinuous increase in V_S 280 km above the CMB, with lower-than-average velocities above the discontinuity. The latter feature is not resolved by the data, but is an artifact of embedding the discontinuity in a smooth reference model. If there is no velocity reduction the discontinuity must be placed about 20 km deeper. Studies of *SdS* and *PdP* provide better resolution of the top of D'' than of the bottom. Models are found through non-unique forward-modeling, and trade-offs exist between model parameters [*Young and Lay*, 1987b]. In matching *SdS* and *PdP* data with synthetics, trade-offs exist between the depth and magnitude of the discontinuity for travel times, and between the magnitude and sharpness of the discontinuity for amplitudes. However, support for these models is strong because of their ability to match both the times and amplitudes of *PdP* and *SdS* data for $\Delta = 65° - 100°$, and of *Pdiff* and *Sdiff* for $\Delta > 100°$.

3.1. Magnitude of the Discontinuity

The magnitude of the D'' discontinuity ranges from about zero (in places where no discontinuity can be resolved) to about 3%, for both V_P and V_S. A small velocity increase would not be detected above noise levels, but studies with good resolution find regions with no evidence of a discontinuity. *Krüger et al.* [1995] find regions with less than a 0.5% possible V_P increase, if the discontinuity is

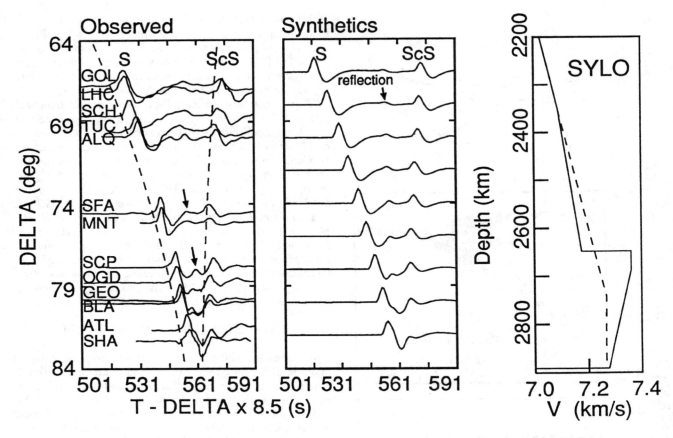

Figure 2. An example showing the forward modeling of *Scd* data for a 9/05/70 earthquake in the Kuril Arc recorded in the United States. The data (left) show a small phase (*Scd*) that arrives in between the *S* and *ScS* arrivals, and is well predicted by the synthetic seismograms (middle) created by the SYLO *S*-velocity model for the base of the mantle (right).

sharp. *Ding and Helmberger* [1997] find no V_P increase beneath Central America to within 1%, and the waveform analyses of *Lay et al.* [1997] find non-observations where the maximum possible undetected increase is 1.5%.

Velocity increases of 2.5-3.0% are consistent with most studies of *SdS* [*Lay and Helmberger*, 1983; *Young and Lay*, 1987; *Young and Lay*, 1990; *Gaherty and Lay*, 1992; *Garnero et al.*, 1993; *Kendall and Shearer*, 1994; *Kendall and Nangini*, 1996; *Matzel et al.*, 1996; *Ding and Helmberger*, 1997; *Garnero and Lay*, 1997; *Lay et al.*, 1997; *Valenzuela and Wysession*, 1997] and *PdP* [*Wright et al.*, 1985; *Baumgardt*, 1989; *Weber and Davis*, 1990; *Weber and Körnig*, 1992; *Houard and Nataf*, 1993; *Weber*, 1993, 1994]. There are also models of lesser V_P increases [1.5%, *Vidale and Benz*, 1993b; 0.7-1.5%, *Krüger et al.*, 1993; 0.5-2.0%, *Krüger et al.*, 1995], and of lesser V_S increases [2.1%, *Lay and Young*, 1991]. Studies of both *PdP* and *SdS* using similar earthquake-station geometries and analysis techniques have found V_P and V_S increases to differ, but not consistently. *Weber* [1993] found regions of D″ beneath the Nansen Basin and Kara Sea with a 3% V_P increase and none

for V_S, and a region of D″ beneath NW Siberia with a 2.3-2.6% V_S increase and none for V_P.

There may be a connection between observations of transverse isotropy at the base of the mantle and the differences between V_{SH} and V_{SV} discontinuities. Observed transverse isotropy in D″ usually (but not always) takes the form of slower V_{SV} relative to V_{SH} (see *Kendall and Silver* [1996], and *Lay et al.*, [1998]), and the magnitudes of V_{SV} discontinuities maybe less than those of V_{SH}. *Garnero and Lay* [1997] found discontinuous V_{SH} increases of 2-3% from *SHcd*, but only 0.5-2% V_{SV} increases for *SVcd* in a region beneath Alaska. Using *SHcd* and *SVcd* also beneath Alaska, *Matzel et al.* [1996] found a 3% V_{SH} increase but no discernible V_{SV} increase. In examinations of *SVdiff*, *Vinnik et al.* [1995] postulate a discontinuous *decrease* in V_S 220 km above the CMB beneath the central Pacific, with the V_{SV} discontinuity is more negative than for V_{SH}. A few observations do report V_{SV} greater than V_{SH} [*Lay et al.*, 1998].

The similarity of the magnitudes of the V_S and V_P discontinuities (about 2.5-3.0) is very unusual, given that V_S variations in D″ tend to be greater than V_P variations by about a factor of 4 [*Robertson and Woodhouse*, 1996; *Bolt-on and Masters*, 1996; *Grand et al.*, 1997]. This result becomes even more extreme when considering that most *SdS* observations are of *SH* waves, and because the *SV* velocities are generally less than for *SH*, the net effect would be that the isotropically-averaged V_S discontinuity increases would actually be less than for V_P. A caveat should be noted that the threshold for observing *PdP* and *SdS* varies greatly between different studies, and a direct comparison may not be valid. For instance, Table 2 lists the results of several comparisons involving observations and non-observations of the D" discontinuity (discussed in detail in a later section). It is suggested that *PdP* non-observations (where a discontinuity was below the threshold of the noise level) represent a greater percentage of the total *P* data set than *SdS* non-observations do for the *S* data (further discussed below). The implication is that the mean V_P discontinuity magnitude may actually be less than for V_S. But in any case, there is a strong suggestion that the mechanism for the D" discontinuity is not the same as the mechanisms for the large-scale lateral velocity variations at the CMB.

3.2. Sharpness of the Discontinuity

Using near-vertical *PdP* reflections off of the top of D", L. Breger and H.-C. Nataf [personal communication] were able constrain the width of the D" discontinuity transition to be less than 8 km in one location. The *PdP* arrivals had a dominant period of 2.5 s, as compared to a *PcP* dominant period of 1 s. While this suggests that the D" discontinuity is not as sharp as the CMB, it does limit the width to being not much greater than λ/4, or 8 km [*Richards*, 1972]. The effects of topography and 3D scattering on this estimate of thickness have not been fully investigated. Long-period *SdS* arrivals cannot constrain the width of the transition, as a velocity increase distributed over about 100 km in depth matches the data as well as a sharp discontinuity [*Young and Lay*, 1987b; *Gaherty and Lay*, 1992; *Garnero et al.*, 1993]. However, a gradual transition greatly decreases pre-critical short-period *PdP* amplitudes [*Young and Lay*, 1987b], placing a maximum transition thickness limit at ≈ 75 km [*Weber et al.*, 1996]. *Revenaugh and Jordan* [1991a,b], using the reflection coefficients of multiple *ScS* bounces, limit the width of the transition to be ≈ 50 km, and modeling of broadband *SdS* signals under Alaska indicates a transition zone width of ≤ 30 km [*Lay and Young*, 1990].

3.3 Velocity Gradients

As seen in Figure 4, models often show a decrease in velocity from reference model values in the regions above the D" discontinuity, though *Ding and Helmberger* [1997] and *Kendall and Nangini* [1996] model structure under the Caribbean with no velocity reduction above the discontinuity. Below the discontinuity, however, the models differ significantly. Some have a zero or slightly positive gradient,

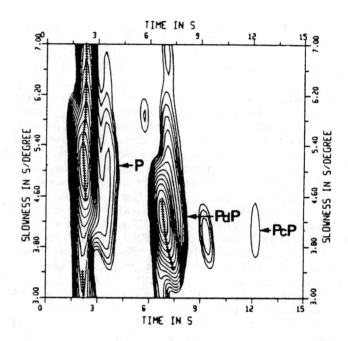

Figure 3. Figure from *Weber et al.* [1996] showing a vespagram for a Kuril earthquake recorded at the Gräfenberg (GRF) Array in Germany. This stacking procedure provides the difference between both slowness and time of the *P* and *PdP* arrivals, which can be converted into the depth of the D" discontinuity. The *PdP* arrival is actually much stronger than the *PcP* arrival at the distances of these seismograms (about 75°-79°).

like SLHO, SLHA, SLHE [*Lay and Helmberger*, 1983], PWDK, SWDK [*Weber and Davis*, 1990], and SKNA2 [*Kendall and Nangini*, 1996]. Others have a strong negative gradient, like SYL1 [*Young and Lay*, 1987b], SGHE [*Garnero et al.*, 1988], and SKNA1 [*Kendall and Nangini*, 1996]. The models of *Valenzuela and Wysession* [1997] and *Ritsema et al.* [1997] use the waveforms of core-diffracted waves, sensitive to the gradients within D", finding strong negative gradients just above the CMB in the regions examined. Still other models have a zero gradient at the top of D" with a negative gradient below, like SYLO [*Young and Lay*, 1990], SGLE [*Gaherty and Lay*, 1992], and SGHP [*Garnero et al.*, 1993]. Because the rate of velocity decrease above the discontinuity can vary, the effect on travel times of phases like *ScS* and *PKP* differ. Models like SGHP (Pacific) and SYL1 (Indian Ocean) are on average slower than PREM, even including the discontinuity. Others like SGLE (Eurasia) and SYLO (Alaska) have faster-than-average velocities. *SdS* and *PdP* arrivals do not require the existence of faster-than-average velocities, but may involve structure superimposed upon both positive and negative anomalies.

A velocity model that is significantly different from others is LAM+ [*Weber*, 1994], that consists of a narrow zone (20 km thick) of very fast velocities atop D". This provides waveforms very similar to conventional models at $\Delta \leq 85°$, but not the secondary arrivals observed beyond about 85°.

4. GEOGRAPHICAL ANALYSIS

Figures 5 and 6, Plates 1 and 2 and Table 2 show a synthesis of studies of the D″ discontinuity, including correlations between observations and non-observations, observations with average D″ velocities, correlations between P and S wave studies, etc. The results are presented with the caveat that studies of many types of data with many different techniques are incorporated. They differ in age, and may differ in quality. For the sake of manageability, all available studies were grouped within Table 2 into 5 different sets of discontinuity *observations* (S16, KSN, LGYG, P40, WK) and *non-observations* (NoS9, NoKSN, NoLGYG, NoP67, NoWK), with the inevitable result of comparing apples and oranges. For instance, the KSN set combines 147 geographically binned values from *Kendall and Shearer* [1994], which are based on 217 individual seismograms, with 65 individual data from *Kendall and Nangini* [1996]. LGYG is mostly individual waveforms from the compilation of *Lay et al.* [1997], but includes the multi-waveform discontinuity determinations of *Garnero et al.* [1993]. The discontinuity data in S16 and P40 are the results of analyses of large numbers of waveforms (*e.g., Vidale and Benz* [1993a,b] used over 1000), but the number and geographical extent of the data vary. Nevertheless, some interesting results emerge.

4.1. Observations and Non-Observations

Plates 1 and 2 show the geographical distribution of locations of observed D″ discontinuities for V_P and V_S, as well as locations where a discontinuity could not be observed. These non-observations are where signal is below noise level, and do not exclude the possibility of small discontinuities or of distributed transition zones that are weak reflectors of short period signals. D″ discontinuities are observed in many parts of the globe, and there has been much discussion about the possibility of it being a global feature [*Nataf and Houard*, 1993]. If the discontinuity is global, then other mechanisms such as topography are required to explain the many non-observations. It is difficult to determine the confidence level of regions that consistently show the presence or absence of a discontinuity, as different studies have differing resolutions, document non-observations to differing degrees, and use different quality data. However, some D″ regions like Franz Josef Land (Arctic Sea) show consistent observations for both V_P and V_S, and some like NE Canada and Greenland show consistent non-observations (data mostly from *Krüger et al.* [1995]).

Estimates for the lateral extent of coherent D″ reflectors vary greatly, with regions as small as a few hundred kilometers [*Weber*, 1994] and as large as 1500 km [*Young and Lay*, 1990] being characterized with relatively uniform

structure. The nature of lateral transitions in the discontinuity structure is unknown: the fast layer may pinch out, or the discontinuity may fade laterally.

An interesting observation in Table 2 is the difference between the KSN and LGYG sets of the ratios of observations to non-observations (212/179 for KSN/NoKSN, and 508/61

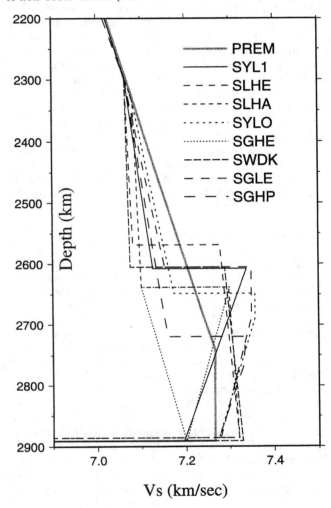

Figure 4. An example of eight variations of shear velocity models for the base of the mantle displaying the D″ discontinuity. These models are all roughly similar in that they have about a 2.5-3.0% discontinuous velocity increase, but differ in several ways including the depth of the discontinuity, the gradient below the discontinuity, the degree to which velocities are decreased above the discontinuity, and the mean vertical velocity. Several other models (not shown) represent the discontinuity as a gradual transition over 50-120 km. The range of models is similar for P velocities. The models shown are SLHA and SLHE [*Lay and Helmberger*, 1983], SYL1 [*Young and Lay*, 1987b], SGHE [*Garnero et al.*, 1988], SYLO [*Young and Lay*, 1990], SWDK [*Weber and Davis*, 1990], SGLE [*Gaherty and Lay*, 1992], and SGHP [*Garnero et al.*, 1993]. Model PREM [*Dziewonski and Anderson*, 1981] is also shown.

Table 2. Analysis of data sets revealing a velocity increase at the base of the mantle.

Obs. Type	Data Source	# of Data	Mean ±1σ	Grand et al. [1997] V_S vs. discontinuity Ave ΔV_S (%)	R^a	Dziewonski et al. [1996] V_S vs. discontinuity Ave ΔV_S (%)	R^a	Van der Hilst et al. [1997] V_P vs. discontinuity Ave ΔV_S (%)	R^a	Wysession [1996] V_P vs. discontinuity Ave ΔV_S (%)	R^a
S	S16[b]	16	257 ± 50	0.80 ± 0.59	0.49	1.34 ± 1.40	0.66	-0.02 ± 0.28	-0.08	-0.32 ±0.51	-0.45
No S	NoS9[c]	9	---	0.85 ± 0.51	---	1.39 ± 1.13	---	0.07 ± 0.41	---	-0.09 ± 0.64	---
S	KSN[d]	212	251 ± 67	0.58 ± 0.87	0.04	0.44 ± 0.82	0.10	-0.01 ± 0.29	-0.01	-0.10 ± 0.44	-0.10
No S	NoKSN[d]	179	---	0.50 ± 0.94	---	0.69 ± 0.82	---	-0.01 ± 0.30	---	-0.02 ± 0.46	---
S	LGYG[e]	508	258 ± 23	0.93 ± 0.63	0.40	0.87 ± 0.88	0.37	-0.07 ± 0.32	0.56	-0.12 ± 0.51	0.01
No S	NoLGYG[e]	61	---	0.89 ± 0.42	---	0.83 ± 0.86	---	0.05 ± 0.31	---	0.02 ± 0.55	---
P	P40[f]	40	265 ± 56	0.86 ± 0.79	-0.04	1.26 ± 1.13	0.41	0.07 ± 0.50	0.09	-0.12 ± 0.57	-0.14
No P	NoP67[g]	67	---	0.64 ± 0.69	---	1.36 ± 1.31	---	0.07 ± 0.41	---	-0.09 ± 0.64	---
P	WK[h]	120	252 ± 35	0.86 ± 0.78	0.01	0.90 ± 0.87	-0.09	0.09 ± 0.28	0.13	-0.06 ± 0.42	-0.00
No P	NoWK[h]	398	---	0.38 ± 0.99	---	0.71 ± 0.66	---	0.02 ± 0.30	---	-0.03 ± 0.38	---

[a] Correlation Coefficient, R, is between the thickness of the discontinuity and the model velocity at the location of the discontinuity.

[b] Includes data from DH97, VW97, W93, SP96, WD90, and OPM97.

[c] Includes data from WD90, W93, and OPM97.

[d] Includes data from KS94 and KN96.

[e] Includes data from LGYG97 and GHG93. Discontinuity depths had 5 possible values: 243 km beneath Alaska, 286 km beneath Eurasia, 283 km beneath the Indian Ocean, 280 km beneath the west Pacific, and 180 beneath the mid-Pacific.

[f] Includes data from HN92, HN93, YN96a,b, NH91, NB96, VB93b, W93, WD90, TW97, KWSS93, KWSS95, SKW97, and STKH93.

[g] Includes data from WD90, NW91, VB93a, W93, KWSS95, DH97, and STKH93.

[h] Includes data from WK90, WK92.

for LGYG/NoLGYG). KSN is dominated by the global study of *Kendall and Shearer* [1994], while LGYG is dominated by sub-Eurasian and sub-Alaskan regional studies. While the particular numbers are not meaningful because of the inclusion of binned data (see above), the ratio for the globally-oriented KSN is about equal to 1, while the more regional LGYG ratio is close to 10. This is likely a result of differences in analysis, as KSN studies use a phase-stripping technique with long-period GDSN data while LGYG studies use synthetic waveform modeling with intermediate-period WWSSN data. The sub-Eurasian and sub-Alaskan regions examined by *Lay et al.* [1997] (and preceding studies) may also be significantly different than the global norm.

4.2. Variations in the Depth of the Discontinuity

Figures 5a,b show the distribution of D" V_P and V_S thicknesses, as determined by the distance from the CMB to the D" discontinuity. The size of the symbols are scaled to the thickness. Table 2 shows that mean thicknesses are in the range of 250 - 265 km, but scatter is large, and the means have 1σ standard deviations of about 50 km. Almost all studies of *SdS* and *PdP* have observed a variable depth of

the discontinuity. The range of thicknesses was 100 - 440 km for *Kendall and Shearer* [1994]. *Lay et al.* [1997] found that ± 50 km of topography could exist over lateral distances of 200-500 km or less beneath Alaska, even in regions that showed generally stable discontinuity structure over 1500-2000 km lengths. *Thomas and Weber* [1997] found topography of 10-100 km over distances of 400-1200 km. *Kendall and Shearer* [1995] identified coherent topography variations over long distances beneath the southwest Pacific. *Schimmel and Paulssen* [1996] found that topography of ± 30-50 km over scale lengths of 1200-1600 km explained the large variation in *SdS* reflection amplitudes at short distances. This focussing-defocussing mechanism may explain alternations of observations and non-observations, even in the case of a global discontinuity.

4.3. Correlations with Averaged Lowermost Mantle Velocities

A natural question is whether the locations of the D" discontinuity observations and non-observations correlate with large-scale models of lowermost mantle velocities. The velocity variations of these models are often attributed to

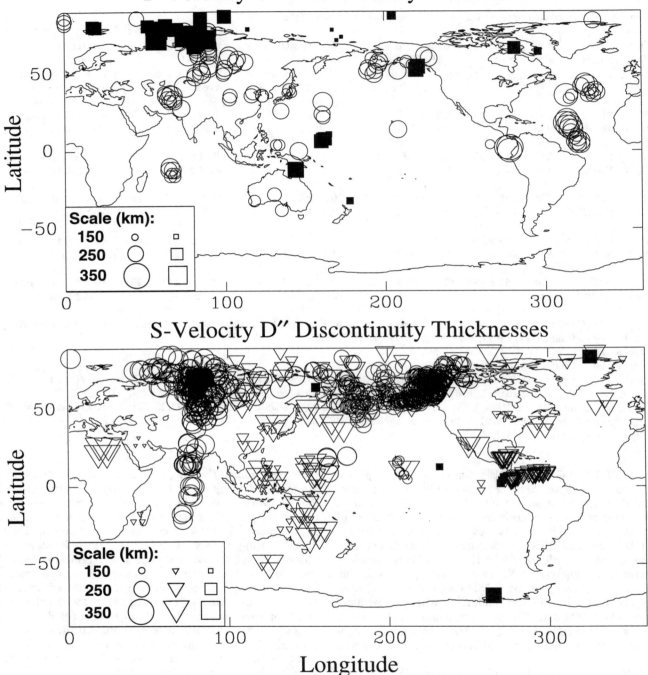

Figure 5. (Top) The geographical distribution of the inferred thickness of the D″ discontinuity for *P* velocities. Circles are for the ISC data of *Weber and Körnig* [1990, 1992], and squares are for the waveform studies of *Weber and Davis* [1990], *Neuberg and Wahr* [1991], *Houard and Nataf* [1992, 1993], *Weber* [1993], *Vidale and Benz* [1993b], *Krüger et al.* [1993, 1995], *L. Breger and H.-C. Nataf* [personal communication], *Yamada and Nakanishi* [1996], *Ding and Helmberger* [1997], *Thomas and Weber* [1997], and *Sherbaum et al.* [1997]. (Bottom) The geographical distribution of the inferred thickness of the D″ discontinuity for *S* velocities. Circles are for the waveform data of *Lay and Helmberger* [1983], *Young and Lay* [1987b], *Young and Lay* [1990], *Gaherty and Lay* [1992], *Garnero et al.* [1993], and *Lay et al.*, [1997], triangles are for the waveform data of *Kendall and Shearer* [1994] and *Kendall and Nangini* [1996], and squares are for the waveform studies of *Weber and Davis* [1990], *Weber* [1993], *Schimmel and Paulssen* [1996], *Ding and Helmberger* [1997], *Valenzuela and Wysession* [1997], and *Olivieri et al.* [1997].

P-Velocity Discontinuity Depths vs. Velocity Models

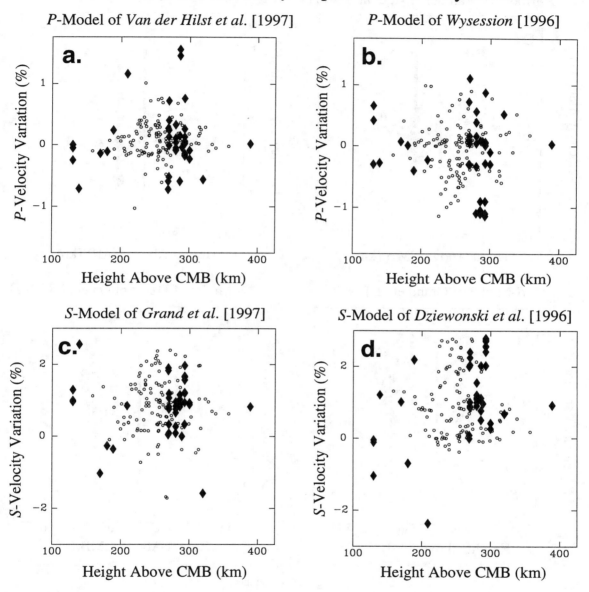

Figure 6. Plots showing the correlations between the variations in inferred D″ thickness, for both *P* and *S* velocities, and variations in the large-scale seismic velocities at the base of the mantle from four different seismic studies. In (a)-(d) the plots show V_P discontinuity thickness vs. the V_P velocity models of (a) *van der Hilst et al.* [1997], and (b) *Wysession* [1996a], and vs. the V_S velocity models of (c) *Grand et al.* [1997], and (d) *Dziewonski et al.* [1996]. In (e)-(h) the plots show V_S discontinuity thickness vs. the V_P velocity models of (e) *van der Hilst et al.* [1997], and (f) *Wysession* [1996a], and vs. the V_S velocity models of (g) *Grand et al.* [1997], and (h) *Dziewonski et al.* [1996]. Note that the correlations are poor. In (a)-(d), data shown is for *Weber and Körnig* [1992] (open circles) and P40 from Table 2 (solid diamonds). In (e)-(h), data shown is for *Lay et al.* [1997] (open circles), *Kendall and Shearer* [1994] and *Kendall and Nangini* [1996] (open triangles), and S16 from Table 2 (solid diamonds).

S-Velocity Discontinuity Depths vs. Velocity Models

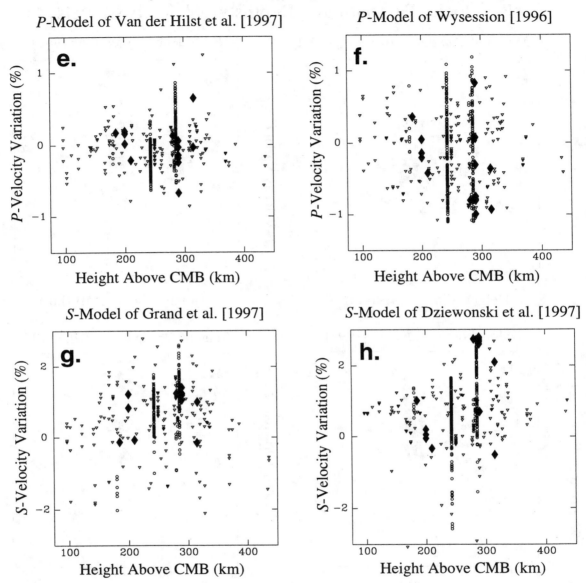

Figure 6. (continued)

temperature variations, as well as the locations of subducted slabs and hotspot origins [i.e., *Richards and Engebretson*, 1992; *Ricard et al.*, 1993]. Any correlation with the locations or topography of discontinuities could have important geodynamic and mineral physics implications. Table 2 and Figures 6*a-h* show these correlations, using a small sample of current velocity models of the lowermost mantle [*Grand et al.*, 1997; *Dziewonski et al.*, 1996; *van der Hilst et al.*, 1997; *Wysession*, 1996a]. While two models are of V_S [*Grand et al.*, 1997; *Dziewonski et al.*, 1996] and two are of V_P [*van der Hilst*, 1997; *Wysession*, 1996a], we show all possible correlations between these and the discontinuity parameters, with the strong caveat that correlations between *P* and *S* velocities at the base of the mantle may be poor [*Wysession et al.*, 1992; *Bolton and Masters*, 1996; *Robertson and Woodhouse*, 1996].

At first glance in Table 2 there seems to be an indication that D″ discontinuities might correlate with regions of fast velocity. For the locations of the 5 groups of observations (3 for V_S, and 2 for V_P), the *Grand et al.* [1997] model predicts increased mean V_S perturbations of 0.4 - 1.0%, and *Dziewonski et al.* [1996] predicts mean V_S values of 0.4 - 1.4%. However, *van der* Hilst [1997] predicts V_P perturbations of about 0%, and *Wysession* [1996a] predicts slightly negative V_P perturbations. It is more likely that these mean velocity values are the result of the limited geographical distributions of our discontinuity observations. Most relevant data sample D" at very high latitudes, which tend to display faster V_S velocities than in V_P models [*Wysession*, 1996a].

There is also no significant difference between the average velocities of locations with and without discontinuities. The mean D" velocities are the same in regions of discontinuities as in regions of non-observations (an average of +0.41% for both in the discontinuity-vs.-velocity model pairings in Table 2). The presence of a D" discontinuity bears no indication of the presence of fast or slow average D" velocities. One is just as likely to find a discontinuity or not in either fast or slow regions of D".

There is also no significant correlation between the depths of D" discontinuities and velocities at the base of the mantle from these 4 models. In Figures 6a-d, neither the V_P discontinuity depth data of P40 (diamonds) nor of *Weber and Körnig* [1992] (circles) correlate well with any of the velocity models. If Figures 6e-h, comparing V_S discontinuity depths with the velocity models, most of the correlations are small, with the exception of the S16 set, though this only has 16 values. However, correlations between the V_S discontinuity depths and the V_S models are all positive, suggesting a tendency for faster-than-average velocities at the base of the mantle in regions of elevated discontinuities. This would be expected if an increase in discontinuity elevation represented an increase in the total amount of fast D" rock. No similar correlation is observed for the V_P discontinuities.

4.4. Differences between P and S Discontinuities

The D" discontinuities for P and S waves exhibit similar characteristics. They are both found across wide geographical distributions, and observations are interspersed with non-observations over a range of spatial wavelengths (Plates 1 and 2). Both have similar mean depths, at about 240-260 km above the CMB. Direct comparisons are difficult because the different techniques required for working with P waves (e.g., stacking) and S waves (e.g., waveform modeling) result in few studies that examine both, though such studies find differences. *Weber* [1993] identified a region beneath NW Siberia with a similar discontinuity for both V_P and V_S, but also found a region beneath NW Siberia that had a discontinuity for V_S but not V_P, and two regions (Nansen Basin and Kara Sea) that had discontinuities for V_P but not V_S. *Ding and Helmberger* [1997] found a 3% discontinuity for V_S beneath Central America, but none for V_P.

One possibility is that the bulk and shear moduli change rapidly enough that D" discontinuities for V_P and V_S actually appear and disappear over the small lateral scales observed in *Weber* [1993]. This would require lateral gradients comparable in strength to the radial gradients, effectively requiring blobs of chemically distinct material. Another possibility is that the different wavelengths for observed *PdP* and *SdS* phases are focussed differently by the discontinuity topography (or scale of volumetric heterogeneities). *Weber* [1993] estimates the Fresnel zones for 1 Hz *PdP* arrivals and lower frequency *SdS* arrivals to be 130 x 260 km and 230 x 460 km, respectively. If D" topography is on the order of 100's of kilometers in scale, it could cause significant changes in the relative amplitudes of *PdP* and *SdS*, occasionally bringing one or both below noise levels.

An important test for this would come with the determination of a relevant comparison between the ratio of observations to non-observations for both V_P and V_S. There is an indication in Table 2 that there are many more V_S discontinuity observations than non-observations, and the opposite holds true for V_P. This observation is likely not to be significant because of the large differences between the studies. Not only are V_S and V_P data and analysis techniques different, but so are the degrees to which different studies have pursued non-observations, which are not as inherently interesting as discontinuity observations. Some data are also very close to each other and should perhaps count as single data. There are many reasons why such a comparison is not valid at this time. Nonetheless, an inference that the V_S discontinuity is in some way more robust or common than the V_P discontinuity would place an important constraint on the dominant wavelength of discontinuity topography or nature of a chemical discontinuity. If factors like the differences between V_S and V_P signal-to-noise thresholds can be assessed, differences in the ratio of observations to non-observations will be geophysically important.

There is likely to be a contribution to differences in the V_P and V_S discontinuities from actual differences between P and S velocities at the base of the mantle. *Grand et al.* [1997] find that while current mantle tomographic P and S velocity models agree very well throughout most of the mantle, this agreement significantly breaks down at the base of the mantle. This drop in correlation between P and S velocities in D" was also observed by *Robertson and Woodhouse* [1996] and *Bolton and Masters* [1996]. *Wysession et al.* [1997], using *Pdiff* and *Sdiff* along identical paths, find the Poisson ratio drops from about 0.31 beneath the mid-Pacific to about 0.25 beneath Alaska. If D" contains within it a chemical boundary layer [*Lay*, 1989], then it is likely that the chemical composition of this layer would vary laterally [*Davies and Gurnis*, 1986], resulting in lateral differences in the ratio between P and S velocities [*Wysession et al.*, 1992]. However, this will not be understood until more regions are examined with both P and S waves, and until we have a better sense of the effects of D" discontinuity topography, which could distinguish between a chemical and thermal origin to D" velocity anomalies.

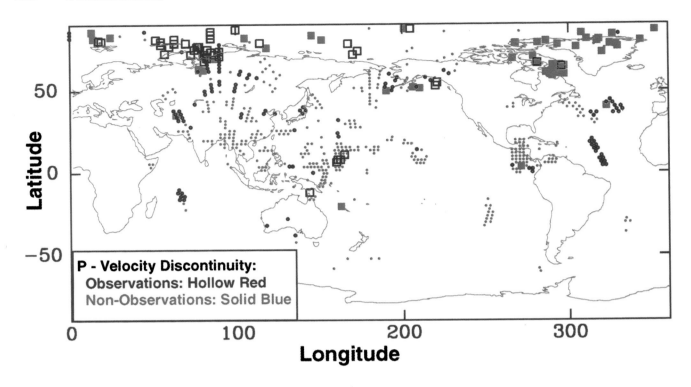

Plate 1. The geographical distributions of D″ V_P discontinuity observations and non-observations. Open red symbols are the locations of discontinuity observations, and closed blue symbols are non-observations. Circles are for ISC bulletin data from *Weber and Körnig* [1990, 1992]. Squares are a combination of studies, including *Weber and Davis* [1990], *Neuberg and Wahr* [1991], *Houard and Nataf* [1992, 1993], *Weber* [1993], *Vidale and Benz* [1993b], *Krüger et al.* [1993, 1995], *L. Breger and H.-C. Nataf* [personal communication], *Yamada and Nakanishi* [1996], *Ding and Helmberger* [1997], *Thomas and Weber* [1997], and *Sherbaum et al.* [1997], and represent composites of many waveforms.

4.5. Correlations with Levels of CMB Scattering

In searching for the origin of the *PdP* and *SdS* phases it is difficult to distinguish between a D″ discontinuity with topography and a distribution of 3D volumetric heterogeneities. With the former, the lower mantle-D″ impedance contrast is assumed to be constant, but topography causes focussing and defocussing that can turn the *PdP* or *SdS* phases on and off [e.g., *Schimmel and Paulssen*, 1996]. With the latter, horizontally varying impedance contrasts can turn *PdP* or *SdS* on and off, and scattering from heterogeneities within D″ can mimic the varying depths of discontinuity topography [*Sherbaum et al.*, 1997]. Conceptually the two models are very different, with a discontinuity appropriate for a sharp chemical or phase boundary, and 3D volumetric heterogeneities suggesting a process such as the accumulation of buckled oceanic crust or laminated CMB reaction products, or 3D phase transition variations due to localized thermochemical variations.

In some cases 1D models cannot adequately explain *PdP* and *SdS* observations. *Sherbaum et al.* [1997] show that *PdP* energy beneath the Arctic Sea does not always arrive along the earthquake-station great circle path, and suggest scatterers at a variety of depths. This was also found by *O'Mongáin and Neuberg* [1996] for *PdP* beneath Central America. However, CMB topography of the sort modeled by *Schimmel and Paulssen* [1996] could also provide *PdP* and *SdS* reflections detected at anomalous back-azimuths. *Sherbaum et al.* [1997] found that reflections from off-azimuth scatterers travel longer paths that look like deep 1D discontinuities, in some cases even from beneath the CMB, though *Lay and Young* [1996] migrated large data sets of *SdS* observations and found general compatibility with a horizontally extending discontinuity structure under Alaska. Further imaging efforts combined with synthetic 3D modeling [e.g., *Thomas and Weber* [1997] may reduce the wide range of distances above the CMB (100-440 km) assumed for 1D discontinuities.

Examining *Scd* phases beneath Alaska, Eurasia and India, *Lay et al.* [1997] found it difficult to distinguish between shear velocity heterogeneity of ± 4% within a 50 km thick region at the top of D″ and ± 50 km of discontinuity topography. In either case, a lack of correlation between *Scd* and *ScS* travel time residuals suggest that heterogeneity is

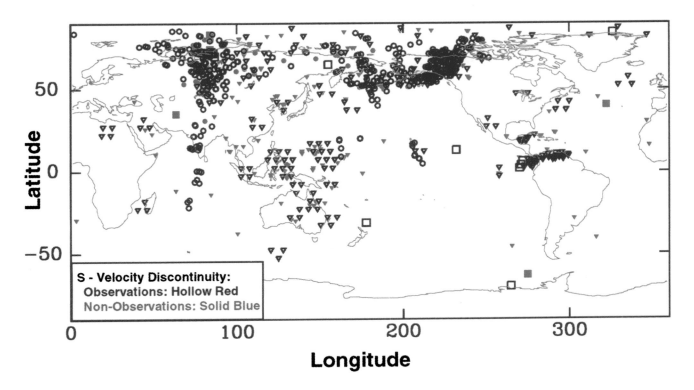

Plate 2. The geographical distributions of D″ V_S discontinuity observations and non-observations. Open red symbols are the locations of discontinuity observations, and closed blue symbols are non-observations. Triangles are for waveform data from *Kendall and Shearer* [1994] and *Kendall and Nangini* [1996]. Circles are for waveform data from *Lay and Helmberger* [1983], *Young and Lay* [1987b], *Young and Lay* [1990], *Gaherty and Lay* [1992], *Garnero et al.* [1993], and *Lay et al.*, [1997]. Squares are a combination of other studies, including *Weber and Davis* [1990], *Weber* [1993], *Schimmel and Paulssen* [1996], *Ding and Helmberger* [1997], *Valenzuela and Wysession* [1997], and *Olivieri et al.* [1997], and represent composites of many waveforms.

concentrated at the top of D″, and low levels of *ScS* fluctuation suggest that most of the rest of D″ is smoothly varying. If the ULVZ at the base of the mantle represents partial melting [*Williams and Garnero*, 1996], then reduced viscosities in the lower part of D″ may not be able to maintain the level of heterogeneity sustainable at the top. This model is consistent with model LAM+ of *Weber* [1994], originating from the accumulation of post-eclogite ocean crust brought down with paleoslabs. Additional support comes from the observations of *Lay et al.* [1997] that heterogeneity at the top of D″ seems to be decoupled from heterogeneities both within D″ and in the overlying mantle.

The scattering of seismic waves from the CMB region has been observed for a long time using *PcP* [*Vinnik and Dashkov*, 1970; *Frasier and Chowdhury*, 1974] and *PKP* [*Haddon and Cleary*, 1974; *Husebye et al.*, 1976], and more recently the codas of *Pdiff* [*Bataille et al.*, 1990; *Bataille and Lund*, 1996]. While these observations support lateral heterogeneity at scales as small as 10 km, they can be explained as either CMB topography of about 300 m or 3D volumetric heterogeneities of 1% distributed throughout a

200-km thick D″ [*Bataille et al.*, 1990]. Comparisons between regions of inferred D″ discontinuities and regions of increased *PKP* scattering is difficult due to incomplete geographical examinations of the latter. However, *Bataille and Flatté* [1988] looked at global *PKP* precursors and found no coherent model of regional scattering variations, concluding that the strength of these inhomogeneities is the same globally to within a factor of three. *Bataille and Flatté* [1988] could not distinguish between CMB topography and 3D volumetric heterogeneities within D″. It is possible that all *PKP* precursors do not have the same source. *Hedlin et al.* [1997] found that global heterogeneity is weak and distributed throughout the lower mantle, while *Vidale and Hedlin* [1998] found regional cases where strong *PKP* precursors may originate in a 60-km-thick layer at the base of D″.

5. IMPLICATIONS FOR MINERAL PHYSICS AND GEODYNAMICS

However numerous the speculations are about the CMB regions, there is only one reality, even if that reality involves a complicated set of regionally varying factors. Some combination of rock and/or liquid exists accounting for all of the unusual seismic observations. The presence of discontinuities, anisotropy, ultra-low velocity layers, and broad regional seismic velocity changes all have explanations, but because geological processes at 136 GPa and 3000-4000 K are likely to differ from those at STP conditions, these explanations remain uncertain. There are several possible factors that likely play major roles. There is either a discontinuity at the top of D″ or a distribution of heterogeneities within (though likely at the top of) D″. While the majority of the CMB is unsampled, the D″ discontinuity is likely globally extensive at a mean height of 250 km above the CMB, but with lateral variations of a strength comparable to the discontinuity itself. There is no correlation between the average lowermost mantle velocity and the occurrence of the discontinuity, and only a hint of one between average velocity and discontinuity thickness. There are no real differences between V_P and V_S. The D″ discontinuity seems to be superimposed upon regional D″ velocities, as it is found in hot (seismically slow) as well as cold (fast) regions, with some indications of systematic regional variations.

5.1. Discontinuity as an Isochemical Thermal Boundary

Because the CMB is much less dense than the underlying core, vertical heat transfer across the CMB in through conduction across D″, which thus contains (or is) a thermal boundary layer (TBL) with a large temperature gradient that remains poorly constrained [*Elsasser et al.*, 1979; *Stacey and Loper*, 1983; *Doornbos et al.*, 1986; *Boehler*, 1993]. An increase in temperature is compatible with the many recent seismic observations that support early suggestions of a decrease in velocities just above the CMB [*Kumagai et al.*, 1992; *Ritsema et al.*, 1997; *Valenzuela and Wysession*, 1997]. However, a simple TBL would not explain an *increase* in seismic-wave velocity. The CMB TBL likely plays only a moderate role in driving broad-scale mantle convection, as estimates of heat flux out of the core are typically 10-20% of that out of surface flux [*Davies*, 1988; *Sleep*, 1990]. This makes it difficult to explain large D″ variations with thermal effects alone, but it is possible for the remnants of slabs to accumulate at the CMB, causing a seismic velocity increase due to their cold signature superimposed upon TBL effects. The positive correlation between the height of the discontinuity above the CMB and corresponding tomographic-model D″ velocity for that location, though slight and not seen for all models, favors a thermal origin for the discontinuity over a chemical one. A chemical boundary layer would be pushed away and thinned in regions of downwelling, the opposite of what seems to be observed. It is also possible that the mechanisms creating the apparent D″ discontinuity differ in regions with and without recent subducted slab, and in paleoslab regions it is primarily of a thermal origin. Slabs may also provide a chemical signature (discussed later).

Could a strong decrease in temperature be the cause of the observed velocity increases atop D″? The major concerns are the magnitude of the required temperature increase and the steepness of the thermal gradient. Much experimental and theoretical work has been been pursued to estimate the seismic-wave velocities in D″, but as neither the exact composition nor prevalent phases of D″ are well established, this is a difficult problem. Guesses involving equations of state for likely lower mantle mineral assemblages have suggested that a 3% increase in seismic velocities would require a drop in temperature of about 700-1100°C [*Wysession et al.*, 1992]. Estimates by *Ding et al.* [1997] are at about 1100°C, and modeling by *Yuen et al.* [1994] (with data of *Chopelas* [1992], *Chopelas and Boehler* [1992], and *Chopelas et al.* [1993]) gives even higher estimates.

These numbers are large, but the cores of subducted slabs may be very cold compared to ambient temperatures at the base of the mantle. Three-dimensional numerical simulations predict that fossil slabs at the base of the mantle could have thermal anomalies exceeding 1000°C below the ambient lower mantle temperature [*Honda et al.*, 1993; *Steinbach et al.*, 1994], provided that the slab material reaches the base of the mantle in fast episodes of flushing [*Weinstein*, 1993; *Tackley et al.*, 1993]. *Ding et al.* [1997] suggest 700°C as the temperature decrease of ponded fossil slab at the CMB relative to the lower mantle adiabat. The rate of descent of slabs through the highly-viscous lower mantle is a key factor in the prediction of lower mantle slab temperatures. Seismic tomography shows a broadening of the slab signature across the 660 km discontinuity in some regions [*Grand et al.*, 1997], and this thickening or buckling suggests a decrease in slab velocities, perhaps a result of increased viscosity in the lower mantle. Slower descent of slabs means greater thermal assimilation of the outer part of the slab by the time it reaches the CMB, but a greater distance for heat to conduct from the core of the slab. The effect would be a broader boundary zone across which a large ΔT might be maintained.

The thickness of this boundary transition is important because of the constraints placed by seismic observations. S-velocity studies suggest a maximum thickness of 50 km [*Revenaugh and Jordan*, 1991a,b] to 75 km [*Young and Lay*, 1987b; *Weber et al.*, 1996], but P-velocity studies may demand the transition to be less than 8 km [*L. Breger and H.-C. Nataf*, personal communication]. It is hard to envision a ΔT of 700-1100°C across less than 8 km at the top of D″, a distance having a thermal diffusion time of less than 5 million years. Distributed across a 50-75 km distance, such a thermal boundary would be more plausible. Certain geodynamic mechanisms might aid in this. If cold slab rock ponded laterally across the CMB, it could displace lower-viscosity hot rock from the base of D″ up and over the ponded slab [*Stevenson*, 1993], enhancing the thermal transition. Seismic observations may support this, modeling velocities above the D″ discontinuity as slower than expected for several hundreds of km into the mantle.

Ding et al. [1997] use numerical modeling to predict that the thermal anomalies resulting from the subduction of the Farallon slab can adequately explain both the existence of the D" discontinuity as well as large-scale tomographic seismic images. The temperature through a vertical profile across their modeled slab at the CMB shows a decrease in temperature of around 700°C at the coldest point (a depth of about 2700, the location of the D" discontinuity) from ambient lower mantle temperatures, and then an increase at the base of D" to about 3300°K. *Ding et al.* [1997] use the approach of *Zhao and Anderson* [1994] to interpret the temperature decrease as a shear velocity increase of 2.5%, distributed over a vertical range of a few hundred kilometers. Even though the seismic modeling still invokes a first-order discontinuity at the top of D", not found in the geodynamic model, the *Scd* data is best fit by a seismic model where most of the velocity increase occurs gradually over a vertical range of 120 km. *Ding et al.* [1997] suggest a thermal origin to the D" discontinuity, as the *SdS* triplication has traditionally been best found beneath regions of convective downwellings [*Lay*, 1995], and that a chemical boundary layer would be dispersed by such downwellings.

A concern with a purely thermal explanation for the D" discontinuity is its lateral extent, as it has been observed in regions with no recent subduction and in regions presumed to be anomalously hot. Is it possible for cold slab to entirely cover the CMB and retain a strong enough thermal signature to provide the discontinuity? Lithosphere is subducting at the current rate of about 3 km^2/yr (or 240 km^3/yr assuming an 80 km thick oceanic lithosphere) so the circulation time required to fill up and empty a 250 km thick D" layer, assuming complete full-mantle circulation, would be about 170 Ma. Since the rock slab is likely to take at least this long to reach the CMB from the surface (not counting mantle avalanche processes), we would require the cold thermal anomaly to last in the mantle for over 350 Ma, unrealistic because the characteristic distance for thermal diffusion is 100 km for such time-scales, and thermal equilibration is hastened by flux of core heat across larger thermal gradients.

Due to the effects of thermal equilibration, the amount of material which is sufficiently cold to generate a 3% velocity contrast is likely to be a relatively small volumetric portion of the slab at CMB depths, and even this portion will be intercalated with warmer (and thus less seismically anomalous) material. Therefore, to produce a relatively broadly distributed laminar boundary with a typical height above the CMB of 250 km through purely thermal effects requires a number of rather specific geodynamic phenomena to occur above the CMB: 1) preservation of sufficient cold slab material to produce the notable seismic velocity contrasts, including possible segregation of cold slab material from surrounding thermally equilibrated material; 2) emplacement of the cold slab material at a height near 250 km above the CMB; and 3) broadly horizontal emplacement of the cold material. Perhaps the most difficult of these conditions to satisfy from a thermal perspective is the predominance of

discontinuity depths in the 200-300 km range (Figure 5). It is unclear why colder (and hence denser) material should be emplaced at this height above the CMB rather than at a range of depths down to the mantle's base, particularly if segregation of cold material from warm material can readily occur at these depths. Although the regional temperature fields of slabs descending to the core-mantle boundary would be anticipated to generate local velocity heterogeneities, there are difficulties with the 250 km discontinuity produced only thermally, unless very rapid mantle avalanches bring large amounts of slab material to the CMB.

It may also be possible that the seismological results are pointing toward a thicker TBL at the base of the mantle. The seminal study of *Stacey and Loper* [1983] used the seismic model PREM [*Dziewonski and Anderson*, 1981] to derive a quasi-exponential thermal profile with a scale height of 73 km and a maximum thermal gradient of 11.2 K/km. The PREM model represented D" as a 150 km layer in which the bulk modulus pressure derivative was unusually low (1.64 instead of about 3.2 in the overlying mantle), and the vertical shear velocity gradient was very slightly negative. Recent work suggests a more dramatic TBL with very strong negative shear velocity gradients extending over the bottom 200-250 km of the mantle. This seems to hold true both beneath discontinuities in fast areas [*Young and Lay*, 1990; *Gaherty and Lay*, 1992] and in the slow velocities of the Pacific [*Ritsema et al.*, 1997; *Valenzuela and Wysession*, 1997], suggesting that the D" thermal boundary layer is much larger than 75 km. Such a thick TBL would require higher viscosities or densities at the base of the mantle to suppress instabilities that could break up the TBL.

5.2. Discontinuity as an Isochemical Phase Boundary

Mineralogical phase changes play an important role in upper mantle geodynamics, and it is possible that the same holds for the lower mantle [*Bina*, 1991]. Even though the lower mantle is viewed as being largely homogeneous and adiabatic, there have been identifications of seismic discontinuities at depths of 710, 900, and 1200 km [*Revenaugh and Jordan*, 1991b; *Wicks and Richards*, 1993], and this may also be the case for D". A 3% D" velocity increase is significant compared to upper mantle discontinuities (the 660 km discontinuity is 4.7% for *P* and 7.0% for *S*), as the high pressures at the CMB cause minerals to be much more incompressible, and might only occur from a breakdown of a major lower mantle phase like perovskite. *Stixrude and Bukowinski* [1992] demonstrated that the breakdown of $(Mg,Fe)SiO_3$ perovskite to its constituent oxides is unlikely to explain the D" discontinuity, but other transformations, perhaps involving Ca and Al, may contribute.

The multi-component phase equilibria of the mantle under CMB conditions remains uncertain, although the approximate mineralogic end-members assumed to predominate under deep mantle conditions have each been examined to pressures approaching those of the CMB. In particular, $(Mg,Fe)SiO_3$-perovskite, $CaSiO_3$-perovskite and $(Mg,Fe)O$ have been probed statically to pressures of 112, 134 and 95 GPa, with $(Mg,Fe)O$ having been examined under shock to pressures exceeding those of the CMB [*Knittle and Jeanloz*, 1987; *Mao et al.*, 1989; *Mao and Bell*, 1979; *Vassiliou and Ahrens*, 1982]. None of these end-members undergoes a resolvable phase transition at pressures corresponding to the lowermost mantle, but there are indications that the partitioning behavior of iron between perovskite and magnesiowüstite could shift depending on the valence state of iron and the aluminum content of deep mantle materials [*Wood and Rubie*, 1996; *McCammon*, 1997]. Aluminum produces enhancements in the amount of iron (particularly as Fe^{3+}) entering into the perovskite structure relative to that which would be inferred from results in the MgO-FeO-SiO_2 system [*Ito et al.*, 1994; *Fei et al.*, 1991; *Kesson and Fitz Gerald*, 1991]. Whether enhanced iron contents in deep mantle perovskite could generate exsolution of Fe or Al-rich phases (possibly with free silica) remains unclear.

All the upper mantle phase transitions, driven by pressure increases, involve increases in seismic velocities. Densities increase, but are over-compensated by increases in incompressibility and rigidity. The same would have to hold for a D″ phase transition. Lateral variations in composition as well as temperature might create topography on such a phase change, the way the 660 km discontinuity is depressed by the cold temperatures of subducting slabs [*Shearer and Masters*, 1992]. However, the seismic observations suggest changes in topography of several hundred kilometers. This would require a very steep Clapyron slope for the phase transition, with very small changes in temperature resulting in very large changes in the pressure at which the transition occurs. If perovskite, the most abundant mineral structure within the Earth, were to break down into an assemblage containing phases like $(Mg,Fe)O$, SiO_2 and Al_2O_3, would this cause a 3% increase in seismic velocities? The high seismic velocities of SiO_2 stishovite [*Sherman*, 1993] might be able to over-compensate for the slower seismic velocities of $(Mg,Fe)O$ to provide an overall increase in seismic velocity [*Wysession*, 1996b]. The transformation would also have to explain why the D″ discontinuity is roughly the same (about 3%) for both V_P and V_S, as lateral S-velocity variations at the base of the mantle are greater than for P-velocities ($\delta \ln V_S / \delta \ln V_P \cong 4$ [*Grand et al.*, 1997; *Wysession and Kuo*, in preparation]). The largest concern with a phase change as the cause of D″ is that it has yet to be experimentally observed, and so remains speculative.

5.3. Discontinuity as a Thermo-Chemical Boundary

Whether or not the thermal signature of slabs or a mineralogical phase change is responsible for the D″ discontinuity, it is probable that chemical heterogeneities play a significant part. The high temperature contrast across the core-mantle boundary required by the high temperatures in the outer core [*Williams et al.*, 1991] should lead to vigorous development of upwellings from the thermal boundary layer, yet the contribution of the core-mantle boundary to the surface heat budget is relatively minor. This suggests that a stable chemical boundary layer (CBL) is present, reducing heat flow across the associated TBL and makes instabilities across the CMB less important to the Earth's surface heat flow. *Bullen* [1949] postulated a chemically inhomogeneous boundary layer as the cause for the seismic D″, and numerous studies have discussed the implications for chemical heterogeneity or a CBL at the base of the mantle [i.e., *Davies and Gurnis*, 1986; *Gurnis*, 1986; *Gurnis and Davies*, 1986; *Ahrens and Hager*, 1987; *Schubert et al.*, 1987; *Zhang and Yuen*, 1987, 1988; *Hansen and Yuen*, 1988, 1989; *Silver et al.*, 1988; *Sleep*, 1988; *Stevenson*, 1988; *Christensen*, 1989; *Knittle and Jeanloz*, 1989, 1991; *Lay*, 1989; *Buffett et al.*, 1990; *Olson and Kincaid*, 1991; *Revenaugh and Jordan*, 1991b; *Ringwood and Hibberson*, 1991; *Stacey*, 1991; *Goarant et al.*, 1992; *Wysession et al.*, 1992, 1993; *Boehler*, 1993; *Buffett*, 1993; *Jeanloz*, 1993; *Kellogg and King*, 1993; *Poirier*, 1993; *Yuen et al.*, 1993, 1994; *Christensen and Hofmann*, 1994, *Weber*, 1994; *Loper and Lay*, 1995; *Kendall and Silver*, 1996; *Lay et al.*, 1997, 1998; and others].

There is a difficulty, however, in finding a mineral assemblage that is both dense enough to be stable at the base of the mantle and seismically fast enough to provide the observed 2-3% increase over ambient lower mantle velocities. For example, increasing density by increasing the iron content of the assemblage, either by a decreased silicate magnesium number or by added iron alloys, will likely reduce seismic velocities. Purely compositional explanations for the possible genesis of layering in D″ can be loosely divided into four categories: 1) descent and segregation of subduction-related geochemical heterogeneities into D″, in the form of subducted basaltic crust and its complementary harzburgite [*Christensen and Hofmann*, 1994]; 2) interaction of the lowermost mantle with the outer core, producing iron enrichment of the lowermost mantle through chemical reactions with outer core material [*Knittle and Jeanloz*, 1989; *Goarant et al.*, 1992]; 3) preservation of primordial stratification of the planet either through enrichment of refractory oxides in a modified heterogeneous accretion scenario [*Ruff and Anderson*, 1980] or through the preservation of fossilized remnants of magma ocean solidification; or 4) descent of dense, possibly molten material into D″ over time, which could produce progressive geochemical enrichment of this region in incompatible elements [*Rigden et al.*, 1988]. Of these explanations, any primordial stratification which persisted over the age of the planet would only be preserved if it were denser than the overlying mantle: such a density difference is not consistent with the intermittent presence of this discontinuity in a hot, inviscid thermal boundary layer. Core-mantle reactions present a more complex set of possibilities. Iron enrichment typically produces a decrease in seismic velocities, and the products of core-mantle reactions likely follow this trend [e.g., *Williams and Garnero*, 1996]. However, these reactions produce both iron-free $MgSiO_3$-perovskite and SiO_2 as reaction products (which are anticipated to be comparatively high velocity constituents), in addition to FeO and FeSi. If the core and mantle have reacted extensively over time, and pervasive segregation of iron-free from iron-rich reaction products can occur, then a possible (although somewhat restrictive) mechanism exists for producing an enrichment of D" in relatively high velocity components.

Considerable attention has been focussed on shifts in silica content with depth in the mantle in the past [e.g., *Stixrude et al.*, 1992], partially motivated by the differences in chemistry between peridotitic and chondritic compositions. Silica enrichment also has the tendency to increase seismic velocity, providing possible chemical explanations for the 250-km D" discontinuity. The observation that basaltic material contains free SiO_2 at deep mantle pressures is significant [*Kesson et al.*, 1994], and provides an additional means by which silica could be enriched in D". It has been proposed that small degrees of partial melting may be present in portions of the lower mantle characterized by high values of $d\ln V_S/d\ln V_P$ [*Duffy and Ahrens*, 1992]. While the melting behavior of the deep mantle remains uncertain, any descent of deep mantle melts into D" could alter the chemistry of this region relative to the overlying mantle, so the existence of moderate SiO_2 enrichment of the lowermost mantle cannot be precluded. Whether SiO_2 enrichment can generate the 250 km discontinuity hinges on its elastic and phase properties under CMB conditions [*Tsuchida and Yagi*, 1989; *Dubrovinsky et al.*, 1997].

Accepting whole mantle convection, there is a possibility of a delamination and accumulation of post-eclogitic ocean crust at the base of the mantle [*Gurnis*, 1986; *Gurnis and Davies*, 1986; *Silver et al.*, 1988; *Christensen*, 1989; *Olson and Kincaid*, 1991]. *Gurnis and Davies* [1986] demonstrated that the recycling of oceanic crust could play an important role in generating the observed geochemical heterogeneity of the mantle. However, unless the crustal component is at least neutrally buoyant compared to the rest of the subducting slab, it may be difficult to separate the crustal component from the rest on the time scale required for transport of a subducting slab to the lowermost mantle [*Richards and Davies*, 1989]. *Christensen and Hofmann* [1994] used numerical simulations to model the delamination of basaltic crust from subducting slabs, the ponding of the former basalt at the base of the mantle, and the re-entrainment of the material in hotspot plumes, providing an explanation for why more enriched isotope ratios are linked to mantle plumes. They estimated that the density of the subducted crust was large enough (by 1.5-2.3%) to accumulate in D", but not too large to be entrained in the upflow of mantle plumes. *Christensen and Hofmann* [1994] predict that MORB anticrust at the CMB could be up to 25 wt% SiO_2 stishovite, and the resulting anticrust compositions could

provide the observed 2-3% seismic velocity increase [*Wysession*, 1996b]. The stabilization of Ca- (+/- Al-) bearing silicate perovskite in basaltic compositions may also lead to enhanced velocities [*Funamori and Jeanloz*, unpublished work]. Because of the return of anticrust to the surface (on the basis of OIB isotope ratios), *Christensen and Hofmann* [1994] suggest that anticrust occupies only 3-15% of D" at any given time, but *Weber* [1994] showed that a thin layer of anomalously fast laminates (model LAM+) at the top of D" could explain many properties of the *PdP* observations. This is also in keeping with results of *Lay et al.* [1997] that *Scd* anomalies, more pronounced than for *ScS*, accumulated near the top of D" and were compatible with ±4% V_S variations within a 50 km thick region (and/or ± 50 km of discontinuity topography). *Lay et al.* [1997] found the uppermost portion of D" to be more heterogeneous than the deeper parts of D".

The other common candidate for the origin of D" heterogeneity is the by-product of core-mantle chemical reactions, as observed in laser-heated diamond anvil cells [*Knittle and Jeanloz*, 1989, 1991; *Goarant et al.*, 1992; *Jeanloz*, 1993], where (Mg,Fe)SiO_3 perovskite reacts with liquid iron to form $MgSiO_3$, SiO_2, and FeO and FeSi metallic alloys. The metallic alloys will not provide any fast velocities, but they will also be dense enough to resist being brought to the top of D". If the SiO_2 stishovite were swept to the top of D" it would be a likely candidate for the D" discontinuity. Further experimentation must be done, as there is an indication that dry CMB conditions would favor reactions between iron and MgO rather than iron and perovskite [*Boehler*, 1993].

Because we have a poor understanding of the TBL and viscosity structure of D", it is impossible to predict exactly how heterogeneities would be transported within D", though many studies have given an indication of the possible modes of transport [*Davies and Gurnis*, 1986; *Gurnis*, 1986; *Gurnis and Davies*, 1986; *Schubert et al.*, 1987; *Zhang and Yuen*, 1987, 1988; *Hansen and Yuen*, 1988, 1989; *Sleep*, 1988; *Stevenson*, 1988; *Christensen*, 1989; *Olson and Kincaid*, 1991; *Kellogg and King*, 1993; *Christensen and Hofmann*, 1994]. The dynamics will be controlled largely by the density contrast created by any existent heterogeneity. Most numerical and laboratory models indicate that a density contrast ≥ 3-6% is required to maintain a stable layer; at lower densities, heterogeneities are gradually entrained in mantle plumes, so the CBL must be resupplied if it is to persist for long. Thus the residence time for heterogeneities in a CBL depends critically on the density contrast, which is yet unknown.

The structure of a CBL also depends on how buoyant it is. A very stable CBL is largely isolated from lower mantle circulation and will likely exhibit little topography or lateral variability. Whether the CBL is made up of subducted oceanic crust, CMB reaction products, or both, it is more likely to be marginally stable (with a $\Delta\rho \approx 2\%$). Numerical simulations indicate that a marginally stable layer tends to pile up under upwellings and thin beneath downwellings [e.g *Hansen and Yuen*, 1988, 1989]. Internal circulation within the

CBL complicates the structure so that upwellings do not necessarily correlate with the thickest CBL regions [e.g. *Hansen and Yuen*, 1989; *Kellogg and King*, 1993].

If motions are dominated by large-scale flow driven by the ponding of slabs and/or the generation of large mantle plumes, laminar heterogeneities would be stretched horizontally in most places except the base of plumes where they may take a more vertical orientation. Such structures have been proposed to explain the unusual nature of D″ anisotropy, with transverse isotropy dominating in D″ regions of mantle slab downwellings but azimuthal anisotropy observed in D″ beneath the mid-Pacific plume groups [*Lay et al.*, 1998]. We would expect a well-defined CBL to be moved about laterally by the mantle flow, thinned in regions of downwelling and thickened beneath upwellings. As no obvious correlation is observed between the height of the D″ discontinuity and regions of downwellings and upwellings (as represented by large-scale velocity models), if such a stable CBL exists, this is likely not what *PdP* and *SdS* waves are detecting. Internal circulation can be driven by the overlying flow (upwelling plumes and downwellings) or by convection within the CBL itself. The CBL is not likely to convect internally if its thermal structure is determined primarily by the presence of cold slabs. On the other hand, if the viscosity in D″ is low enough, there could be internal convection within D″ that would create local pockets of variable composition within the D″ layer but isolate the layer from the overlying mantle. Internal convection creates a secondary thermal boundary layer at the top of the CBL as well as a TBL at the core-mantle boundary. The resulting TBL's above and below a layer with a quasi-adiabatic interior are not, however, in agreement with seismic observations that favor a strong negative gradient throughout D″.

The upshot is that chemical heterogeneities are likely to exist in D″, whether they are a primitive differentiated layer, core-mantle reaction products, delaminated oceanic crust, or the result of ongoing mid-mantle differentiation, and are likely to be related to the anomalous seismic observations seen in the form of seismic anisotropy and the D″ discontinuity. Just as continental crust influences convection patterns of the upper mantle, a CBL at the CMB will influence flow in the overlying mantle. Plumes rising from the top of a combined CBL and TBL tend to be hotter [*Farnetani*, 1997] and more stable in position [*Kellogg*, 1997] than plumes rising from an isochemical TBL.

6. CONCLUSIONS

The D″ discontinuity remains one of the most enigmatic seismic features of the Earth. Since the postulation [*Wright*, 1973; *Wright and Lyons*, 1975, 1981] and discovery [*Lay and Helmberger*, 1983] of the feature, over 40 publications have presented observations and interpretations of seismic waves interacting with the D″ discontinuity. And yet there is still no consensus as to what form the discontinuity takes and what is causing it. The height of the discontinuity is

seen to range from 100-450 km above the CMB, though the mean height for both *P* and *S* velocities is about 250-265 km above the CMB. A vertical discontinuity transition width of 50-75 km is compatible with most data, though some work with high frequency waves suggests it might be thinner in some places. The velocity increase is generally about 2-3% above that of the overlying mantle (again, the same for both *P* and *S* velocities), when the discontinuity is observed. The geographical distribution of the D″ discontinuity is such that it is both observed and not observed in most places of the CMB where it has been pursued. The discontinuity can sometimes turn on and off over very short distances (10-100 km), and other times behave coherently over much larger distances (> 1000 km). While many of the observations occur beneath Alaska and the Caribbean, which are the likely repositories of Mesozoic subducted slabs [*Ricard et al.*, 1993; *Grand et al.*, 1997], there is no difference between the average D″ velocity (from tomographic models) for locations of discontinuity observations and locations of non-observations. However, there is a slight positive correlation between the height of the inferred D″ discontinuity above the CMB and the regional velocity in the vicinity of the discontinuity as determined from tomographic models. This might suggest a preference toward a thermal rather than chemical origin for the discontinuity.

The source of the D″ discontinuity could be either a true surface, with large amounts of topography to explain both the range in depths observed as well as the the ability for the discontinuity to turn on and and off through focusing and defocussing effects, or a distribution of heterogeneities distributed within D″. Such a surface could be (1) thermal, representing the top of ponded subducted slab material, (2) mineralogical, representing a phase boundary between two different sets of phase assemblages, or (3) chemical, representing the separation between two different heterochemical regions. The alternate model, of scattered heterogeneities, could represent chemical or phase anomalies (triggered by slight chemical anomalies) distributed by convective patterns at the base of the mantle. These chemical anomalies may come from core-mantle reactions, the delamination of oceanic crust, or whole-mantle differentiation (either primordial or concurrent). It is likely that thermal, phase and chemical anomalies are all present, and even possible that the D″ discontinuity could have different causes in different locations. It is also likely that the D″ discontinuity is intimately related to the process of seismic anisotropy within D″. More work is needed to determine the geographical extent of the discontinuity, to model the seismic observations in three-dimensions, to determine the thermal structure of D″, to identify the phase relations of minor as well as major minerals at CMB conditions, and understand the possible mechanisms of convective mixing under these conditions.

Acknowledgments. This research was supported by NSF grants EAR-9305894 (TL), EAR-9418643 (TL), EAR-93-05892 (LK), EAR-9417542 (MW), and the Packard Foundation (MW). CMB workshops organized under EAR-9305894 helped develop inter-disciplinary coordination on this topic. Contribution 338 of the Institute of Tectonics and W. M. Keck Seismological Laboratory.

REFERENCES

Ahrens, T. J., and B. H. Hager, Heat transport across D" (abstract), *Eos Trans. AGU, 68*, 1493, 1987.

Bataille, K., and S. M. Flatté, Inhomogeneities near the core-mantle boundary inferred from short-period scattered *PKP* waves recorded at the Global Digital Seismograph Network, *J. Geophys. Res., 93*, 15,057-15,064, 1988.

Bataille, K., and F. Lund, Strong scattering of short-period seismic waves by the core-mantle boundary and the *P*-diffracted wave, *Geophys. Res. Lett.*, 23, 2413-2416, 1996.

Bataille, K., R. S. Wu, and S. M. Flatté, Inhomogeneities near the core-mantle boundary evidenced from scattered waves: A review, *Pure Appl. Geophys., 132*, 151-174, 1990.

Baumgardt, D. R., Evidence for a *P*-wave velocity anomaly in D", *Geophys. Res. Lett., 16,* 657-660, 1989.

Bina, C. R., Mantle Discontinuities, *Rev. Geophys., U.S. National Rep. to International Union of Geodesy and Geophysics 1987-1990*, 783-793, 1991.

Boehler, R., Core-mantle reactions? (abstract), *Eos Trans. AGU,* 74(43), Fall Meeting Supp., 415, 1993.

Bolton, H., and G. Masters, A region of anomalous *dlnVs/dlnVp* in the deep mantle (abstract), *Eos Trans. AGU,* 77(46), Fall Meeting Supp., F697, 1996.

Buffett, B. A., Influence of a toroidal magnetic field on the nutations of Earth, *J. Geophys. Res., 98*, 2105-2117, 1993.

Buffett, B. A., T. A. Herring, P. M. Matthews, and I. I. Shapiro, Anomalous dissipation in Earth's forced nutations: Inferences on the electrical conductivity and magnetic energy at the core-mantle boundary (abstract), *Eos Trans. AGU, 71*, 496, 1990.

Bullen, K. E., Compressibility-pressure hypothesis and the Earth's interior, *Month. Not. R. Astr. Soc., Geophys. Suppl., 5*, 355-368, 1949.

Chopelas, A., Sound velocities of MgO to very high compression, *Earth Planet. Sci. Lett., 114*, 185-192, 1992.

Chopelas, A., and R. Boehler, Thermal expansivity in the lower mantle, *Geophys. Res. Lett.*, 19, 1983-1986, 1992.

Chopelas, A., H. J. Reichmann, and L. Zhang, Sound velocities of four minerals to very high compression: constraints on δlnρ/δlnV in the lower mantle, *EOS, Trans. Am. Geophys. Un., 74* (16), 41, 1993.

Christensen, U. R., Models of mantle convection: one or several layers, *Philos. Trans. R. Soc. London A, 328*, 417-424, 1989.

Christensen, U. R., and A. W. Hofmann, Segregation of subducted oceanic crust in the convecting mantle, *J. Geophys. Res., 99*, 19,867-19,884, 1994.

Cleary, J. R., The D" region, *Phys. Earth Planet. Inter., 30*, 13-27, 1974.

Creager, K. C., and T. H. Jordan, Aspherical structure of the core-mantle boundary from *PKP* travel times, *Geophys. Res. Lett., 13*, 1497-1500, 1986.

Dahm, C. G., New values for dilatational wave-velocities through the Earth, *Trans., Am. Geophys. Union*, 80-83, 1934.

Davies, G. F., Ocean bathymetry and mantle convection, 1, Large-scale flow and hotspots, *J. Geophys. Res., 93*, 10,447-10,480, 1988.

Davies, G. F., and M. Gurnis, Interaction of mantle dregs with convection: lateral heterogeneity at the core-mantle boundary, *Geophys. Res. Lett., 13*, 1517-1520, 1986.

Ding, X., and D. V. Helmberger, Modeling D" structure beneath Central America with broadband seismic data, *Phys. Earth Planet. Inter., 101*, 245-270, 1997.

Ding, X., Sidorin, I., Helmberger, D. V., Gurnis, M., and Grand, S., D" seismic structure in down-welling regions, *Geophys. Res. Lett.*, submitted, 1997.

Doornbos, D. J., S. Spiliopoulos, and F. D. Stacey, Seismological properties of D" and the structure of a thermal boundary layer, *Phys. Earth Planet. Int., 41*, 225-239, 1986.

Dubrovinsky, L. S., S. K. Saxena, P. Lazor, R. Ahuja, O. Eriksson, J. M. Wills, and B. Johansson, Experimental and theoretical identification of a new high-pressure phase of silica, *Nature, 388*, 362-365, 1997.

Duffy, T. S., and T. J. Ahrens, Sound velocities at high pressure and temperature and their geophysical implications, *J. Geophys. Res., 97*, 4503-4520, 1992.

Dziewonski, A. M., and D. L. Anderson, Preliminary reference earth model, *Phys. Earth Planet. Inter., 25,* 297-356, 1981.

Dziewonski, A. M., G. Ekström, and X.-F. Liu, Structure at the top and bottom of the mantle, in *Monitoring a Comprehensive Test Ban Treaty*, Kluwer Academic Publishers, pp. 521-550, 1996.

Elsasser, W. M., P. Olson, and B. D. Marsh, The depth of mantle convection, *J. Geophys. Res., 84*, 147-155, 1979.

Farnetani, C. G., Excess temperature of mantle plumes: The role of chemical stratification across D", *Geophys. Res. Lett, 24*, 1583-1586, 1997.

Fei, Y., H.-K. Mao, and B. O. Mysen, Experimental determination of element partitioning and calculation of phase relations in the *MgO-FeO-SiO₂* system at high pressure a high temperature, *J. Geophys. Res., 96*, 2157-2170, 1991.

Frasier, C., and D. K. Chowdhury, Effect of scattering of *PcP/P* amplitude ratios at LASA from 40° to 84° distance, *J. Geophys. Res., 79*, 5469-5477, 1974.

Gaherty, J. B., and T. Lay, Investigation of laterally heterogeneous shear velocity structure in D" beneath Eurasia, *J. Geophys. Res., 97*, 417-435, 1992.

Garnero, E. J., and T. Lay, Lateral variations in lowermost mantle shear wave anisotropy beneath the north Pacific and Alaska, *J. Geophys. Res., 102*, 8121-8135, 1997.

Garnero, E. J., D. V. Helmberger, and G. Engen, Lateral variations near the core-mantle boundary, *Geophys. Res. Lett., 15,* 609-612, 1988.

Garnero, E. J., D. V. Helmberger, and S. Grand, Preliminary evidence for a lower mantle shear wave velocity discontinuity beneath the central Pacific, *Phys. Earth Planet. Int., 79*, 335-347, 1993.

Garnero, E. J., J. Revenaugh, Q. Williams, T. Lay, and L. Kellogg, Ultra-low velocity zone at the core-mantle boundary, in *Observational and Theoretical Constraints on he Core-Mantle Boundary Region*, ed. by M. Gurnis, M. Wysession, B. Buffett and E. Knittle, AGU, Washington, D. C., this volume, 1998.

Goarant, F. F. Guyot, J. Peyronneau, and J.-P. Poirier, High-pressure and high-temperature reactions between silicates and liquid iron alloys in the diamond anvil cell, studied by analytical electron microscopy, *J. Geophys. Res., 97*, 4477-4487, 1992.

Grand, S. P., R. D. van der Hilst, and S. Widiyantoro, Global seismic tomography: A snapshot of convection in the Earth, *GSA Today, 7*, 1-7, 1997.

Gurnis, M., The effects of chemical density differences on convective mixing in the Earth's mantle, *J. Geophys. Res., 91*, 11,407-11,419, 1986.

Gurnis, M., and G. F. Davies, The effect of depth-dependent viscosity on convective mixing in the mantle and the possible survival of primitive mantle, *Geophys. Res. Lett., 13*, 541-544, 1986.

Gutenberg, B., Über die Konstitution des Erdinnern, erschlossen aus Erdbebenbeobachtungen, *Phys. Zeitschrift, 14*, 1217-1218, 1913.

Gutenberg, B., and C. F. Richter, On seismic waves, *Beitr. Geophys., 54*, 94-136, 1939.

Haddon, R. A. W., and J. R. Cleary, Evidence for scattering of seismic *PKP* waves near the core-mantle boundary, *Phys. Earth Planet. Int., 8*, 211-234, 1974.

Hansen, U., and D. A. Yuen, Numerical simulation of thermal chemical instabilities at the core-mantle boundary, *Nature, 334*, 237-240, 1988.

Hansen, U., and D. A. Yuen, Dynamical influences from thermal-chemical instabilities at the core-mantle boundary, *Geophys. Res. Lett., 16*, 629-632, 1989.

Hedlin, M. A. H., P. M. Shearer, and P. S. Earle, Seismic evidence for small-scale heterogeneity throughout the Earth's mantle, *Nature, 387*, 145-150, 1997.

Honda, S., D. A. Yuen, S. Balachandar, and D. Reuteler, Three-dimensional instabilities of mantle convection with multiple phase transitions, *Science, 259*, 1308-1311, 1993.

Houard, S., and H.-C. Nataf, Further evidence for the 'Lay discontinuity' beneath northern Siberia and the North Atlantic from short-period *P*-waves recorded in France, *Phys. Earth Planet. Int., 72*, 264-275, 1992.

Houard, S., and H.-C. Nataf, Laterally varying reflector at the top of D″ beneath northern Siberia, *Geophys. J. Int., 115*, 168-182, 1993.

Husebye, E. S., D. W. King, and R. A. W. Haddon, Precursors to *PKIKP* and seismic wave scattering near the core-mantle boundary, *J. Geophys. Res., 81*, 1870-1882, 1976.

Inoue, H., Y. Fukao, K. Tanabe, and Y. Ogata, Whole mantle *P*-wave travel time tomography, *Phys. Earth Planet. Int., 59*, 294-328, 1990.

Ito, E., E. Takahashi, and Y. Matsui, The mineralogy and chemistry of the lower mantle: an implication of the ultrahigh-pressure phase relations in the system $MgO-FeO-SiO_2$, *Earth Planet. Sci. Lett., 67*, 238-248, 1984.

Jeanloz, R., Chemical reactions at Earth's core-mantle boundary: Summary of evidence and geomagnetic implications, in *Relating Geophysical Structures and Processes: The Jeffreys Volume, Geophys. Monogr. Ser.*, Vol. 76, edited by K. Aki and R. Dmowska, pp. 121-127, AGU, Washington, D. C., 1993.

Jeffreys, H., The times of P, S, and SKS and the velocities of P and S, *Mon. Not. R. Astron. Soc., 4*, 498-533, 1939.

Kellogg, L. H., Growing the Earth's D″ layer: effect of density variations at the core-mantle boundary, *Geophys. Res. Lett., 24*, 2749-2752, 1997.

Kellogg, L. H., and S. D. King, Effect of mantle plumes on the growth of D″ by reaction between the core and mantle, *Geophys. Res. Lett., 20*, 379-382, 1993.

Kendall, J. M., and C. Nangini, Lateral variations in D″ below the Caribbean, *Geophys. Res. Lett., 23*, 399-402, 1996.

Kendall, J. M., and P. M. Shearer, Lateral variations in D″ thickness from long-period shear-wave data, *J. Geophys. Res., 99*, 11,575-11,590, 1994.

Kendall, J. M., and P. M. Shearer, On the structure of the lowermost mantle beneath the southwest Pacific, southeast Asia and Australasia, *Phys. Earth Planet. Int., 92*, 85-98, 1995.

Kendall, J. M., and P. G. Silver, Constraints from seismic anisotropy on the nature of the lowermost mantle, *Nature, 381*, 409-412, 1996.

Kennett, B. L. N., E. R. Engdahl, and R. Buland, Constraints on seismic velocities in the Earth from traveltimes, *Geophys. J. Int., 122*, 108-124, 1995.

Kesson, S. E., and J. D. Fitz Gerald, Partitioning of MgO, FeO, NiO, MnO, and Cr_2O_3 between magnesian silicate perovskite and magnesiowüstite: Implications for the origin of inclusions in diamond and the composition of the lower mantle, *Earth Planet. Sci. Lett., 111*, 229-240, 1991.

Kesson, S. E., J. D. Fitz Gerald, and J. M. G. Shelley, Mineral chemistry and density of subducted basaltic crust at lower-mantle pressures, *Nature, 372*, 767-769, 1994.

Knittle, E., and R. Jeanloz, Synthesis and equation of state of *(Mg, Fe)SiO₃* perovskite to over 100 GPa, *Science, 235*, 668-670, 1987.

Knittle, E., and R. Jeanloz, Simulating the core-mantle boundary: an experimental study of high-pressure reactions between silicates and liquid iron, *Geophys. Res. Lett., 16*, 609-612, 1989.

Knittle, E., and R. Jeanloz, The high pressure phase diagram of $Fe_{0.94}O$: A possible constituent of the Earth's core, *J. Geophys. Res., 96*, 16169-16180, 1991.

Krüger, F., M. Weber, F. Scherbaum, and J. Schlittenhardt, Double beam analysis of anomalies in the core-mantle boundary region, *Geophys. Res. Lett., 20*, 1475-1478, 1993.

Krüger, F., M. Weber, F. Scherbaum, and J. Schlittenhardt, Normal and inhomogeneous lowermost mantle and core-mantle boundary under the Arctic and Northern Canada, *Geophys. J. Int., 122*, 637-658, 1995.

Krüger, F., F. Scherbaum, M. Weber, and J. Schlittenhardt, Analysis of asymmetric multipathing with a generalization of the double beam method, *Bull. Seismol. Soc. Am., 86*, 737-749, 1996.

Kumagai, H., Y. Fukao, N. Suda, and N. Kobayashi, Structure of the D″ layer inferred from the Earth's free oscillations, *Phys. Earth Planet. Inter., 73*, 38-52, 1992.

Kuo, B. Y., and K. Y. Wu, Global shear velocity heterogeneities in the D″ layer: Inversion from *Sd-SKS* differential travel times, *J. Geophys. Res., 102*, 11775-11788, 1997.

Lavely, E. M., D. W. Forsyth, and P. Friedmann, Scales of heterogeneity near the core-mantle boundary, *Geophys. Res. Lett., 13*, 1505-1508, 1986.

Lay, T., Structure of the core-mantle transition zone: a chemical and thermal boundary layer, *Eos Trans. AGU, 70*, 49, 54-55, 58-59, 1989.

Lay, T., Seismology of the lowermost mantle and core-mantle boundary, *Rev. of Geophys., Suppl.*, 325-328, 1995.

Lay, T., and D. V. Helmberger, A lower mantle *S*-wave triplication and the velocity structure of D″, *Geophys. J. R. astron. Soc., 75*, 799-837, 1983.

Lay, T., and C. J. Young, Analysis of seismic *SV* waves in the core's penumbra, *Geophys. Res. Lett., 18*, 1373-1376, 1991.

Lay, T., and C. J. Young, Imaging scattering structures in the lower mantle by migration of long-period *S* waves, *J. Geophys. Res., 101*, 20,023-20,040, 1996.

Lay, T., E. J. Garnero, C. J. Young, and J. B. Gaherty, Scale-lengths of shear velocity heterogeneity at the base of the mantle from *S* wave differential travel times, *J. Geophys. Res., 102*, 9887-9910, 1997.

Lay, T., E. J. Garnero, Q. Williams, B. Romanowicz, L. Kellogg, and M. E. Wysession, Seismic wave anisotropy in the D″ region and its implications, in *Observational and Theoretical Constraints on he Core-Mantle Boundary Region*, ed. by M. Gurnis, M. Wysession, B. Buffett and E. Knittle, AGU, Washington, D. C., this volume, 1998.

Li, X. D. and B. Romanowicz, Global mantle shear velocity model developed using nonlinear asymptotic coupling theory, *J. Geophys. Res., 101*, 22245-22272, 1996.

Loper , D. E., and T. Lay, The core-mantle boundary region, *J. Geophys. Res., 100*, 6397-6420, 1995.

Mao, H. K., and P. M. Bell, Equations of state of MgO and epsilon Fe under static pressure conditions, *J. Geophys., Res., 84*, 4533-4536, 1979.

Mao, H. K., C. Chen, R. J. Hemley, A. P. Jephcoat, Y. Wu, and W. A. Bassett, Stability and equation of state of $CaSiO_3$-perovskites to 134 GPa, *J. Geophys. Res., 94*, 17,889-17,894, 1989.

Masters, G., S. Johnson, G. Laske, and H. Bolton, A shear-velocity model of the mantle, *Phil. Trans. R. Soc. Lond. A, 354*, 1385-1411, 1996.

Matzel, Eric, Mrinal K. Sen, and Stephen P. Grand, Evidence for anisotropy in the deep mantle beneath Alaska, Geophys. Res. Lett., 23, 2417-2420, 1996.

Maupin, V., On the possibility of anisotropy in the D" layer as inferred from the polarisation of diffracted S waves, *Phys. Earth Planet. Int., 87*, 1-32, 1994.

McCammon, C. A., Perovskite as a possible sink for ferric iron in the lower mantle, Nature, 387, 694-696, 1997.

McCammon, C. A., J. W. Harris, B. Harte, and M. T. Hutchison, Partitioning of ferric iron between lower mantle phases: Results from natural and synthetic samples, *J. Conf. Abstr., 1*, 390, 1996.

McSweeney, T. J., and K. C. Creager, Global core-mantle boundary structure inferred from *PKP* differential travel times, (abstract), *Eos Trans. AGU Supplement, 75*, 663, 1994.

Mitchell, B. J., and D. V. Helmberger, Shear velocities at the base of the mantle from observations of S and ScS, *J. Geophys. Res., 78*, 6009-6020, 1973.

Nataf, H.-C., and S. Houard, Seismic discontinuity at the top of D": A world-wide feature?, *Geophys. Res. Lett., 20*, 2371-2374, 1993.

Neuberg, J., and J. Wahr, Detailed investigation of a spot on the core-mantle boundary using digital *PcP* data, *Phys. Earth planet. Int., 68*, 132-143, 1991.

O'Mongáin, A., and J. Neuberg, D" topography: Fact or fiction (abstract), *Eos Trans. AGU Supplement, 77*, F679, 1996.

Olivieri, M., N. A. Pino, and A. Morelli, Evidence for an S-velocity discontinuity in the lowermost mantle beneath the South Eastern Pacific Basin, *Geophys. Res., Lett., 24*, 2617-2620, 1997.

Olson, P. L., and C. Kincaid, Experiments on the interaction of thermal convection and compositional layering at the base of the mantle, *J. Geophys. Res., 96*, 4347-4354, 1991.

Poirier, J.-P., Core-infiltrated mantle and the nature of the D" layer, *J. Geomagn. Geoelectr., 45*, 1221-1227, 1993.

Pulliam, R. J., D. W. Vasco, and L. R. Johnson, Tomographic inversions for mantle *P* wave velocity structure based on the minimization of l^2 and l^1 norms of International Seismological Centre travel time residuals, *J. Geophys. Res., 98*, 699-734, 1993.

Revenaugh, J., and T. H. Jordan, Mantle layering from *ScS* Reverberations, 3, The transition zone, *J. Geophys. Res., 96*, 19763-19780, 1991a.

Revenaugh, J., and T. H. Jordan, Mantle layering from *ScS* Reverberations, 4, The lower mantle and core-mantle boundary, *J. Geophys. Res., 96*, 19811-19824, 1991b.

Ricard, Y., M. Richards, C. Lithgow-Bertelloni, and Y. Le Stunff, A geodynamic model of mantle density heterogeneity, *J. Geophys. Res., 98*, 21,895-21,909, 1993.

Richards, P. G., Seismic waves reflected from velocity gradient anomalies within the Earth's upper mantle, *Z. Geophys., 38*, 517-527, 1972.

Richards, M. A. and Davies, G. F., On the separation of relatively buoyant components from subducted lithosphere, *Geophys. Res. Lett., 16*, 831-834, 1989.

Richards, M. A., and D. C. Engebretson, The history of subduction , and large-scale mantle convection, *Nature, 355*, 437-440, 1992.

Rigden, S. M., T. J. Ahrens, E. M. Stolper, Shock compression of molten silicates; results for a model basaltic composition, *J. Geophys. Res., 93*, 367-382, 1988.

Ringwood, A. E., and W. Hibberson, Solubilities of mantle oxides in molten iron at high pressures and temperatures: Implications for the composition and formation of Earth's core, *Earth Planet. Sci. Lett., 102*, 235-251, 1991.

Ritsema, J., E. Garnero, and T. Lay, A strongly negative shear velocity gradient and lateral variability in the lowermost mantle beneath the Pacific, *J. Geophys. Res., 102*, 20,395-20,411, 1997.

Robertson, G. S., and J. H. Woodhouse, Ratio of relative S to P velocity heterogeneity in the lower mantle, *J. Geophys. Res., 101*, 20,041-20,052, 1996.

Ruff, L., and D. L. Anderson, Core formation, evolution, and convection: A geophysical model, *Phys. Earth Planet. Int., 21*, 181-201, 1980.

Schimmel, M., and H. Paulssen, Steeply reflected *ScSH* precursors from the D" region, *J. Geophys. Res., 101*, 16,077-16,088, 1996.

Schlittenhardt, J., Investigation of the velocity- and Q-structure of the lowermost mantle using *PcP/P* amplitude ratios from arrays at distances of 70°-84°, *J. Geophys., 60*, 1-18, 1986.

Scherbaum, F., F. Krüger, and M. Weber, Double beam imaging: Mapping lower mantle heterogeneities using combinations of source and receiver arrays, *J. Geophys. Res., 102*, 507-522, 1997.

Schubert, G., P. Olson, C. A. Anderson, and P. Goldman, Mantle plumes, solitons and the D" layer (abstract), *Eos Trans. AGU, 68*, 1488, 1987.

Shearer, P. M., and T. G. Masters, Global mapping of topography on the 660-km discontinuity, *Nature, 355*, 791-796, 1992.

Sherman, D. M., Equation of state and high-pressure phase transitions of Stishovite (SiO_2): Ab initio (Periodic Hartree-Fock) results, *J. Geophys. Res, 98*, 11,865-11,873, 1993.

Shibutani, T., A. Tanaka, M. Kato, and K. Hirahara, A study of *P*-wave velocity discontinuity in D" layer with J-array records: Preliminary results, *J. Geomag. Geoelectr., 45*, 1275-1285, 1993.

Silver, P. G., R. W. Carlson, and P. Olson, Deep slabs, geochemical heterogeneity, and the large-scale structure of mantle convection: investigation of an enduring paradox, *Ann. Rev. Earth Planet. Sci., 16*, 477-541, 1988.

Sleep, N. H., Gradual entrainment of a chemical layer at the base of the mantle by overlying convection, *Geophys. J. R. astron. Soc., 95*, 437-447, 1988.

Sleep, N. H., Hotspots and mantle plumes: Some phenomenology, *J. Geophys. Res., 95*, 6715-6736, 1990.

Song, X., and D. V. Helmberger, Effect of velocity structure in D" on *PKP* phases, *Geophys. Res. Lett., 20*, 285-288, 1993.

Souriau, A. and G. Poupinet, Lateral variations in *P* velocity and attenuation in the D" layer, from diffracted *P* waves, *Phys. Earth Planet. Int., 84*, 227-234, 1994.

Stacey, F. D., Effects on the core of structure within D", *Geophys. Astrophys. Fluid Dyn., 60*, 157-163, 1991.

Stacey, F. D., and D. E. Loper, The thermal boundary-layer interpretation of D" and its role as a plume source, *Phys. Earth Planet. Inter., 33*, 45-55, 1983.

Steinbach, V. C., D. A. Yuen, G. Y. A. Bussod, H. Paulssen, Effects of depth-dependent properties on the thermal anomalies produced in flush instabilities from phase transitions, *Phys. Earth Planet. Int.*, 86, 165-183, 1994.

Stevenson, D., Infiltration, dissolution and underplating: Rules for mixing core-mantle cocktails (abstract), Eos Trans. AGU, 69, 1404, 1988.

Stevenson, D. J., Why D" is unlikely to be caused by core-mantle interactions (abstract), *Eos Trans. AGU Supplement, 74*, 51, 1993.

Stixrude, L., and M. S. T. Bukowinski, Stability of $(Mg,Fe)SiO_3$ perovskite and the structure of the lowermost mantle, *Geophys. Res. Lett., 19*, 1057-1060, 1992.

Stixrude, L., R. J. Hemley, Y. Fei, and H. K. Mao, Thermoelasticity of silicate perovskite and magnesiowüstite and stratification of the Earth's mantle, *Science, 257*, 1099-1101, 1992

Su, W., R. L. Woodward, and A. M. Dziewonski, Degree 12 model of shear velocity heterogeneity in the mantle, *J. Geophys. Res., 99*, 6945-6981, 1994.

Sylvander, M. and A. Souriau, P-velocity structure of the core-mantle boundary region inferred from *PKP(AB)-PKP(BC)* differential travel times, *Geophys. Res. Lett., 23*, 853-856, 1996.

Tackley, P. J., D. J. Stevenson, G. A. Glatzmaier, and G. Schubert, Effects of an endothermic phase transition at 670 km depth in a spherical model of convection in the Earth's mantle, *Nature, 361*, 699-704, 1993.

Tanimoto, T., Long-wavelength S-wave velocity structure throughout the mantle, *Geophys. J. Int., 100*, 327-336, 1990.

Thomas, C., and M. Weber, P velocity heterogeneities in the lower molten determined with the German regional seismic network: Improvement of previous models and results of 2D modeling, *Phys. Earth Planet. Int., 101*, 105-117, 1997.

Tsuchida, Y., and T. Yagi, A new, post-stishovite high-pressure polymorph of silica, *Nature, 340*, 217-220, 1989.

Valenzuela, R. W., and M. E. Wysession, Illuminating the base of the mantle with core-diffracted waves, in *Observational and Theoretical Constraints on he Core-Mantle Boundary Region*, ed. by M. Gurnis, M. Wysession, B. Buffett and E. Knittle, AGU, Washington, D. C., this volume, 1998.

Van der Hilst, R. D., S. Widiyantoro, and E. R. Engdahl, Evidence for deep mantle circulation from global tomography, *Nature, 386*, 578-584, 1997.

Vasco, D. W., L. R. Johnson, R. J. Pulliam, and P. S. Earle, Robust inversion of IASP91 travel time residuals for mantle P and S velocity structure, earthquake mislocations, and station corrections, *J. Geophys. Res., 99*, 13,727-13,755, 1994.

Vassiliou, M. S., and T. J. Ahrens, The equation of state of $Mg_{0.6}Fe_{0.4}O$ to 200 GPa, *Geophys. Res. Lett., 9*, 127-130, 1982.

Vidale, J. E., and H. M. Benz, A sharp and flat section of the core-mantle boundary, *Nature, 359*, 627-629, 1993a.

Vidale, J. E., and H. M. Benz, Seismological mapping of fine structure near the base of the Earth's mantle, *Nature, 361*, 529-532, 1993b.

Vidale, J. E., and M. A. H. Hedlin, Intense scattering at the core-mantle boundary north of Tonga: Evidence for partial melt, *Nature*, in press, 1998.

Vinnik, L. P., and G. G. Dashkov, *PcP* waves from atomic explosions and the nature of the core-mantle boundary, *Izv., Earth Physics, 1*, 7-16, 1970.

Vinnik, L., B. Romanowicz, Y. Le Stunff, and L. Makeyeva, Seismic anisotropy in the D" layer, *Geophys. Res. Lett., 22*, 1657-1660, 1995.

Weber, M., P and S wave reflections from anomalies in the lowermost mantle, *Geophys. J. Int., 115*, 183-210, 1993.

Weber, M., Lamellae in D"? An alternative model for lower mantle anomalies, *Geophys. Res. Lett., 21*, 2531-2534, 1994.

Weber, M., and J. P. Davis, Evidence of a laterally variable lower mantle structure from P- and S-waves, *Geophys. J. Int., 102*, 231-255, 1990.

Weber, M., and M. Körnig, Lower mantle inhomogeneities inferred from *PcP* precursors, *Geophys. Res. Lett., 17*, 1993-1996, 1990.

Weber, M., and M. Körnig, A search for anomalies in the lowermost mantle using seismic bulletins, *Phys. Earth Planet. Int., 73*, 1-28, 1992.

Weber, M., J. P. Davis, C. Thomas, F. Krüger, F. Scherbaum, J. Schlittenhardt, and M Körnig, The structure of the lowermost mantle as determined from using seismic arrays, in *Seismic Modelling of the Earth's Structure*, Edited by E. Boschi, G Ekström, and A. Morelli, pp. 399-442, Istit. Naz. di Geophys., Rome, 1996.

Weinstein, S. A., Catastrophic overturn in the Earth's mantle driven by multiple phase changes and internal heat generation, *Geophys. Res. Lett., 20*, 101-104, 1993.

Wicks, C. W., and M. A. Richards, Seismic evidence for the 1200 km discontinuity (abstract), *Eos Trans. AGU, 74(43)*, Spring Meeting suppl., 550, 1993.

Williams, Q., and E. J. Garnero, On the possible origin of a seismically thin boundary layer at the base of the mantle, *Science, 273*, 1528-1530, 1996.

Williams, Q., E. Knittle, and R. Jeanloz, The high-pressure melting curve of iron: A technical discussion, *J. Geophys. Res., 96*, 2171-2184, 1991.

Witte, H., Beiträge zur Berechnung der Geschwindigkeit der Raumwellen im Erdinnern, *Nachrichten der Gesellschaft der Wissenschaften zu Göttingen, Heft 2*, pp. 199-241, 1932.

Wood, B. J., and D. C. Rubie, The effect of alumina on phase transformations at the 660-kilometer discontinuity from Fe-Mg partitioning experiments, *Science, 273*, 1522-1524, 1996.

Woodward, R. L., and G. Masters, Lower-mantle structure from *ScS-S* differential travel times, *Nature, 352*, 231-233, 1991.

Wright, C., Array studies of P phases and the structure of the D" region of the mantle, *J. Geophys. Res., 78*, 4965-4982, 1973.

Wright, C., and J. A. Lyons, Seismology, dT/dΔ and deep mantle convection, *Geophys. J. R. Astron. Soc., 40*, 115-138, 1975.

Wright, C., and J. A. Lyons, Further evidence for radial velocity anomalies in the lower mantle, *Pure Appl. Geophys., 119*, 137-162, 1981.

Wright, C., Muirhead, K. J., and A. E. Dixon, The P wave velocity structure near the base of the mantle, *J. Geophys. Res., 90*, 623-634, 1985.

Wysession, M. E., Large-scale structure at the core-mantle boundary from core-diffracted waves, *Nature, 382*, 244-248, 1996a.

Wysession, M. E., Imaging cold rock at the base of the mantle: The sometimes fate of Slabs?, in *Subduction: Top to Bottom*, edited by G. E. Bebout, D. Scholl, S. Kirby, and J. P. Platt, AGU, Washington, D. C., pp. 369-384, 1996b.

Wysession, M. E., E. A. Okal and C. R. Bina, The structure of the core-mantle boundary from diffracted waves, *J. Geophys. Res., 97*, 8749-8764, 1992.

Wysession, M. E., C. R. Bina and E. A. Okal, Constraints on the temperature and composition of the base of the mantle, in *Dynamics of the Earth's Deep Interior and Earth Rotation, Geophys. Monogr. Ser.,* edited by J.-L. LeMoüel et al., AGU, Washington, D.C., 181-190, 1993.

Wysession, M. E., L. Bartkó, and J. Wilson, Mapping the lowermost mantle using core-reflected shear waves, *J. Geophys. Res., 99,* 13,667-13,684, 1994.

Wysession, M. E., S. Robertson, G. I. Al-eqabi, A. Langenhorst, M. J. Fouch, K. M. Fischer, and T. J. Clarke, D" lateral, vertical and anisotropic structure from MOMA core-diffracted waves, *Eos Trans. AGU, 78*(46), Fall Meeting suppl., F1, 1997.

Yamada, A., and I. Nakanishi, Detection of P wave reflector in D" beneath the southwestern Pacific using double-array stacking, *Geophys. Res. Lett., 23,* 1553-1556, 1996.

Young, C. J., and T. Lay, The core mantle boundary, *Ann. Rev. Earth Planet. Sci., 15,* 25-46, 1987*a*.

Young, C. J., and T. Lay, Evidence for a shear velocity discontinuity in the lower mantle beneath India and the Indian Ocean, *Phys. Earth Planet. Inter., 49,* 37-53, 1987*b*.

Young, C. J., and T. Lay, Multiple phase analysis of the shear velocity structure in the D" region beneath Alaska, *J. Geophys. Res., 95,* 17,385-17,402, 1990.

Yuen, D. A., O. P. Cadek, A. Chopelas, and C. Matyska, Geophysical inferences of thermal-chemical structures in the lower mantle, *Geophys. Res. Lett., 20,* 899-902, 1993.

Yuen, D. A., O. P. Cadek, R. Boehler, J. Moser, and C. Matyska, Large cold anomalies in the deep mantle and mantle instability in the Cretaceous, *Terra Nova, 6,* 238-245, 1994.

Zhu, A.-N., and M. E. Wysession, Mapping global D" *P* velocity structure from ISC *PcP-P* differential travel times, *Phys. Earth Planet. Int., 99,* 69-82, 1997.

Zhang, S., and D. A. Yuen, Deformation of the core-mantle boundary induced by spherical shell, compressible convection, *Geophys. Res. Lett., 14,* 899-902, 1987.

Zhang, S., and D. A. Yuen, Dynamical effects on the core-mantle boundary from depth-dependent thermodynamical properties of the lower mantle, *Geophys. Res. Lett., 15,* 451-454, 1988.

Zhao, Y., and D. L. Anderson, Mineral physics constraints on the chemical composition of the Earth's lower mantle, *Phys. Earth Planet. Int., 85,* 273-292, 1994.

M. E. Wysession, Department of Earth and Planetary Sciences, Washington University, St. Louis, MO 63130

T. Lay, J. Revenaugh, and Q. Williams, Institute of Tectonics and Earth Sciences Department, University of California, Santa Cruz, CA 95064

E. J. Garner and R. Jeanloz, Department of Geology and Geophysics, University of California, Berkeley, CA 94720

L. H. Kellogg, Geology Department, University of California, Davis, CA 95616

Seismic Wave Anisotropy in the D" Region and its Implications

Thorne Lay, Quentin Williams

Earth Sciences Department and Institute of Tectonics, University of California, Santa Cruz California

Edward J. Garnero

Department of Geology and Geophysics and Seismological Laboratory, University of California, Berkeley, California

Louise Kellogg

Department of Geology, University of California, Davis, California

Michael E. Wysession

Department of Earth and Planetary Sciences, Washington University, St. Louis, Missouri

Seismological observations provide evidence for localized regions of aniso-tropic splitting of shear waves that graze through the D" region at the base of mantle. Beneath Alaska and the Caribbean *S*, *Sdiff* (diffracted *S*), and *ScS* phases exhibit shear wave splitting with the longitudinal component (*SV*, in the standard great-circle reference frame) delayed relative to the transverse component (*SH*) by as much as 4-9 s. This can be explained by the presence of 1-3% transverse isotropy (hexagonal symmetry with a vertical axis) within the D" region. These circum-Pacific areas have higher-than-average lower mantle shear velocities as well as localized *S* velocity discontinuities (but no detectable *P* velocity discontinuities) at the top of a 200-300 km thick D" layer. It appears that the anisotropy exists at and below the D" discontinuity, resulting in a stronger velocity contrast for *SH* than for *SV*. The magnitude of anisotropy varies laterally, and can decrease to zero over scale lengths of about 1000 km. Beneath the central Pacific Ocean, where the lowermost mantle appears to have lower-than-average shear velocities, little or no D" discontinuity, and strongly negative velocity gradients within D", there are in-termittent observations of *Sdiff*, *S*, and *ScS* splitting with relatively early arri-vals on either SH or SV components. The variable splitting suggests the presence of non-vertical hexagonal symmetry axis or general anisotropy in localized patches, apparently concentrated toward the core-mantle boundary.

INTRODUCTION

Determination of the dynamical processes occurring in the thermal and chemical boundary layer at the base of the mantle is of great importance to fundamental geophysical problems such as the nature of mantle convection, the

The Core-Mantle Boundary Region
Geodynamics 28
Copyright 1998 by the American Geophysical Union.

Various mechanisms for generating anisotropy in D" are considered, including lattice preferred orientation (LPO) of anisotropic minerals in the D" boundary layer, anisotropic structures of slab remnants that have sunk to the base of the mantle, and shape preferred orientation (SPO) induced by shearing of chemical heterogeneities and/or zones of partial melt resulting in horizontally or vertically laminated fabrics. Constraining D" dynamical properties from shear wave anisotropy has potential, but the observational and modeling challenges remain formidable.

process of core formation, and the nature of coupling between the core and mantle [e.g., *Stacey and Loper*, 1983; *Lay*, 1989; *Loper and Lay*, 1995]. Seismological investigations provide the strongest constraints on the core-mantle transition zone, with substantial complexity of the deep elastic velocity structure having been revealed by seismic waves [see reviews by *Young and Lay*, 1987; *Lay*, 1995]. There is much progress in mapping of large-scale patterns of volumetric heterogeneity [e.g., *Su et al.*, 1994; *Li and Romanowicz*, 1996; *Grand et al.*, 1997], with higher velocity regions often accompanied by *P* and *S* velocity discontinuities at the top of the D" region (the lowermost 200-300 km of the mantle) [see *Wysession et al.*, 1998]. There is also evidence for a thin, laterally varying ultra-low velocity zone (ULVZ), with *P* and *S* velocity reductions of 10% or more right above the core-mantle boundary (CMB) [see *Garnero et al.*, 1998]. These seismological investigations suggest the presence of chemical and thermal heterogeneity, and possibly partial melt, in a dynamic boundary layer at the base of the mantle [*Lay et al.*, 1998].

While resolving the elastic *P* and *S* wave velocity structures in D" is an important undertaking, interpretation of isotropic seismological models in terms of boundary layer processes is very poorly constrained. Many trade-offs exist between plausible thermal and chemical contributions to the observed structures, and inferred dynamical processes are correspondingly non-unique; what is needed is a seismic observable more directly influenced by dynamical processes. For lithospheric investigations, observations of anisotropic seismological properties have proved to be significant for dynamic interpretations, because the mineralogical preferred orientation and cracking processes that produce seismic anisotropy in the lithosphere appear to be directly linked to large-scale deformations of the medium [e.g., *Kawasaki*, 1986; *Nataf et al.*, 1986; *Karato*, 1989; *Silver*, 1996]. Anisotropic structures in the asthenosphere are also valuable for inferring mantle processes, although the lack of direct constraint on the flow regime complicates the interpretations [e.g., *Karato*, 1989].

A substantial number of observations are consistent with the presence of laterally varying seismic anisotropy with variable symmetry orientation in the D" region. These seismological observations require geodynamical and mineral physics interpretations. We conduct a multidisciplinary examination of the current status of this topic to provide impetus for further work on a difficult, but promising approach to enhancing our understanding of the dynamical processes in the core-mantle transition zone.

CHALLENGES TO OBSERVING DEEP MANTLE ANISOTROPY

Seismic anisotropy can be detected in *P* wave travel times if extensive raypath coverage is available and if the effects can be localized spatially to a particular region. While this has proved to be the case in some investigations of *Pn* and *P* wave travel time anomalies for studying the crust and uppermost mantle [e.g., *Morris et al.*, 1969; *Dziewonski and Anderson*, 1983], neither criterion can be readily satisfied for studies of the base of the mantle. Anisotropy may arise from fine layering of materials with strong contrasts in physical properties, and the presence of lamination in the deep mantle may possibly be detected by high resolution imaging of velocity structure [e.g., *Weber*, 1994]. However, there are many difficulties in recovering deterministic models for high frequency energy traversing all the way to the D" region. Patterns of small scale heterogeneities that produce scattering of *PKP* phases have led to some speculations on anisotropic fabrics in D" [e.g., *Haddon*, 1982; *Cormier*, 1995], but the models are very poorly resolved and there are strong trade-offs with CMB topography, which at present is poorly known. Seismic anisotropy can also be detected in the behavior of surface waves or free oscillations [e.g., *Nishimura and Forsyth*, 1988; *Tanimoto and Anderson*, 1985], but as yet these waves provide little constraint on anisotropic structure near the CMB. Mantle models need to improve before normal modes can be exploited as they are for the lithosphere. Given the limitations of *P* waves and normal modes, the most effective probe of anisotropic structure in the deep mantle is shear wave splitting.

Shear wave splitting occurs when an *S* wave traverses a seismically anisotropic region, with the *S* wave energy partitioning into two polarizations that propagate with different velocities. The overall polarization of the *S* wave is thus modified, and can be used to infer characteristics of the anisotropic structure. If the path length is known, the magnitude of anisotropy can be inferred from the time shift between the split *S* arrivals, and the orientation can be determined from the polarization angles of the fast and slow waves.

In general, the lower mantle does not exhibit significant shear wave splitting for teleseismic S wave signals [e.g., *Kaneshima and Silver*, 1992; *Meade et al.*, 1995b], despite the intrinsic anisotropy of major lower mantle minerals such as perovskite [e.g., *Mao et al.*, 1991] and periclase, and the likely presence of shear flows associated with up-wellings and downwellings. Most observations of shear wave splitting in teleseismic signals have been explained in terms of shallow mantle anisotropy beneath seismic stations and in the vicinity of subducting slabs [e.g., *Silver*, 1996]. Indeed, the upper 100-300 km of the mantle has significant anisotropy that can affect body wave signals that traverse the deep mantle.

The main strategy for identifying shear wave splitting caused by deep mantle anisotropy has been analysis of polarization of S waveforms. Comparisons of S and ScS or SKS and $Sdiff$ (S wave energy diffracted by the core) are made to bound source and receiver contributions to the overall S waveforms and polarization. Recent models of lithospheric anisotropy can account for shallow anisotropic effects explicitly, either by selecting stations that are known to lack lithospheric anisotropy or by correcting for the splitting effects predicted by lithospheric models. However, correction for shallow effects is not an error-free procedure, and several studies [e.g., *Garnero and Lay*, 1997; *Vinnik et al.*, 1998] have found that corrections based on existing models are not always correct; improved lithospheric corrections must be determined. It is always necessary to account for lithospheric anisotropic effects.

For earthquake sources located within subducting slabs one must also consider the possibility that near-source slab structure is anisotropic, as this is expected given the observed anisotropy of oceanic lithosphere [e.g., *Anderson*, 1987; *Kendall and Thomson*, 1993]. In this case, S wave splitting may have complex variations with azimuth and take-off angle with respect to the dipping slab, making it difficult to even detect the near-source effect. Existing anisotropic slab models are inadequate for making corrections, and empirical arguments are usually invoked to rule out near-source contributions; however, none of the studies considered here convincingly eliminate near-source contributions.

Correcting for shallow mantle anisotropy is not the only challenge for quantifying deep mantle anisotropy. Body waves that reach the core-mantle transition zone have long paths through the heterogeneous mantle, and relatively short paths within the D" region, reducing sensitivity to any deep anisotropy. Increasing the proportion of path length within the deep mantle mandates the study of grazing waves, which have much greater sensitivity to the overall velocity structure than do the steeply incident waves used in most studies of lithospheric anisotropy. The depth extent over which D" anisotropy may occur is not known independently, and it is essential to account for the precise raypath when estimating path length within a possibly anisotropic zone [*Garnero and Lay*, 1998]. For core-grazing

geometries, distance dependence of anisotropic observations may involve either depth dependence or lateral variations in the anisotropic structure [*Ritsema et al.*, 1998].

Accounting for propagation effects on S wave polarization is also essential. Layered structure imparts distinct effects on longitudinally (SV) and transversely (SH) polarized S wave motions, as defined relative to the vertical plane containing the source and receiver. Differences in the frequency dependent interactions of SV and SH polarizations with the CMB, with radial gradients in the D" region, and with any discontinuities in the medium, add complexity to the S wavefield that will superimpose on any effects of anisotropic splitting. If splitting occurs, the fast and slow polarizations will each have SV and SH components of motion relative to any radial layering that is encountered as the grazing wave turns, possibly obscuring the effects of the splitting. The reflection coefficient for SV and SH components of ScS causes non-linear particle motion at some distances, as does diffraction, even for a standard Earth model lacking strong gradients in D" [e.g., *Maupin*, 1994]. Finally, small-scale lateral heterogeneity of isotropic structure can produce waveform complexity easily misinterpreted as the result of transmission through an anisotropic zone [e.g., *Grechka and McMechan*, 1995]. Indeed, a laminated region with many thin layers of isotropic structure is indistinguishable from a medium with intrinsic mineralogical anisotropy for certain geometries and wavelengths [*Backus*, 1962]. Strong velocity gradients and multi-scale heterogeneity do exist in D" [e.g., *Lay*, 1989] as well as thin layers [e.g., *Garnero and Helmberger*, 1995; *Garnero et al.*, 1993, 1998], thus the observational challenges of quantifying D" anisotropy should not be underestimated.

SEISMOLOGICAL OBSERVATIONS OF ANISOTROPY IN THE D" REGION

The observational challenges noted above have slowed development of anisotropic models for the D" region compared to the more extensive progress in mapping out isotropic structures [see *Wysession et al.*, 1998]. Another factor is that well-calibrated digital broadband instrumentation is particularly valuable for making subtle S wave polarization measurements, and such data have only become abundant in the past decade. There has been a flurry of papers on D" anisotropy in the past 3 years, but these build on important earlier work, which we describe first.

Early studies of shear wave splitting

Systematic waveform differences between S and ScS arrivals in long-period World Wide Standardized Seismograph Network (WWSSN) recordings were first observed by *Mitchell and Helmberger* [1973] and *Lay and Helmberger* [1983b]. $ScSV$ components often have slightly late peaks relative to $ScSH$ components, in contrast to more consistent relative times of S wave components. *Mitchell and*

Helmberger [1973] studied paths beneath the Caribbean, with *ScSV* peaks ranging from 0 to 3 s behind *ScSH*. *Lay and Helmberger* [1983b] extended those observations, and added paths beneath Alaska, where the peak of *ScSV* is sometimes as much as 4 s late relative to *ScSH*. Neither study corrected for shallow mantle anisotropy (due to the lack of obvious *S* splitting), and both performed isotropic waveform modeling, achieving satisfactory fits to the observations by introducing thin (20 to 40 km) high velocity layers (as much as 5% higher than standard values) at the CMB. *SH* and *SV* component interactions with such a thin layer cause an apparent shift of the *ScS* peak on the two components in the long-period WWSSN passband (internal multiples within the layer must be accounted for).

The WWSSN data have a positive 1 s baseline shift (with a ±1 s range of data) in the *ScSV-ScSH* observations from 40° to 75° for the Caribbean and Alaskan paths, and the delay of *ScSV* increases rapidly with distance from 75° to 80° beneath Alaska. The rapid increase (but not the baseline shift) can be explained by the thin high velocity layer model. Explanation of the baseline shift and rapid increase in terms of anisotropy requires lateral variation in the magnitude of anisotropy beneath Alaska. *Cormier* [1986] attempted to explain these data with D" models with transverse isotropy (hexagonal symmetry with a vertical symmetry axis, for which the *SV* and *SH* components are not coupled) in a 150 km thick zone with anisotropy increasing with depth to a maximum at the CMB. His synthetics produce phase shifts of the overall *ScS* waveform relative to *S*, but, in detail, do not match the shift of the peaks of *ScSV* and *ScSH* that was previously measured.

Doornbos et al. [1986] sought to reconcile thermal boundary layer models with *P* and *S* wave velocity structures inferred from diffracted waves by invoking transverse isotropy D". However, this approach is predicated on the existence of globally representative velocity structures for D", which is inconsistent with the strong heterogeneity in the region (most published *Sdiff* decay spectra exhibit huge scatter [e.g., *Doornbos and Mondt*, 1979]).

Diffracted wave analysis gained attention with the report of broadband *SVdiff* signals at large distances into the core shadow zone by *Vinnik et al.* [1989]. A few low amplitude longitudinal signals were found near the time of strong tangential *Sdiff* arrivals at distances near 118° for paths bottoming beneath the central Pacific, with a possible 1/4 period phase shift of the *SV* component of the *Sdiff* energy. For Earth models with mild velocity gradients in D" *SV* energy does not diffract efficiently relative to *SH*, predicting that *Sdiff* become predominantly transversely polarized beyond distances of 105°. While noting that a strong negative velocity gradient in D" can allow *SVdiff* energy to propagate further than expected for a standard Earth model [e.g., *Kind and Müller*, 1977], *Vinnik et al.* [1989] invoked azimuthal anisotropy in D" which would couple the *SV* and *SH* energy, allowing strong (quasi-) *SH* diffractions to couple into weak *SV* diffractions. No corrections were

made for lithospheric anisotropy in this study, but subsequent correction of the same data by *Maupin* [1994] did not eliminate the non-linear polarization. There is now evidence for a strong negative shear velocity gradient beneath the central Pacific in the distance decay of *SVdiff* energy, and the polarization of *Sdiff* waveforms for corresponding isotropic structures is complicated [*Ritsema et al.*, 1997], thus, interpretation of *SVdiff* observations deep in the shadow zone requires knowledge of the structure in D" along the entire path, which is very difficult to achieve.

Profiles of long-period WWSSN *SV* waveforms were considered by *Lay and Young* [1991], with amplitude and waveform data at distances from 70° to 105° used to estimate D" shear velocity structure beneath Alaska and the northern Pacific. A shear velocity discontinuity exists 200-300 km above the CMB in this region [*Lay and Helmberger*, 1983a; *Young and Lay*, 1990; *Matzel et al.*, 1996], which predicts non-linear *S* wave polarizations in the distance range 90° to 105°. The *SV* amplitudes decay rapidly beyond 95°, consistent with positive, or only slightly negative velocity gradients in D". Some data exhibit shifts in the apparent onset of *SV* and *SH* at distances near 95°, with *SV* being delayed by from 0 to 4.5 s, which cannot be accounted for by isotropic structure. The delayed *SV* arrivals bottom in the same vicinity as the delayed *ScSV* arrivals observed by *Lay and Helmberger* [1983b], but the thin high velocity layer models that had been invoked to explain the latter data do not predict a shift of *S* arrivals near 95°. This suggests that laterally varying anisotropy at the base of the mantle is a more reasonable explanation for both data sets. Application of receiver lithospheric anisotropy corrections to both data sets [*Garnero and Lay*, 1997] supports this conclusion.

Recent studies of shear wave splitting

The studies mentioned above established the basic approaches to detecting lower mantle anisotropy. Recent work has utilized higher quality data sets to examine shear waves in the three key distances ranges (*ScS* at distances from 50° to 80°; *S* grazing D" at distances of from 90 to 105°; and *Sdiff* in the core shadow beyond about 105°). Figure 1 shows a map summarizing the regions that have now been sampled, keyed to Table 1, which identifies associated references and seismic phases analyzed. The map also indicates the nature of the anisotropy detected in each region, characterized by whether the *SH* velocity (Vsh) or for the *SV* velocity (Vsv) is larger. In general, shear wave splitting is most evident when fast and slow quasi-shear wave polarizations are aligned with the great-circle reference frame, but there is no physical requirement that this be the case. Generic raypaths of each of the three principal data types are illustrated in Figure 2. Examples of waveforms exhibiting splitting for each geometry are shown in Figure 3.

Only a small portion of the lowermost mantle has been examined, and recent work has emphasized the same basic regions studied in the early studies discussed above; beneath

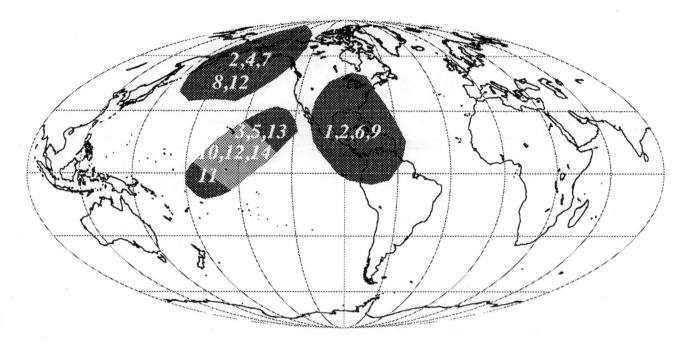

Figure 1. Global map showing regions of D" that have at least some reported evidence of shear wave splitting, with the number indicating the corresponding reference in Table 1. The darker regions under the Caribbean and Alaska are consistent with laterally varying transverse isotropy from 0 to 2%, producing relative early tangential component S, ScS and $Sdiff$ signals. The region under the Pacific has variable anisotropy, with the northeasternmost and southwesternmost areas (darker shading) possibly having transverse isotropy, while the central area (lighter shading) has either no anisotropy or patchy azimuthal anisotropy.

Table 1. Observations of Shear Velocity Anisotropy in D"

No.	Reference	S Phases Used	Sense of Splitting
1	*Mitchell and Helmberger* [1973]	ScS	Vsh>Vsv
2	*Lay and Helmberger* [1983b]	ScS	Vsh>Vsv
3	*Vinnik et al.* [1989]	$Sdiff$	Azimuthal
4	*Lay and Young* [1991]	ScS, S	Vsh>Vsv
5	*Vinnik et al.* [1995]	$Sdiff$	Vsh>Vsv
6	*Kendall and Silver* [1996]	S	Vsh>Vsv
7	*Matzel et al.* [1996]	S	Vsh>Vsv
8	*Garnero and Lay* [1997]	ScS, S, $Sdiff$	Vsh>Vsv
9	*Ding and Helmberger* [1997]	ScS	Vsh>Vsv
10	*Pulliam and Sen* [1998]	S	Vsv>Vsh
11	*Vinnik et al.* [1998]	$Sdiff$	Vsh>Vsv
12	*Kendall and Silver* [1998]	S, $Sdiff$	Vsh>Vsv/None
13	*Ritsema et al.* [1998]	$Sdiff$	Vsh>=<Vsv
14	*Russell et al.* [1998]	ScS	Vsh>=<Vsv

the Caribbean, Alaska and the Central Pacific. While there is a need to extend the coverage as much as possible, it has been essential to first establish the strength and variability of the anisotropic observations (or lack of anisotropy) in each region with higher quality data and modeling. We summarize recent observations in each of the three regions.

Caribbean. *Kendall and Silver* [1996] analyzed high quality digital S and $Sdiff$ waves in the distance range 90°

to 110° for paths sampling D" below the Caribbean. They corrected for relatively strong lithospheric anisotropy effects under the Canadian seismic stations used. Delays of 3-9 s in the apparent onset of the SV components are consistent with transverse isotropy with a vertical axis, but there is little azimuthal coverage. Absence of splitting of direct S waves at closer distances indicates that the anisotropy is located in the deep mantle. The sense of the delay observed by *Kendall and Silver* [1996] is similar to that seen in ScS splitting for paths with shorter lengths in D" by *Mitchell and Helmberger* [1973] and *Lay and Helmberger* [1983b] for the same region, but the magnitude is larger, consistent with either an effect that accumulates with distance or possibly a northward increase in the magnitude of anisotropy. *Ding and Helmberger* [1997] show further examples of late $ScSV$ in broadband recordings for paths under the Caribbean.

An appropriate reference model is needed to constrain the depth range of anisotropy. The Caribbean region has a laterally varying shear velocity discontinuity about 250±50 km above the CMB [e.g., *Lay and Helmberger*, 1983a; *Kendall and Nangini*, 1996; *Ding and Helmberger*, 1997], which imparts complexity to the S waveforms that must be distinguished from anisotropy. *Ding and Helmberger* [1997] note that there is evidence for SV reflections from

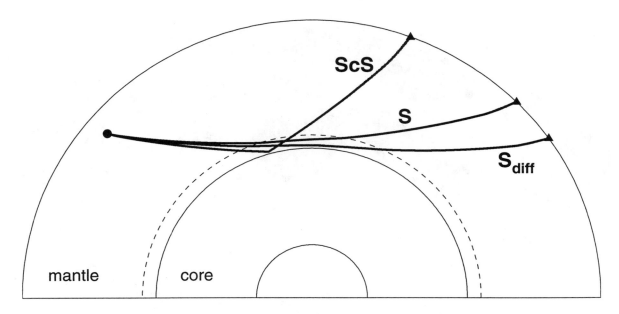

Figure 2. Raypath for the three basic classes of seismic shear wave used to examine anisotropy in the lowermost mantle: *ScS*, *S* near 95° and *Sdiff* in the core shadow zone.

the top of D" under the Caribbean, so the *S* wave discontinuity is not manifested only in Vsh. Any *P* wave discontinuity beneath the Caribbean region is much smaller than the 3% *S* wave discontinuity, and the *P* velocity structure may be smoothly varying with depth. *Ding and Helmberger* [1997] show isotropic broadband *SV* synthetics for distances beyond 90° with strong secondary arrivals caused by energy traveling above the discontinuity, while weak initial motions are associated with energy that actually penetrates into D". If the latter is obscured by noise or weakened by strong attenuation, one could infer an erroneous splitting by comparison of the strong onsets for *SH* and *SV* signals.

At distances of 90°-95° the *SV* arrival sometimes has an initial weak motion which appears to have opposite polarity to what is expected, and this energy is seldom picked as the onset of the *SVdiff* signal if there is a later impulsive onset. If the weak motion is picked, the magnitude of inferred shear wave splitting is reduced significantly. No isotropic or anisotropic structure has yet been shown to account for this reversed *SV* onset (which is intermittently observed beneath Alaska as well), and it may be a manifestation of scattering or heterogeneous general anisotropy.

Kendall and Silver [1996] assumed that the anisotropic region extends over the full 250 km thickness of D" beneath the Caribbean, and inferred 0.5-2.8% anisotropy over regions of 1000-2800 km in length. This can be compared with the estimate by *Doornbos et al.* [1986] that the *ScS* splitting of up to 2 s observed by *Mitchell and Helmberger* [1973] requires 2% anisotropy over a 75 km thick layer. The observed shift of the onset of *S* at 90° requires that the anisotropy is present at least several hundred kilometers

above the CMB, not concentrated at the bottom of the thermal boundary layer as in the models of *Cormier* [1986]. In fact, for *S* waves that graze a shear velocity discontinuity, it is possible to account for the anisotropic effects by localizing anisotropic structure near the discontinuity rather than throughout the D" layer [*Garnero and Lay*, 1997; *Lay et al.*, 1997]. Absence of *SKS* splitting apart from that explained by receiver structure indicates that transverse isotropy is more likely than azimuthal anisotropy in this region.

Alaska. The region under Alaska and the northern Pacific has recently been studied by *Matzel et al.* [1996] and *Garnero and Lay* [1997]. *Matzel et al.* [1996] considered essentially the same WWSSN data as *Lay and Young* [1991], emphasizing the *S* observations in the distance range around 95°. Beyond 93° the apparent *SV* arrival onset is late relative to *SH* by 3 to 5 s (in 21 of 35 cases), as noted by *Lay and Young* [1991], and the *SH* waveforms show evidence of multiple arrivals associated with a shear velocity discontinuity near 243 km above the CMB, while the *SV* waveforms are simpler. Various isotropic models were explored but could not replicate these features. Given that the *SH* complexity is generated near the depth of the discontinuity, this suggests that anisotropy is present near that depth, as under the Caribbean. The primary contribution of this study was the computation of transverse isotropy synthetic seismograms using a reflectivity technique, with good agreement being achieved in the basic waveform behavior of the *SH* and *SV* signals near 95°. In a successful model (Figure 4b), the Vsh velocity structure has a 3% shear velocity discontinuity at a depth of 2600 km, with no

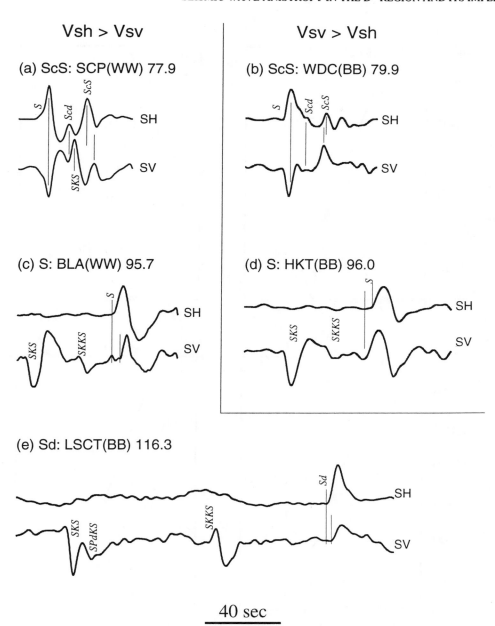

Figure 3. Representative shear waveform examples showing splitting between *SH* and *SV* components for each of the three phase geometries in Figure 2. (a) and (b) show relative shifts of *ScSH* and *ScSV* for paths under Alaska (a) and under the Pacific (b). The seismogram in (a) is the most extreme case for Alaska, with the first peak of *ScSH* (indicated by the vertical stroke) arriving more than 4 s ahead of the first peak of *ScSV*. The seismogram in (b) (kindly provided by Sara Russell), for the Pacific, shows the opposite behavior, but this appears to be fairly rare.   (c) and (d) show examples of *S* waves near 95°, where the onset of the *SH* pulse is either earlier than *SV* (c), as under Alaska and the Caribbean, or later than *SV* (d), as intermittently observed under the Pacific (seismogram (d) was kindly provided by Jay Pulliam).   (e) shows a long distance *Sdiff* observation traversing under the Pacific with Shdiff arriving ahead of *SVdiff*.

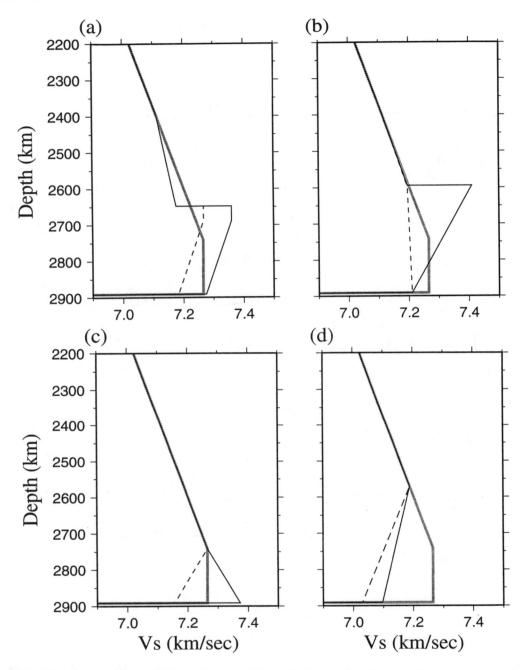

Figure 4. Classes of transverse anisotropy D" shear velocity models that have been proposed In various studies. The solid line indicates Vsh, the dotted line indicates Vsv, and the gray line indicates the shear velocity structure in PREM [*Dziewonski and Anderson*, 1981], which is the reference mantle model in each case. (a) is from study of the mantle under Alaska by *Garnero and Lay* [1997], and has a uniform anisotropy throughout the D" layer, along with a small Vsv discontinuity at the same depth as the strong Vsh discontinuity. (b) is from a study of the mantle under Alaska by *Matzel et al.* [1996], and reduces anisotropy within the D" layer. (c) is from *Maupin* [1994], and shows a class of models that was explored to model data under the Pacific observed by *Vinnik et al.* [1989], although not all signals could be matched with the same model. Anisotropy is concentrated in D". (d) is one of the models from *Vinnik et al.* [1995] which was used to model stacked waveforms from a cluster of Fiji events, with paths under the Pacific.

discontinuity in the Vsv structure. Anisotropy of 1.5-3% is needed to match the 3-5 s time lags. The D" discontinuity in this region thus appears to be stronger for Vsh than for Vsv, and there is evidence for either a small Vp discontinuity in this region [*Vidale and Benz*, 1993] or smooth radial structure [*Young and Lay*, 1989]. This led *Matzel et al.* [1996] to propose that the D" discontinuity is itself entirely an anisotropic transition in this region, primarily manifested in the Vsh structure.

Garnero and Lay [1997] further examine WWSSN and digital data traversing D" beneath Alaska, extending the mapping of laterally varying transverse isotropy initiated by *Lay and Young* [1991]. The resulting image obtained from a combination of path-length weighted *ScS* observations and *S* observations near 95°, is shown in Figure 5, with spatial smoothing applied to simulate the Fresnel zone effects of the long-period data. The anisotropy magnitude varies from 1.5% to zero over lateral scale lengths of 1000 km. Transverse isotropy models that successfully predict the waveforms of many (but not all) of the split signals are shown in Figure 4a, with a small, but non-zero shear velocity discontinuity in the Vsv structure being preferred due to observation of *SV* waveform complexities consistent with triplication by a small velocity increase [*Lay and Young*, 1991]. The anisotropy must exist near the discontinuity, but its variation with depth into D" is not constrained.

Garnero and Lay [1998] explore the effects of transverse isotropy in D" under Alaska on *SKS-SH* differential times. Anomalies of several seconds relative to reference structures are easily generated. It is notable that isotropic modeling of the *SKS-S* and *SKKS-SKS* differential time anomalies of several seconds for this region leads to incorporation of a low velocity layer in the outermost core to slightly delay the *SKS* or *SKKS* waves [*Lay and Young*, 1990], while including transverse isotropy in D" provides a more self-consistent explanation [*Garnero and Lay*, 1998].

Central Pacific. Recent studies of D" structure under the central Pacific have revealed substantial complexity, with a mix of observations of relatively slow *SV* (consistent with transverse isotropy), relatively fast *SV* (consistent with azimuthal anisotropy), and little or no shear wave splitting. This region is underlain by slower-than-average large scale shear velocity in most global tomography models [e.g. *Su et al.*, 1994; *Li and Romanowicz*, 1996], and anomalously large amplitudes of *SVdiff* observations traversing the lower mantle there indicate negative shear velocity gradients over the lowermost 200-300 km of the mantle [*Ritsema et al.*, 1997; *Valenzuela and Wysession*, 1998]. There is some evidence for a localized shear velocity discontinuity 175-185 km above the CMB [*Garnero et al.*, 1993; *Valenzuela and Wysession*, 1998], but it is likely that no discontinuity is present in much of the region. There is little evidence for a *P* velocity discontinuity in D" in this region [e.g. *Young and Lay*,

1989; *Wysession et al.*, 1998], while a thin very low velocity layer exists at the base of the mantle in extensive regions [see *Garnero et al.*, 1998]. These attributes differ from the circum-Pacific regions described previously.

Maupin [1994] presented a thorough, and sobering, theoretical study of effects of anisotropy in D", motivated by the observations of *Vinnik et al.* [1989]. She showed that the mere presence of *SVdiff* with amplitudes 30% of *SHdiff*, or isolated observations of complex *Sdiff* polarization with poorly defined onset times, does not uniquely resolve whether isotropic structure, transverse isotropy or azimuthal anisotropy is responsible. Using multiple events one can reliably detect azimuthal anisotropy if the horizontal particle motion of *Sdiff* proves independent of focal mechanism polarization (as noted above, there are some instances where the SV polarity near 95° is the same as that of SKS, which is inconsistent with the focal mechanism). Maupin favored the interpretation of azimuthal anisotropy for the *Sdiff* observations near 118°, but such a model could not fully explain all of the 3 higher quality observations of *Vinnik et al.* [1989], nor could models of transverse isotropy. In considering a suite of transverse isotropy models with different Vsv and Vsh gradients in D" (anisotropy was taken to increase with depth down to the CMB over a 150 km thick layer (Figure 4c)), Maupin found that shifts in the onset and first peak by 2.5 to 6 s can be achieved, as necessary to account for the observations near 95° in other regions by *Lay and Young* [1989] and *Kendall and Silver* [1996]. However, observed waveforms under the Pacific exhibit greater variability than predictions for any laterally uniform model, suggesting the importance of heterogeneous structure.

Vinnik et al. [1995] extended the *Sdiff* data set traversing the central Pacific, using recordings from 12 events at two east coast digital stations (HRV and WFM) at epicentral distances from 114° to 121°. The relative amplitude of motions identified as *SVdiff* to the usually clear *SHdiff* arrivals varied substantially, indicating little coupling between the two, and there was good correlation between *SVdiff* and *SKKS* amplitudes, indicating that source radiation pattern influences the *SVdiff* arrivals. This suggests that negative velocity gradients in D" are present, in order to have *SV* energy penetrate deep into the shadow zone. A 3 s shift of onset time of *SVdiff* relative to *SHdiff* was also measured for a stack of signals from 6 events from 116° to 120°, although one has to be concerned that the focal mechanisms and distances differ for the stacked traces.

Reflectivity synthetics [*Kind and Müller*, 1975] were used to separately model the *SH* and *SV* components under the hypothesis of transverse isotropy, and *Vinnik et al.* [1995] were able to explain three features of the stacked data: (1) anomalously late *SHdiff* arrivals (up to 10 s later than the reference IASP91 model), (2) the 3 s delay of the onset of *SVdiff* with respect to *SHdiff*, and (3) different spectral content of *SVdiff* and *SHdiff*, with *SVdiff* peaking

D" Anisotropy
5 deg cap averages

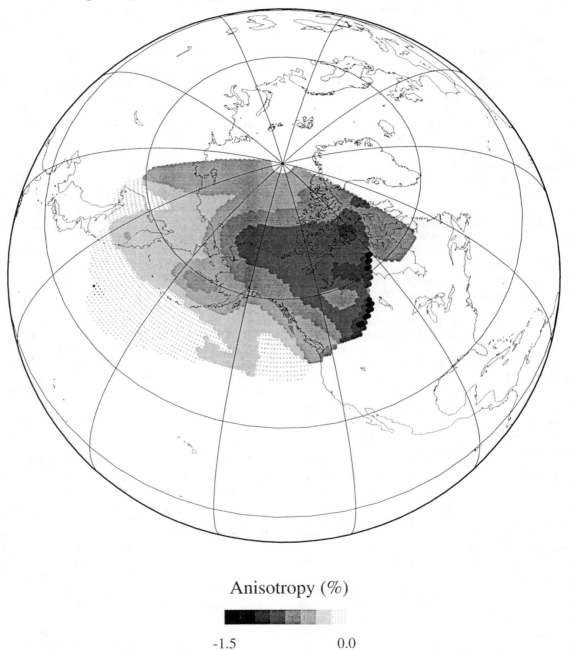

Anisotropy (%)

-1.5   0.0

Figure 5. Map of lateral gradients in anisotropy in the lowermost mantle beneath Alaska, showing how the transverse isotropy appears to increase from west to east. [from *Garnero and Lay*, 1997].

at a higher frequency. The observations can be explained by a model with a low Vsh layer with a thickness of 250-350 km with Vsv even slower by about 1%. Alternative models featured negative gradients in the same depth range with equivalent velocity reductions, which are more compatible with profiles of *SVdiff* data modeled by *Ritsema et al.* [1997]. Figure 4d indicates successful anisotropic models with negative gradients that fit the *SVdiff/SHdiff* spectral ratios and the 3 s travel time shift [*Vinnik et al.*, 1995]. Four events from 114° to 115° showed no splitting, indicating that anisotropy is laterally varying in this region, and there is little constraint on the vertical distribution of anisotropy within the D" region (it need not be distributed over the entire depth range in which the shear velocity decreases). It appears that transverse isotropy may be present in lower mantle with either faster-than-average or slower-than-average velocities.

Garnero and Lay [1997] considered a profile of broadband recordings in North America from a Fiji event, with no apparent splitting at 95°, about 1 s delay of the onset of *SVdiff* relative to *SHdiff* at 102°, and 3.3 s relative delay of *SVdiff* at 116.3° (see Figure 3e). The latter number is consistent with the 3 s delay seen by *Vinnik et al.* [1995] at east coast stations. *Kendall and Silver* [1998] report negligible splitting at Canadian stations for paths under the central Pacific, with the exception of small amounts of splitting at distances larger than 110° (to Central and Eastern Canada), which is consistent with the profile of *Garnero and Lay* [1997]. However, observations in Northwestern Canada do show increasing delays of *SVdiff* onset with distance [*Ritsema et al.*, 1998]. It is possible that a lateral gradient of increasing transverse isotropy structure toward the North American continent is sensed by these data.

Pulliam and Sen [1998] considered a small data set at closer ranges, finding 3 records with *SV* advances of up to 2.2 s relative to *SH* at distances from 93.9° to 96° at a station in Texas (corrected for receiver anisotropy), one of which is shown in Figure 3d. Some stations at closer distances show no splitting for the same events, so it was inferred that the anisotropy is near the base of the mantle. An azimuthal anisotropy model with Vsv having a positive velocity increase in the lowermost 150 km of the mantle, while Vsh is nearly constant, gives good waveform matches to the data. The independent evidence for a regionally extensive negative gradient in the sub-Pacific Vsv structure casts some doubts on the specific model used, but the observation of intermittent early *SVdiff* arrivals is supported by other work [*Vinnik et al.*, 1998; *Ritsema et al.*, 1998].

Analysis of broadband *ScS* recordings in western North America (see Figure 3b) indicates small scale variations in anisotropy with the fast polarization sometimes being close to the longitudinal component and sometimes close to the transverse component [*Russell et al.*, 1998]. The

data vary systematically over scale lengths of only a few hundred kilometers, indicating that is unwise to make generalizations about anisotropic structure from sparse data distributions.

While a better azimuthal distribution of seismic data will be required before azimuthal anisotropy in the lowermost mantle can be constrained, certain *ScS* and *SHdiff* profiles are compatible with this possibility. *Winchester and Creager* [1997] infer azimuthal anisotropy from azimuthal patterns in transverse *ScS-S* travel time differentials for three different D" locations beneath the western Pacific. This interpretation assumes that *ScS* is the anomalous phase, and that the anomalies arise from the same lower mantle volume. *Valenzuela and Wysession* [1998] find azimuthal variation in *SHdiff* profiles in D" beneath northeastern Siberia. Most profiles trending SE-NW are 1-3% faster than for PREM [*Dziewonski and Anderson*, 1981], while most profiles trending SW-NE are 1-2% slower than for PREM. Given the differences in path in a heterogeneous region, the interpretation in terms of azimuthal anisotropy is very tentative.

Vinnik et al. [1998] consider profiles of observations at individual stations in North America in the range 92° to 122° from Fiji events to explore anisotropic structure under the Pacific. Station profiles of differential travel time residuals for *SKS-S* and *SKKS-S* show increases with distance indicative of accumulating delay of *S* in a low velocity layer at the base of the mantle. Changes in the rate of increase with distance are attributed to strong lateral gradients of Vsh in a layer about 300 km thick, with velocities as much as 10% lower than PREM. In the southwestern region of the model, Vsh returns to "normal" over a distance of only several hundred km, while Vsv remains low by about 10%, suggesting very large anisotropy. Several stations display small increases in *SVdiff-SHdiff* differential arrival times with increasing distance, supporting some increase in isotropy toward the southwest. The strength of the lateral gradients and magnitude of the anisotropy are highly dependent on interpretation of the distance trends, particularly the assumption of a 300 km thick zone of anomalous structure. The very low Vsv in the southwestern area is compatible with observed *SKKS-SKS* travel time residuals of 2-3 s that likely originate in the same region [*Schweitzer and Müller*, 1986; *Garnero and Helmberger*, 1993]. In the central region where Vsh is very low, a few observations of early *SVdiff* are reported, consistent with *Pulliam and Sen* [1998] and *Russell et al.* [1998].

Clearly, more work is needed, but a preliminary synthesis of observations under the Pacific is that there are strong lateral gradients in isotropic and anisotropic structure in the low velocity central region, fringed by regions to the northeast and southwest that may have transverse isotropy (see Figure 1). The spatial gradients are poorly resolved, as are the overall extent and magnitude of anisotropy.

GEODYNAMICAL AND MINERAL PHYSICS CONSIDERATIONS OF D" ANISOTROPY

The underlying physical origin of anisotropy in D" must be addressed by considering a suite of scenarios involving physical properties under conditions not yet extensively explored in the laboratory. At the most basic level, that D" anisotropy is detectable in the regions shown in Figure 1 and (apparently) not observed in the bulk of the lower mantle indicates that fundamental differences in length scales, structure, deformation style and/or chemistry exist between the lowermost mantle and the overlying material. As noted by *Karato* [1989], identifying the dominant deformation mechanisms (dislocation motions, grain boundary diffusion, melt alignment) and controlling factors (strain versus stress) that can give rise to anisotropy requires knowledge of the flow regime or vice versa. Current ignorance of the detailed chemistry, physical properties and flow regime in D" limits our ability to interpret anisotropy in the region.

It is probable that a hot, low viscosity thermal boundary layer (TBL) is present at the base of the mantle [e.g., *Loper and Lay*, 1995]. Conventional estimates of the TBL thickness suggest 50-75 km [e.g., *Stacey and Loper*, 1983], however, the evidence described above for much thicker zones of negative velocity gradient in D" beneath the central Pacific, Alaska, and the Caribbean suggests that the TBL may actually be 200-250 km thick. The TBL dynamic regime is embedded the within larger scale circulation of the lower mantle induced by downwellings (maybe involving slabs), internal heating, and upwellings in "superplumes" or plume provinces. Strong horizontal shear flows are likely in the low viscosity TBL, and thermal instabilities may give rise to small-scale upwellings [e.g., *Olson et al.*, 1987]. Upwelling plumes will tend to entrain material up to a few percent denser than the overlying mantle [e.g., *Davies and Gurnis*, 1986; *Sleep*, 1988; *Hansen and Yuen*, 1988; *Olson and Kincaid*, 1991]. In numerical simulations, the entrained material tends to form vertically oriented "cusps".

Another distinct attribute of the lowermost mantle is the proximity of the huge density contrast at the CMB. Given the long history of chemical differentiation of the planet, D" is a likely depository for chemical heterogeneities with densities intermediate to those of the lower mantle and core. D" chemical heterogeneities may arise from many processes, including chemical reactions between the core and mantle, chemical distinctions associated with relic slabs, and chemical fractionation associated with partial melting at the base of the mantle. The chemical buoyancy and physical characteristics of any chemical heterogeneity will determine how it interacts with the TBL flow regime.

The probable thermal and chemical complexity of D", coupled with the lack of experimental constraints on lower mantle deformation processes at relevant pressure and temperature conditions, has generated an abundance of possible causes for the anisotropy in D" (Figure 6) [*Wysession*,

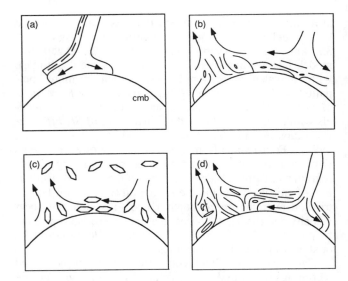

Figure 6. Schematic illustrations of possible dynamically induced anisotropic structure in D". (a) Subducted slabs, retaining anisotropic fabrics, or having melting/deformation of crustal components accumulate in D" and produce transverse isotropy associated with the D" discontinuity in faster-than-average areas. (b) Core-mantle chemical reaction products and/or partial melt components from the thermal boundary layer are entrained in large-scale boundary layer flow and sheared to make lamellae with strong velocity contrasts. This can produce both general and transverse isotropy concentrated near the CMB. (c) Lattice preferred orientation (LPO) in primary lower mantle minerals is organized by strong boundary layer shear flows. A transition in deformational mechanism, perhaps under the high temperatures of the boundary layer is responsible for the enhanced anisotropic fabric developed in the D" region. (d) Chemical heterogeneities and/or partial melt components are sheared into vertical lamellae near the CMB beneath upwellings and ride up over the top of downwelling slab structures to produce transverse isotropy near the D" discontinuity.

1996; *Karato*, 1997]. Two major families of mechanisms are those involving large-scale structural anisotropy and those involving lattice preferred orientation (LPO) of minerals within this region. The former family postulates a broad range of possible heterogeneities in chemistry or phase incorporated within D"; the latter family most likely involves either a change in chemistry, phase, or deformation style from the overlying mantle. Accordingly, the basic observation of anisotropy within D" reemphasizes the anomalous character of this zone relative to the overlying mantle.

Structural Anisotropy in D"

Proposals of structurally-induced anisotropy in D" have emphasized mechanisms that give rise to finely laminated structures with strong contrasts in material properties.

Horizontally laminated structures will yield *SH* velocities faster than *SV* [*Backus*, 1962; *Kendall and Silver*, 1996], in general accord with most shear wave splitting in D". Mechanisms which could generate such laminated structures within a TBL at the base of the mantle include entrainment of core-mantle reaction products in the lowermost mantle [*Knittle and Jeanloz*, 1991]; injection of subducted oceanic crustal material into the boundary layer, perhaps associated with partial melt [*Kendall and Silver*, 1996]; and diking induced by either the presence of partial melt or of solidified partial melt near the CMB [e.g. *Williams and Garnero*, 1996]. Even small pockets of partial melt will tend to be sheared by flow, creating a fabric with shaped preferred orientation (SPO), but the chemical buoyancy and distribution of the melt will play an important role in whether such a fabric is sustained or whether the melt drains out (either upward or downward). In short, such laminations could be: generated by interactions with the core below; derived from geochemical heterogeneities produced from tectonic processes; a manifestation of the presence of melt in this region; or some combination of these.

Core-mantle chemical reactions have been demonstrated as viable at the CMB based on both experimental data [*Knittle and Jeanloz*, 1989; 1991; *Goarant et al.*, 1992] and thermochemical calculations [*Song and Ahrens*, 1994]. Notably, CMB reactions appear to be significantly enhanced by the presence of partial melt [*Ito et al.*, 1995; *Williams*, 1998]. The distribution of reaction products within laminar structures is inferred from their being generated at the CMB, with subsequent entrainment into the overlying mantle. Two types of laminations can plausibly arise. First, unreacted mantle could be juxtaposed with lamellae containing a mixture of the reaction products: this possibility has been examined by *Kendall and Silver* [1996] who noted that iron-enrichment and reaction within D" could produce the observed anisotropy of D", but would likely lower the shear velocity of the iron-enriched zone by up to 6%. Second, any physical separation between the relatively high seismic velocity (and lower density) iron-free perovskite and silica reaction products and the iron alloy reaction products would give rise to three possible combinations of materials which could produce laminar features: seismically fast iron-depleted reaction products, markedly slow (by approximately 40% in shear [*Kendall and Silver*, 1996; *Williams and Garnero*, 1996]) iron alloys, and unreacted mantle. *Pulliam and Sen* [1998] suggest that azimuthal anisotropy beneath the central Pacific is caused by entrainment of chemical heterogeneities, perhaps from CMB chemical reactions, in upwelling flows, or possibly in a shear zone between flows with opposing directions. A horizontal symmetry axis for the resulting lateral 'lamination' can account for the shear wave splitting. Although the precise mechanisms of, and scale-lengths at which, such reaction-induced geochemical heterogeneities are entrained into the overlying D" layer remain conjectural, the recognition that D" may be associated with low

viscosity, partially molten and plausibly rapidly flowing material provides a number of possible entrainment and emplacement mechanisms for such reaction products.

If lithospheric slabs sink to the D" region (Figure 6a) [e.g. *Christensen and Hofmann*, 1994; *Grand et al.*, 1997] their influence on anisotropy in D" depends greatly on their thermal, mechanical and chemical structure. Numerical simulations [e.g., *Tackley*, 1995] indicate that slabs could accumulate in the transition zone before descending into the lower mantle in large blobs, sustaining strong thermal anomalies all the way to the CMB. Thus, slabs could enhance lateral temperature variations in D". Slabs may also undergo significant internal deformation as they descend; the eclogitic component may delaminate from the residual slab. The separate components may break up and stretch to form lamellae whose orientation would depend not only the local flow regime, but on the integrated flow history [e.g., *Gurnis*, 1986]. Subducted slabs approaching the CMB will tend to form laterally-aligned tendrils as the slab folds and spreads out along the base of the mantle, possibly disrupting any compositionally distinct layer and perhaps contributing slab derived heterogeneities as well (Figure 6d).

The chemical properties of slabs also influence whether anisotropy results. The basaltic chemistry material is likely to be juxtaposed with mantle of approximately peridotitic chemistry and, if the slab retains its initial stratification, a harzburgitic (olivine-enriched) layer near its former base (Figure 6a). The degree to which purely solid-state elastic differences between basalt and either peridotitic or harzburgitic chemistry can generate differences in shear velocity is unclear. The basaltic layer is anticipated to be enriched in $CaSiO_3$-perovskite and to have an enhanced iron and aluminum content present within $(Mg,Fe)SiO_3$-perovskite relative to the surrounding material; the formerly basaltic material may also contain small (~10%) amounts of free SiO_2 [*Irifune and Ringwood*, 1987; *Irifune*, 1994; *Kesson et al.*, 1994]. Unfortunately, the elastic properties of Al- and Fe-enriched magnesium silicate perovskite are ill-constrained, as is the shear modulus of calcium silicate perovskite (and, of equivalent importance for the conditions of the CMB, the pressure derivative of the shear modulus of both calcium and magnesium silicate perovskites). As a result, it is difficult to assess the magnitude of solid-state shear velocity contrasts between layers of basaltic chemistry, their complementary harzburgitic layers, and peridotitic mantle under CMB conditions.

Kendall and Silver [1996] propose that D" anisotropy may be associated with the presence of melt inclusions within the basaltic layer: such melting could be associated with a depression of the eutectic temperature associated with enhanced Ca, Al and Fe contents or, speculatively, even with the retention of volatiles within the subducted oceanic crust to ultra-high pressures. Aligned inclusions of melt can readily produce *S* wave anisotropy, with the magnitude of anisotropy at a given melt fraction hinging crucially on the aspect ratios of the inclusions [*Kendall and*

Silver, 1996]. For small aspect ratio inclusions (0.01), as little as 0.5% partial melting could give rise to the observed anisotropy. *Karato* [1997] argues that melting of slab components seems unlikely because the downwelling regions are likely to be the coldest portions of D". Perhaps a stronger argument is that it is unclear what flow regime in the downwelling could produce the multiple oscillating, or anastamozing layers of material with few kilometer scale lengths required to produce a seismically anisotropic fabric (a single, or even a few layers will not suffice).

Partial melt in D" need not involve slabs; *Williams and Garnero* [1996] interpret the ULVZ in terms laterally varying partial melt concentrations of between 5 and 30% in the lowermost 5-40 km of the mantle. Such large-scale melting in the lowermost mantle could easily give rise to very large values of S wave anisotropy if the orientation of the melted material is only marginally aligned in a horizontal direction. Although the distribution of partial melt as a function of depth within D" is not clear, the existence of large-scale melting in the lowermost regions of D" raises the crucial issue of the depth dependence of anisotropy within this zone. That is, depending on the degree of preferred alignment of melt in the lowermost mantle, even a modest sampling of the lowermost mantle could result in the observation of S wave anisotropy. In short, if liquid is present in abundance within the lowermost mantle, and this region experiences significant strain, preferred alignment of melt inclusions would be expected.

As an additional effect, depending upon the detailed geographic distribution and thickness of the partially molten ULVZ over time and the overall stability of the D" region with respect to deformation and entrainment in the overlying flow of the lower mantle [e.g., *Sleep*, 1988; *Kellogg and King*, 1993], regions which contained a ULVZ in the past may retain a structural anisotropy produced by the presence of laminations of solidified partial melt. In particular, the thickness of the ULVZ through geologic time remains obscure: if, as is likely, the mantle has undergone secular cooling, a thicker and more widespread ULVZ would be expected to have been present in the past. The level at which fossil remnants of a more pervasive partially molten zone are preserved depends on both their detailed chemistry (which controls the buoyancy of such features) and the dynamics of the lowermost mantle: simulations suggest that an augmentation of only a few percent in density is needed to stably stratify the lowermost mantle for extended periods of geologic time [*Sleep*, 1988]. While speculative, the role of solidified dike- or laccolith-like features in generating anisotropy in D" hinges critically on how different the composition of melt generated near the CMB is from the composition of normal mantle. That is, the degree to which the eutectic composition in the CMB differs from the bulk composition of the lower mantle represents a crucial parameter in determining whether such fossilized melt regions could have markedly different shear wave velocities from normal mantle.

These disparate mechanisms of producing structural laminations; swept-up CMB reaction products, emplacement of subducted oceanic crust, and preferred orientation of partially molten material (or its fossilized remnants) at the CMB, can each fulfill the fundamental requirement of producing an intercalation of material of differing shear modulus within the material of the lowermost mantle. Yet, each mechanism does have specific implications, requirements and first-order uncertainties associated with it. It is also important to note that these mechanisms need not operate to the exclusion of one another.

Intrinsic Mineralogical Anisotropy in D"

Any LPO that develops in D" will depend on the strain experienced by grains, the grain size, and the dominant deformation mechanism(s) active there. *Kellogg and Turcotte* [1990] showed that the amount of strain on a marker in mantle flow is largely controlled by the time it spends undergoing pure shear. Thus, LPO is likely to be greatest in the vicinity of strong horizontal flows and near active upwellings and downwellings, where the strain rates are highest (Figure 6c). Strain patterns within D" may not correspond directly to flow in the overlying mantle, as small scale circulation within D" can decouple the strain patterns from the overlying upwellings and downwellings, so it is very difficult to infer any particular strain regime in D". Given that the entire mantle is straining, it is important to account for LPO being concentrated at the base of the mantle.

The primary means by which anisotropy could be generated by LPO in D", but not in the overlying mantle (without invoking a change in chemistry) are through a dramatic difference in the strain history of the lowermost mantle from the overlying material, a phase change of one of the constituents of the lowermost mantle, and/or a shift in deformational mechanism within (or just above) D". The viability of the first of these mechanisms is difficult to assess, given the profound uncertainties in the flow field of the lowermost (or for that matter, lower) mantle. The second of these mechanisms, the presence of a phase change in the lowermost mantle which could in turn give rise to LPO within a new assemblage is extremely speculative: although possible phase changes in deep mantle constituents have been sporadically reported based on both theoretical and experimental grounds [*Cohen et al.*, 1997; *Cohen*, 1992; *Meade et al.*, 1995a; *Saxena et al.*, 1996; *Stixrude et al.*, 1996], each of these proposed transitions suffer from one of the following difficulties: 1) they are not necessarily coincident with the pressures at the top of D", but rather may occur in the mid-mantle [*Meade et al.*, 1995a; *Saxena et al.*, 1996; *Stixrude et al.*, 1996]; 2) they are not observed in shock experiments on compositions expected to be similar to those in the lowermost mantle [*Cohen et al.*, 1997; *Vassiliou and Ahrens*, 1982]; or 3) they hinge on the presence of abundant SiO_2 in the deep mantle [*Cohen*, 1992], a mineralogy which is unlikely to occur in an isochemical mantle unless perovskite is destabilized at high pressures: a possibility that is highly controversial [e.g., *Knittle and Jeanloz*, 1987; *Kesson et al.*, 1994]. A chemically distinct D" layer could enable phase transitions to occur that would not be expected within a homogeneous lower mantle, but this possibility is almost unconstrained. At present, we consider a shift in deformational mechanism to be the most viable for generating LPO in D", given the low pressure rheologic behavior of materials, speculations on the rheologic behavior of the lower mantle, and the role of D" as a TBL.

In particular, the lack of anisotropy in the lower mantle has been attributed to the deformational mechanism of perovskite not giving rise to preferred orientation based on ambient temperature, high-pressure experiments on perovskite deformation [*Meade et al.*, 1995b]. The observed average shear wave elastic anisotropy of perovskite is in excess of 6% [*Meade et al.*, 1995b; *Yeganeh-Haeri*, 1994], with a peak shear velocity anisotropy of 16% [*Yeganeh-Haeri*, 1994]; other possible lower mantle phases (MgO, SiO_2) have similar magnitudes of maximum anisotropy [*Meade et al.*, 1995b; *Jackson and Niesler*, 1982; *Weidner et al.*, 1982]. Notably, the degree to which these elastic anisotropies will change at the conditions of the CMB is unclear: studies of the pressure-dependence of anisotropy at ultra-high pressures remain in their infancy, but there are both theoretical and experimental indications that the level of anisotropy in lower mantle minerals could be quite different at high pressures [e.g., *Duffy et al.*, 1995; *Karki et al.*, 1997]. Based on ambient temperature results, *Meade et al.* [1995b] observed essentially no formation of anisotropic textures within sheared silicate perovskite, and proposed that the available slip systems of perovskite are such that they do not give rise to anisotropic textures under deformation. Whether this result remains robust at high temperatures remains unclear [e.g., *Karato et al.*, 1995].

For comparison, *Karato et al.* [1995] proposed, based on high temperature and low pressure (0.3 GPa) deformation results on an analogue material (CaTiO3), that deformation of the lower mantle could occur within a superplastic regime, in which grain boundary sliding provides the dominant means of deformation. In general, such superplastic deformation does not produce anisotropic textures. In a number of materials (including olivine, ice and silicon), increasing temperature while holding shear stress constant can yield a rheologic transition from such plastic deformation into a power-law deformation regime [e.g., *Frost and Ashby*, 1982]. Such a transition would be expected to yield a transition from largely isotropic textures to anisotropic textures: as present estimates of the temperature jump across D" are typically in excess of 800 K [*Williams and Jeanloz*, 1990; *Boehler*, 1993; *Williams*, 1998], homologous temperature changes of the order of 0.3 are expected across D". Such large shifts in homologous temperature lie

in a range which could readily drive the behavior of the lowermost mantle from predominantly plastic flow into a power-law creep regime. This change in deformation mechanism would be expected if (1) the boundary between the plastic and power-law flow regimes exists under conditions of the deep mantle, and has a positive slope in the shear-stress/homologous temperature plane; and 2) if the lower mantle does actually deform in a superplastic regime [e.g. *Karato et al.*, 1995]. As a result, a shift from plastic to power-law flow could plausibly alter the rheology of the lowermost mantle in a manner that could result in anisotropy being generated within D", but not within the overlying mantle. *Karato* [1997] favors such a possibility for D" anisotropy, with strong elongation of MgO minerals in a high viscosity perovskite matrix. The anisotropy of D" could be intimately associated with the temperature jump across this zone, and be a fundamental manifestation of a different deformational process occurring within D" relative to the overlying lower mantle.

DISCUSSION AND CONCLUSIONS

We do not yet have a full understanding of anisotropy in D", and many fundamental issues need to be resolved. However, it is clear that observations of anisotropy hold potential for constraining processes taking place at the base of the mantle. Distinguishing between various scenarios like those in Figure 6 will be challenging, as each has attendant uncertainties at the heart of the mechanism for generating anisotropy. One source of guidance is provided by relating the observations of anisotropy to large-scale patterns in mantle structure. This is considered in Figure 7, which compares observed anisotropic characteristics (Figure 7a) with the spatial distribution of ULVZ in D", a model of subducted slab that may have penetrated to D", and buoyancy-flux weighted hot spots (Figure 7b), and with spatial patterns in deep mantle shear velocity structure from a high resolution tomographic inversion (Figure 7c). There is a general spatial association of regions with transverse isotropy and areas of subducted slab accumulation and absence of a detectable ULVZ. Transverse isotropy tends to be observed in faster than average lower mantle, while the slow region of the central Pacific exhibits small-scale general anisotropy. Extending these spatial correlations is one important direction for future research.

The most plausible mechanisms for generating anisotropy near the base of the mantle appear to be either the development of sheared lamellae of partial melt or chemical heterogeneities or the development of LPO if there is a transition in predominant deformation mechanism in the hot boundary layer. Multiple mechanisms for generating anisotropy may be operative; for example, it is difficult to reconcile all observations with downwelling slab structures. However, the latter may play an important role if they displace chemical heterogeneities in D" as they sink to

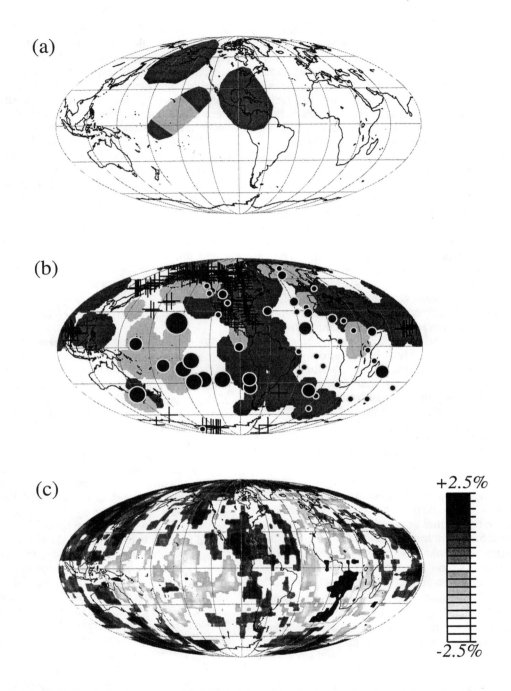

Figure 7. (a) Locations where shear wave has been observed in the D" region from Figure 1. Darker areas are consistent with transverse isotropy while the mid-Pacific lighter area has a mix of general and no anisotropy. (b) Locations where a thin very low velocity region at the base of the mantle (light shaded areas) has been observed or not (dark areas), along with hotspot locations indicated by circles with sizes proportional to estimated buoyancy flux (from *Garnero et al.* [1998]). Also shown are the regions of subducted slab that may have reached the base of the mantle (plus signs) from the model of *Lithgow-Bertelloni and Richards* [1997]. (c) Lateral variation in the shear velocity structure at a depth of 2750 m from the tomographic model reported in *Grand et al.* [1997].

the CMB, with the in situ material riding up horizontally on top of the ponding slab to produce anisotropic structure several hundred kilometers above the CMB [e.g. *Wysession*, 1996]. This could reconcile observations of anisotropy at the CMB in hot areas and well above it in colder areas.

Important research efforts need to be undertaken. Seismological work needs to expand the spatial sampling of D", to establish any systematic relationships between D" anisotropy and large scale structure in D" and the overlying mantle, and to investigate localized regions to establish the variability of anisotropy on small scale lengths. These efforts should include attempts to detect *P* velocity anisotropy, particularly in areas like the mid-Pacific where general anisotropy is suggested by the *S* wave splitting results. Continued characterization of shallow mantle and near source contributions to anisotropic measurements must be sustained as well. Geodynamical topics include boundary layer calculations for materials with strong viscosity contrasts, such as in the case of partial melt, to assess fabric development in the presence of shear flow. With the evidence for a thick thermal boundary layer (200-300 km thick) in the Central Pacific, along with regions of partial melt underlying this zone, dynamical models used to assess entrainment and boundary layer instability need to be reconsidered. Experimental constraints on deformation mechanisms operating under D" pressure and temperature conditions are essential for assessing whether LPO in perovskite and magnesiowustite can actually develop in strong shear flows in a hot boundary layer. It is also important to assess the physical properties of possible partial melt components in D", along with those of slab remnants to assess whether SPO of sufficient strength to explain the seismic data can actually develop or not. The stability of orthorhombic perovskite and the possibility of enhanced stishovite content in the D" layer must be further explored. All of these topics are at the leading edge of research in each discipline, but it appears that by understanding anisotropy in D" we will significantly advance our understanding of deep Earth processes.

Acknowledgments. GMT mapping software of *Wessel and Smith* [1991] was used in figures. C. Lithgow-Bertelloni and S. Grand provided slab and seismic tomography models, respectively. We thank J. Pulliam for providing some of his waveform data in advance of publication. B. Romanowicz, J. Ritsema, D. Helmberger and an anonymous reviewer provided comments on the manuscript. This research was supported by NSF grants EAR 9305894, EAR 9418643 (T.L.) and 9896047 (E.J.G.). Workshops on the core-mantle boundary organized under EAR 9305894 helped develop interdisciplinary coordination. Contribution 337 of the Institute of Tectonics and the W. M. Keck Seismological Laboratory.

REFERENCES

Anderson, D. L., Thermally induced phase changes, lateral heterogeneity of the mantle, continental roots, and deep slab anomalies, *J. Geophys. Res., 92*, 13968-13980, 1987.

Backus, G. E., Long-wave elastic anisotropy produced by horizontal layering, *J. Geophys. Res., 67*, 4427-4440, 1962.

Boehler, R., Temperatures in the Earth's core from melting-point measurements of iron at high static pressures, *Nature, 363*, 534-536, 1993.

Christensen, U. R., and A. W. Hofmann, Segregation of subducted oceanic crust in the convecting mantle, *J. Geophys. Res., 99*, 19867-19884, 1994.

Cohen, R. E., First principles predictions of elasticity and phase transitions in high pressure SiO_2 and geophysical implications, in *High Pressure Research: Application to Earth and Planetary Sciences*, Eds. Y. Syono and M. H. Manghnani, pp. 425-431, American Geophysical Union, Washington, D.C., 1992.

Cohen, R. E., I. I. Mazin, and D. G. Isaak, Magnetic collapse in transition metal ions at high pressure: Implications for the Earth, *Science, 275*, 654-657, 1997.

Cormier, V. F., Synthesis of body waves in transversely isotropic Earth models, *Bull. Seism. Soc. Am., 76*, 231-240, 1986.

Cormier, V. F., Anisotropically distributed heterogeneity in D", *EOS Trans. AGU, Fall Meeting Suppl., 76 (46)*, F402, 1995.

Davies, G. F., and M. Gurnis, Interaction of mantle dregs with convection: Lateral heterogeneity at the core-mantle boundary, *Geophys. Res. Lett., 13*, 1517-1520, 1986.

Ding, X., and D. V. Helmberger, Modeling D" structure beneath Central America with broadband seismic data, *Phys. Earth Planet. Int., 101*, 245-270, 1997.

Doornbos, D. J., and J. C. Mondt, P and S waves diffracted around the core and the velocity structure at the base of the mantle, *Geophys. J. Roy. Astron. Soc., 57*, 381-395, 1979.

Doornbos, D. J., S. Spiliopoulos and F. D. Stacey, Seismological properties of D" and the structure of a thermal boundary layer, *Phys. Earth Planet. Inter., 41*, 225-239, 1986.

Duffy, T. S., R. J. Hemley, and H. K. Mao, Equation of state and shear strength at multimegabar pressures: Magnesium oxide to 227 Gpa, *Phys. Rev. Lett., 74*, 1371-1374, 1995.

Dziewonski, A. M., and D. L. Anderson, Preliminary reference Earth model, *Phys. Earth Planet. Inter., 25*, 297-356, 1981.

Dziewonski, A. M., and D. L. Anderson, Travel times and station corrections for *P* waves at teleseismic distances, *J. Geophys. Res., 88*, 3296-3314, 1983.

Frost, H. J., and M. F. Ashby, *Deformation Mechanism Maps*, Pergamon Press, Oxford, 1982.

Garnero, E., and D. V. Helmberger, Travel times of *S* and *SKS*: Implications for three-dimensional lower mantle structure beneath the central Pacific, *J. Geophys. Res., 98*, 8225-8241, 1993.

Garnero, E. J., and D. V. Helmberger, A very slow basal layer underlying large-scale low-velocity anomalies in the lower mantle beneath the Pacific: evidence from core phases, *Phys. Earth Planet. Inter., 91*, 161-176, 1995.

Garnero, E. J., and T. Lay, Lateral variations in lowermost mantle shear wave anisotropy beneath the north Pacific and Alaska, *J. Geophys. Res., 102*, 8121-8135, 1997.

Garnero, E. J., and T. Lay, Effects of D" anisotropy on seismic velocity models of the outermost core, *Geophys. Res. Lett.*, submitted, 1998.

Garnero, E. J., D. V. Helmberger, and S. Grand, Preliminary evidence for a lower mantle shear wave velocity discontinuity beneath the central Pacific, *Phys. Earth Planet. Inter., 79*, 335-347, 1993.

Garnero, E. J., J. Revenaugh, Q. Williams, T. Lay, and L. Kellogg, Ultra-low velocity zone at the core-mantle boundary, *This volume*, 1998.

Goarant, F., F. Guyot, J. Peyronneau, and J.-P. Poirier, High-pressure and high-temperature reactions between silicates and liquid iron alloys, in the diamond anvil cell, studied by analytical electron microscopy, *J. Geophys. Res., 97*, 4477-4487, 1992.

Grand, S. P., R. D. van der Hilst, S. Widiyantoro, Global seismic tomography: A snapshot of convection in the Earth, *GSA Today, 7*, 1-7, 1997.

Grechka, V. Y., and G. A. McMechan, Anisotropy and nonlinear polarization of body waves in exponentially heterogeneous media, *Geophys. J. Int., 123*, 959-965, 1995.

Gurnis, M., Stirring and mixing in the mantle by plate-scale flow; large persistent blobs and long tendrils coexist, *Geophys. Res. Lett., 13*, 1474-1477, 1986.

Haddon, R. A. W., Evidence for inhomogeneities near the core-mantle boundary, *Philos. Trans. R. Soc. London A, 306*, 61-70, 1982.

Hansen, U., and D. A. Yuen, Numerical simulations of thermal-chemical instabilities at the core-mantle boundary, *Nature, 334*, 237-240, 1988.

Irifune, T., Absence of an aluminous phase in the upper part of the Earth's lower mantle, *Nature, 370*, 131-133, 1994.

Irifune, T., and A. E. Ringwood, Phase transformations in primitive MORB and pyrolite compositions to 25 GPa and some geophysical implications, in *High Pressure Research in Mineral Physics,* Eds. Y. Syono and M.H. Manghnani, pp. 231-242, American Geophysical Union, Washington, D.C., 1987.

Ito, E., K. Morooka, O. Ujike and T. Katsura, Reactions between molten iron and silicate melts at high pressure: Implications for the chemical evolution of Earth's core, *J. Geophys. Res., 100*, 5901-5910, 1995.

Jackson, I., and H. Niesler, The elasticity of periclase to 3 GPa and some geophysical implications, in *High Pressure Research in Geophysics,* Eds. S. Akimoto and M. H. Manghnani, pp. 93-113, Center for Academic Publications, Tokyo, 1982.

Kaneshima, S., and P. G. Silver, A search for source-side anisotropy, *Geophys. Res. Lett., 19*, 1049-1052, 1992.

Karato, S.-I., Seismic anisotropy: mechanisms and tectonic implications, In: *Rheology of Solids and of the Earth*, edited by S. Karato and M. Toriumi, Oxford University Press, Oxford, pp. 393-422, 1989.

Karato, S.-I., Seismic anisotropy in the deep mantle, boundary layers and the geometry of mantle convection, *PAGEOPH*, in press, 1997.

Karato, S.-I., S. Zhang, and H.-R. Wenk, Superplasticity in Earth's lower mantle: Evidence from seismic anisotropy and rock physics, *Science, 270*, 458-461, 1995.

Karki, B. B., L. Stixrude, S. J. Clark, M. C. Warren, G. J. Ackland and J. Crain, Elastic properties of orthorhombic MgSiO3 perovskite at lower mantle pressures, *Am. Mineral., 82*, 635-638, 1997.

Kawasaki, I., Azimuthally anisotropic model of the oceanic upper mantle, *Phys. Earth Planet. Inter., 43*, 1-21, 1986.

Kellogg, L. H., and D. L. Turcotte, Mixing and the distribution of heterogeneities in a chaotically convecting mantle, *J. Geophys. Res., 95*, 421-432, 1990.

Kellogg, L. H., and S. D. King, Effect of mantle plumes on the growth of D" by reaction between the core and mantle, *Geophys. Res. Lett., 20*, 379-382, 1993.

Kendall, J.-M., and C. Nangini, Lateral variations in D" below the Caribbean, *Geophys. Res. Lett., 23*, 399-402, 1996.

Kendall, J.-M., and P. G. Silver, Constraints from seismic anisotropy on the nature of the lowermost mantle, *Nature, 381*, 409-412, 1996.

Kendall, J.-M., and P. G. Silver, Investigating causes of D" anisotropy, *This volume*, 1998.

Kendall, J.-M., and C. J. Thomson, Seismic modeling of subduction zones with inhomogeneity and anisotropy - 1. Teleseismic *P*-wavefront tracking, *Geophys. J. Int., 112*, 39-66, 1993.

Kesson, S., J. D. FitzGerald and J. M. G. Shelley, Mineral chemistry and density of subducted basaltic crust at lower-mantle pressures, *Nature, 372*, 767-769, 1994.

Kind, R., and G. Müller, Computations of *SV* waves in realistic Earth models, *J. Geophys., 41*, 142-172, 1975.

Kind, R., and G. Müller, The structure of the outer core from SKS amplitudes, *Bull. Seism. Soc. Am., 67*, 1541-1554, 1977.

Knittle, E. and R. Jeanloz, Synthesis and equation of state of (Mg,Fe)SiO3 perovskite to over 100 GPa, *Science, 235*, 668-670, 1987.

Knittle, E., and R. Jeanloz, Simulating the core-mantle boundary: An experimental study of high-pressure reactions between silicates and liquid iron, *Geophys. Res. Let., 16*, 609-612, 1989.

Knittle, E., and R. Jeanloz, Earth's core-mantle boundary: Results of experiments at high pressures and temperatures, Sciences, 251, 1438-1443, 1991.

Lay, T., Structure of the core-mantle transition zone: A chemical and thermal boundary layer, *EOS Trans. AGU, 70*, 49, 54-55, 58-59, 1989.

Lay, T., Seismology of the lower mantle and core-mantle boundary, *Rev. of Geophys., Suppl.*, 325-328, 1995.

Lay, T., and D. V. Helmberger, A lower mantle *S*-wave triplication and the shear velocity structure of D", *Geophys. J. R. Astron. Soc., 75*, 799-838, 1983a.

Lay, T., and D. V. Helmberger, The shear-wave velocity gradient at the base of the mantle, *J. Geophys. Res., 88*, 8160-8170, 1983b.

Lay, T., and C. J. Young, The stably stratified outermost

core revisited, *Geophys. Res. Lett.*, *17*, 2001-2004, 1990.

Lay, T., and C. J. Young, Analysis of seismic SV waves in the core's penumbra, *Geophys. Res. Lett.*, *18*, 1373-1376, 1991.

Lay, T., Q. Williams, and E. J. Garnero, The core-mantle boundary layer and deep earth dynamics, *Nature*, in press, 1998.

Lay, T., E. J. Garnero, C. J. Young, and J. B. Gaherty, Scale lengths of shear velocity heterogeneity at the base of the mantle from S wave differential travel times, *J. Geophys. Res.*, *102*, 9887-9909, 1997.

Li, X. D., and B. Romanowicz, Global mantle shear velocity model developed using nonlinear asymptotic coupling theory, *J. Geophys. Res.*, *101*, 22245-22272, 1996.

Lithgow-Bertelloni, C., and M. Richards, The dynamics of Cenozoic and Mesozoic plate motions, *Rev. of Geophys.*, in press, 1997.

Loper, D. E., and T. Lay, The core-mantle boundary region, *J. Geophys. Res.*, *100*, 6397-6420, 1995.

Mao, H. K., R. J. Hemley, Y. Fei, J. F. Shu, L. C. Chen, A. P. Jephcoat, Y. Wu, and W. A. Bassett, Effect of pressure, temperature, and composition on lattice parameters and density of $(Fe, Mg)SiO_3$-Perovskites to 30 GPa, *J. Geophys. Res.*, *96*, 8069-8079, 1991.

Matzel, E., M. K. Sen, and S. P. Grand, Evidence for anisotropy in the deep mantle beneath Alaska, *Geophys. Res. Lett.*, *23*, 2417-2420, 1996.

Maupin, V., On the possibility of anisotropy in the D" layer as inferred from the polarization of diffracted S waves, *Phys. Earth Planet. Inter.*, *87*, 1-32, 1994.

Meade, C., H. K. Mao, and J. Z. Hu, High-temperature phase transition and dissociation of $(Mg,Fe)SiO_3$ perovskite at lower mantle pressures, *Science*, *268*, 1743-1745, 1995a.

Meade, C., P. G. Silver, and S. Kaneshima, Laboratory and seismological observations of lower mantle isotropy, *Geophys. Res. Lett.*, *22*, 1293-1296, 1995b.

Mitchell, B. J., and D. V. Helmberger, Shear velocities at the base of the mantle from observations of S and ScS, *J. Geophys. Res.*, *78*, 6009-6020, 1973.

Morris, E. M., R. W. Raitt and G. G. Shor, Velocity anisotropy and delay time maps of the mantle near Hawaii, *J. Geophys. Res.*, *74*, 4300-4316, 1969.

Nataf, H.-C., I. Nakanishi, and D. L. Anderson, Measurement of mantle wave velocities and inversion for lateral heterogeneities and anisotropy, 3. Inversion, *J. Geophys. Res.*, *91*, 7261-7307, 1986.

Nishimura, C. E. and D. W. Forsyth, Rayleigh wave phase velocities in the Pacific with implications for azimuthal anisotropy and lateral heterogeneity, *Geophys. J. Roy. Astron. Soc.*, *94*, 479-501, 1988.

Olson, P., and C. Kincaid, Experiment on the interaction of thermal convection and compositional layering at the base of the mantle, *J. Geophys. Res.*, *96*, 4347-4354, 1991.

Olson, P., G. Schubert, and C. Anderson, Plume formation in the D" layer and the roughness of the core-mantle boundary, *Nature*, *327*, 409-413, 1987.

Pulliam, J., and M. K. Sen, Seismic anisotropy in the core-mantle transition zone, *Geophys. J. Int.*, in press, 1997.

Ritsema, J., E. Garnero, and T. Lay, A strongly negative shear velocity gradient and lateral variability in the lowermost mantle beneath the Pacific, *J. Geophys. Res.*, *102*, 20,395-20,411, 1997.

Ritsema, J., T. Lay, E. J. Garnero, and H. Benz, Seismic anisotropy in the lowermost mantle beneath the Pacific, *Geophys. Res. Lett.*, in press, 1998.

Russell, S., T. Lay, and E. J. Garnero, Small-scale boundary layer structure in the lowermost mantle: Evidence for in-flow at the root of the Hawaiian plume, *Nature*, in press, 1998.

Saxena, S.K. et al., Stability of perovskite $(MgSiO_3)$ in the Earth's mantle, *Science*, *274*, 1357-1359, 1996.

Schweitzer, J., and G. Müller, Anomalous difference trav-eltimes and amplitude ratios of SKS and SKKS from Tonga-Fiji events, *Geophys. Res. Lett.*, *13*, 1529-1532, 1986.

Silver, P. G., Seismic anisotropy beneath the continents: Probing the depths of geology, *Annu. Rev. Earth Planet. Sci.*, *24*, 385-432, 1996.

Sleep, N. H., Gradual entrainment of a chemical layer at the base of the mantle by overlying convection, *Geophys. J.*, *95*, 437-447, 1988.

Song, Y., and T.J. Ahrens, Pressure-temperature range of re-actions between liquid iron in the outer core and mantle silicates, *Geophys. Res. Lett.*, *21*, 153-156, 1994.

Stacey, F. D., and D. E. Loper, The thermal boundary layer interpretation of D" and its role as a plume source, *Phys. Earth Planet. Inter.*, *33*, 45-55, 1983.

Stixrude, L., R. E. Cohen, R. Yu and H. Krakauer, *Am. Mineral.*, *81*, 1293-1296, 1996.

Su, W-J., R. Woodward, and A.M. Dziewonski, Degree 12 model of shear velocity heterogeneity in the mantle, *J. Geophys. Res.*, *99*, 6945-6980, 1994.

Tackley, P. J., On the penetration of an endothermic phase transition by upwellings and downwellings, *J. Geophys. Res.*, *100*, 15,477-15,488, 1995.

Tanimoto, T., and D. L. Anderson, Lateral heterogeneity and azimuthal anisotropy of the upper mantle: Love and Rayleigh waves 100-250 s, *J. Geophys. Res.*, *90*, 1842-1858, 1985.

Valenzuela, R., and M. E. Wysession, The base of the mantle as illuminated by core-diffracted waves, *This volume*, 1998.

Vassiliou, M. S. and T. J. Ahrens, The equation of state of $Mg_{0.6}Fe_{0.4}O$ to 200 GPa, *Geophys. Res. Lett.*, *9*, 127-130, 1982.

Vidale, J. E., and H. M. Benz, Seismological mapping of fine structure near the base of Earth's mantle, *Nature*, *361*, 529-532, 1993.

Vinnik, L., L. Breger, and B. Romanowicz, Anisotropic structures at the base of the oceanic mantle, *Nature*, in press, 1998.

Vinnik, L. P., V. Farra, and B. Romanowicz, Observational evidence for diffracted SV in the shadow of the Earth's core, *Geophys. Res. Lett.*, *16*, 519-522, 1989.

Vinnik, L., B. Romanowicz, Y. Le Stunff, and L. Makeyeva, Seismic anisotropy in the D" layer, *Geophys. Res. Lett.*, *22*, 1657-1660, 1995.

Weber, M., Lamellae in D"? An alternative model for lower

mantle anomalies, *Geophys. Res. Lett., 21*, 2531-2534, 1994.

Weidner, D. J., J. D. Bass, A. E. Ringwood and W. Sinclair, The single-crystal elastic moduli of stishovite, *J. Geophys. Res., 87*, 4740-4746, 1982.

Wessel, P., and W. H. F. Smith, Free software helps map and display data, *EOS, 72*, 441, 445-446, 1991.

Williams, Q.,The temperture contrast across D", *This volume*, 1998.

Williams, Q., and E. J. Garnero, On the possible origin of a seismically thin boundary layer at the base of the mantle, *Science, 273*, 1528-1530, 1996.

Williams, Q., and R. Jeanloz, Melting relations in the iron-sulfur system at high pressures: Implications for the thermal state of the Earth, *J. Geophys. Res., 95*, 19299-19310, 1990.

Winchester, J. P., and K. C. Creager, Azimuthal anisotropy and abrupt transitions from slow to fast anomalies in the velocity structure of D", *Proc. of 1997 IRIS Meeting*, abstract.

Wysession, M. E., Continents of the core, *Nature, 381*, 373-374, 1996.

Wysession, M. E., T. Lay, E. J. Garnero, J. Revenaugh, Q. Williams, D. V. Helmberger, R. Jeanloz, and L. Kellogg, Large-scale structure of D": Lateral gradients and discontinuity variations, *This volume*, 1998.

Yeganeh-Haeri, Y., Synthesis and re-investigation of the elastic properties of single-crystal magnesium silicate perovskite, *Phys. Earth Planet. Inter., 87*, 111-121, 1994.

Young, C. J., and T. Lay, The core-mantle boundary, *Annu. Rev. Earth Planet. Sci., 15*, 25-46, 1987.

Young, C. J., and T. Lay, The core shadow zone boundary and lateral variations of the P velocity structure of the lowermost mantle, *Phys. Earth Planet. Int., 54*, 64-81, 1989.

Young, C. J., and T. Lay, Multiple phase analysis of the shear velocity structure in the D" region beneath Alaska, *J. Geophys. Res., 95*, 17385-17402, 1990.

Edward J. Garnero, Department of Geology and Geophysics, University of California, Berkeley, CA 94720.

Louise Kellogg, Department of Geology, University of California, Davis, CA 95616.

Thorne Lay, Earth Sciences Department, University of California, Santa Cruz, CA 96064.

Quentin Williams, Earth Sciences Department, University of California, Santa Cruz, CA 95064.

Michael E. Wysession, Department of Earth and Planetary Sciences, Washington University, St. Louis, MO, 63130.

Ultralow Velocity Zone at the Core-Mantle Boundary

Edward J. Garnero

Berkeley Seismological Laboratory, University of California, Berkeley, CA

Justin Revenaugh, Quentin Williams, and Thorne Lay

Institute of Tectonics and Earth Sciences Department, University of California, Santa Cruz, CA

Louise H. Kellogg

Department of Geology, University of California, Davis, CA

Accumulating seismic evidence for a laterally varying thin ultralow velocity zone (ULVZ) at the base of the mantle is summarized, and shown to have far reaching implications. Anomalous *SPdKS* delays and amplitudes, and precursory energy to *PcP*, *ScP*, and *PKP* require seismic velocity reductions of at least 10% in some regions of the boundary layer. Estimates of ULVZ thickness depend on poorly constrained parameters such as the shear velocity and density, but it is likely less than 20-40 kilometers, where present. The most probable cause of the ULVZ is partial melting of the lowermost mantle, with chemical heterogeneity and dynamic effects playing an important role as well. Strong spatial correlations between regions with detected ULVZ zones and surface positions of buoyancy flux-weighted hot spots, as well as between regions lacking a detectable ULVZ and areas of accumulated subducted material, suggest a direct relationship between the top and bottom boundary layers of the mantle, strongly supporting the idea of circulation between the upper and lower mantle.

1. INTRODUCTION

Over the past several decades, deep Earth research has yielded first-order information on properties of the core-mantle boundary (CMB) region, such as large-scale lateral variations and strong radial gradients and discontinuities in mantle seismic velocities, and inferred large-scale flow properties (see *Loper and Lay* [1995] for a review). More recently, details of structure in the lowermost 200-300 kilometers of the mantle (the D″ region) have emerged, revealing the presence of seismic shear wave anisotropy, coherent small scale scattering, thin low-velocity basal layering, and probable ongoing

chemical reactions between the solid silicate mantle and liquid iron-alloy core; the emerging picture of the lowermost mantle is as complex as for the Earth's lithosphere, with both regions being dynamic thermal and chemical boundary layers [see references contained in *Loper and Lay*, 1995; *Wysession*, 1996a; *Wysession et al.*, 1997; *Lay et al.*, 1998].

Resolving structure of the Earth's boundary layers at the top and bottom of the mantle is crucial to understanding large scale mantle flow, since the dynamics of the boundary layers is expected to be closely coupled to upwelling and downwelling motions of the convective system. Determining the configuration of seismic velocity layering, heterogeneity and anisotropy provides direct information on the thermal and chemical processes in both boundary layers.

Recently, evidence for a laterally varying thin layer at the base of the mantle with large velocity reductions has been discovered using different methods [e.g., *Mori and*

The Core-Mantle Boundary Region
Geodynamics 28

Copyright 1998 by the American Geophysical Union.

Helmberger, 1995; *Garnero and Helmberger*, 1996; *Vidale and Hedlin*, 1998]. This ultralow velocity zone (ULVZ) has been imaged as having variable thickness (5-40 kilometers) and *P* wave velocity reductions of at least 10%. Explanation of the ULVZ as a result of partial melt [*Williams and Garnero*, 1996; *Holland and Ahrens*, 1997; *Revenaugh and Meyer*, 1997; *Vidale and Hedlin*, 1998; *Wen and Helmberger*, 1998a] implies approximately a 3:1 reduction in *S* wave velocities (V_S) compared to *P* wave velocities (V_P) in the layer (thus at least a 30% shear velocity reduction). Other possible explanations for the ULVZ, such as chemical heterogeneity, have different implications for shear velocity reductions. Better understanding of this layer of ultralow velocities in relation to lower mantle mineralogy (e.g., composition and state), mantle dynamics (e.g., small and large scale circulation, including plumes and downwellings), core dynamics (possible influences on magnetic reversals and the geodynamo) and other seismic phenomena (e.g., scattering, anisotropy) are the focus of this paper. In what follows, we first review relevant studies of lower mantle *P* waves, since ULVZ work to date has primarily imaged *P* wave velocity structure, followed by a detailed discussion of present seismic methods used for ULVZ imaging (and their uncertainties). Finally, mineral physics and geodynamics considerations demonstrate far reaching implications. Diverse observables at the Earth's surface such as hot spots and subduction zones, and observed magnetic field reversal paths, show strong spatial correlations with the ULVZ distribution.

2. LOWER MANTLE *P* VELOCITY STRUCTURE

Patterns of *P* wave velocity heterogeneity in the lowermost few hundred kilometers of the mantle have been imaged globally at scale lengths of 3000-5000 kilometers and greater [e.g., *Dziewonski*, 1984; *Inoue et al.*, 1990; *Pulliam et al.*, 1993; *Wysession*, 1996b; *Sylvander et al.*, 1997], as well as locally at intermediate to short wavelength ($\approx$ 3000 kilometers to less than 100 kilometers) [e.g., *Young and Lay*, 1989; *Wysession and Okal*, 1989; *Weber and Körnig*, 1990; *Weber*, 1993; *Souriau and Poupinet*, 1994; *Krüger et al.*, 1995; *Wysession et al.*, 1995; *Sylvander and Souriau*, 1996; *Scherbaum et al.*, 1997]. Many studies use seismic waves and methods that average the radial V_P structure over a minimum of several hundred kilometers, i.e., throughout D″ (e.g., travel time analyses of *PcP*, *PKP*, and diffracted *P* waves, *Pd*) and thus do not resolve the velocity structure in the lowermost few tens of kilometers of the mantle where the ULVZ has been detected. Thus any strong velocity variations in an unresolved thin transition zone structure may have been erroneously mapped into models of V_P heterogeneity or velocity gradients in the lowermost few hundred kilometers of the mantle. For

example, *Song and Helmberger* [1995] argued for moderately reduced V_P in the lowermost 200-300 kilometers of the mantle to explain observed separations of the DF and AB branches of *PKP* phases at large distances. Alternatively, the data may be explained by stronger velocity reductions in the lowermost 5-40 kilometers of the mantle.

Some studies have noted that the presence of a thin heterogeneous mantle basal layer helps in explaining various seismic observations, even if the thin zone is not uniquely resolved by the data under consideration. For example, in an analysis of *PcP*, *PKP* and *PKKP* arrival times (as recorded by the International Seismological Centre), *Doornbos and Hilton* [1989] considered a model parameterization with a 20 kilometers thick layer at the CMB. While noting a strong trade-off between the velocity reduction and thickness of the layer, they preferred a layer having a 20% reduction in V_P. More recently, it has been shown that a ULVZ basal layer can explain anomalous *PKP* AB-BC separations [see *Sylvander and Souriau*, 1996]. Also, perturbations in *Pd* times have been attributed to a basal layer with ±10% V_P heterogeneity [*Sylvander et al.*, 1997]. As we discuss below, the presence of a thin transition zone can cause detectable waveform (and associated travel time) perturbations in particular long period and broadband seismic arrivals, as well as producing additional precursory arrivals in short period *P* wave recordings [e.g., *Wysession*, 1996b; *Scherbaum et al.*, 1997; *Revenaugh and Meyer*, 1997]. Thus it is possible to directly establish the presence and parameters of a thin basal layer rather than inferring it from a parameterized tomographic inversion.

3. SEISMIC IMAGING OF ULTRALOW VELOCITY ZONES

3.1 Previous Work

Detailed waveform modeling studies have revealed evidence for a laterally varying low velocity layer right above the CMB. A map of lowermost mantle regions that have been previously investigated for this ULVZ structure is displayed in Figure 1. Fresnel zones are shown, and cover roughly 44% of the CMB surface area. Regions having a ULVZ are denoted by light shading, and those showing no evidence for a ULVZ have dark shading (recognizing that the ULVZ may be present, but too thin to detect in these areas). The Fresnel zones provide an estimate of the region which kinematically can contribute to the first quarter wavelength of the seismic arrivals used to map the ULVZ. The actual wavefield involves lateral averaging over these dimensions, however, it is important to keep in mind that variations at scale lengths smaller than the Fresnel zone are not precluded or constrainable, and in fact may significantly

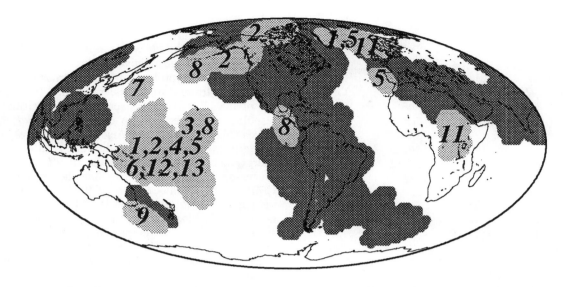

Figure 1. Shaded areas correspond to CMB regions investigated for ULVZ structure. CMB Fresnel zones are shown. Lighter shading corresponds to regions where evidence for ULVZ structure has been found, the numbers correspond to specific studies, as listed in Table 1. The darker shading indicates CMB regions lacking evidence for ULVZ structure, and is based on data in studies 2, 5, and 9 in Table 1 (see text for more details).

contribute to any observed anomaly. The large ULVZ patch in the southwest Pacific (Figure 1), for example, may contain lateral variations within it, and any single one-dimensional model for the region applies only in an average sense. Nonetheless, the Fresnel zone representation provides a first-order characterization of spatial extent of detectable ULVZ at a lateral scale that allows comparisons with other geophysical observations.

The ULVZ regions are numbered corresponding to the studies listed in Table 1. The table also lists the seismic phases and data type used. Since seismology is the most direct means for investigating this structure, we discuss the different seismic phases that constrain the ULVZ characteristics.

3.2 SPdKS Data

The seismic phase *SPdKS* is an *SKS* wave with additional segments of mantle-side CMB *P* wave diffraction (*Pd*) at the *SKS* core entry and exit locations (Figure 2a). The nomenclature "*SPdKS*" is generically used to represent a composite of seismic energy having contributions from *(i)* an *SKS* wave with a segment of *Pd* on the source-side of the wavepath (i.e., *SPdKS*, Figure 2b); *(ii)* an *SKS* wave with a segment of *Pd* on the receiver-side of the wavepath (i.e., *SKPdS*); *(iii)* various combinations of *(i)* and *(ii)*, see *Choy* [1975] and *Garnero et al.* [1993]; and *(iv)* if a ULVZ is present, energy from an *SV*-to-*P* mode-conversion at the CMB followed by an internally reflected *P* wave ("*SpKS*", Figure 2b, see also *Helmberger et al.* [1996]). Unless

specifically indicated otherwise, the term "*SPdKS*" refers to the composite energy arrival. We note that our synthetic modeling of *SPdKS* incorporates all of the separate contributions to the arrival.

SPdKS wavepaths through the lower mantle are very close to *SKS*, typically separated by less than 200 kilometers at the CMB [*Garnero et al.*, 1993]. Thus, referencing *SPdKS* to *SKS* in time and amplitude eliminates source dependence and provides sensitivity to very localized CMB regions. This is especially true when using *SPdKS* data with short *Pd* segments, i.e., at distances close to the inception of *SPdKS* (near 108°, a model dependent distance). It is noteworthy that anomalously low *P* velocities can affect amplitude ratios involving *SKS* [*Silver and Bina*, 1993] since the amount of *SKS* energy transformed into *SPdKS* arrivals is a model dependent phenomena [*Helmberger et al.*, 1996].

An uncertainty in *SPdKS* modeling is that the composite energy arriving as *SPdKS* has *Pd* contributions from possibly differing CMB structures on the source and receiver sides of the wavepath. In this case, however, the energy contributing to *SPdKS* from the more anomalous of the two ULVZ regions will dominate the arrival, due to more effective SV-to-P mode conversions in the ULVZ (e.g., *SpKS*) and better trapping of *Pd* (of *SPdKS*) energy in the boundary layer. Additional information, such as criss-crossing geometries, is needed to resolve the uncertainty of which side of the wavepath contains the ULVZ anomaly (as demonstrated by *Garnero and Helmberger* [1996] for the case of the central Pacific ULVZ).

Table 1. Past ULVZ studies

#	Study	Seismic Phase	Data Type[a]
1	*Garnero et al.* [1993]	SPdKS	LP
2	*Garnero, Helmberger* [1995]	SPdKS	LP
3	*Mori, Helmberger* [1995]	PcP	SPA
4	*Vidale et al.* [1995]	ScP	SPA,BB
5	*Garnero, Helmberger* [1996]	SPdKS	LP
6	*Helmberger et al.* [1996]	SPdKS	LP,SPW
7	*Fischer et al.* [1996]	SPdKS	BB
8	*Revenaugh, Meyer* [1997]	PcP	SPA
9	*Williams et al.* [1998]	SPdKS,PcP	LP,SPA
10	*Helmberger et al.* [1998]	SPdKS	LP
11	*Wen, Helmberger* [1998b]	SPdKS	LP
12	*Wen, Helmberger* [1998a]	PKP prec.	BB
13	*Vidale, Hedlin* [1998]	PKP prec.	SPA,BB

[a] LP, Long period WWSSN data; SPW, Short period WWSSN data; SPA, short period array data; BB broadband data

The threshold for detection of ULVZ structure using *SPdKS* data is dependent on both the wavelength of seismic data and ULVZ properties. For a ULVZ with V_P and V_S reductions of 10% and 30%, respectively, layer thicknesses greater than 3-5 kilometers are detectable with long period World Wide Standardized Seismographic Network (WWSSN) data. Smaller ULVZ thicknesses are detectable with broadband data and/or larger velocity reductions in the layer. Thus the areas with dark shading in Figure 1, many of which are sampled by WWSSN data, may actually have an undetectably thin ULVZ (< 3-5 kilometers).

Data that sample anomalous ULVZ structure beneath the southwest Pacific and Iceland are displayed in Figure 3. The first arrival is *SKS*, and the secondary pulse, appearing as a shoulder on *SKS* at smaller distances (e.g., 110°), is *SPdKS*. Predictions for the delay time of the peak of *SPdKS* relative to *SKS* for the Preliminary Reference Earth Model (PREM) [*Dziewonski and Anderson*, 1981] are indicated by a dashed line (as measured from synthetic seismograms generated by the reflectivity method [e.g., see *Fuchs and Müller*, 1971]). Readily apparent are observed *SPdKS* delays of up to 5 s relative to *SKS*. These delays are often accompanied with anomalously large *SPdKS* amplitudes [see *Garnero and Helmberger*, 1998]. Waveforms with particularly large and delayed *SPdKS* are indicated by small filled circles. The observed *SPdKS* peaks arrive well after the PREM prediction (dashed line). A dotted line which fits the average of the anomalous *SPdKS* arrivals is delayed by several seconds relative to PREM, with large variations about this trend indicating the presence of strong small scale lateral variations [*Garnero and Helmberger*, 1996, 1998].

3.3 PcP and ScP Data

ScP and *PcP* waveforms (typically having measurements made relative to direct *P*, see Figure 2a) are also effective probes for ULVZ structure. Any large contrast in lowermost mantle properties (such as at the top of the ULVZ) can reflect significant seismic energy, resulting in a precursor to core-reflected *ScP* and *PcP* waves (Figure 2c and 2d). Also, *ScP* can have precursory energy from an *S*-to-*P* mode conversion in down-going *S* energy encountering the top of the ULVZ (see Figure 2c). Precursors to *ScP* waves sampling the southwest Pacific CMB can be explained by a ULVZ having strong lateral variations [*Vidale et al.*, 1995]. Short-period array data are useful for analysis of *PcP* and *ScP* waveforms when stacking methods are employed [*Mori and Helmberger*, 1995; *Vidale et al.*, 1995; *Revenaugh and Meyer*, 1997]. Also, issues such as sharpness of the discontinuity at the top of the ULVZ can be addressed with short-period data.

Unlike *SPdKS* phase anomalies which can be observed in individual seismograms, short-period precursors to *PcP* reflected from the top of the ULVZ can only be seen in stacks of many records. Figure 4 is a profile of stacked traces of nearly 1500 recordings from the Northern and Southern California Seismic Networks for nearly 100 intermediate and deep earthquakes in the Tonga subduction zone. Prior to stacking, the records were deconvolved using source-specific average *P* waveforms on an event-by-event basis. The stack combines records within small (0.4 s) calculated *PcP-P* delay time bins; the total range of delay times corresponds to a distance range

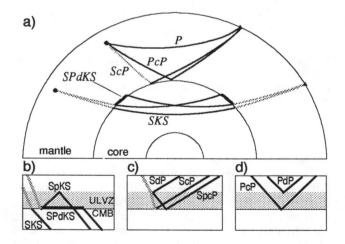

Figure 2. (a) Cross-section of the Earth showing geometrical ray paths for the seismic phases *SKS* and *SPdKS* at 120°, and *PcP*, *ScP*, and *P* at 60°. *S* wave particle motions are indicated by gray lines, *P* wave motion by black lines. Zoom in at the CMB is shown for (b) *SPdKS*, (c) *ScP*, and (d) *PcP*, displaying additional arrivals due to the ULVZ.

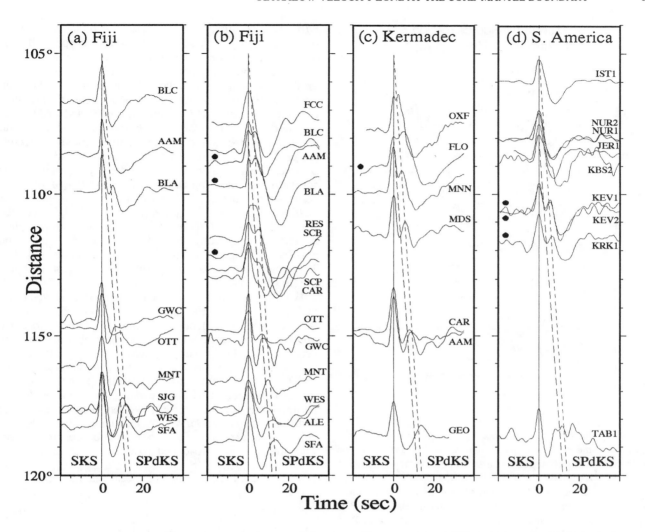

Figure 3. Longitudinal component long-period WWSSN recordings of *SKS* and *SPdKS*, the first and second arrivals, respectively. The first (solid) and second (dashed) lines correspond to *SKS* and *SPdKS* predictions, respectively, of the PREM model. The third small dashed line is drawn to aid in identifying delayed *SPdKS* peaks. Many records show extreme *SPdKS* time and amplitude anomalies, and are indicated by dots on the left of the trace. All times and amplitudes are normalized to SKS.

of ≈78° to 90°. A standard reference Earth model is used to predict the differential times and data are aligned on actual *P* arrivals. The asymmetry of the *PcP* wavelets (relative to the stacked *P* signals) is evidence of a reverse-polarity precursor, referred to as *PdP*.

A curved-wavefront stacking algorithm using 5509 seismograms was computed for events in the Tonga-Fiji subduction complex for target reflectors spaced every two kilometers in depth in the lowermost mantle [*Revenaugh and Meyer*, 1997]. Figure 5a displays the result as computed *PdP/P* stack amplitude ratio versus the corresponding reflector depth relative to the CMB. In this case, a single midpoint bin was used, ignoring

structural variation near the bounce points of *PdP*. Nonetheless, a peak identified as *PcP* is non-zero well above the 95% confidence level, as is a precursory (shallower) trough that *Revenaugh and Meyer* [1997] associate with reverse-polarity reflection from a discontinuity approximately 16 kilometers above the CMB. (Delay of *PcP* due to the presence of a ULVZ, should be manifested as a reflector below the CMB in such a representation.) Comparison with a synthetic data stack illustrates the need for a second arrival. Although the stacked synthetic waveform is slightly asymmetric, the magnitude of the secondary shallow trough is much less than observed. It is not possible to mimic the trough

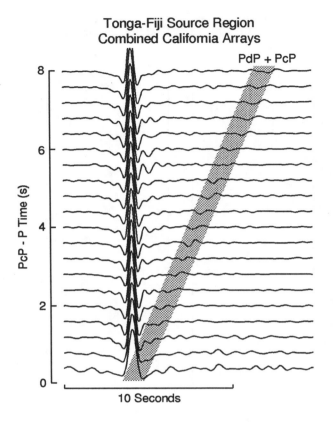

Figure 4. Stacked seismograms of Tonga-Fiji earthquakes recorded by the combined California regional arrays. *P* is the dominant signal. *PcP* appears as a small positive peak following *P* by the delay time shown on the left-hand axis (see shaded area). A small negative polarity precursor to *PcP* is apparent (*PdP*). The asymmetry of the *PcP* waveform relative to *P* is evidence of a low-amplitude, reverse polarity precursor to PcP, interpreted as a reflection from the top of the ULVZ.

simply by summing variably delayed *PcP* phases, i.e., the stack cannot be adequately modeled without a reversed polarity precursor to *PcP*, in accord with *Mori and Helmberger* [1995].

Figure 5b shows the aligned bin stacks of the same source area. Bins measuring 3° on a side were tiled over the region of densest midpoint coverage. The composite stack consists of 5 adjacent bins and 4217 seismograms. Peak amplitude depths range from 27 to 11 kilometers below the CMB in the five bin stacks, signifying a delayed *PcP* arrival and lower-than-average velocities in D″. The peak aligned stack amplitude of *PcP* and *PdP* is increased over that of Figure 5a and there is noticeable compaction of the stack peaks, signifying the reduction in travel time variability obtained by binning.

The mean reflector depth, reflection coefficient and remaining travel time variability of *PcP* and its precursor

PdP were estimated by Monte Carlo methods with synthetic data stacks. Models were accepted if the observed minus predicted stack resulted in a 80% or greater variance reduction, a value chosen with reference to the bootstrap-derived confidence intervals. For the stacks in Figures 5a and 5b, all acceptable models have reversed polarity *PdP* (i.e., opposite of *P* and *PcP*). The amplitude of *PdP* relative to *P* can be used to bound acceptable velocity and density contrasts of the reflector. A grid search was conducted over shear velocity, compressional velocity, and density _perturbations, computing a mean reflection coefficient ($\overline{R}_{PdP}$) for each perturbation triplet _by matching the ray-parameter distribution of data. $\overline{R}_{PdP}$ predictions within 1.5 times the observed range of *PdP* reflection strengths were accepted. A further constraint, that the average *PcP* reflection coefficient be positive and less than 0.4, was included. The upper limit (0.4) is 3 times the largest accepted Monte Carlo value.

We find that both shear and compressional velocities must drop across the layer to produce the observed reversed polarity reflections (Figure 6). Furthermore, we observe a ≈3:1 ratio of shear to compressional velocity decrease in the model solution space. Density is relatively unimportant unless allowed to vary greatly (> 15%), in which case certain trade-offs exist as discussed in the next section. Here we present results for density perturbations less than ±10%. We explored the effects of a diffuse top of the ULVZ by calculating the acceptable velocity perturbations for a 5 kilometer transition zone. The effect of the broadened transition is to increase the

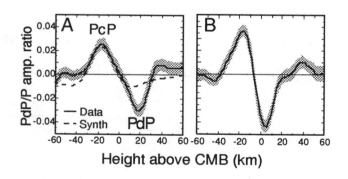

Figure 5. (a) Curved wavefront stack of observed (solid line) and synthetic (dashed line) *PdP-PcP* data for Fiji-Tonga events. Amplitudes normalized to *P*. Gray shading denotes 95% confidence intervals of the data stack. Synthetic stack contains only *P* and *PcP* arrivals and fails to match the precursory downswing to the *PcP* peak centered at -20 kilometers (negative depths imply slow lowermost mantle and a delayed *PcP* arrival, and should not be construed as CMB displacements). (b) As in (a) except data were separately stacked in small bounce point bins and aligned before stacking [see *Revenaugh and Meyer*, 1997].

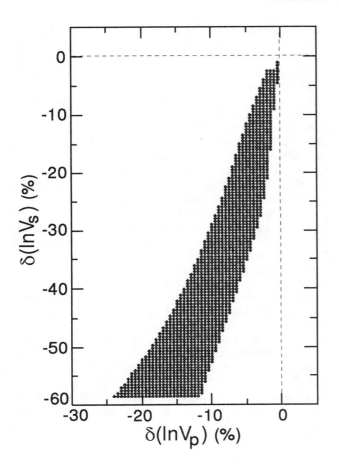

Figure 6. Acceptable velocity variations across an assumed first-order discontinuity responsible for the *PdP* precursor to *PcP* for Fiji-Tonga events. Negative values signify a velocity decrease across the discontinuity. All acceptable models have large V_S decreases. [After *Revenaugh and Meyer*, 1997].

magnitude of the minimum acceptable velocity reductions (i.e., greater velocity reduction) and to increase the range of acceptable models at greater velocity decrease, but the near 3:1 ratio of shear to compressional velocity decrease persists. Similar results were obtained from *PdP* phases of South American and western Pacific earthquakes recorded in California [*Revenaugh and Meyer*, 1997]. Note that the absolute values of the velocity drops are not well constrained by this simple modeling approach, and require a more sophisticated waveform modeling approach such as that taken by *Mori and Helmberger* [1995]. Adopting $\delta V_P = -10\%$ [*Mori and Helmberger*, 1995], consistent with *SPdKS* modeling, suggests shear velocity drops of 25 to 40% when we average our stacks of all regions [see *Revenaugh and Meyer*, 1997]. These numbers are in excellent agreement with predictions for a partial melt layer [*Williams and Garnero*, 1996].

3.4 PKP Data

As with *PcP* data, strong contrasts in seismic properties at the top of the ULVZ (including shear velocity) can affect *PKP* waves, giving rise to precursors [*Vidale and Hedlin*, 1998; *Wen and Helmberger*, 1998a]. In fact, in the southwest Pacific, anomalously large *PKP* precursor amplitudes have been attributed to small scale heterogeneity with 10 to 15% RMS V_P variations over a 60 km layer, having an origin of partial melt [*Vidale and Hedlin*, 1998]. *P* heterogeneity of 8% RMS with an 8 km correlation length was given by *Wen and Helmberger* [1998a], with identical implications. Thus future work analyzing and modeling *PKP* precursors on a global scale holds promise for further ULVZ characterization.

4. ULVZ MODELING TRADE-OFFS

Past modeling efforts have mapped V_P reductions in the ULVZ of 10% [e.g., *Garnero and Helmberger*, 1996]. Subsequently, *Williams and Garnero* [1996] predicted that associated V_S reductions should be on the order of 30%, if the origin of the ULVZ is partial melt. The *PdP-PcP* modeling of *Revenaugh and Meyer* [1997] appears to corroborate such 3:1 δV_S:δV_P reductions. In order to accommodate the effects on *SPdKS-SKS* timing of such a large V_S drop and still explain the observations, *Garnero and Helmberger* [1998] showed that the ULVZ layer thickness must be reduced by approximately one-half of that in *Garnero and Helmberger* [1996], i.e., a ULVZ maximum thickness of 20 kilometers as compared to 40 kilometers. Also, large density (ρ) increases in the ULVZ can retard the peak time of the *SPdKS* arrival, by increasing the amplitude of the internal reflection (*SpKS*, Figure 1b), leading to further reduction of modeled ULVZ layer thickness [*Garnero and Helmberger*, 1998].

Trade-offs in modeling a given *SPdKS* record are schematically illustrated in Figure 7. The shape of the solution space is represented as a plane, but may more appropriately be a volume (possibly with curvature) given uncertainties in the data. This figure can be used to make several important points. First, there exists a minimum ULVZ velocity reduction (and associated thickness) that can explain any given anomaly. This is especially apparent in those data with very short *Pd* segments in *SPdKS* that have several second anomalies, e.g., to model a 4 s *SPdKS* delay accrued over a 100 kilometers *Pd* segment requires a minimum V_P reduction of around 10%. Predictions for models having milder reductions over a thicker layer are unable to explain the anomalous data. As the figure shows, greater velocity reductions can be accommodated for smaller ULVZ thicknesses. Thus the trade-off goes in the direction of thinner boundary layers with more extreme contrasts in properties. If density is allowed to increase, as might be expected for iron enrichment, then ULVZ thickness is reduced. We

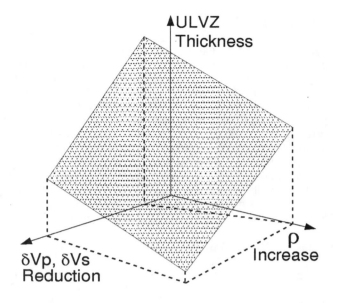

Figure 7. Schematic illustration showing trade-offs in *SPdKS* modeling space. Larger ULVZ velocity reductions can be accommodated if layer thickness is reduced. Increasing density in the ULVZ layer must also be accompanied with a layer thickness reduction (see text for more details).

note that large density increases (e.g., ≥50%) with no associated velocity reductions can produce *SPdKS* anomalies, due to increasing the *SpKS* amplitude. The actual scales on the axes in Figure 7 depend on the record being modeled, as well as on the $\delta V_S : \delta V_P$ ratio assumed. Modeling distance profiles of seismic data, including analyses of different phases sampling the same region, help to reduce these trade-offs. Uncertainties in ULVZ modeling are many, but some bounds can be placed on maximum thickness, minimum velocity reduction, and their variations [*Garnero and Helmberger*, 1998].

SPdKS modeling typically assumes a sharp discontinuity at the top of the ULVZ. While this is supported by observation of short period *PdP* [*Mori and Helmberger*, 1995; *Revenaugh and Meyer*, 1997], the depth interval over which the change in properties occurs from the ULVZ to the overlying mantle certainly requires further analysis.

5. MINERAL PHYSICS CONSIDERATIONS

Constraints on the physical origins of the ULVZ can be derived from the families of solutions for the *P*- and *S*-wave velocities and density within such a layer (e.g., see Figure 7). From the available seismic constraints, it is apparent that the ULVZ is remarkable in at least two major respects. First, the negative sign of the *P*- and *S*-

velocity changes implies the ULVZ is fundamentally different from most other mantle discontinuities, with only the "low velocity zone" and the actual core-mantle boundary having similar characteristics. We do not view it as coincidence that each of these other boundaries has either been proposed to be produced by partial melting or, for the CMB, is due to both melting and a profound change in chemistry. The second notable aspect of the ULVZ is the magnitude of its velocity contrasts. These exceed those present at any other horizontal boundary in the planet aside from the surface and the CMB (Figure 8). The absolute change in V_P is about a factor of two greater than that observed at the seismic discontinuities observed near 400 and 670 kilometers depth. The inferred V_S discontinuity associated with the ULVZ dwarfs those present at the major mantle discontinuities (including the that at the top of D″) by a factor of five or more. As a result, it appears that the ULVZ is the largest seismic anomaly in the mantle, with shifts in seismic velocity more reminiscent of a magma chamber than those generally associated with simple changes in phase or moderate alterations in chemistry. Again, we do not view this similarity between the properties of the ULVZ and those of partially molten systems as coincidental [e.g., *Williams and Garnero*, 1996]. However, because of the considerably different phases and chemical behavior which exist at the ultra-high pressure and temperature conditions of D″ relative to the uppermost mantle conditions, it is important to evaluate different mechanisms that might give rise to such a basal layer.

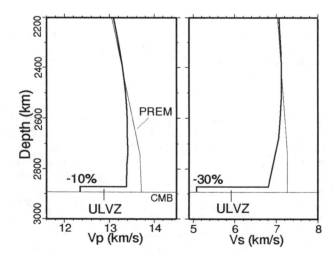

Figure 8. Lower mantle velocity profiles, displaying a ULVZ with 10% V_P and 30% V_S reductions, 2-3% reductions in D″ above the ULVZ is also shown. These profiles are characteristic of the Pacific region, where there are reduced velocities in the lowermost several hundred kilometers of the mantle overlying the ULVZ.

The primary mechanisms for generating a layer at the base of the mantle (or indeed, anywhere in the mantle) are through a change in chemistry and/or a change in phase at depth: in the case of melting, these options obviously need not be exclusive. Several mechanisms exist which could alter the chemistry of the mantle at depth: these include *i)* enrichment in material associated with subduction (whether formerly basaltic crust or its complementary harzburgitic material [*Christensen and Hofmann*, 1994]; *ii)* core-mantle boundary reaction products [*Knittle and Jeanloz*, 1989, 1991]; and *iii)* negatively buoyant descent of melt from the overlying lower mantle [e.g., *Rigden et al.*, 1984]. The first of these options is unlikely to explain the magnitude of ULVZ velocity anomalies without the occurrence of large-scale melting. Harzburgite, which is expected to be enriched in $(Mg,Fe)_2SiO_4$, should have rather similar elastic behavior to the lower mantle; similarly, the primary mineralogic difference between subducted basalt and an approximately peridotitic mantle is in the increased calcium and aluminum (and, to a lesser degree, iron) content of the basalt. Therefore, any material of basaltic chemistry is expected to contain relatively large amounts of $CaSiO_3$-perovskite. From the known elastic behavior of silicate perovskites, it appears that $CaSiO_3$-perovskite is nearly indistinguishable in elastic properties from magnesium silicate perovskite-rich assemblages [e.g., *Mao et al.*, 1989], demonstrating it is extremely unlikely that calcium enrichment could generate the very large difference in seismic properties observed in the ULVZ. The role of aluminum in altering the elastic properties of the silicate perovskites is ill-constrained, but densely packed aluminous phases (such as Al_2O_3) tend to have elastic properties which lie generally close to silicate perovskite [*Gieske and Barsch*, 1968; *Bass*, 1995; *Knittle*, 1995].

Williams and Garnero [1996] have calculated the seismic velocity contrast associated with core-mantle boundary reactions, and found that a maximum V_P depression of about 4% can be produced from the mixture which results from such reactions: far short of that necessary to generate the ULVZ. However, if segregation of solid FeO or FeSi occurs from the pure $MgSiO_3$-perovskite and SiO_2 produced by such core-mantle boundary reactions [e.g. *Manga and Jeanloz*, 1996], a nearly pure FeO or FeSi enriched layer could have V_P anomalies which approach 10%. Using parameters from *Williams and Garnero* [1996], the shear velocity anomaly associated with such a layer would be slightly over 20%. Ignoring dynamical issues associated with invoking a relatively dense ULVZ, we note that the V_S anomaly of such a solid FeO/FeSi layer is anticipated to fall short of that implied by Figure 6. This does not preclude the ULVZ from being significantly iron-enriched, potentially through CMB reactions: as *Ito et al.* [1995] have demonstrated, chemical reactions between liquid silicates and iron occur quite readily.

To explain the ULVZ utilizing solid->solid phase transitions requires that dramatic decreases in shear modulus occur across (or associated with) the transition. Indeed, a 30% decrease (or more) in shear wave velocity in this zone implies (if constant density is assumed) that the shear modulus of this layer is approximately half that of the overlying material. Therefore, transitions which involve a softening of the shear modulus provide particularly attractive explanations for low velocity features. However, the primary transition which has been proposed to involve shear softening under CMB conditions is the shift from the stishovite structure of SiO_2 to the $CaCl_2$ structure [*Cohen*, 1992]. Such shear softening typically occurs in the low pressure phase, prior to the occurrence of the phase transition. If the stishovite -> $CaCl_2$ transition in SiO_2 actually generates the ULVZ, a large abundance of SiO_2 at depth (of order 30%) would be required, which necessitates essentially complete breakdown of the silicate perovskites under deep mantle conditions to their constituent oxides. As no compelling seismic evidence exists for a mid-lower mantle discontinuity in material properties that could be associated with the dissociation of silicate perovskite, SiO_2-dependent scenarios for generating the ULVZ require *i)* a dramatic change in mineralogy at depth in the lower mantle; *ii)* the intersection of the Clapeyron slope of the stishovite -> $CaCl_2$ transition with the conditions of the lowermost mantle; and *iii)* a smearing out of the shear softening produced by this transition over the lowermost portion of the mantle, with relatively little (high shear velocity) $CaCl_2$-structure generated at ULVZ depths.

In contrast to the restrictive conditions required for generating the ULVZ via mineralogic or chemical mechanisms, partial melting provides a rather natural means for explaining this feature [*Williams and Garnero*, 1996]. The ability of differing geometry-dependent amounts of partial melts to explain the velocity perturbations associated with the ULVZ are shown in Figure 9. The advantages of partial melting in explaining the ULVZ include *i)* quantitative agreement with large V_P and V_S depressions; *ii)* explanation of why this feature is generally correlated with zones of hot upwelling in the overlying mantle; and *iii)* providing both thermally-based and density-based rationales for this region lying at the absolute base of the mantle. The density of basic melts at ultra-high pressures is anticipated to be either comparable or exceed that of their coexisting solids [*Rigden et al.*, 1984; 1989]. Therefore, if melting is initiated at depth, the melt is likely to either be produced at, or descend to, the base of the mantle. Thus, the partial melting scenario involves a minimum of required special circumstances, but also involves a number of first order uncertainties: these include the geometry of melt distribution, and thus the melt fraction, and the precise chemistry of the melt.

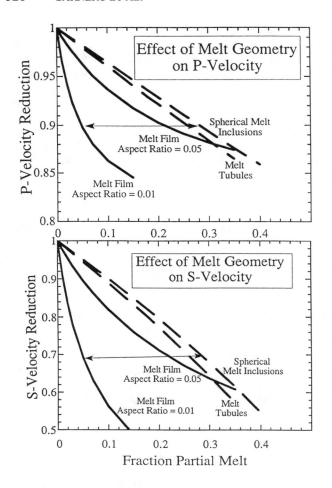

Figure 9. (a) The ratio of *P* wave velocity of melt bearing mantle to that of solid mantle for varying melt fractions and geometries. Arrow indicates the range of melt fractions that produce a 10% V_P reduction. (b) As in (a), except for shear waves. Arrow corresponds to the variation in melt fraction implied by the inferred 10% V_P depression, coupled with results of (a). [After *Williams and Garnero*, 1997.]

6. GEODYNAMICS CONSIDERATIONS

The existence of a thermal boundary layer at the base of the mantle is quite probable [*Stacey and Loper*, 1983], with the hottest temperatures in the mantle existing just above the core. The temperature increase in the CMB thermal boundary layer is uncertain, but has been estimated by comparing mantle adiabats and core melting temperatures to be on the order of 800° or more [e.g., *Jeanloz and Morris*, 1986] The increase in temperature with depth into the thermal boundary layer provides a natural mechanism for concentrating partial melt in a ULVZ, especially if densities are elevated. The deepest part of the boundary layer will be the lowest viscosity region as well, with strong horizontal flows being likely. The CMB is essentially isothermal due to the low viscosity of the core, thus a ULVZ might be globally present if it represents partial melting of a bulk component of the lower mantle (such as MgO). However, regions with mantle downwellings should have thinner thermal boundary layers than regions with upwellings, due to dynamical effects on a low viscosity basal layer and also radial temperature gradients, giving rise to lateral variations in thickness of the ULVZ. ULVZ thicknesses are at most a few tens of kilometers, and possibly modulated down to unobservable levels beneath regions of cold downwellings. Hotter temperatures early in Earth history could have produced a thicker global ULVZ in this scenario.

A purely thermal cause for the ULVZ, with associated partial melting of some bulk mantle component, predicts interesting dynamical effects that we explore. Survival of the zone of partial melt requires a significant density increase to resist entrainment or upward drainage, especially given the effects of reduced viscosity and probably small-scale convection. In fact, the partially molten region may convect separately from the large scale overlying mantle flow, modifying heat flux out of the core in thick ULVZ regions, resulting in the hottest areas of D″, with small radial thermal gradients. Such enhanced efficiency of heat transport across the CMB could influence core flow as well as mantle upwelling, perhaps increasing the lateral dimension of the latter relative to the concentrated upwelling expected for hot subsolidus boundary layers. Numerical calculations are needed to fully explore these ramifications.

If chemical heterogeneity is also an important factor for the existence of any lateral variations in the ULVZ, the dynamical issues are even more complex. If the origin of the ULVZ is partially (or completely) due to compositional uniqueness relative to the lower mantle, then it must be denser than the lower mantle to be gravitationally stable. Stability is determined by a buoyancy ratio, the ratio of the compositional to thermal buoyancy, $\Delta\rho_{ULVZ}/\rho\alpha\Delta T$ where $\Delta\rho_{ULVZ}$ is the density anomaly of the ULVZ, ρ is the density of the lower mantle, α is the coefficient of thermal expansion, and ΔT is the temperature difference across the layer (see *Hansen and Yuen* [1990] for a detailed discussion of the physics of thermo-chemical convection.) Generally, a buoyancy number close to 1 results in a compositionally distinct layer through time; this corresponds to a minimum increase in $\Delta\rho_{ULVZ}$ of around 6% [*Sleep*, 1988]. Such a layer will pile up under upwellings [*Davies and Gurnis*, 1986]. A large viscosity drop due to partial melt can change this result, as low-viscosity dense material is not as easily entrained in mantle plumes [*Sleep*, 1988].

Several studies have used numerical or laboratory models of thermo-chemical convection to determine the conditions for maintaining a stable, dense layer, and the structure of that layer [e.g., *Davies and Gurnis* 1986; *Hansen and Yuen*, 1988, 1989; *Olson and Kincaid*, 1991; *Kellogg and King*, 1993]. In time-dependent flow models, a stable layer often forms complex features due to internal circulation and mixing between the dense layer and the overlying layer. The structure of a thermo-chemical boundary layer at the base of the mantle is influenced primarily by the intrinsic density of the layer, but the vigor of the convection throughout the mantle, the rheology within the layer and in the overlying mantle, and the thermal conductivity of the layer relative to the lower mantle also play a role.

The most complex structures are observed when the layer is marginally stable [*Hansen and Yuen*, 1990; *Kellogg and King*, 1993]. At moderate buoyancy ratios (up to about 1), the layer tends to be stable but can vary drastically in thickness from nearly invisible where it has been pushed away by downwellings to relatively thick under upwellings (see Figure 10). As the buoyancy ratio increases, so does the stability of the ULVZ layer, and at higher buoyancy numbers the resulting layer is rather stagnant, not varying much in thickness from place to place [*Kellogg and King*, 1993; *Kellogg*, 1997].

However it originates, a ULVZ layer at the base of the mantle can significantly influence the thermal structure of the lower mantle (especially upwelling plumes) and the heat flux across the CMB. Maintaining a stable ULVZ layer at the base of the convecting mantle results in an increased temperature gradient across the lower boundary layer; hence the temperature at the base of the mantle may be relatively high (Figure 10a). Upwelling mantle plumes in this case may not arise directly from the core-mantle boundary and so one possible consequence of a stable layer at the base of the mantle could be cooler plumes [*Farnetani*, 1997]. An iron-rich layer at the base of the mantle would also be a good heat conductor [*Manga and Jeanloz*, 1996]. The consequences of high heat flow though the ULVZ (and thus D″) layer may also include more stable and hotter plumes [*Manga and Jeanloz*, 1996]. The layer itself may stablize plumes, as piles of dense material provide an "anchor" for upwellings [*Manga and Jeanloz*, 1996; *Montague et al.*, 1996].

The detailed thermal structure of the lower mantle will be determined in part by whether the ULVZ layer circulates internally. In order to attempt answering to this question, we must estimate the Rayleigh number in this layer. In principle, if the Rayleigh number is supercritical, we would expect that the layer is convecting internally. However, uncertainties in important ULVZ properties, such as ΔT across the layer, and viscosity preclude any constrained estimate of the Rayleigh number at present. Thus we cannot rule out free convection; nor can we assert that there must be vigorous convection within the layer. The flow patterns within the layer may

be strongly influenced by the overall pattern of circulation within the lower mantle. Convection within a thick ULVZ may result in an additional thermal boundary layer at both the base and the top of the ULVZ (Figure 10c). On the other hand, the steep thermal gradient at the base of the mantle might account for the ULVZ.

7. DISCUSSION

We now compare the ULVZ distribution of Figure 1 to various other geophysical phenomena. Figure 11a shows the distributions of buoyancy flux weighted hot spots (white circles, from *Sleep* [1992]) and calculated locations of subducted material in the D″ layer (crosses, from *Lithgow-Bertelloni and Richards* [1998]) superimposed on the ULVZ map. The hot spot spatial pattern is correlated with the pattern of ULVZ detections. For the hotspot catalog of Sleep (1990), the probability of chance correlation exceeding the observed is ~1%. When hotspots are weighted by flux (Sleep, 1990), the likelihood of chance correlation drops to 0.37% [*Williams et al.*, 1998]. Also evident in the figure, is that projected locations of subducted material most strongly coincide with areas where a ULVZ has not been detected, as well as where hot spots are absent (as expected in regions of downwelling, e.g., see *Richards and Engebretson* [1992]). This suggests thermal and/or dynamical effects on ULVZ structure and distribution from overlying convection.

A partially molten iron rich layer can have higher electrical conductivity [e.g., see *Jeanloz*, 1990]. Therefore the presence of a ULVZ potentially influences the Earth's magnetic field. Figure 11b shows magnetic field reversal paths and ULVZ distribution The calculated distribution of reversal paths are most strongly correlated

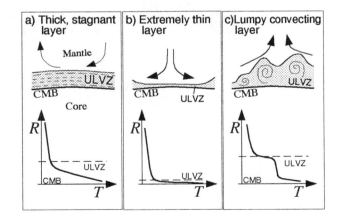

Figure 10. Schematic cross-sections and temperature profiles for (a) a thick stagnant ULVZ layer; (b) an extremely thin ULVZ; and (c) a thick, contorted convecting ULVZ layer (vertical dimension exaggerated).

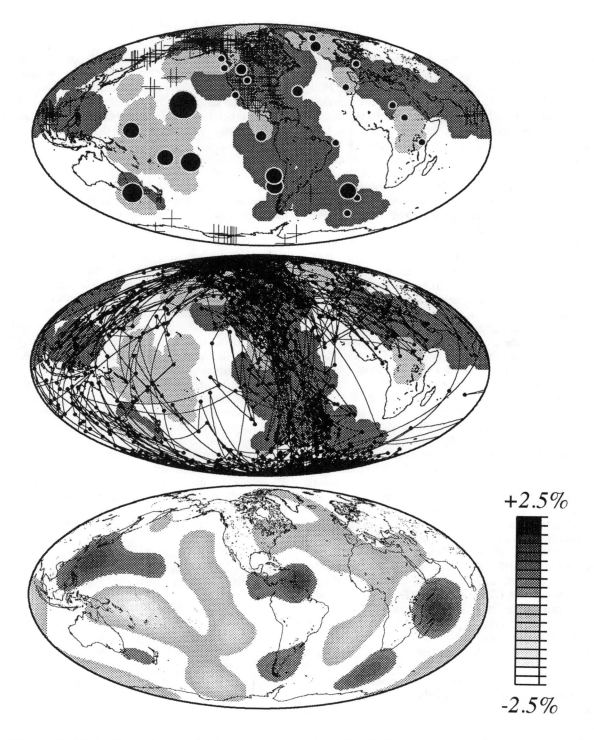

Figure 11. Mollweide projection of globe showing: (top) ULVZ distribution as in Figure 1: light shading corresponds to Fresnel zone regions where a ULVZ has been detected, dark regions are where no ULVZ is detected, no shading corresponds to no coverage. Black-filled circles are hot spot locations (where we have ULVZ coverage), scaled to buoyancy flux estimates of *Sleep* [1992]. Crosses are locations of calculated lower mantle density anomalies due to subducted material [*Lithgow-Bertelloni and Richards, 1998*]. (middle) ULVZ distribution with VGP reversal paths of *Laj et al.* [1991]. (bottom) *P* wave velocity heterogeneity in the lowermost 200-300 kilometers of the mantle, as calculated by *Wysession* [1996a]. Light and dark regions correspond to low and high velocity perturbations, respectively, with the ±0.5% range padded white.

with regions where a ULVZ has not been detected (e.g., under South and North America). If the ULVZ is iron-rich, and has relatively high electrical conductivity, regions where the ULVZ is thick and easily detected may be avoided by the poles of the magnetic field during reversals [e.g., see *Aurnou et al.*, 1996].

A recent inversion for large scale V_P structure in the lowermost few hundred km of the mantle is presented in Figure 11c for comparison the ULVZ distribution [from *Wysession*, 1996b]. High and low velocities (dark and light shadings, respectively) outside the -0.5% < δV_P < 0.5% range are contoured. In most of the areas studied, the ULVZ is present in regions of large scale *P* velocity reductions (Figure 11c). However, some areas lack this correlation, such as under parts of northern Africa, off the west coast of Central America, or just off the east coast of Japan. This may be the result of poor resolution in the tomographic image due to the long D″ paths of the *Pd* phase used in that study. On the other hand, the depth scales sampled by ULVZ investigations and lower mantle tomography studies are quite different (i.e., 5-40 kilometers compared to 200-400 kilometers, respectively), and thus may be sampling different length scales that need not be strongly correlated spatially. Alternatively, we cannot preclude a thin ULVZ (< 5 km) in these regions, which might go undetected by the long-period WWSSN data.

The large scale low-velocity feature in the southwest Pacific [e.g., see *Su et al.*, 1994; *Wysession*, 1996b; *Masters et al.*, 1996] has been interpreted as a large scale lower mantle thermal upwelling (often referred to as a "superplume"), and coincides with a large ULVZ patch. A ULVZ that laterally extends over a large area may act as the source to many plumes (and thus hot spots), which may effectively heat the lower mantle over large scale lengths, giving rise to low seismic velocities at long wavelengths. However, inversions at shorter scale lengths, such as that of *Grand et al.* [1997], show more variability in the southwest Pacific D″ structure, with alternating high and low velocities. Future investigations should attempt to resolve to what extent previous superplume interpretations are dependent upon scale length of resolvability, as well as spatial correlation of reduced lower mantle velocities and ULVZ distribution. *Lay et al.* [1997] detect significant small scale heterogeneities within the D″ layer in circum-Pacific regions, thus an intermittent ULVZ in these regions may be a possibility.

As previously mentioned, analyses of seismic phases other than *SPdKS* and precursors to core reflected energy may not show direct evidence for ULVZ structure, especially considering the strong ULVZ lateral variations [*Garnero and Helmberger*, 1996, 1998]. Long range diffracted energy, however, can allow us to place constraints on ULVZ structure allowable in any average structure. In an analysis of broadband amplitudes, travel times, and waveforms of diffracted shear waves *Ritsema et al.* [1997] show that a ULVZ can be tolerated in a one-dimensional reference structure for the region beneath the central Pacific if ULVZ thickness is ≤ 5 km (assuming δV_P=-10% and δV_S=-30%). Lateral variations with much thicker localized ULVZ regions [as in *Garnero and Helmberger*, 1996] cannot be precluded by the data. Intense small scale lateral variations in ULVZ topography and/or heterogeneity [*Wen and Helmberger*, 1998; *Helmberger et al.*, 1998; *Garnero and Helmberger*, 1998] can however give rise to spatial patterns in the amplitudes of *PKP* precursor energy correlating with the ULVZ distribution [see *Vidale and Hedlin*, 1998]. Future seismological efforts should include predictions through more detailed ULVZ structures, such as variable ULVZ topography [as in *Wen and Helmberger*, 1998b].

It is possible that systematic layering of ULVZ melt material, either solidified or liquid, may be related to the cause of D″ anisotropy [see *Lay et al.*, 1998]. A better understanding of lower mantle mineralogy along with more detailed seismic information are necessary to establish the feasibility of such a scenario. Future ULVZ modeling must include exploring the whole parameter space of ULVZ thickness, V_P, V_S, ρ, as well as possible strong attenuation due to partial melt, to determine the range of acceptable models, which will ultimately help in determining plausible ULVZ as well as lowermost mantle mineralogy.

8. CONCLUSIONS

Evidence for a thin boundary layer of ultralow velocities is provided by several seismic phases analyzed by different methods. The resulting distribution of ULVZ layering, where we have coverage, correlates strongly with hot spot locations (especially flux-weighted hot spots) and lower mantle velocity reductions. Regions where a ULVZ has not been detected (therefore, absent or thinner than our 3-5 kilometers threshold of detection thickness) spatially correlate to predicted locations of subducted slab material, higher seismic velocities in the overlying mantle, as well as virtual geomagnetic pole reversal paths. The preferred explanation of the cause of the ULVZ is partial melting of lowermost mantle rock, with probable chemical variations (e.g., from core-mantle chemical reaction products) producing very short wavelength heterogeneity. The emerging picture is thus one in which the layering in the lowermost mantle is strongly coupled to the dynamics of mantle circulation: low viscosity partially molten ULVZ regions may support elevated heat flow from the core, forming the root of mantle plumes; and such a layer is suppressed in regions where cold downwelling material has fallen to the CMB. The hypothesis of extensive material exchange between the upper and lower mantles is strongly supported by this ULVZ analysis.

Acknowledgements. We thank Carolina Lithgow Bertelloni for the lower mantle slab calculations, Michael Wysession for the lower mantle V$_P$ model, Daniel Brito for the paleomagnetic reversal paths, Julie Zaslow and Xiaoming Ding for data, and the authors of GMT software with which all figures were constructed [*Wessel and Smith*, 1991]. Thanks also to Steve Grand, Don Helmberger, Barbara Romanowicz, and John Vidale for helpful discussions. Also, reviews by M. Wysession and an anonymous reviewer improved the manuscript. Contribution #334 of the W. M. Keck Seismological Laboratory and Institute of Tectonics. This research was partially supported by NSF grant EAR9305894. E.J.G. was supported by NSF grant EAR-9896046 and EAR9418643. Workshops on the core-mantle boundary organized under EAR9305894 were instrumental in developing interdisciplinary coordination on this topic. The 1996 Fall AGU Meeting held a special session on the core-mantle boundary that further focussed attention on this topic.

9. REFERENCES

Aurnou, J.M., J.L Buttles, G.A., Neumann, and P.L. Olson, Electromagnetic core-mantle coupling and paleomagnetic reversal paths, *Geophys. Res. Lett.*, 23, 2705-2708, 1996.

Bass, J.D., Elasticity of minerals, glasses and melts, Handbook of Physical Constants, Vol. 2, pp. 45-63, Ed. T.J. Ahrens, American Geophysical Union, Washington, D.C., 1995.

Choy, G. L., Theoretical seismograms of core phases calculated by frequency-dependent full wave theory, and their interpretation, *Geophys. J. R. astr. Soc.*, 51, 275-312, 1975.

Christensen, U.R., and A.W. Hofmann, Segregation of subducted oceanic crust in the convecting mantle, *J. Geophys. Res.*, 99, 19867-19884, 1994.

Cohen, R.E., First principles predictions of elasticity and phase transitions in high pressure SiO$_2$ and geophysical implications, in High Pressure Research: Application to Earth and Planetary Sciences, Eds. Y. Syono and M.H. Manghnani, pp. 425-431, American Geophysical Union, Washington, D.C., 1992.

Davies, G.F., and M. Gurnis, Interaction of mantle dregs with convection: lateral heterogeneity at the core-mantle boundary, *Geophys. Res. Lett.*, 13, 1517-1520, 1986.

Doornbos, D. J., and T. Hilton, Models of the core-mantle boundary and the travel times of internally reflected core phases, *J. Geophys. Res.*, 94, 15,741-15,751, 1989.

Dziewonski, Mapping the lower mantle: Determination of lateral heterogeneity in *P* velocity up to degree and order 6, *J. Geophys. Res.*, 89, 5929-5952, 1984.

Dziewonski, A. M., and D. L. Anderson, Preliminary reference Earth model (PREM), *Phys. Earth Planet. Int.*, 25, 297-356, 1981.

Farnetani, C.G., Excess temperature of mantle plumes; The role of chemical stratification across D″, *Geophys. Res. Lett.*, 24, 1583-1586, 1997.

Fischer, K.M., J.M. Zaslow, E.J.Garnero, M.J. Fouch, M.E. Wysession, T.J. Clarke, and G.I. Al-eqabi, SpdKS constraints on a thin slow layer at the base of the mantle beneath the northwestern Pacific, *Eos Trans. AGU*, 77, no. 48, 1996.

Fuchs, K. and G. Müller, Computation of synthetic seismograms with the reflectivity method and comparison with observations, *Geophys. J. R. Astron. Soc.*, 23, 417-433, 1971.

Garnero E. J., S. P. Grand, and D. V. Helmberger, Low *P* wave velocity at the base of the mantle, *Geophys. Res. Lett.*, 20, 1843-1846, 1993.

Garnero, E.J., and D.V. Helmberger, Seismic detection of a thin laterally varying boundary layer at the base of the mantle beneath the central-Pacific, *Geophys. Res. Lett.*, 23, 977-980, 1996.

Garnero, E.J., and D.V. Helmberger, Further constraints and uncertainties in modeling a thin laterally varying ultralow velocity layer at the base of the mantle, *J. Geophys. Res.* (in press), 1998.

Gieske, J.H., and G.R. Barsch, Pressure dependence of the elastic constants of single crystalline aluminum oxide, *Physica Status Solidi*, 29, 121-31, 1968.

Grand, S.P., R.D. van der Hilst, and S. Widiyantoro, Global seismic tomography: a snapshot of convection in the Earth, *G.S.A. Today*, 7, 1-7, 1997.

Hansen, U, and Yuen, D.A., Numerical simulations of thermal-chemical instabilities at the core-mantle boundary, *Nature*, 334, 237-240, 1988.

Hansen, U, and Yuen, D.A., Dynamical influences from thermal-chemical instabilities at the core-mantle boundary, *Geophys. Res. Lett.*, 16, 629-632, 1989.

Hansen, U. and Yuen, D.A., Nonlinear physics of double-diffusive convection in geological systems, *Earth-Sci. Rev.*, 29, 385-399, 1990.

Helmberger, D.V., E.J. Garnero, and X.-M. Ding, Modeling two-dimensional structure at the core-mantle boundary, *J. Geophys. Res.*, 101, 13,963-13,972, 1996.

Helmberger, D.V., L. Wen, and X. Ding, Ultra low velocity zone beneath Iceland, *Nature*, (in review) 1998.

Holland, K.G., and T.J. Ahrens, Melting of (Mg,Fe)$_2$SiO$_4$ at the core-mantle boundary of the Earth, *Science*, 275, 1623-1625, 1997.

Inoue H., Y. Fukao, K. Tanabe, and Y. Ogata, Whole mantle *P* wave travel time tomography, *Phys. Earth Planet. Int.*, 59, 294-328, 1990.

Ito, E., K. Morooka, O. Ujike and T. Katsura, Reactions between molten iron and silicate melts at high pressure: Implications for the chemical evolution of Earth's core, *J. Geophys. Res.*, 100, 5901-5910, 1995.

Jeanloz, R., The nature of the Earth's core, *Ann. Rev. Earth Planet. Sci.*, 18, 357-386, 1990.

Jeanloz, R., and S. Morris, Temperature distribution in the crust and mantle, *Ann. Rev. Earth Planet. Sci.*, 14, 377-415, 1986.

Kellogg, L.H. and S.D. King, Effect of mantle plumes on the growth of D″ by reaction between the core and mantle, *Geophys. Res. Lett.*, 20, 379-382, 1993.

Kellogg, L.H., Growing the Earth's D″ layer: Effect of density variations at the core-mantle boundary, *Geophys. Res. Lett.*, 24, 2749-2752, 1997.

Knittle, E., Static compression measurements of equations of state, Handbook of Physical Constants, Vol. 2, pp. 98-142, Ed. T.J. Ahrens, American Geophysical Union, Washington, D.C., 1995.

Knittle, E. and R. Jeanloz, Simulating the core-mantle boundary: an experimental study of high-pressure reactions between silicates and liquid iron, *Geophys. Res. Lett.*, 16, 609-612, 1989.

Knittle, E. and R. Jeanloz, Earth's core-mantle boundary: results of experiments at high pressures and temperatures, *Science, 251,* 1438-1443, 1991.

Krüger, F., M. Weber, F. Scherbaum, and J. Schlittenhardt, Normal and inhomogeneous lowermost mantle and core-mantle boundary structure under the Arctic and northern Canada, *Geophys. J. Int., 122,* 637-658, 1995.

Laj, C., A. Mazaud, R. Weeks, M. Fuller, and E. Herrero-Berverra, Geomagnetic reversal paths, *Nature, 351,* 447, 1991.

Lay, T., E.J. Garnero, C.J. Young, and J.B. Gaherty, Scale lengths of shear velocity heterogeneity at the base of the mantle from *S* wave differential travel times, *J. Geophys. Res., 102,* 9887-9909, 1997.

Lay, T., E.J. Garnero, Q. Williams, R. Jeanloz, B. Romanowicz, L. Kellogg, and M.E. Wysession, Shear wave anisotropy in the D″ region and its implications, AGU Geophysical Monograph on the Core-Mantle Region, (this volume) 1998.

Lithgow-Bertelloni, C., and M.A. Richards, The dynamics of cenezoic and mesozoic plate motions, *Ann. Rev. Earth Planet. Sci., 36,* 27-78, 1998.

Loper, D.E., and T. Lay, The core-mantle boundary region, *J. Geophys. Res., 100,* 6397-6420, 1995.

Manga, M. and Jeanloz, R. Implications of a metal-bearing chemical boundary layer in D″ for mantle dynamics, *Geophys. Res. Lett., 23,* 3091-3094, 1996.

Mao, H.K., Chen, L.C., Hemley, R.J., and A.P. Jephcoat, Stability and equation of state of CASiO3-perovskite to 134-GPA, *J. Geophys. Res., 94,* 17889-17894, 1989.

Montague, M., M. Manga, L.H. Kellogg, R. Jeanloz, Models of a metal-bearing chemical boundary layer in the D″ region: Implications for mantle and core dynamics, *EOS, Trans. AGU, 77,* 709-710, 1996.

Mori, J., and D.V. Helmberger, Localized boundary layer below the mid-Pacific velocity anomaly identified from a *PcP* precursor, *J. Geophys. Res., 100* 20,359-20,365, 1995.

Olson, P. and C. Kincaid, Experiments on the interaction of thermal convection and compositional layering at the base of the mantle, *J. Geophys. Res., 96,* 4347-4354, 1991.

Pulliam, R.J., D.W. Vasco, and L.R. Johnson, Tomographic inversions for mantle *P* wave velocity structure based on the minimization of l^2 and l^1 norms of International Seismic Centre travel time residuals, *J. Geophys. Res., 98,* 699-734, 1993.

Revenaugh J.S., and R. Meyer, Seismic evidence of partial melt within a possibly ubiquitous low-velocity layer at the base of the mantle, *Science, 277,* 670-673, 1997.

Richards, M.A., and D.C. Engebretson, Large-scale mantle convection and the history of subduction, *Nature, 355,* 437-440, 1992.

Rigden, S.M., T.J. Ahrens, and E.M. Stolper, Densities of liquid silicates at high pressures, *Science, 226,* 1071-1074, 1984.

Rigden, S.M., T.J. Ahrens, and E.M. Stolper, High-pressure equation of state of molten anorthite and diopside, *J. Geophys. Res., 94,* 9508-9522, 1989.

Ritsema, J.E., E.J. Garnero, and T. Lay, A strongly negative shear velocity gradient and lateral variability in the lowermost mantle beneath the Pacific, *J. Geophys. Res, 102,* 20,395-20,411, 1997.

Scherbaum, F., F. Krüger, and M. Weber, Double beam imaging: mapping lower mantle heterogeneities using combinations of source and receiver arrays, *J. Geophys. Res., 102,* 507-522, 1997.

Silver, P., and C. R. Bina, An anomaly in the amplitude ratio of *SKKS/SKS* in the range 100-108° from portable teleseismic data, *Geophys. Res. Lett., 20,* 1135-1138, 1993.

Sleep, N.H., Gradual entrainment of a chemical layer at the base of the mantle by overlying convection, *Geophys. J., 95,* 437-447, 1988.

Sleep, N.H., Time dependence of mantle plumes: some simple theory, *J. Geophys. Res., 97,* 20,007-20,019, 1992.

Song, X.-D., and D.V. Helmberger, A *P* wave velocity model of Earth's core, *J. Geophys. Res., 100,* 9817-9830, 1995.

Souriau, A., and G. Poupinet, Lateral variations in *P* velocity and attenuation in the D″ layer, from diffracted *P* waves, *Phys. Earth Planet. Int., 84,* 227-234, 1994.

Sylvander, M. and A. Souriau, Mapping S-velocity heterogeneities in the D″ region, from *SmKS* differential travel times, *Phys. Earth Planet. Int., 94,* 1-21, 1996.

Sylvander, M., B. Ponce, and A. Souriau, Seismic velocities at the core-mantle boundary inferred from *P* waves diffracted around the core, *Phys. Earth Planet. Int., 101,* 189-202, 1997.

Stacey, F.D. and D.E. Loper, The thermal boundary-layer interpretation of D″ and its role as a plume source, *Phys. Earth Planet. Int., 33,* 45-55, 1983.

Su, W.-J., R.L. Woodward, and A.M. Dziewonski, Degree 12 model of shear velocity heterogeneity in the mantle, *J. Geophys. Res., 99,* 6945-6980, 1994.

Vidale, J.E., and M.A.H. Hedlin, Intense scattering at the core-mantle boundary north of Tonga: evidence for partial melt, *Nature* (in press), 1998.

Vidale, J.E., E.J. Garnero, and L.S.L. Kong, Sounding the base of the mantle by core reflections, *Eos Trans. AGU, 76,* no. 46, 404, 1995.

Weber, M., *P* and *S* wave reflections from anomalies in the lowermost mantle, *Geophys. J. Int., 115,* 183-210, 1993.

Weber, M., and M. Körnig, Lower mantle inhomogeneities inferred from *PcP* precursors, *Geophys. Res. Lett., 17,* 1993-1996, 1990.

Wen, L., and D.V. Helmberger, Ultra low velocity zones near the core-mantle boundary from broadband *PKP* precursors, *Science,* (in press) 1998a.

Wen, L., and D.V. Helmberger, A 2-D P-SV hybrid method and its application to modeling localized structures near the core-mantle boundary, *J. Geophys. Res.,* (in press) 1998b.

Williams, Q., and E.J. Garnero, Seismic evidence for partial melt at the base of Earth's mantle, *Science, 273,* 1528-1530, 1996.

Williams, Q., J. Revenaugh, and E.J. Garnero, A correlation between the hot spot distribution and ultra-low basal velocities in the mantle, *Nature,* (in review), 1998.

Wysession, M. E., Imaging cold rock at the base of the mantle: The sometimes fate of slabs?, in *Subduction: Top to Bottom,* edited by G.E. Bebout, D. Scholl, S. Kirby, and J.P. Platt, AGU, Washington, D.C., pp. 369-384, 1996a.

Wysession, M. E., Large-scale structure at the core-mantle boundary from diffracted waves, *Nature, 382,* 244-248, 1996b.

Wysession, M. E., and E. A. Okal, Regional analysis of D″ velocities from the ray parameters of diffracted *P* profiles, *Geophys. Res. Lett., 16,* 1417-1420, 1989.

Wysession, M.E., R.W. Valenzuela, A.-N. Zhu, and L. Bartko,

Investigating the base of the mantle using differential travel times, *Phys. Earth Planet. Int., 92,* 67-84, 1995.

Wysession, M.E., T. Lay, J. Revenaugh, Q. Williams, E.J. Garnero, R. Jeanloz, L. Kellogg, The D'' discontinuity and its implications, AGU Geophysical Monograph on the Core-Mantle Region, (this volume) 1997.

Young, C. J., and T. Lay, The core shadow zone boundary and lateral variations of the *P* velocity structure of the lowermost mantle, *Phys. Earth Planet. Int., 54,* 64-81, 1989.

E. Garnero, Berkeley Seismological Laboratory, University of California, 475 McCone Hall, Berkeley, CA 94720

J. Revenaugh, Q. Williams, and T. Lay, Institute of Tectonics and Earth Sciences Board, University of California, Santa Cruz, CA 95064

L. Kellogg, Department of Geology, University of California, Davis, CA 95616